Student Solutions Manual
for Hirsch and Goodman's
UNDERSTANDING
INTERMEDIATE ALGEBRA
A Graphing Approach

Cheryl Roberts
Northern Virginia Community College

Brooks/Cole Publishing Company
IT**P**® An International Thomson Publishing Company

Pacific Grove • Albany • Belmont • Bonn • Boston • Cincinnati • Detroit • Johannesburg • London
Madrid • Melbourne • Mexico City • New York • Paris • Singapore • Tokyo • Toronto • Washington

Sponsoring Editors: Denise Bayko, Linda Row
Production Coordinator: Dorothy Bell
Printing and Binding: West Publishing

For more information, contact:

BROOKS/COLE PUBLISHING COMPANY
511 Forest Lodge Rd.
Pacific Grove, CA 93950
USA

International Thomson Publishing Europe
Berkshire House 168-173
High Holborn
London WC1V 7AA
England

Thomas Nelson Australia
102 Dodds Street
South Melbourne, 3205
Victoria, Australia

Nelson Canada
1120 Birchmount Road
Scarborough, Ontario
Canada M1K 5G4

International Thomson Editores
Seneca 53
Col. Polanco
11560 México, D. F., México

International Thomson Publishing GmbH
Königswinterer Strasse 418
53227 Bonn
Germany

International Thomson Publishing Asia
221 Henderson Road
#05-10 Henderson Building
Singapore 0315

International Thomson Publishing Japan
Hirakawacho Kyowa Building, 3F
2-2-1 Hirakawacho
Chiyoda-ku, Tokyo 102
Japan

Printed in the United States of America

10 9 8 7 6 5 4 3 2 1

ISBN 0-314-21006-7

TABLE OF CONTENTS

CHAPTER 1

1.1 Exercises

1. {3, 4, 5, 6, 7, 8, 9, 10, 11, 12, 13}

3. ∅

5. {41, 43, 47}

7. {6, 12, 18, 24, ... }

9. {1, 2, 3, 4, 6, 9, 12, 18, 36}

11. $A \cap B$ = {3, 6}

13. A = {0, 1, 2, 3, 4, 5, 6}
    B = {3, 6, 9, 12, 15, 18, 21, 24, 27, 30, 33}

    $A \cup B$ = {0, 1, 2, 3, 4, 5, 6, 9, 12, 15, 18, 21, 24, 27, 30, 33}

15. A = {0, 1, 2, 3, 4, 5, 6}
    D = {7, 8, 9, 10, 11, 12, 13}

    $A \cup D$ = {0, 1, 2, 3, 4, 5, 6, 7, 8, 9, 10, 11, 12, 13}

17. $66 = 2 \cdot 33$
    $= 2 \cdot 3 \cdot 11$

19. $128 = 2 \cdot 64$
    $= 2 \cdot 2 \cdot 32$
    $= 2 \cdot 2 \cdot 2 \cdot 16$
    $= 2 \cdot 2 \cdot 2 \cdot 2 \cdot 8$
    $= 2 \cdot 2 \cdot 2 \cdot 2 \cdot 2 \cdot 4$
    $= 2 \cdot 2 \cdot 2 \cdot 2 \cdot 2 \cdot 2 \cdot 2$

21. Prime

23. $91 = 7 \cdot 13$

25. True, −8 is an integer.

27. True, $\frac{8}{4} = 2$ is an integer.

29. True, 1.8 is a rational number.

31. False, $\sqrt{19}$ is irrational.

33. False, $0 \in W$, but $0 \notin N$.

35. True, the rational numbers are a subset of the real numbers.

37. $6 \cdot 2 \qquad 6 + 2$
    $\quad 12 \qquad\quad 8$
    $\qquad >, \geq, \neq$

39. $3 \cdot 0 \qquad 3 + 0$
    $\quad 0 \qquad\quad 3$
    $\qquad <, \leq, \neq$

41. $\{x \mid x < 4\}$

43. $\{a \mid a \leq -3\}$

45. $\{y \mid y \geq -4\}$

47. $\{y \mid -2 < y < 5, y \in Z\}$

49. $\{r \mid 5 \leq r \leq 12\}$

51. $\{z \mid -3 < z \leq 0, z \in Z\}$

53. No solution; There are no numbers greater than −4 and less than or equal to −7.

55. $\{a \mid -4 < a < 4\}$

57. $C = \{-4, -3, -2, -1, 0, 1, 2, 3, 4, 5, 6\}$
$D = \{1, 2, 3, 4, 5, 6, 7, 8\}$

$C \cup D = \{-4, -3, -2, -1, 0, 1, 2, 3, 4, 5, 6, 7, 8\}$
$= \{x \mid -4 \leq x \leq 8, x \in Z\}$

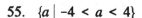

59. $A = \{-3, -2, -1, 0, 1, \dots\}$
$D = \{1, 2, 3, 4, 5, 6, 7, 8\}$

$A \cap D = \{1, 2, 3, 4, 5, 6, 7, 8\} = D$

61. $A \cup B = \{x \mid x \geq -3 \text{ or } x < -3\} = R$

63. $C = \{x \mid -4 \leq x \leq 6\}$

$D = \{x \mid 1 \leq x < 9\}$

$C \cap D$ is where the two graphs overlap.

$C \cap D = \{x \mid 1 \leq x \leq 6\}$

65. $A \cup C = \{x \mid x \geq -3 \text{ or } -4 \leq x \leq 6\}$
$= \{x \mid x \geq -4\}$

67. True, associative property for addition

69. True, commutative property for addition

71. True, distributive property

73. False

75. True, distributive property

77. False

79. True, associative property for multiplication

81. True, associative property for multiplication

83. True, additive inverse property

85. True, commutative property of addition

87. True, distributive property

89. True, closure for multiplication

91. False

93. Let $x = 0.674\overline{674}$
$1000x = 674.674\overline{674}$

$\begin{array}{r} 1000x = 674.674\overline{674} \\ - \quad x = 0.674\overline{674} \\ \hline 999x = 674 \end{array}$

$x = \dfrac{674}{999}$

Let $x = 0.92\overline{92}$
$100x = 92.92\overline{92}$

$\begin{array}{r} 100x = 92.92\overline{92} \\ -x = 0.92\overline{92} \\ \hline 99x = 92 \end{array}$

$x = \dfrac{92}{99}$

95. No, subtraction is not commutative.
For example, $6 - 3 \neq 3 - 6$.

No, subtraction is not associative.
For example, $6 - (4 - 2) \neq (6 - 4) - 2$.

97. 1. Additive identity property
    2. Distributive property
    4. Additive inverse property
    5. Associative property for addition
    6. Additive inverse property
    7. Additive identity property

1.2 Exercises

1. $-3 + 8 = + (8 - 3) = 5$

3. $-3 - 8 = -3 + (-8)$
   $ = -11$

5. $-3(-8) = 24$

7. $1.692 - 3.965 + 8.754$
   $= 1.692 + (-3.965) + 8.754$
   $= -2.273 + 8.754$
   $= 6.481$

9. $-3 - 4 - 5 = -3 + (-4) + (-5)$
   $ = -7 + (-5)$
   $ = -12$

11. $-3 - 4(5) = -3 - 20$
    $ = -3 + (-20)$
    $ = -23$

13. $-3(-4)(-5) = 12(-5)$
    $ = -60$

15. $3(-4 - 5) = 3(-9)$
    $ = -27$

17. $8 - 4 \cdot 3 - 7 = 8 - 12 - 7$
    $ = -4 - 7$
    $ = -11$

19. $8 - (4 \cdot 3 - 7) = 8 - (12 - 7)$
    $ = 8 - 5$
    $ = 3$

21. $(8 - 4)(3 - 7) = (4)(-4)$
    $ = -16$

23. $(2.6)^2 - |7.8 - 13.69| = 6.76 - |-5.89|$
    $ = 6.76 - 5.89$
    $ = 0.87$

25. $\dfrac{-20}{-5} = 4$

27. $\dfrac{-5 - 11}{-9 + 4} = \dfrac{-16}{-5}$

    $\phantom{\dfrac{-5 - 11}{-9 + 4}} = \dfrac{16}{5}$

29. $\dfrac{-10 - 2 - 4}{-2} = \dfrac{-12 - 4}{-2}$

    $\phantom{\dfrac{-10 - 2 - 4}{-2}} = \dfrac{-16}{-2}$

    $\phantom{\dfrac{-10 - 2 - 4}{-2}} = 8$

31. $\dfrac{-10 - (2 - 4)}{-2} = \dfrac{-10 - (-2)}{-2}$

    $\phantom{\dfrac{-10 - (2 - 4)}{-2}} = \dfrac{-10 + 2}{-2}$

    $\phantom{\dfrac{-10 - (2 - 4)}{-2}} = \dfrac{-8}{-2}$

    $\phantom{\dfrac{-10 - (2 - 4)}{-2}} = 4$

33. $\dfrac{-10 - 2(-4)}{-2} = \dfrac{-10 - (-8)}{-2}$

    $\phantom{\dfrac{-10 - 2(-4)}{-2}} = \dfrac{-10 + 8}{-2}$

    $\phantom{\dfrac{-10 - 2(-4)}{-2}} = \dfrac{-2}{-2}$

    $\phantom{\dfrac{-10 - 2(-4)}{-2}} = 1$

35. $\dfrac{-4(-3)(-6)}{-4(-3) - 6} = \dfrac{12(-6)}{12 - 6}$

$\qquad\qquad\qquad = \dfrac{-72}{6}$

$\qquad\qquad\qquad = -12$

37. $8 - 3(5 - 1) = 8 - 3(4)$

$\qquad\qquad\quad = 8 - 12$

$\qquad\qquad\quad = -4$

39. $-7 - 2(4 - 6) = -7 - 2(-2)$

$\qquad\qquad\qquad = -7 - (-4)$

$\qquad\qquad\qquad = -7 + 4$

$\qquad\qquad\qquad = -3$

41. $7 + 2[4 + 3(4 + 1)] = 7 + 2[4 + 3(5)]$

$\qquad\qquad\qquad\qquad = 7 + 2(4 + 15)$

$\qquad\qquad\qquad\qquad = 7 + 2(19)$

$\qquad\qquad\qquad\qquad = 7 + 38$

$\qquad\qquad\qquad\qquad = 45$

43. $7 - 2[4 - 3(4 - 1)] = 7 - 2[4 - 3(3)]$

$\qquad\qquad\qquad\qquad = 7 - 2(4 - 9)$

$\qquad\qquad\qquad\qquad = 7 - 2(-5)$

$\qquad\qquad\qquad\qquad = 7 - (-10)$

$\qquad\qquad\qquad\qquad = 7 + 10$

$\qquad\qquad\qquad\qquad = 17$

45. $8 - \left(\dfrac{10}{-5}\right) = 8 - (-2)$

$\qquad\qquad\qquad = 8 + 2$

$\qquad\qquad\qquad = 10$

47. $8\left(\dfrac{10}{-5}\right) = 8(-2)$

$\qquad\qquad\quad = -16$

49. $\dfrac{12}{-4} - \left(\dfrac{10}{-2}\right) = -3 - (-5)$

$\qquad\qquad\qquad\quad = -3 + 5$

$\qquad\qquad\qquad\quad = 2$

51. $\dfrac{-8 + 2}{4 - 6} - \dfrac{6 - 11}{-3 - 2} = \dfrac{-6}{-2} - \dfrac{-5}{-5}$

$\qquad\qquad\qquad\qquad\qquad = 3 - 1$

$\qquad\qquad\qquad\qquad\qquad = 2$

53. $-12 - \dfrac{6 - 2(-3)}{-3} = -12 - \dfrac{6 + 6}{-3}$

$\qquad\qquad\qquad\qquad = -12 - \dfrac{12}{-3}$

$\qquad\qquad\qquad\qquad = -12 - (-4)$

$\qquad\qquad\qquad\qquad = -12 + 4$

$\qquad\qquad\qquad\qquad = -8$

55. $(-6)^2 = 36$

57. $2^2 + 3^2 + 4^2 = 4 + 9 + 16$

$\qquad\qquad\qquad\quad = 13 + 16$

$\qquad\qquad\qquad\quad = 29$

59. $2(5)^2 = 2(25)$

$\qquad\qquad = 50$

61. $-3^4 = -81$

63. $-8 - 2(-4)^2 = -8 - 2(16)$

$\qquad\qquad\qquad = -8 - 32$

$\qquad\qquad\qquad = -40$

65. $2(-5)(-6)^2 = 2(-5)(36)$

$\qquad\qquad\qquad = (-10)(36)$

$\qquad\qquad\qquad = -360$

67. $2(-5) - 6^2 = 2(-5) - 36$

$\qquad\qquad\qquad = -10 - 36$

$\qquad\qquad\qquad = -46$

69. $\dfrac{5[-8 - 3(-2)^2]}{-6 - 6 - 2} = \dfrac{5[-8 - 3(4)]}{-12 - 2}$

$\qquad\qquad\qquad\qquad = \dfrac{5(-8 - 12)}{-14}$

$\qquad\qquad\qquad\qquad = \dfrac{5(-20)}{-14}$

$\qquad\qquad\qquad\qquad = \dfrac{-100}{-14}$

$\qquad\qquad\qquad\qquad = \dfrac{50}{7}$

71. $-3 - 2[-4 - 3(-2 - 1)]$
$= -3 - 2[-4 - 3(-3)]$
$= -3 - 2(-4 + 9)$
$= -3 - 2(5)$
$= -3 - 10$
$= -13$

73. $-3\{5 - 3[2 - 6(3 - 5)]\}$
$= -3\{5 - 3[2 - 6(-2)]\}$
$= -3[5 - 3(2 + 12)]$
$= -3[5 - 3(14)]$
$= -3(5 - 42)$
$= -3(-37)$
$= 111$

75. $|3 - 8| - |3| - |-8|$
$= |-5| - |3| - |-8|$
$= 5 - 3 - 8$
$= 2 - 8$
$= -6$

77. $|-3 - 2 - 4| - (2 - 3)^3$
$= |-5 - 4| - (-1)^3$
$= |-9| - (-1)$
$= 9 + 1$
$= 10$

79. $(-3 - 2)^2 - (-3 + 2)^2$
$= (-5)^2 - (-1)^2$
$= 25 - 1$
$= 24$

81. $\dfrac{42.2}{1.63 - (2.1)(5.8)} = \dfrac{42.2}{1.63 - 12.18}$
$= \dfrac{42.2}{-10.55}$
$= -4$

83. $6\left(\dfrac{3}{7}\right)^2 + 4\left(\dfrac{3}{7}\right) - 3 = 6\left(\dfrac{9}{49}\right) + 4\left(\dfrac{3}{7}\right) - 3$
$= \dfrac{54}{49} + \dfrac{12}{7} - 3$
$= \dfrac{54}{49} + \dfrac{84}{49} - \dfrac{147}{49}$
$= -\dfrac{9}{49}$

85. $x + y + z = -2 + (-3) + 5$
$= -5 + 5$
$= 0$

87. $xyz = (-2)(-3)(5)$
$= 6(5)$
$= 30$

89. $-x^2 - 4x + 2 = -(-2)^2 - 4(-2) + 2$
$= -4 - 4(-2) + 2$
$= -4 + 8 + 2$
$= 4 + 2$
$= 6$

91. $|xy - z| = |(-2)(-3) - 5|$
$= |6 - 5|$
$= |1|$
$= 1$

93. $\dfrac{3x^2y - x^3y^2}{3x - 2y} = \dfrac{3(-2)^2(-3) - (-2)^3(-3)^2}{3(-2) - 2(-3)}$
$= \dfrac{3(4)(-3) - (-8)(9)}{-6 + 6}$
$= \dfrac{12(-3) - (-72)}{0}$
$= \dfrac{-36 + 72}{0}$
$= \dfrac{36}{0}$
$=$ undefined

95. $s_e = s_y\sqrt{1 - r^2}$
$= 2.3\sqrt{1 - (0.74)^2}$
$= 2.3\sqrt{1 - 0.5476}$
$= 2.3\sqrt{0.4524}$
$= 2.3(0.6726)$
$= 1.55$

97. (a) $\sigma_r = \sqrt{\dfrac{1 - \rho^2}{n - 1}}$

$= \sqrt{\dfrac{1 - (0.72)^2}{100 - 1}}$

$= 0.070$

(b) $\sigma_r = \sqrt{\dfrac{1 - \rho^2}{n - 1}}$

$= \sqrt{\dfrac{1 - (0.64)^2}{50 - 1}}$

$= 0.110$

99. $(-3)(4) = [(-1)(3)](4)$ additive

$= (-1)[(3)(4)]$ inverse

$= (-1)[12]$ multiplication

$= -12$ multiplication

1.3 Exercises

1. $6x + 2x = (6 + 2)x$
   $= 8x$

3. $6x(2x) = (6)(2)x \cdot x$
   $= 12x^2$

5. $2x - 6x = (2 - 6)x$
   $= -4x$

7. $2x(-6x) = (2)(-6)x \cdot x$
   $= -12x^2$

9. $3m - 4m - 5m = (3 - 4 - 5)m$
   $= -6m$

11. $3m(-4m)(-5m) = (3)(-4)(-5)m \cdot m \cdot m$
    $= 60m^3$

13. $-2t^2 - 3t^2 - 4t^2 = (-2 - 3 - 4)t^2$
    $= -9t^2$

15. $-2t^2(-3t^2)(-4t^2) = (-2)(-3)(-4)t^2t^2t^2$
    $= -24t^6$

17. $2x + 3y + 5z$

19. $2x(3y)(5z) = (2)(3)(5)xyz$
    $= 30xyz$

21. $x^3 + x^2 + 2x$

23. $x^3(x^2)(2x) = 2x^3x^2x$
    $= 2x^6$

25. $-5x(3xy) - 2x^2y = (-5)(3)x \cdot x \cdot y - 2x^2y$
    $= -15x^2y - 2x^2y$
    $= (-15 - 2)x^2y$
    $= -17x^2y$

27. $-5x(3xy)(-2x^2y)$
    $= (-5)(3)(-2)x \cdot x \cdot x^2y \cdot y$
    $= 30x^4y^2$

29. $2x^2 + 3x - 5 - x^2 - x - 1$
    $= (2 - 1)x^2 + (3 - 1)x + (-5 - 1)$
    $= 1x^2 + 2x - 6$
    $= x^2 + 2x - 6$

31. $10x^2y - 6xy^2 + x^2y - xy^2$
    $= (10 + 1)x^2y + (-6 - 1)xy^2$
    $= 11x^2y - 7xy^2$

33. $3(m + 3n) + 3(2m + n)$
$= 3m + 9n + 6m + 3n$
$= (3 + 6)m + (9 + 3)n$
$= 9m + 12n$

35. $6(a - 2b) - 4(a + b)$
$= 6a - 12b - 4a - 4b$
$= (6 - 4)a + (-12 - 4)b$
$= 2a - 16b$

37. $8(2c - d) - (10c + 8d)$
$= 16c - 8d - 10c - 8d$
$= (16 - 10)c + (-8 - 8)d$
$= 6c - 16d$

39. $x(x - y) + y(y - x) = x^2 - xy + y^2 - xy$
$= x^2 - 2xy + y^2$

41. $a^2(a + 3b) - a(a^2 + 3ab)$
$= a^3 + 3a^2b - a^3 - 3a^2b$
$= (1 - 1)a^3 + (3 - 3)a^2b$
$= 0a^3 + 0a^2b$
$= 0 + 0$
$= 0$

43. $5a^2bc(-2ab^2)(-4bc^2)$
$= (5)(-2)(-4)a^2abb^2bcc^2$
$= 40a^3b^4c^3$

45. $(2x)^3(3x)^2$
$= (2x)(2x)(2x)(3x)(3x)$
$= (2 \cdot 2 \cdot 2 \cdot 3 \cdot 3)(x \cdot x \cdot x \cdot x \cdot x)$
$= 72x^5$

47. $2x^3(3x)^2 = 2x^3 \cdot 3x \cdot 3x$
$= 2 \cdot 3 \cdot 3 \cdot x^3 \cdot x \cdot x$
$= 18x^5$

49. $(-2x)^5(x^6)$
$= (-2x)(-2x)(-2x)(-2x)(-2x)(x^6)$
$= (-2)(-2)(-2)(-2)(-2)(x \cdot x \cdot x \cdot x \cdot x)(x^6)$
$= -32x^{11}$

51. $(-2x)^4 - (2x)^4$
$= (-2x)(-2x)(-2x)(-2x) - (2x)(2x)(2x)(2x)$
$= (-2)(-2)(-2)(-2)(x \cdot x \cdot x \cdot x)$
$\quad - (2)(2)(2)(2)(x \cdot x \cdot x \cdot x)$
$= 16x^4 - 16x^4$
$= (16 - 16)x^4$
$= 0$

53. $(-2x)^3 - (2x)^3$
$= (-2x)(-2x)(-2x) - (2x)(2x)(2x)$
$= (-2)(-2)(-2)(x \cdot x \cdot x) - (2)(2)(2)(x \cdot x \cdot x)$
$= -8x^3 - 8x^3$
$= (-8 - 8)x^3$
$= -16x^3$

55. $4b - 5(b - 2) = 4b - 5b + 10$
$= (4 - 5)b + 10$
$= -1b + 10$
$= -b + 10$

57. $8t - 3[t - 4(t + 1)]$
$= 8t - 3(t - 4t - 4)$
$= 8t - 3(-3t - 4)$
$= 8t + 9t + 12$
$= 17t + 12$

59. $a - 4[a - 4(a - 4)]$
$= a - 4(a - 4a + 16)$
$= a - 4(-3a + 16)$
$= a + 12a - 64$
$= 13a - 64$

61. $x + x[x + 3(x - 3)] = x + x(x + 3x - 9)$
$= x + x(4x - 9)$
$= x + 4x^2 - 9x$
$= 4x^2 - 8x$

63. $x - \{y - 3[x - 2(y - x)]\}$
$= x - [y - 3(x - 2y + 2x)]$
$= x - [y - 3(3x - 2y)]$
$= x - (y - 9x + 6y)$
$= x - (-9x + 7y)$
$= x + 9x - 7y$
$= 10x - 7y$

65. $3x + 2y[x + y(x - 3y) - y^2]$
$= 3x + 2y(x + xy - 3y^2 - y^2)$
$= 3x + 2y(x + xy - 4y^2)$
$= 3x + 2xy + 2xy^2 - 8y^3$

67. $6s^2 - [st - s(t + 5s) - s^2]$
$= 6s^2 - (st - st - 5s^2 - s^2)$
$= 6s^2 - (-6s^2)$
$= 6s^2 + 6s^2$
$= 12s^2$

69. (a) $3x - 15y - 11x$
$= 3(2.82) - 15(7.25) - 11(2.82)$
$= 8.46 - 108.75 - 31.02$
$= -100.29 - 31.02$
$= -131.31$

(b) $3x - 15y - 11x$
$= -8x - 15y$
$= -8(2.82) - 15(7.25)$
$= -22.56 - 108.75$
$= -131.31$

71. (a) $2(3x - 4) - 5(3x + 8)$
$= 2[3(2.82) - 4] - 5[3(2.82) + 8]$
$= 2(8.46 - 4) - 5(8.46 + 8)$
$= 2(4.46) - 5(16.46)$
$= 8.92 - 82.3$
$= -73.38$

(b) $2(3x - 4) - 5(3x + 8)$
$= 6x - 8 - 15x - 40$
$= -9x - 48$
$= -9(2.82) - 48$
$= -25.38 - 48$
$= -73.38$

73. $A = (12)(8) - (3)(x)$
$A = 96 - 3x$

75. $A = x(x + 5) - (2)(x)$
$A = x^2 + 5x - 2x$
$A = x^2 + 3x$

1.4 Exercises

1. number: n

8   more than   a number
8   + n
     $8 + n$

3. number: n

3 less than   twice a number
     $2n - 3$

5. number: n

4   more than   3 times a number   is

$4 + 3n =$ 7 less than   the number.
                     $n - 7$
     $4 + 3n = n - 7$

7. 1st number: n
2nd number: m

sum of two numbers is 1 more than their product
$n + m$ $= 1$ + nm
     $n + m = 1 + nm$

9. smaller number: n
larger number: 5 more than twice the smaller
       5   + $2n$
       $5 + 2n$

11. smallest number: n
middle number: 3 times the smallest
       $3n$
largest number: 12 more than the middle number
       12   + $3n$
       $12 + 3n$

13. 1st integer: n
2nd consecutive integer: $n + 1$

15. 1st even integer: n
2nd consecutive even integer: $n + 2$
3rd consecutive even integer : $n + 4$

17. 1st integer: n
2nd integer: $n + 1$
cube of 1st integer: n^3
cube of 2nd integer: $(n + 1)^3$
sum of the cubes: $n^3 + (n + 1)^3$

19. 1st number: n
2nd number: $40 - n$

21. 1st number: n
2nd number: $2n$
3rd number: $100 - (n + 2n) = 100 - 3n$

23. width: n
length: $3n$

$A = $ width $\cdot$ length
$A = n \cdot 3n$
$A = 3n^2$

$P = 2 \cdot$ width $+ 2 \cdot$ length
$P = 2(n) + 2(3n)$
$P = 2n + 6n$
$P = 8n$

25. 2nd side's length: n
1st side's length: $2n$
3rd side's length: $4 + n$

P = 1st side's length + 2nd side's length
         + 3rd side's length
$P = 2n + n + 4 + n$
$P = 4n + 4$

27. (a) $12 + 9 + 10 = 31$ coins

(b)   value of nickels: $12(0.05) = \$0.60$
value of dimes: $9(0.10) = \$0.90$
value of quarters: $10(0.25) = \$2.50$
Total value = $\$0.60 + \$0.90 + \$2.50 = \$4.00$

29. (a) $(n + d + q)$ coins

(b)   value of nickels: $5n$ cents
value of dimes: $10d$ cents
value of quarters: $25q$ cents
Total value = $5n + 10d + 25q$ cents

31. width: w
length: $3w$

(a)   regular fence = $2 \cdot$ width = $2w$ meters
(b)   cost of regular fence = $(2w$ meters$)(\$2$ per meter$)$
                                          = $4w$ dollars

(c)   heavy-duty fence = $2 \cdot$ length = $2(3w) = 6w$ meters
(d)   cost of heavy-duty fence = $(6w$ meters$)(\$5$ per meter$)$
                                                = $30w$ dollars

(e)   Total cost = cost of regular fence
                            + cost of heavy-duty fence
                        = $4w + 30w$
                        = $34w$ dollars

33. number of nickels: n cents
number of dimes: $20 - n$ cents
value of nickels: $5n$ cents
value of dimes: $10(20 - n) = 200 - 10n$ cents
Total value = value of nickels + value of dimes
                      = $5n + 200 - 10n$ cents
                      = $200 - 5n$ cents

35. (a) $A = 1500(1.06) = \$1590$

(b) $A = 1590(.06)^2 = \$1685.40$

(c) $A = 1500(1.06)^n$

(d) $A = 1500(1.06)^{15} = \$3594.84$

1.5 Exercises

1. $5(x - 3) - (x + 2) = -5$

Test $x = 0$:
$5(0 - 3) - (0 + 2) = -5$
$5(-3) - 2 = -5$
$-15 - 2 = -5$
$-17 = -5$
No

Test $x = 3$:
$5(3 - 3) - (3 + 2) = -5$
$5(0) - 5 = -5$
$0 - 5 = -5$
$-5 = -5$
Yes

Test $x = 5$:
$5(5 - 3) - (5 + 2) = -5$
$5(2) - 7 = -5$
$10 - 7 = -5$
$3 = -5$
No

3. $3(a - 5) + (1 - a) = -11$

Test $a = -3$:
$3(-3 - 5) + 2[1 - (-3)] = -11$
$3(-8) + 2(4) = -11$
$-24 + 8 = -11$
$-16 = -11$
No

Test $a = 0$:
$3(0 - 5) + 2(1 - 0) = -11$
$3(-5) + 2(1) = -11$
$-15 + 2 = -11$
$-13 = -11$
No

Test $a = 2$:
$3(2 - 5) + 2(1 - 2) = -11$
$3(-3) + 2(-1) = -11$
$-9 - 2 = -11$
$-11 = -11$
Yes

5. $x^2 - 4x = 5$

    Test $x = -1$:
$$(-1)^2 - 4(-1) = 5$$
$$1 + 4 = 5$$
$$5 = 5$$
             Yes

    Test $x = 5$:
$$5^2 - 4(5) = 5$$
$$25 - 20 = 5$$
$$5 = 5$$
             Yes

    Test $x = 2$:
$$2^2 - 4(2) = 5$$
$$4 - 8 = 5$$
$$-4 = 5$$
             No

7. $a(a + 8) = a + 8$

    Test $a = -1$:
$$-1(-1 + 8) = -1 + 8$$
$$-1(7) = 7$$
$$-7 = 7$$
             No

    Test $a = 1$:
$$1(1 + 8) = 1 + 8$$
$$1(9) = 9$$
$$9 = 9$$
             Yes

    Test $a = 3$:
$$2(3 + 8) = 3 + 8$$
$$3(11) = 11$$
$$33 = 11$$
             No

9. $$2x - 7 = 5$$
$$2x - 7 + 7 = 5 + 7$$
$$2x = 12$$
$$\frac{2x}{2} = \frac{12}{2}$$
$$x = 6$$

11. $$4y + 2 = -1$$
$$4y + 2 - 2 = -1 - 2$$
$$4y = -3$$
$$\frac{4y}{4} = \frac{-3}{4}$$
$$y = -\frac{3}{4}$$

13. $$m + 3 = 3 - m$$
$$m + 3 + m = 3 - m + m$$
$$2m + 3 = 3$$
$$2m + 3 - 3 = 3 - 3$$
$$2m = 0$$
$$\frac{2m}{2} = \frac{0}{2}$$
$$m = 0$$

15. $$3t - 5 = 5t - 13$$
$$3t - 5 - 3t = 5t - 13 - 3t$$
$$-5 = 2t - 13$$
$$-5 + 13 = 2t - 13 + 13$$
$$8 = 2t$$
$$\frac{8}{2} = \frac{2t}{2}$$
$$4 = t$$

17. $$11 - 3y = 38$$
$$11 - 3y - 11 = 38 - 11$$
$$-3y = 27$$
$$\frac{-3y}{-3} = \frac{27}{-3}$$
$$y = -9$$

19. $$5s + 2 = 3s - 7$$
$$5s + 2 - 3s = 3s - 7 - 3s$$
$$2s + 2 = -7$$
$$2s + 2 - 2 = -7 - 2$$
$$2s = -9$$
$$\frac{2s}{2} = \frac{-9}{2}$$
$$s = -\frac{9}{2}$$

21. $$3.24x - 5.2 = 7.74x + 0.3$$
$$3.24x - 5.2 - 3.24x = 7.74x + 0.3 - 3.24x$$
$$-5.2 = 4.5x + 0.3$$
$$-5.2 - 0.3 = 4.5x + 0.3 - 0.3$$
$$-5.5 = 4.5x$$
$$\frac{-5.5}{4.5} = \frac{4.5x}{4.5}$$
$$-\frac{11}{9} = x$$

23.
$$5 - 21x = 3x - 16$$
$$5 - 21x + 21x = 3x - 16 + 21x$$
$$5 = 24x - 16$$
$$5 + 16 = 24x - 16 + 16$$
$$21 = 24x$$
$$\frac{21}{24} = \frac{24x}{24}$$
$$\frac{7}{8} = x$$

25.
$$3x - 12 = 12 - 3x$$
$$3x - 12 + 3x = 12 - 3x + 3x$$
$$6x - 12 = 12$$
$$6x - 12 + 12 = 12 + 12$$
$$6x = 24$$
$$\frac{6x}{6} = \frac{24}{6}$$
$$x = 4$$

27.
$$3x - 12 = -12 - 3x$$
$$3x - 12 + 3x = -12 - 3x + 3x$$
$$6x - 12 = -12$$
$$6x - 12 + 12 = -12 + 12$$
$$6x = 0$$
$$\frac{6x}{6} = \frac{0}{6}$$
$$x = 0$$

29.
$$2t + 1 = 7 - t$$
$$2t + 1 + t = 7 - t + t$$
$$3t + 1 = 7$$
$$3t + 1 - 1 = 7 - 1$$
$$3t = 6$$
$$\frac{3t}{3} = \frac{6}{3}$$
$$t = 2$$

31.
$$3(x - 1) = 2(x + 1)$$
$$3x - 3 = 2x + 2$$
$$3x - 3 - 2x = 2x + 2 - 2x$$
$$x - 3 = 2$$
$$x - 3 + 3 = 2 + 3$$
$$x = 5$$

33.
$$3(x + 1) + x = 2(x + 3)$$
$$3x + 3 + x = 2x + 6$$
$$4x + 3 = 2x + 6$$
$$4x + 3 - 2x = 2x + 6 - 2x$$
$$2x + 3 = 6$$
$$2x + 3 - 3 = 6 - 3$$
$$2x = 3$$
$$\frac{2x}{2} = \frac{3}{2}$$
$$x = \frac{3}{2}$$

35.
$$0.06x + 0.0725(22.500 - x) = 1{,}500$$
$$0.06x + 1631.25 - 0.0725x = 1{,}500$$
$$-0.0125x + 1631.25 = 1{,}500$$
$$-0.0125x + 1631.25 - 1631.25 = 1{,}500 - 1631.25$$
$$-0.0125x = -131.25$$
$$\frac{-0.0125x}{-0.0125} = \frac{-131.25}{-0.0125}$$
$$x = 10{,}500$$

37.
$$\frac{x}{2} - 1 = 5$$
$$2\left(\frac{x}{2} - 1\right) = 2(5)$$
$$\frac{2}{1} \cdot \frac{x}{2} - 2(1) = 2(5)$$
$$x - 2 = 10$$
$$x - 2 + 2 = 10 + 2$$
$$x = 12$$

39.
$$\frac{2x}{3} + 2 = \frac{x}{2}$$
$$6\left(\frac{2x}{3} + 2\right) = 6\left(\frac{x}{2}\right)$$
$$\frac{6}{1} \cdot \frac{2x}{3} + 6(2) = \frac{6}{1} \cdot \frac{x}{2}$$
$$4x + 12 = 3x$$
$$4x + 12 - 4x = 3x - 4x$$
$$12 = -x$$
$$\frac{12}{-1} = \frac{-x}{-1}$$
$$-12 = x$$

41.

$$5x - \frac{2}{3} = \frac{x}{4}$$

$$12\left(5x - \frac{2}{3}\right) = 12\left(\frac{x}{4}\right)$$

$$12(5x) - \frac{12}{1} \cdot \frac{2}{3} = \frac{12}{1} \cdot \frac{x}{4}$$

$$60x - 8 = 3x$$
$$60x - 8 - 60x = 3x - 60x$$
$$-8 = -57x$$

$$\frac{-8}{-57} = \frac{-57x}{-57}$$

$$\frac{8}{57} = x$$

43.

$$6x - \frac{2}{5} = \frac{3x}{4}$$

$$20\left(6x - \frac{2}{5}\right) = 20\left(\frac{3x}{4}\right)$$

$$20(6x) - \frac{20}{1} \cdot \frac{2}{5} = \frac{20}{1} \cdot \frac{3x}{4}$$

$$120x - 8 = 15x$$
$$120x - 8 - 120x = 15x - 120x$$
$$-8 = -105x$$

$$\frac{-8}{-105} = \frac{-105x}{-105}$$

$$\frac{8}{105} = x$$

45.

$$\frac{3}{5}x - \frac{2}{3} = 5x$$

$$15\left(\frac{3}{5}x - \frac{2}{3}\right) = 15(5x)$$

$$\frac{15}{1} \cdot \frac{3}{5}x - \frac{15}{1} \cdot \frac{2}{3} = 15(5x)$$

$$9x - 10 = 75x$$
$$9x - 10 - 9x = 75x - 9x$$
$$-10 = 66x$$

$$\frac{-10}{66} = \frac{66x}{66}$$

$$-\frac{5}{33} = x$$

47.

$$\frac{3x - 2}{4} = 5$$

$$\frac{4}{1} \cdot \left(\frac{3x - 2}{4}\right) = 4(5)$$

$$3x - 2 = 20$$
$$3x - 2 + 2 = 20 + 2$$
$$3x = 22$$

$$\frac{3x}{3} = \frac{22}{3}$$

$$x = \frac{22}{3}$$

49.

$$\frac{7x - 1}{2} = \frac{2}{3}$$

$$\frac{6}{1} \cdot \left(\frac{7x - 1}{2}\right) = \frac{6}{1} \cdot \frac{2}{3}$$

$$3(7x - 1) = 4$$
$$21x - 3 = 4$$
$$21x - 3 + 3 = 4 + 3$$
$$21x = 7$$

$$\frac{21x}{21} = \frac{7}{21}$$

$$x = \frac{1}{3}$$

51.

$$5x + 7y = 4$$
$$5x + 7y \cdot 7y = 4 - 7y$$
$$5x = 4 - 7y$$

$$\frac{5x}{5} = \frac{4 - 7y}{5}$$

$$x = \frac{4 - 7y}{5}$$

53.

$$2x - 9y = 11$$
$$2x - 9y - 2x = 11 - 2x$$
$$-9y = 11 - 2x$$

$$\frac{-9y}{-9} = \frac{11 - 2x}{-9}$$

or

$$y = \frac{2x - 11}{9}$$

55.
$$2(x - y) = 3x + 4$$
$$2x - 2y = 3x + 4$$
$$2x - 2y - 2x = 3x + 4 - 2x$$
$$-2y = x + 4$$
$$-2y - 4 = x + 4 - 4$$
$$-2y - 4 = x$$

57.
$$2x - 5 < 7x - 2$$
$$2x - 5 - 2x < 7x - 2 - 2x$$
$$-5 < 5x - 2$$
$$-5 + 2 < 5x - 2 + 2$$
$$-3 < 5x$$

$$\frac{-3}{5} < \frac{5x}{5}$$

$$-\frac{3}{5} < x$$

or

$$x > -\frac{3}{5}$$

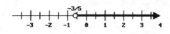

59.
$$4x - 25 < x - 8$$
$$4x - 25 - x < x - 8 - x$$
$$3x - 25 < -8$$
$$3x - 25 + 25 < -8 + 25$$
$$3x < 17$$

$$\frac{3x}{3} < \frac{17}{3}$$

$$x < \frac{17}{3}$$

61.
$$3 + (2x - 1) < 4x - 6$$
$$2x + 2 < 4x - 6$$
$$2x + 2 - 4x < 4x - 6 - 4x$$
$$-2x + 2 < -6$$

$$-2x + 2 - 2 < -6 - 2$$
$$-2x < -8$$

$$\frac{-2x}{-2} > \frac{-8}{-2}$$

$$x > 4$$

63.
$$3x - (2x - 7) \leq 4x + 6$$
$$3x - 2x + 7 \leq 4x + 6$$
$$x + 7 \leq 4x + 6$$
$$x + 7 - x \leq 4x + 6 - x$$
$$7 \leq 3x + 6$$
$$7 - 6 \leq 3x + 6 - 6$$
$$1 \leq 3x$$

$$\frac{1}{3} \leq \frac{3x}{3}$$

$$\frac{1}{3} \leq x$$

or

$$x \geq \frac{1}{3}$$

65.
$$6x - (3x + 1) \geq 2x + (2x - 3)$$
$$6x - 3x - 1 \geq 4x - 3$$
$$3x - 1 \geq 4x - 3$$
$$3x - 1 - 4x \geq 4x - 3 - 4x$$
$$-x - 1 \geq -3$$
$$-x - 1 + 1 \geq -3 + 1$$
$$-x \geq -2$$

$$\frac{-x}{-1} \leq \frac{-2}{-1}$$

$$x \leq 2$$

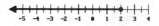

67. $3.4 - (2x - 5.8) \leq 2.6x + 4$

$3.4 - 2x + 5.8 \leq 2.6x + 4$

$-2x + 9.2 \leq 2.6x + 4$

$-2x + 9.2 + 2x \leq 2.6x + 4 + 2x$

$9.2 \leq 4.6x + 4$

$9.2 - 4 \leq 4.6x + 4 - 4$

$5.2 \leq 4.6x$

$$\frac{5.2}{4.6} \leq \frac{4.6x}{4.6}$$

$$\frac{26}{23} \leq x$$

or

$$x \geq \frac{26}{23}$$

CHAPTER 1 REVIEW EXERCISES

1. $A = \{1, 2, 3, 4\}$

3. $C \cap D = \{b\}$

5. $A = \{1, 2, 3, 4\}$
   $B = \{6, 7, 8, \dots\}$

   $A \cup B = \{1, 2, 3, 4, 6, 7, 8, \dots\}$

7. $A = \{1, 2, 3, 4, 6, 12\}$

9. $A = \{1, 2, 3, 4, 6, 12\}$
   $B = \{12, 24, 36, \dots\}$

   $A \cap B = \{12\}$

11. $B = \{12, 24, 36, \dots\}$
    $C = \{6, 12, 18, 24, \dots\}$

    $B \cap C = \{12, 24, 36, \dots\} = B$

13. $A = \{-1, 0, 1, 2, 3, 4\}$
    $B = \{3, 4, 5, 6, 7, 8, 9, 10, 11, 12\}$

    $A \cap B = \{3, 4\} = \{x \mid 3 \leq x \leq 4, x \in Z\}$

15. $\{x \mid x \leq 4\}$

17. $\{b \mid -8 \leq b \leq 5\}$

19. $\{a \mid -2 < a < 4\}$

21. False, $\frac{1}{2} \in Q$.

23. True, it is a repeating decimal.

25. False, $\pi \in I$.

27. True

29. True, commutative property of addition.

31. True, distributive property

33. True, multiplicative inverse property

35. False

37. $(-2) + (-3) - (-4) + (5) = -5 + 4 + (-5)$
    $= -1 + (-5)$
    $= -6$

39. $6 - 2 + 5 - 8 - 9 = 4 + 5 - 8 - 9$
    $= 9 - 8 - 9$
    $= 1 - 9$
    $= -8$

41. $(-2)(-3)(-5) = 6(-5)$
    $= -30$

43. $(-2)^6 = (-2)(-2)(-2)(-2)(-2)(-2)$
    $= 64$

45. $(-2) - (-3)^2 = (-2) - 9$
    $= -11$

47. $(-6 - 3)(-2 - 5) = (-9)(-7)$
$$= 63$$

49. $5 - 3[2 - (4 - 8) + 7]$
$= 5 - 3[2 - (-4) + 7]$
$= 5 - 3(6 + 7)$
$= 5 - 3(13)$
$= 5 - 39$
$= -34$

51. $5 - \{2 + 3[6 - 4(5 - 9)] - 2\}$
$= 5 - \{2 + 3[6 - 4(-4)] - 2\}$
$= 5 - [2 + 3(6 + 16) - 2]$
$= 5 - [2 + 3(22) - 2]$
$= 5 - (2 + 66 - 2)$
$= 5 - (66)$
$= -61$

53. $\dfrac{4[5 - 3(8 - 12)]}{-6 - 2(5 - 6)} = \dfrac{4[5 - 3(-4)]}{-6 - 2(-1)}$

$$= \dfrac{4(5 + 12)}{-6 + 2}$$

$$= \dfrac{4(17)}{-4}$$

$$= \dfrac{68}{-4}$$

$$= -17$$

55. $x^2 - 2xy + y^2 = (-2)^2 - 2(-2)(-1) + (-1)^2$
$= 4 - 4 + 1$
$= 1$

57. $|x - y| - (|x| - |y|)$

$= |-2 - (-1)| - (|-2| - |-1|)$

$= |-1| - (2 - 1)$
$= 1 - 1$
$= 0$

59. $\dfrac{2x^2y^3 + 3y^2}{zx^2y} = \dfrac{2(-2)(-1)^3 + 3(-1)^2}{0(-2)^2(-1)}$

$$= \dfrac{2(-2)(-1) + 3(1)}{0(4)(-1)}$$

$$= \dfrac{4 + 3}{0}$$

$$= \dfrac{7}{0}$$

undefined

61. $t = \dfrac{\overline{X} - a}{\dfrac{sx}{\sqrt{n}}}$

$= \dfrac{100 - 95}{\dfrac{7.1}{\sqrt{30}}}$

$= 3.86$

63. $(2x + y)(-3x^2y) = -6x^3y - 3x^2y^2$

65. $(2xy^2)^2(-3x)^2$
$= (2xy^2)(2xy^2)(-3x)(-3x)$
$= (2)(2)(-3)(-3)x \cdot x \cdot x \cdot x \cdot y^2 \cdot y^2$
$= 36x^4y^4$

67. $3x - 2y - 4x + 5y - 3x$
$= (3 - 4 - 3)x + (-2 + 5)y$
$= -4x + 3y$

69. $-2r^2s + 5rs^2 - 3sr^2 - 4s^2r$
$= (-2 - 3)r^2s + (5 - 4)rs^2$
$= -5r^2s + 1rs^2$
$= -5r^2s + rs^2$

71. $-2x - (3 - x) = -2x - 3 + x$
$= (-2 + 1)x - 3$
$= -x - 3$

73. $3a(a - b + c) = 3a^2 - 3ab + 3ac$

75. $2x - 3(x - 4) = 2x - 3x + 12$
$= (2 - 3)x + 12$
$= -x + 12$

77. $3a - [5 - (a - 4)] = 3a - (5 - a + 4)$
$= 3a - (9 - a)$
$= 3a - 9 + a$
$= 4a - 9$

79. $5x - \{3x + 2[x - 3(5 - x)]\}$
    $= 5x - [3x + 2(x - 15 + 3x)]$
    $= 5x - [3x + 2(4x - 15)]$
    $= 5x - (3x + 8x - 30)$
    $= 5x - (11x - 30)$
    $= 5x - 11x + 30$
    $= -6x + 30$

81. 1st number: n
    2nd number: m

    Five less than the product of two numbers   is   3 more than   their sum.
    $\qquad\qquad nm - 5 \qquad\qquad\qquad = 3 \qquad + \qquad n + m$

    $nm - 5 = 3 + n + m$

83. 1st integer: n
    2nd consecutive odd integer: $n + 2$
    3rd consecutive odd integer: $n + 4$

    sum of first two is 5 less than the third
    $n + n + 2 \qquad\qquad = \qquad\qquad n + 4 - 5$

    $n + n + 2 = n + 4 - 5$

85. width: n
    length: $4n - 5$

    $A = $ width $\cdot$ length
    $A = n(4n - 5)$
    $A = 4n^2 - 5n$

    $P = 2 \cdot$ width $+ 2 \cdot$ length
    $P = 2(n) + 2(4n - 5)$
    $P = 2n + 8n - 10$
    $P = 10n - 10$

87. 1st number: n
    2nd number: m

    sum of the squares is 8 more than the product of the number
    $n^2 + m^2 \qquad\quad = 8 \qquad + \qquad\qquad nm$
    $n^2 + m^2 = 8 + nm$

89. number of dimes: n
    number of nickels: $40 - n$
    value of dimes: $10n$
    value of nickels: $5(40 - n)$

    Total value = value of dimes + value of nickels
    Total value $= 10n + 5(40 - n)$
    $\qquad\qquad\quad = 10n + 200 - 5n$
    $\qquad\qquad\quad = 5n + 200$ cents

91. $\qquad 5x - 2 = -2$
    $5x - 2 + 2 = -2 + 2$
    $\qquad\quad 5x = 0$

    $\qquad \dfrac{5x}{5} = \dfrac{0}{5}$

    $\qquad\quad x = 0$

93. $\qquad 3x - 5 = 2x + 6$
    $3x - 5 - 2x = 2x + 6 - 2x$
    $\qquad\quad x - 5 = 6$
    $x - 5 + 5 = 6 + 5$
    $\qquad\qquad x = 11$

95. $\qquad 11x + 2 = 6x - 3$
    $11x + 2 - 6x = 6x - 3 - 6x$
    $\qquad\quad 5x + 2 = -3$
    $\quad 5x + 2 - 2 = -3 - 2$
    $\qquad\qquad 5x = -5$

    $\qquad\qquad \dfrac{5x}{5} = \dfrac{-5}{5}$

    $\qquad\qquad x = -1$

97. $\qquad 5(a - 3) = 2(a - 4)$
    $\qquad 5a - 15 = 2a - 8$
    $5a - 15 - 2a = 2a - 8 - 2a$
    $\qquad 3a - 15 = -8$
    $3a - 15 + 15 = -8 + 15$
    $\qquad\qquad 3a = 7$

    $\qquad\qquad \dfrac{3a}{3} = \dfrac{7}{3}$

    $\qquad\qquad a = \dfrac{7}{3}$

99. $6(q - 4) + 2(q + 5) = 8q - 19$
$6q - 24 + 2q + 10 = 8q - 19$
$8q - 14 = 8q - 19$
$8q - 14 - 8q = 8q - 19 - 8q$
$-14 = -19$
$\emptyset$

101. $3x - 5x = 7x - 4x$
$-2x = 3x$
$-2x + 2x = 3x + 2x$
$0 = 5x$

$$\frac{0}{5} = \frac{5x}{5}$$

$$0 = x$$

103. $x - \dfrac{2}{3} = 2x + 4$

$$3\left(x - \frac{2}{3}\right) = 3(2x + 4)$$

$$3x - \frac{3}{1} \cdot \frac{2}{3} = 6x + 12$$

$3x - 2 = 6x + 12$
$3x - 2 - 3x = 6x + 12 - 3x$
$-2 = 3x + 12$
$-2 - 12 = 3x + 12 - 12$
$-14 = 3x$

$$\frac{-14}{3} = \frac{3x}{3}$$

$$-\frac{14}{3} = x$$

105. $\dfrac{x - 3}{5} = x + 1$

$$\frac{5}{1} \cdot \frac{x - 3}{5} = 5(x + 1)$$

$x - 3 = 5x + 5$
$x - 3 - x = 5x + 5 - x$
$-3 = 4x + 5$
$-3 - 5 = 4x + 5 - 5$
$-8 = 4x$

$$\frac{-8}{4} = \frac{4x}{4}$$

$$-2 = x$$

107. $7x + 8y = 22$
$7x + 8y - 7x = 22 - 7x$
$8y = 22 - 7x$

$$\frac{8y}{8} = \frac{22 - 7x}{8}$$

$$y = \frac{22 - 7x}{8}$$

109. $3x - 6y = 2 - 4x + 2y$
$3x - 6y - 2y = 2 - 4x + 2y - 2y$
$3x - 8y = 2 - 4x$
$3x - 8y - 3x = 2 - 4x - 3x$
$-8y = 2 - 7x$

$$\frac{-8y}{-8} = \frac{2 - 7x}{-8}$$

$$y = \frac{2 - 7x}{-8} \quad \text{or} \quad y = \frac{7x - 2}{8}$$

111. $3x + 12 < 2x - 9$
$3x + 12 - 2x < 2x - 9 - 2x$
$x + 12 < -9$
$x + 12 - 12 < -9 - 12$
$x < -21$

113. $5 - (2x - 7) > 3x - 15$
$5 - 2x + 7 > 3x - 15$
$12 - 2x > 3x - 15$
$12 - 2x + 2x > 3x - 15 + 2x$
$12 > 5x - 15$
$12 + 15 > 5x - 15 + 15$
$27 > 5x$

$$\frac{27}{5} > \frac{5x}{5}$$

$$\frac{27}{5} > x \quad \text{or} \quad x < \frac{27}{5}$$

115.
$$3.2 - (5.3x - 1.2) \le 3.4x - 1.8$$
$$3.2 - 5.3x + 1.2 \le 3.4x - 1.8$$
$$4.4 - 5.3x \le 3.4x - 1.8$$
$$4.4 - 5.3x + 5.3x \le 3.4x - 1.8 + 5.3x$$
$$4.4 \le 8.7x - 1.8$$
$$4.4 + 1.8 \le 8.7x - 1.8 + 1.8$$
$$6.2 \le 8.7x$$

$$\frac{6.2}{8.7} \le \frac{8.7x}{8.7}$$

$$0.71 \le x$$

or

$$x \ge 0.71$$

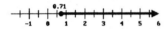

1. $A = \{2, 3, 5, 7\}$
        $B = \{2, 3, 5, 7, 11, 13, 17, 19, 23\}$

   (a) $A \cap B = \{2, 3, 5, 7\}$

   (b) $A \cup B = \{2, 3, 5, 7, 11, 13, 17, 19, 23\} = B$

2. (a) False; $\sqrt{16} = 4 \in Z$

   (b) True; $2 = \dfrac{2}{1} \in Q$

   (c) False; $-\dfrac{3}{4} \in Q$

3. (a) $\{a \mid a > 4\}$

   (b) $\{x \mid -3 \le x < 10\}$

4. (a) False
   (b) True, commutative property of addition

5. (a) $-3 - (-6) + (-4) - (-9)$
          $= -3 + 6 + (-4) + 9$
          $= 3 + (-4) + 9$
          $= -1 + 9$
          $= 8$

   (b) $(-7)^2 - (-6)(-2)(-3)$
          $= 49 - (-6)(-2)(-3)$
          $= 49 - (-36)$
          $= 85$

   (c) $|3 - 8| - |5 - 9| = |-5| - |-4|$
                                $= 5 - 4$
                                $= 1$

   (d) $6 - 5[-2 - (7 - 9)]$
          $= 6 - 5[-2 - (-2)]$
          $= 6 - 5(0)$
          $= 6 - 0$
          $= 6$

6. (a) $(x - y)^2 = [(-2) - (-3)]^2$
                    $= (1)^2$
                    $= 1$

   (b) $\dfrac{x^2 - y^2}{x^2 - 2xy - y^2} = \dfrac{(-2)^2 - (-3)^2}{(-2)^2 - 2(-2)(-3) - (-3)^2}$

   $= \dfrac{4 - 9}{4 - 12 - 9}$

   $= \dfrac{-5}{-17}$

   $= \dfrac{5}{17}$

7. (a) $(5x^3y^2)(-2x^2y)(-xy^2)$
          $= (5)(-2)(-1)x^3x^2xy^2yy^2$
          $= 10x^6y^5$

   (b) $3rs^2 - 5r^2s - 4rs^2 - 7rs$
          $= (3 - 4)rs^2 - 5r^2s - 7rs$
          $= -rs^2 - 5r^2s - 7rs$

   (c) $2a - 3(a - 2) - (6 - a)$
          $= 2a - 3a + 6 - 6 + a$
          $= (2 - 3 + 1)a + (6 - 6)$
          $= 0$

   (d) $7r - \{3 + 2[s - (r - 2s)]\}$
          $= 7r - [3 + 2(s - r + 2s)]$
          $= 7r - [3 + 2(3s - r)]$
          $= 7r - (3 + 6s - 2r)$
          $= 7r - 3 - 6s + 2r$
          $= 9r - 6s - 3$

8. width: n
    length: $3n - 8$

    $P = 2 \cdot \text{width} + 2 \cdot \text{length}$
    $P = 2(n) + 2(3n - 8)$
    $P = 2n + 6n - 16$
    $P = 8n - 16$

9. number of dimes: x
    number of nickels: $34 - x$
    value of dimes: $10x$
    value of nickels: $5(34 - x)$

Total value = value of dimes + value of nickels
Total value $= 10x + 5(34 - x)$
            $= 10x + 170 - 5x$
            $= 5x + 170$ cents

10. (a)
$$3x - 2 = 5x + 8$$
$$3x - 2 - 3x = 5x + 8 - 3x$$
$$-2 = 2x + 8$$
$$-2 - 8 = 2x + 8 - 8$$
$$-10 = 2x$$

$$\frac{-10}{2} = \frac{2x}{2}$$

$$-5 = x$$

(b)
$$2(x - 4) = 5x - 7$$
$$2x - 8 = 5x - 7$$
$$2x - 8 - 2x = 5x - 7 - 2x$$
$$-8 = 3x - 7$$
$$-8 + 7 = 3x - 7 + 7$$
$$-1 = 3x$$

$$\frac{-1}{3} = \frac{3x}{3}$$

$$-\frac{1}{3} = x$$

(c)
$$\frac{2x + 1}{3} = \frac{1}{4}$$

$$\frac{12}{1} \cdot \frac{2x + 1}{3} = \frac{12}{1} \cdot \frac{1}{4}$$

$$4(2x + 1) = 3$$
$$8x + 4 = 3$$
$$8x + 4 - 4 = 3 - 4$$
$$8x = -1$$

$$\frac{8x}{8} = \frac{-1}{8}$$

$$x = -\frac{1}{8}$$

(d)
$$3x - (5x - 4) \le 2x + 8$$
$$3x - 5x + 4 \le 2x + 8$$
$$-2x + 4 \le 2x + 8$$
$$-2x + 4 + 2x \le 2x + 8 + 2x$$
$$4 \le 4x + 8$$
$$4 - 8 \le 4x + 8 - 8$$
$$-4 \le 4x$$

$$\frac{-4}{4} \le \frac{4x}{4}$$

$$-1 \le x$$

or

$$x \ge -1$$

(e)
$$2.1x - 5 > 5.3x - 4.9$$
$$2.1x - 5 - 2.1x > 5.3x - 4.9 - 2.1x$$
$$-5 > 3.2x - 4.9$$
$$-5 + 4.9 > 3.2x - 4.9 + 4.9$$
$$0.1 > 3.2x$$

$$\frac{-0.1}{3.2} > \frac{3.2x}{3.2}$$

$$-0.03125 > x \text{ or } -\frac{1}{32} > x$$

or

$$x < -0.03125 \text{ or } x < -\frac{1}{32}$$

11.
$$3s - 5t = 2t - 4$$
$$3s - 5t + 5t = 2t - 4 + 5t$$
$$3s = 7t - 4$$
$$3s + 4 = 7t - 4 + 4$$
$$3s + 4 = 7t$$

$$\frac{3s + 4}{7} = \frac{7t}{7}$$

$$\frac{3s + 4}{7} = t$$

CHAPTER 2

2.1 Exercises

1. $3x - 5y = 17$
   $3(4) - 5(1) = 17$
   $12 - 5 = 17$
   $7 = 17$
   No

3. $4y - 3x = 7$
   $4(-1) - 3(1) = 7$
   $-4 - 3 = 7$
   $-7 = 7$
   No

5. $2x + 3y = 2$
   $2\left(\dfrac{3}{2}\right) + 3\left(-\dfrac{1}{3}\right) = 2$
   $3 - 1 = 2$
   $2 = 2$
   Yes

7. $\dfrac{2}{3}x - \dfrac{1}{4}y = 1$
   $2\left(\dfrac{3}{2}\right) + 3\left(-\dfrac{1}{3}\right) = 2$
   $4 - 3 = 1$
   $1 = 1$
   Yes

9. $x + y = 8$

   $-1 + y = 8$
   $y = 9$
   $(-1, 9)$

   $0 + y = 8$
   $y = 8$
   $(0, 8)$

   $1 + y = 8$
   $y = 7$
   $(1, 7)$

   $x + (-2) = 8$
   $x = 10$
   $(10, -2)$

$x + 0 = 8$
$x = 8$
$(8, 0)$

$x + 4 = 8$
$x = 4$
$(4, 4)$

11. $5x + 4y = 20$

   $5(-2) + 4y = 20$
   $-10 + 4y = 20$
   $4y = 30$
   $y = \dfrac{15}{2}$
   $\left(-2, \dfrac{15}{2}\right)$

   $5(0) + 4y = 20$
   $4y = 20$
   $y = 5$
   $(0, 5)$

   $5(4) + 4y = 20$
   $20 + 4y = 20$
   $4y = 0$
   $y = 0$
   $(4, 0)$

   $5x + 4(-5) = 20$
   $5x - 20 = 20$
   $5x = 40$
   $x = 8$
   $(8, -5)$

   $5x + 4(0) = 20$
   $5x = 20$
   $x = 4$
   $(4, 0)$

   $5x + 4(4) = 20$
   $5x + 16 = 20$
   $5x = 4$
   $x = \dfrac{4}{5}$
   $\left(\dfrac{4}{5}, 4\right)$

13. $\dfrac{x}{3} + \dfrac{y}{4} = 1$

$12\left(\dfrac{x}{3} + \dfrac{y}{4}\right) = 12(1)$

$4x + 3y = 12$

$4(-3) + 3y = 12$
$-12 + 3y = 12$
$3y = 24$
$y = 8$
$(-3,\ 8)$

$4(0) + 3y = 12$
$3y = 12$
$y = 4$
$(0,\ 4)$

$4(3) + 3y = 12$
$12 + 3y = 12$
$3y = 0$
$y = 0$
$(3,\ 0)$

$4x + 3(-4) = 12$
$4x - 12 = 12$
$4x = 24$
$x = 6$
$(6,\ -4)$

$4x + 3(0) = 12$
$4x = 12$
$x = 3$
$(3,\ 0)$

$4x + 3(4) = 12$
$4x + 12 = 12$
$4x = 0$
$x = 0$
$(0,\ 4)$

15. $x + y = 6$

x-intercept: Let $y = 0$
$x + 0 = 6$
$x = 6$

y-intercept: Let $x = 0$
$0 + y = 6$
$y = 6$

17. $x - y = 6$

x-intercept: Let $y = 0$
$x - 0 = 6$
$x = 6$

y-intercept: Let $x = 0$
$0 - y = 6$
$-y = 6$
$y = -6$

19. $y - x = 6$

x-intercept: Let $y = 0$
$0 - x = 6$
$-x = 6$
$x = -6$

y-intercept: Let $x = 0$
$y - 0 = 6$
$y = 6$

21. $2x + 4y = 12$

x-intercept: Let $y = 0$
$2x + 4(0) = 12$
$2x = 12$
$x = 6$

y-intercept: Let $x = 0$
$2(0) + 4y = 12$
$4y = 12$
$y = 3$

23. $y = -\dfrac{4}{3}x + 4$

x-intercept: Let $y = 0$

$0 = -\dfrac{4}{3}x + 4$

$\dfrac{4}{3}x = 4$

$x = 3$

y-intercept: Let $x = 0$

$y = -\dfrac{4}{3}(0) + 4$

$y = 4$

25. $y = \dfrac{3}{5}x - 3$

   x-intercept: Let $y = 0$

   $0 = \dfrac{3}{5}x - 3$

   $3 = \dfrac{3}{5}x$

   $5 = x$

   y-intercept: Let $x = 0$

   $y = \dfrac{3}{5}(0) - 3$

   $y = -3$

27. $2y - 3x = 7$

   x-intercept: Let $y = 0$

   $2(0) - 3x = 7$

   $\qquad -3x = 7$

   $\qquad x = -\dfrac{7}{3}$

   y-intercept: Let $x = 0$

   $2y - 3(0) = 7$

   $\qquad 2y = 7$

   $\qquad y = \dfrac{7}{2}$

29. $4x + 3y = 0$

   Let $x = 0$: $4(0) + 3y = 0$

   $\qquad\qquad\qquad 3y = 0$

   $\qquad\qquad\qquad y = 0$

   $\qquad\qquad\qquad (0, 0)$

   Let $x = 3$: $4(3) + 3y = 0$

   $\qquad\qquad\quad 12 + 3y = 0$

   $\qquad\qquad\qquad 3y = -12$

   $\qquad\qquad\qquad y = -4$

   $\qquad\qquad\qquad (3, -4)$

Let $x = -3$: $4(-3) + 3y = 0$

$\qquad\qquad\quad -12 + 3y = 0$

$\qquad\qquad\qquad 3y = 12$

$\qquad\qquad\qquad y = 4$

$\qquad\qquad\qquad (-3, 4)$

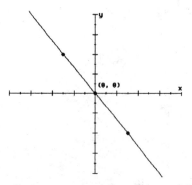

31. $y = x$

   Let $x = 0$: $y = 0$

   $\qquad\qquad\quad (0, 0)$

   Let $x -= 1$: $y = 1$

   $\qquad\qquad\quad (1, 1)$

   Let $x = 2$: $y = 2$

   $\qquad\qquad\quad (2, 2)$

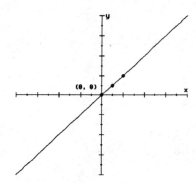

33. $\dfrac{x}{2} - \dfrac{y}{3} = 1$

   $6\left(\dfrac{x}{2} - \dfrac{y}{3}\right) = 6(1)$

   $3x - 2y = 6$

   Let $x = 0$: $3(0) - 2y = 6$

   $\qquad\qquad\qquad -2y = 6$

   $\qquad\qquad\qquad y = -3$

   $\qquad\qquad\qquad (0, -3)$

Let $y = 0$: $3x - 2(0) = 6$
$$3x = 6$$
$$x = 2$$
$$(2, 0)$$

Let $x = 4$: $3(4) - 2y = 6$
$$12 - 2y = 6$$
$$-2y = -6$$
$$y = 3$$
$$(4, 3)$$

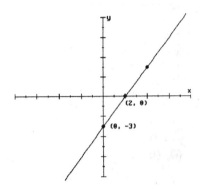

35. $y = 3x - 1$

Let $x = 0$: $y = 3(0) - 1$
$$y = -1$$
$$(0, -1)$$

Let $y = 0$: $0 = 3x - 1$
$$1 = 3x$$

$$\frac{1}{3} = x$$

$$\left(\frac{1}{3}, 0\right)$$

Let $x = 1$: $y = 3(1) - 1$
$$y = 2$$
$$(1, 2)$$

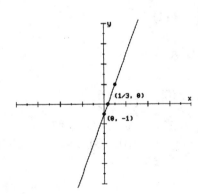

37. $y = -\frac{2}{3}x + 4$

Let $x = 0$: $y = -\frac{2}{3}(0) + 4$
$$y = 4$$
$$(0, 4)$$

Let $y = 0$: $0 = -\frac{2}{3}x + 4$

$$\frac{2}{3}x = 4$$

$$x = 6$$
$$(6, 0)$$

Let $x = 3$: $y = -\frac{2}{3}(3) + 4$

$$y = -2 + 4$$
$$y = 2$$
$$(3, 2)$$

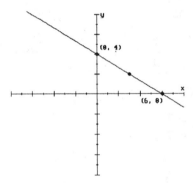

39. $5x - 4y = 20$

Let $x = 0$: $5(0) - 4y = 20$
$$-4y = 20$$
$$y = -5$$
$$(0, -5)$$

Let $y = 0$: $5x - 4(0) = 20$
$$5x = 20$$
$$x = 4$$
$$(4, 0)$$

Let $x = 2$: $5(2) - 4y = 20$
$$10 - 4y = 20$$
$$-4y = 10$$
$$y = -\frac{5}{2}$$

$$\left(2, -\frac{5}{2}\right)$$

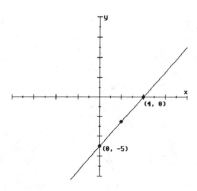

41. $5x - 7y = 30$

Let $x = 0$: $5(0) - 7y = 30$
$$-7y = 30$$
$$y = -\frac{30}{7}$$

$$\left(0, -\frac{30}{7}\right)$$

Let $y = 0$: $5x - 7(0) = 30$
$$5x = 30$$
$$x = 6$$
$$(6, 0)$$

Let $x = 1$: $5(1) - 7y = 30$
$$5 - 7y = 30$$
$$-7y = 25$$
$$y = -\frac{25}{7}$$

$$\left(1, -\frac{25}{7}\right)$$

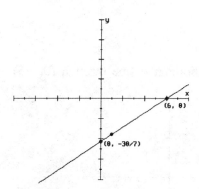

43. $5x + 7y = 30$

Let $x = 0$: $5(0) + 7y = 30$
$$7y = 30$$
$$y = \frac{30}{7}$$

$$\left(0, \frac{30}{7}\right)$$

Let $y = 0$: $5x + 7(0) = 30$
$$5x = 30$$
$$x = 6$$
$$(6, 0)$$

Let $x = 1$: $5(1) + 7y = 30$
$$5 + 7y = 30$$
$$7y = 25$$
$$y = \frac{25}{7}$$

$$\left(1, \frac{25}{7}\right)$$

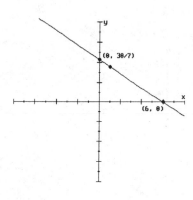

45. $x = 5$

This is a vertical line through $(5, 0)$.

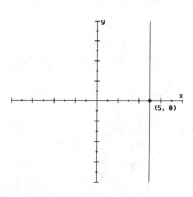

47. $-\dfrac{3}{4}x + y = 2$

    Let $x = 0$: $-\dfrac{3}{4}(0) + y = 2$

                             $y = 2$
                             $(0, 2)$

    Let $y = 0$: $-\dfrac{3}{4}x + 0 = 2$

                       $-\dfrac{3}{4}x = 2$

                         $x = -\dfrac{8}{3}$

$$\left(-\dfrac{8}{3},\, 0\right)$$

    Let $x = -4$: $-\dfrac{3}{4}(-4) + y = 2$

                       $3 + y = 2$
                            $y = -1$
                         $(-4, -1)$

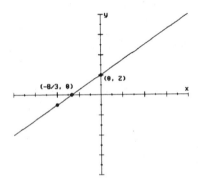

49. $\dfrac{3}{4}x - y = 2$

    Let $x = 0$: $\dfrac{3}{4}(0) - y = 2$

                         $-y = 2$
                         $y = -2$
                         $(0, -2)$

    Let $y = 0$: $\dfrac{3}{4}x - 0 = 2$

                         $\dfrac{3}{4}x = 2$

                         $x = \dfrac{8}{3}$

$$\left(\dfrac{8}{3},\, 0\right)$$

    Let $x = 4$; $\dfrac{3}{4}(4) - y = 2$

                       $3 - y = 2$
                         $-y = -1$
                         $y = 1$
                         $(4, 1)$

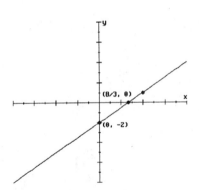

49. $\dfrac{3}{4}x - y = 2$

    Let $x = 0$: $\dfrac{3}{4}(0) - y = 2$

                         $-y = 2$
                         $y = -2$
                         $(0, -2)$

    Let $y = 0$: $\dfrac{3}{4}x - 0 = 2$

                         $\dfrac{3}{4}x = 2$

                         $x = \dfrac{8}{3}$

$$\left(\dfrac{8}{3},\, 0\right)$$

51. $y + 5 = 0$
      $y = -5$

This is a horizontal line through $(0, -5)$.

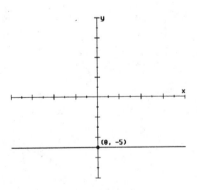

53. $5x - 4y = 0$

Let $x = 0$: $5(0) - 4y = 0$
$$-4y = -$$
$$y = 0$$
$$(0, 0)$$

Let $x = 4$: $5(4) - 4y = 0$
$$20 - 4y = 0$$
$$-4y = -20$$
$$y = 5$$
$$(4, 5)$$

Let $x = -4$: $5(-4) - 4y = 0$
$$-20 - 4y = 0$$
$$-20 = 4y$$
$$-5 = y$$
$$(-4, -5)$$

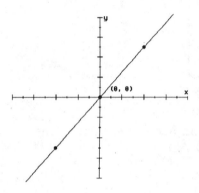

55. $5x - 4 = 0$
$$5x = 4$$
$$x = \frac{4}{5}$$

This is a vertical line through $\left(\frac{4}{5}, 0\right)$.

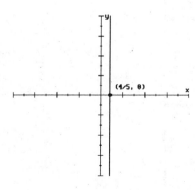

57. $y = 10x - 6$

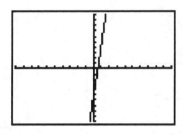

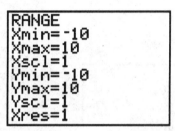

59. $y = -0.2x + 7$

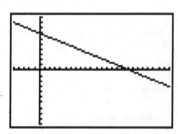

61. $2x + 3y = 180$

$\quad\quad 3y = -2x + 180$

$\quad\quad\; y = -\dfrac{2}{3}x + 60$

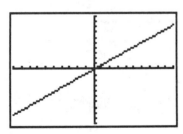

```
RANGE
Xmin=-10
Xmax=100
Xscl=10
Ymin=-10
Ymax=100
Yscl=10
Xres=1
```

65. $y = 0.1x + 2.4$

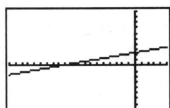

```
RANGE
Xmin=-40
Xmax=10
Xscl=2
Ymin=-10
Ymax=10
Yscl=1
Xres=1
```

67. (a) distance = distance from home   +   rate · time
                at 11:00 AM   since 11:00 AM

$$d = 260 + 52h$$

(b)

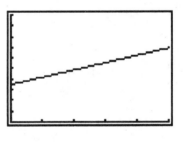

```
RANGE
Xmin=0
Xmax=5
Xscl=1
Ymin=0
Ymax=750
Yscl=75
Xres=1
```

63. $15x - 18y = 1$

$\quad\quad -18y = -15x + 1$

$\quad\quad\quad\; y = \dfrac{5}{6}x - \dfrac{1}{18}$

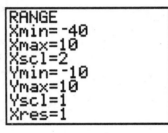

```
RANGE
Xmin=-10
Xmax=10
Xscl=1
Ymin=-10
Ymax=10
Yscl=1
Xres=1
```

(c)

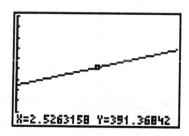

```
RANGE
Xmin=0
Xmax=200
Xscl=10
Ymin=0
Ymax=60
Yscl=5
Xres=1
```

391 miles

(c)

(d)

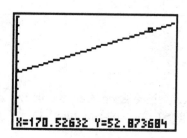

X=120   Y=45.8

120 miles

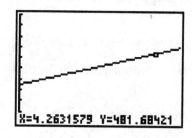

(d)

X=4.2631579 Y=481.68421

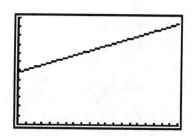

4.3 hours

X=170.52632 Y=52.873604

69. (a)   Rental charge

= daily charge + charge per mile · number of miles

$C = 29 + 0.14n$

(b)

$52.87

2.2 Exercises

1. {(3, 9), (3, 7), (8, 2), (7, 2)}

3. {(3, a), (3, b), (8, b), (−1, b), (−1, c)}

5. domain: {3, 4, 5}
    range: {2, 3}

7. domain: {3, −2, 4}
    range: {−2, −1, 3}

9. $\{x \mid x \neq 0\}$

11. $\{x \mid x \text{ is a real number}\}$

13. $2x + 3 \neq 0$
$$2x \neq -3$$
$$x \neq -\frac{3}{2}$$
$$\left\{x \mid x \neq -\frac{3}{2}\right\}$$

15. $\{x \mid x \text{ is a real number}\}$

17. $x - 4 \geq 0$
$$x \geq 4$$
$$\{x \mid x \geq 4\}$$

19. $5 - 4x \geq 0$
$$5 \geq 4x$$
$$\frac{5}{4} \geq x$$
$$\left\{x \mid x \leq \frac{5}{4}\right\}$$

21. $4x \geq 0$
$$x \geq 0$$
$$\{x \mid x \geq 0\}$$

23. $x - 3 > 0$
$$x > 3$$
$$\{x \mid x > 3\}$$

25. Function, since each element in the domain is assigned only one element in the range.

27. Not a function, since the domain element 6 is assigned two range elements, 3 and 1.

29. Function, since each element in the domain is assigned only one element in the range.

31. Not a function, since the domain element 3 is assigned two range elements, 1 and 2.

33. Function, since each element in the domain is assigned only one element in the range.

35. Function, since each element in the domain is assigned only one element in the range.

37. Not a function, since the domain element 9 is assigned two range elements, −1 and 3.

39. Function, since there corresponds exactly one y-value to each x-value.

41. Not a function. For example, let $x = 0$.
$$0^2 + y^2 = 81$$
$$y^2 = 81$$
$$y = \pm 9$$

There are two range elements corresponding to a domain element.

43. Not a function. For example, let $x = 0$.
$$0 = y^2 - 4$$
$$4 = y^2$$
$$\pm 2 = y$$

There are two range elements corresponding to a domain element.

45. Function, it passes the Vertical Line Test.

47. Not a function, it doesn't pass the Vertical Line Test.

49. Function, it passes the Vertical Line Test.

51. Not a function, it doesn't pass the Vertical Line Test.

53. $y = 0$

55. If $y = -5$, then $x = 7$. The ordered pair is $(7, -5)$.

57. 3 values; −6, −2, and 4

59. 1 value

61. The domain is $\{x \mid -7 \leq x \leq 7\}$.

63. Domain: $\{x \mid -4 \leq x \leq 5\}$
Range: $\{y \mid -4 \leq y \leq 6\}$

65. Domain: $\{x \mid 0 \leq x \leq 4\}$
Range: $\{y \mid -3 \leq y \leq 3\}$

67. Domain: $\{x \mid -5 \leq x \leq -2, 1 \leq x \leq 5\}$
Range: $\{y \mid -3 \leq y \leq 1, 3 \leq y \leq 4\}$

69. $y = 2x - 4$

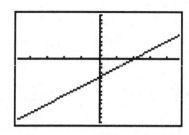

Range: $\{y \mid -13.4 \le y \le 5.4\}$

71. $y = x^3 - 4x$

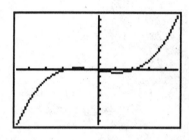

Range: $\{y \mid -85.0 \le y \le 85.0\}$

73. $y = \dfrac{x^2 + 4}{x^2 + 1}$

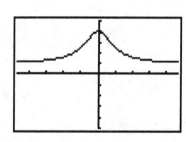

Range: $\{y \mid 1.1 \le y \le 4\}$

75. $A = s^2$

77.

Area of shaded region	=	Total Area	−	Unshaded Area

$$A = (a)(2a) - (5)(3)$$
$$A = 2a^2 - 15$$

$a > 3$ and $2a > 5$

$a > 3$ and $a > \dfrac{5}{2}$

a can be any value greater than 3.

79.

$C =$	fee per day	·	number of days	+	charge per mile	·	number miles

$$C = 29 \cdot 4 + 0.12 \cdot m$$
$$C = 116 + 0.12m$$

81.
$$C = 10 \cdot d + 12 \cdot (d + 5)$$
$$C = 10d + 12d + 60$$
$$C = 22d + 60$$

83.
$$C = 0.02n^2 + 100n + 10000$$
$n = 100$:　$C = 0.02(100)^2 + 100(100) + 10000$
　　　　　　$= 20{,}200$

$20,200

$n = 200$:　$C = 0.02(200)^2 + 100(200) + 10000$
　　　　　　$= 30{,}800$

$30,800

$n = 500$:　$C = 0.02(500)^2 + 100(500) + 10000$
　　　　　　$= 65{,}000$

$65,000

$n = 1000$:　$C = 0.02(1000)^2 + 100(1000) + 10000$
　　　　　　$= 130{,}000$

$130,000

85. (a)

$C =$	flat fee	+	fee per minute	·	number of minutes
$C =$	29.95	+	0.33	·	m

(b) $m = 75$:　$C = 29.95 + 0.33(75)$
　　　　　　$= 54.7$

$54.70

$$m = 180: \quad C = 29.95 + 0.33(180)$$
$$= 89.35$$

$89.35

$$m = 300: \quad C = 29.95 + 0.33(300)$$
$$= 128.95$$

$128.95

87. (a) Yes, it passes the Vertical Line Test.

   (b) Domain: The rent varies between $200 and $400.
   Range: The profit varies between $35,000 and $50,000.

2.3 Exercises

1. $f(0) = 2(0) - 3$
   $= 0 - 3$
   $= -3$

3. $g(2) = 3(2)^2 - 2 + 1$
   $= 3(4) - 2 + 1$
   $= 11$

5. $g(-2) = 3(-2)^2 - (-2) + 1$
   $= 3(4) + 2 + 1$
   $= 15$

7. $h(3) = \sqrt{3 + 5}$
   $= \sqrt{8}$
   $= 2\sqrt{2}$

9. $h(-3) = \sqrt{-3 + 5}$
   $= \sqrt{2}$

11. $h(a) = \sqrt{a + 5}$

13. $f(5) = 5^2 + 2$
    $= 25 + 2$
    $= 27$

    $f(2) = 2^2 + 2$
    $= 4 + 2$
    $= 6$

    $f(5) + f(2) = 27 + 6$
    $= 33$

15. $f(6 - 4) = f(2)$
    $= 2^2 + 2$
    $= 4 + 2$
    $= 6$

17. $g(x + 2) = 2(x + 2) - 3$
    $= 2x + 4 - 3$
    $= 2x + 1$

19. $f(2x) = (2x)^2 + 2$
    $= 4x^2 + 2$

21. $g(3x + 2) = 2(3x + 2) - 3$
    $= 6x + 4 - 3$
    $= 6x + 1$

23. $g(x + 2) = 2(x + 2) - 3$
    $= 2x + 4 - 3$
    $= 2x + 1$

    $g(x) = 2x - 3$

    $g(x + 2) - g(x) = (2x + 1) - (2x - 3)$
    $= 2x + 1 - 2x + 3$
    $= 4$

25. $f(-2) = (-2)^2 + 2(-2) - 3$
    $= 4 - 4 - 3$
    $= -3$

    $f(3) = 3^2 + 2(3) - 3$
    $= 9 + 6 - 3$
    $= 12$

    $f(-2) + f(3) = -3 + 12$
    $= 9$

27. $f(x) - 4 = (x^2 + 2x - 3) - 4$
    $= x^2 + 2x - 7$

29. $f(x) = x^2 + 2x - 3$

    $f(4) = 4^2 + 2(4) - 3$
    $= 16 + 8 - 3$
    $= 21$

    $f(x) - f(4) = (x^2 + 2x - 3) - 21$
    $= x^2 + 2x - 24$

31. $f(3x) = (3x)^2 + 2(3x) - 3$
    $= 9x^2 + 6x - 3$

33. $3f(x) = 3(x^2 + 2x - 3)$
$= 3x^2 + 6x - 9$

35. $g(-3) = \dfrac{-3}{-3 + 5}$

$= -\dfrac{3}{2}$

37. $g\left(\dfrac{1}{4}\right) = \dfrac{\dfrac{1}{4}}{\dfrac{1}{4} + 5}$

$= \dfrac{\dfrac{1}{4}}{\dfrac{1}{4} + 5} \cdot \dfrac{4}{4}$

$= \dfrac{1}{1 + 20}$

$= \dfrac{1}{21}$

39. $g(x + 5) = \dfrac{x + 5}{x + 5 + 5}$

$= \dfrac{x + 5}{x + 10}$

41. (a) $g(-5) = 1$; $(-5, \underline{1})$

(b) $g(-2) = -1$; $(-2, \underline{-1})$

(c) $g(0) = -2$; $(0, \underline{-2})$

(d) $g(1) = -\dfrac{1}{2}$; $\left(1, -\dfrac{1}{\underline{2}}\right)$

(e) $g(3) = 2$; $(3, \underline{2})$

43. $C(m) = $ daily $\cdot$ number $+$ charge $\cdot$ number
$ = $ charge of days per mile miles

$C(m) = 29 \cdot 3 + 0.14 \cdot m$
$C(m) = 87 + 0.14m$

45. length: L
width: $3L - 12$

$A = $ length $\cdot$ width
$A = L(3L - 12)$
$A = 3L^2 - 12L$

47. number of \$30 tickets sold: x
number of \$24 tickets sold: $600 - x$

$C = $ amount $+$ amount
 from \$30 from \$24
 tickets tickets

$C = 30x + 24(600 - x)$

49. number of minutes for machine A: m
number of minutes for machine B: $m + 35$

$N = $ copies made $+$ copies made
 by machine A by machine B

$N = 20m + 22(m + 35)$

51. electrician's hours: h
assistant's hours: $h - 2$

$A = $ amounts earned $+$ amount earned
 by electrician by assistant

$A = 45h + 25(h - 2)$

53. number of hours tutoring: t
number of hours as a clerk: $15 - t$

S = amount earned + amount earned
     as tutor   as clerk

S = $10t$ + $6.35(15 - t)$

2.4 Exercises

1. increasing: $x \geq 0$
   decreasing: $x \leq 0$
   turning point: $(0, 0)$

3. increasing: $-2 \leq x \leq 5$
   decreasing: never
   turning point: none

5. increasing: for all values of x
   decreasing: never
   turning point: none

7. increasing: $x \leq -3$; $x \geq 3$
   decreasing: $-3 \leq x \leq 3$
   turning points: at $x = -3$; at $x = 3$

9. increasing: $-2 \leq x \leq 0$; $x \geq 3$
   decreasing: $x \leq -2$; $0 \leq x \leq 3$
   turning points: $(-2, 0)$; $(0, 1)$; $(3, 0)$

11. increasing: $x \geq 5$
    decreasing: $x \leq -4$
    constant: $-4 \leq x \leq 5$
    turning points: none

13. $f(6.5) = 0.05(6.5)(6.5 - 3)$
         $= 1.14$

15. $h(0.8) = \dfrac{3}{4}(0.8)^3 - \dfrac{1}{2}(0.8) + 4$
         $= 3.98$

17. $g(-14.1) = \dfrac{(-14.1)^2 - 7}{(-14.1)^2 - 14.1 + 3}$

         $= 1.02$

19. $f(10) = 0.05(10)(10 - 3)$
         $= 3.5$

    $f(25) = 0.05(25)(25 - 3)$
         $= 27.5$

    $f(60) = 0.05(60)(60 - 3)$
         $= 171$

    $f(10) + f(25) + f(60) = 3.5 + 27.5 + 171$
                        $= 202$

21. $h\left(\dfrac{1}{2}\right) = \dfrac{3}{4}\left(\dfrac{1}{2}\right)^3 - \dfrac{1}{2}\left(\dfrac{1}{2}\right) + 4$

         $= 3.84$

    $h\left(\dfrac{5}{8}\right) = \dfrac{3}{4}\left(\dfrac{5}{8}\right)^3 - \dfrac{1}{2}\left(\dfrac{5}{8}\right) + 4$

         $= 3.87$

    $h\left(\dfrac{9}{4}\right) = \dfrac{3}{4}\left(\dfrac{9}{4}\right)^3 - \dfrac{1}{2}\left(\dfrac{9}{4}\right) + 4$

         $= 11.42$

    $h\left(\dfrac{1}{2}\right) + h\left(\dfrac{5}{8}\right) - h\left(\dfrac{9}{4}\right) = 3.84 + 3.87 - 11.42$

                        $= -3.71$

23. $y = x^2 + 5x$

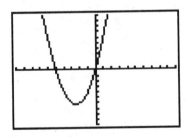

    turning point: $(-2.5, -6.3)$

25. $y = -2x^2 + 3x + 6$

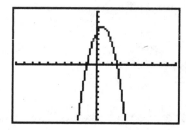

turning point: (0.8, 7.1)

27. $y = \dfrac{x^2 - 2}{x^2 + 4}$

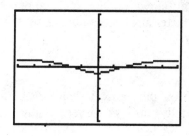

turning point: (0, -0.5)

29. $g(x) = 8 + 5x - x^3$

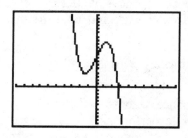

turning points: (-1.3, 3.7), (1.3, 12.3)

31. (a) $s(t) = 40t - 16t^2$

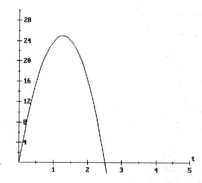

Domain: $0 \le t \le 2.5$

(b) 25 (y-value of turning point)

(c) 25 feet; the maximum value of this function is the maximum height the object reaches.

(d) 1.25 sec (x-value of turning point)

33. (a) $C(n) = 0.05n^2 - 185n + 300000$

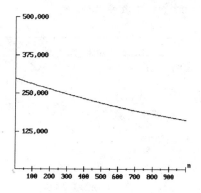

Domain: $0 \le n$

(b) $C(1500) = 0.05(1500)^2 - 185(1500) + 300000$
$= 135000$
$135,000

$C(2000) = 0.05(2000)^2 - 185(2000) + 300000$
$= 130000$
$130,000

(c) 1850 items (x-value of turning point)

(d) $128,875 ($y$-value of turning point)

35. $y = 2x^2 - 9x - 12$

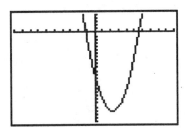

minimum value: −22.125

2.5 Exercises

1. (a) 7.5 thousand = 7500 calls

   (b) 4 a.m. (lowest point on graph)

   (c) 28 thousand = 28,000 calls

   (d) $0 \le t \le 4$; $16 \le t \le 22$;
   This corresponds to: between midnight
   and 4 a.m. and between 4 p.m. and
   10 p.m.

   (e) $10 \le t \le 16$; This corresponds to
   between 10 a.m. and 4 p.m.

   (f) $10 \le t \le 16$; This corresponds to
   between 10 a.m. and 4 p.m.

3. (a) It denies this belief since as the altitude
   increases (go right on horizontal axis),
   the temperature both increases and
   decreases.

   (b) $0 \le a \le 10$; $50 \le a \le -80$

   This corresponds to between 0 and 10
   kilometers and between 50 and 80
   kilometers.

(c) Starting at 45 km, the temperature
increases until it reaches approximately
10° C at an altitude of 50 km, then the
temperature decreased.

5. (a) At $t = 1$, $p = 30$
   Approximately 30%

   (b) Find 25 on vertical axis, go horizontally
   until you intersect the curve, move
   vertically down to the t-axis, $t \approx 4$.
   Approximately 4 days

   (c) As time passes material is forgotten.
   Material is forgotten rapidly during the
   first 3 hours of the given time period
   (more than half the material is forgotten
   during this time), then the rate of
   forgetting slows down. Only about 25%
   of the material is remembered by the
   6th day.

7. (a) Find 10 on the vertical scale, then go
   horizontally until you intersect the
   massed practice curve, move vertically
   down to the horizontal axis, and note
   the number of trials.

   Approximately 14 trials

   (b) Find 10 on the vertical scale, then go
   horizontally until you intersect the spaced
   practice curve, move vertically down to
   the horizontal axis, and note the number
   of trials.

   Approximately 3 trials

   (c) Performance with spaced practice is
   superior to performance with massed
   practice. Learning with spaced practice
   occurs more rapidly than with massed
   practice.

9. **Starting at home.** Kyle starts his trip by traveling 90 miles away during the first 2 hours of the day. His average rate of speed during that 2-hour period was 45 mph. For the next 2 hours Kyle did not travel. Then Kyle travelled closer to his home between hours 4 and 5. He was travelling an average of 30 mph during this time. Kyle stopped again for about 1 hour, then drove home in 1½ hours at an average rate of 40 mph.

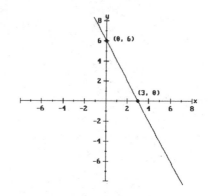

11. The graph shows that children assigned to each group had approximately the same number of aggressive behaviors before treatment. By the end of the treatment period, however, treatments A and C seemed to be the most effective in reducing these aggressive behaviors, where treatment C reduced their behaviors the most quickly.

   After the treatment period, however, the children in treatment C reverted back to their agressive behavior. Hence treatment C ends up being the least effective at the end of the time period given following treatment.

   The children in treatment B slowly reduced their aggressive behaviors even after the treatment ended. Although it seemed the least effective at the end of the treatment, it actually is shown to be the most effective of the three at the end of the time period given following treatment.

3. $2x - 6y = -6$

$x = 0$: $2(0) - 6y = -6$
$$-6y = -6$$
$$y = 1$$
$$(0, 1)$$

$y = 0$: $2x - 6(0) = -6$
$$2x = -6$$
$$x = -3$$
$$(-3, 0)$$

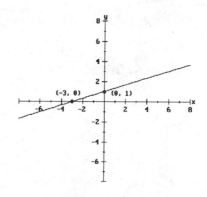

CHAPTER 2 REVIEW EXERCISES

1. $2x + y = 6$

$x = 0$: $2(0) + y = 5$
$$y = 6$$
$$(0, 6)$$

$y = 0$: $2x + 0 = 6$
$$2x = 6$$
$$x = 3$$
$$(3, 0)$$

5. $5x - 3y = 10$

$x = 0$: $5(0) - 3y = 10$
$$-3y = 10$$
$$y = -\frac{10}{3}$$
$$\left(0, -\frac{10}{3}\right)$$

$y = 0$: $5x - 3(0) = 10$
$$5x = 10$$
$$x = 2$$
$$(2, 0)$$

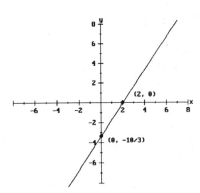

$$y = 0: \quad 3x - 8(0) = 11$$
$$3x = 11$$
$$x = \frac{11}{3}$$

$$\left(\frac{11}{3}, 0\right)$$

7. $2x + 5y = 7$

$$x = 0: \quad 2(0) + 5y = 7$$
$$5y = 7$$
$$y = \frac{7}{5}$$

$$\left(0, \frac{7}{5}\right)$$

$$y = 0: \quad 2x + 5(0) = 7$$
$$2x = 7$$
$$x = \frac{7}{2}$$

$$\left(\frac{7}{2}, 0\right)$$

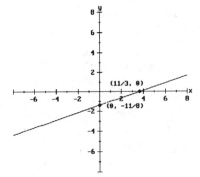

11. $5x + 7y = 21$

$$x = 0: \quad 5(0) + 7y = 21$$
$$7y = 21$$
$$y = 3$$
$$(0, 3)$$

$$y = 0: \quad 5x + 7(0) = 21$$
$$5x = 21$$
$$x = \frac{21}{5}$$

$$\left(\frac{21}{5}, 0\right)$$

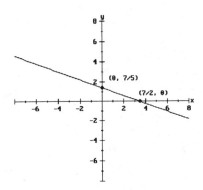

9. $3x - 8y = 11$

$$x = 0: \quad 3(0) - 8y = 11$$
$$-8y = 11$$
$$y = -\frac{11}{8}$$

$$\left(0, -\frac{11}{8}\right)$$

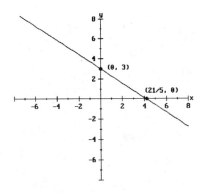

13. $y = x$

    $x = 0$: $y = 0$
              (0, 0)

    $x = 1$: $y = 1$
              (1, 1)

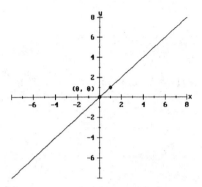

$y = 0$: $0 = \dfrac{2}{3}x + 2$

      $-2 = \dfrac{2}{3}x$

      $-3 = x$
      (-3, 0)

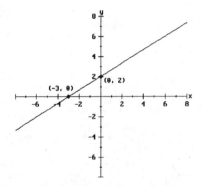

15. $y = -2x$

    $x = 0$: $y = 0$
             (0, 0)

    $x = 1$: $y = -2(1)$
            $y = -2$
           (1, -2)

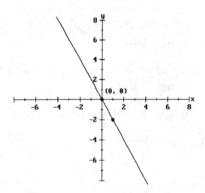

19. $\dfrac{x}{3} + \dfrac{y}{2} = 12$

    $6\left(\dfrac{x}{3} + \dfrac{y}{2}\right) = 6(12)$

       $2x + 3y = 72$

  $x = 0$: $2(0) + 3y = 72$
               $3y = 72$
                $y = 24$
             (0, 24)

  $y = 0$: $2x + 3(0) = 72$
              $2x = 72$
               $x = 36$
            (36, 0)

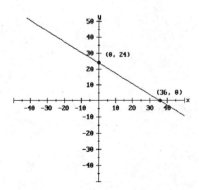

17. $y = \dfrac{2}{3}x + 2$

    $x = 0$: $y = \dfrac{2}{3}(0) + 2$

           $y = 2$
          (0, 2)

21. $x - 2y = 8$

$x = 0$: $\quad 0 - 2y = 8$
$\qquad\qquad -2y = 8$
$\qquad\qquad\quad y = -4$
$\qquad\qquad\quad (0, -4)$

$y = 0$: $\quad x - 2(0) = 8$
$\qquad\qquad\quad\; x = 8$
$\qquad\qquad\quad\; (8, 0)$

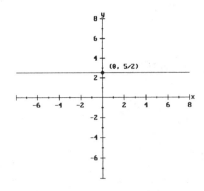

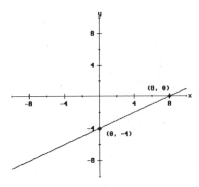

23. $x - 2 = 0$
$\quad\;\; x = 2$

This is a vertical line through (2, 0).

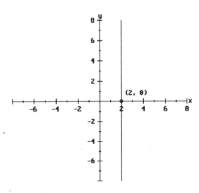

25. $2y = 5$
$\quad\; y = \dfrac{5}{2}$

This is a horizontal line through $\left(0, \dfrac{5}{2}\right)$.

27. $\{x \mid x$ is a real number$\}$

29. $4 - x \geq 0$
$\qquad 4 \geq x$

$\{x \mid x \leq 4\}$

31. $x + 2 \neq 0$
$\qquad x \neq -2$

$\{x \mid x \neq -2\}$

33. Function, since each element in the domain is assigned only one element in the range.

35. Not a function, since the domain element 4 is assigned two range elements, 2 and 7.

37. Function, since there corresponds exactly one y-value to each x-value.

39. Function, since there corresponds exactly one y-value to each x-value.

41. Function, it passes the Vertical Line Test.

43. Not a function, it doesn't pass the Vertical Line Test.

45. Domain: $\{-2, 0, 3, 5\}$
    Range: $\{3, 5, 7, 4\}$
    Function

47. Domain: $\{4, 6, 1, 8\}$
    Range: $\{9\}$
    Function

49. Domain: $\{x \mid -5 \leq x \leq 6\}$
Range: $\{y \mid -2 \leq y \leq 5\}$
Function

51. Domain: $\{x \mid -3 \leq x \leq 3\}$
Range: $\{y \mid -4 \leq y \leq 4\}$
Not a function, doesn't pass the Vertical Line Test.

53. $f(x) = 3x + 5$
$f(-1) = 3(-1) + 5 = 2$
$f(0) = 3(0) + 5 = 5$
$f(1) = 3(1) + 5 = 8$
$f(2) = 3(2) + 5 = 11$

55. $f(x) = 2x^2 - 3x + 2$
$f(-1) = 2(-1)^2 - 3(-1) + 2$
$\qquad = 2(1) - 3(-1) + 2$
$\qquad = 7$

$f(0) = 2(0)^2 - 3(0) + 2$
$\qquad = 2(0) - 3(0) + 2$
$\qquad = 2$

$f(1) = 2(1)^2 - 3(1) + 2$
$\qquad = 2(1) - 3(1) + 2$
$\qquad = 1$

$f(2) = 2(2)^2 - 3(2) + 2$
$\qquad = 2(4) - 3(2) + 2$
$\qquad = 4$

57. $h(x) = \sqrt{x - 5}$

$h(6) = \sqrt{6 - 5}$
$\qquad = \sqrt{1}$
$\qquad = 1$

$h(5) = \sqrt{5 - 5}$
$\qquad = \sqrt{0}$
$\qquad = 0$

4 is not in the domain of $h(x)$.

59. $h(x) = \dfrac{x - 1}{x + 3}$

$h(1) = \dfrac{1 - 1}{1 + 3}$

$\qquad = \dfrac{0}{4}$

$\qquad = 0$

$h(3) = \dfrac{3 - 1}{3 + 3}$

$\qquad = \dfrac{2}{6}$

$\qquad = \dfrac{1}{3}$

−3 is not in the domain of $h(x)$.

61. $f(x) = 2x^2 + 4x - 1$
$f(a) = 2a^2 + 4a - 1$
$f(z) = 2z^2 + 4z - 1$

63. $f(x + 2) = 5(x + 2) + 2$
$\qquad = 5x + 10 + 2$
$\qquad = 5x + 12$

65. $f(x) + 2 = (5x + 2) + 2$
$\qquad = 5x + 4$

67. $f(x) = 5x + 2$
$f(2) = 5(2) + 2 = 12$

$f(x) + f(2) = (5x + 2) + 12$
$\qquad = 5x + 14$

69. $g(x + 2) = 6 - (x + 2)$
$\qquad = 6 - x - 2$
$\qquad = 4 - x$

71. $g(2x) = 6 - 2x$

73. $2g(x) = 2(6 - x)$
$\qquad = 12 - 2x$

75. Rasheed starts his trip by traveling 90 miles away from his home during the first 2 hours of the day. His average rate of speed for the first hour was 40 mph; during the second hour his average rate was 50 mph. During the next hour Rasheed did not travel. Then Rasheed drove closer to his home between hours 3 and 4. He was traveling an average of 60 mph during this time. Rasheed drove another 20 miles away from his home, travelling at 20 mph between the 4th and 5th hour. Finally, during the last hour, Rasheed drove home at a rate of 50 mph.

77. Total   = delivery   +   charge   · number
    charges   fee   per day   days

    C =   30   +   42   · n

    $C = 30 + 42n$

79. (a) $f(-6) = 0$

    (b) $f(0) = 2$

    (c) $f(-4) = -2$

    (d) $f(6) = -3$

    (e) $(3, 5)$

    (f) $f(5)$ is smaller

    (g) $(-6, 0)$; $(-2, 0)$; $(5, 0)$

81. $y = 6x - x^2$

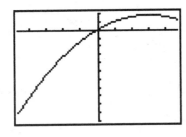

range: $\{y \mid -50.3 \le y \le 9\}$
turning point: $(3, 9)$

83. $y = \sqrt{x - 3}$

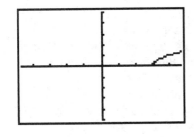

range: $\{y \mid 0 \le y \le 1.3\}$

85. $y = 0.4x^2 - 2x + 5$

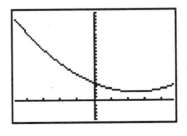

range: $\{y \mid 2.5 \le y \le 23.2\}$
turning point: $(2.5, 2.5)$

87. (a) $C(n) = 450000 - 175n + 0.02n^2$

(b) $P = C(5000) = 450000 - 175(5000) + 0.02(5000)^2$
    $= 75000$

    $\$75,000$

(c) Find $y = 100000$ on the graph; it occurs at
    $x = 3090$ and $x = 5660$.

    Approximately 3090 items or 5660 items.

1. (a) $3x - 5y = 30$

    $x = 0$: $3(0) - 5y = 30$
    $-5y = 30$
    $y = -6$
    $(0, -6)$

    $y = 0$: $3x - 5(0) = 30$
    $3x = 30$
    $x = 10$
    $(10, 0)$

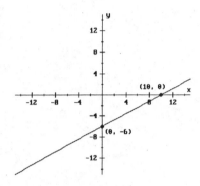

    (b) $x - 7 = 0$
    $x = 7$

    This is a vertical line that passes through $(7, 0)$.

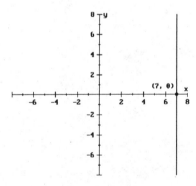

2. (a) domain: $\{2, 3, 4\}$
    range: $\{-3, 5, 6\}$

    (b) domain: $\{x \mid -4 \leq x \leq 5\}$
    range: $\{y \mid -3 \leq y \leq 6\}$

3. (a) $x - 4 \geq 0$
    $x \geq 4$
    $\{x \mid x \geq 4\}$

    (b) $3x - 4 \neq 0$
    $3x \neq 4$
    $x \neq \dfrac{4}{3}$

    $\left\{ x \mid x \neq \dfrac{4}{3} \right\}$

4. (a) Not a function, since the domain element 2 is assigned two range elements, 5 and 4.

    (b) Function, since each element in the domain is assigned only one element in the range.

    (c) Not a function. For example, let $x = 27$.

    $27 = 3y^2$
    $9 = y^2$
    $\pm 3 = y$

    There are two range elements corresponding to a domain element.

    (d) Function, it passes the Vertical Line Test.

5. (a) $g(2) = \sqrt{2(2) + 1}$
    $= \sqrt{5}$

    (b) $f(-3) = 3(-3)^2 - 4$
    $= 3(9) - 4$
    $= 23$

    (c) $h(x - 2) = 5(x - 2) - 3$
    $= 5x - 10 - 3$
    $= 5x - 13$

    (d) $f(x^2) = 3(x^2)^2 - 4$
    $= 3x^4 - 4$

    (e) $f(x) = 3x^2 - 4$
    $f(5) = 3(5)^2 - 4$
    $= 3(25) - 4$
    $= 71$

    $f(x) - f(s) = (3x^2 - 4) - 71$
    $= 3x^2 - 75$

6. (a) $f(-6) = 2$

   (b) $f(3) = -2$

   (c) $f(0) = 0$

   (d) $f(7) = 2$

   (e) $(-4, 6)$; $(3, -2)$

   (f) $f(-4)$

   (g) x-intercepts: $x = 0,\ 5$
       y-intercept: $y = 0$

7. (a) $y = x^2 - 3x - 8$

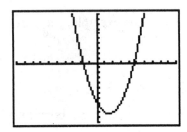

   (b) $(1.5,\ -10.3)$

   (c) $\{y \mid y \geq -10.3\}$

   (d) x-intercepts: $x = -1.7,\ 4.7$
       y-intercept: $y = -8$

8. (a) 30 minutes
   (b) 8 mg
   (c) between 10 and 20 minutes it increases
       $6 - 3 = 3$ mg.

CHAPTER 3

3.1 Exercises

1. final exam score: x
   course grade: y

   $y = 0.25x + 0.75(82)$
   $y = 0.25x + 61.5$

   Using TABLE

   (a) 74
   (b) 114

3. final exam score: x
   course grade: y

   $y = 0.25x + 0.75(78)$
   $y = 0.25x + 58.5$

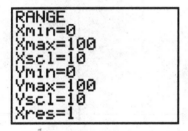

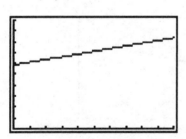

She needs at least a grade of 86.

5. gross sales: x
   salary: y

   $y = 0.09x$

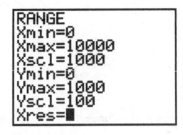

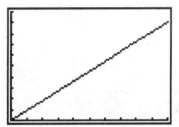

His gross sales need to be at least \$6666.67.

7. gross sales: x
   salary: y

   Plan 1: commission only
   $y = 0.09x$

   Plan 2: salary + commission
   $y = 120 + 0.06x$

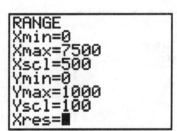

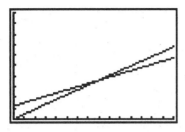

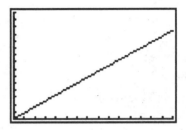

Plan 2 is the best for sales up to $4000 then Plan 1 becomes the better deal.

She should invest $12048.19.

9. width: x
length: $1 + 3x$
perimeter: y

$$y = 2(x) + 2(1 + 3x)$$
$$y = 2x + 2 + 6x$$
$$y = 8x + 2$$

Using TABLE,

width = 9.75
length = $1 + 3x = 1 + 3(9.75) = 30.25$

The dimensions must be 9.75 in. by 30.25 in.

11. amount invested: x
interest earned: y

$$y = 0.083x$$

13. amount in high-risk account: x
remainder of investment: $12500 - x$
interest earned: y

$$y = 0.08x + 0.05(12500 - x)$$
$$y = 0.08x + 625 - 0.05x$$
$$y = 0.03x + 625$$

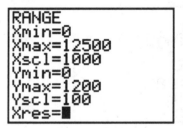

```
RANGE
Xmin=0
Xmax=12500
Xscl=1000
Ymin=0
Ymax=1200
Yscl=100
Xres=█
```

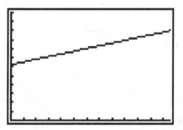

```
RANGE
Xmin=0
Xmax=15000
Xscl=1000
Ymin=0
Ymax=1500
Yscl=100
Xres=1
```

The minimum occurs at $x = 0$. Hence if he doesn't invest in the high-risk account, his interest would be $625. The minimum occurs at $x = 12,500$. Hence if he invests all his money in the high-risk account, he will earn $1000.

15. amount in low-risk account: x
    remainder of investment: $24000 - x$
    interest earned: y

    $y = 0.042x + 0.078(24000 - x)$
    $y = 0.042x + 1872 - 0.078x$
    $y = -0.036x + 1872$

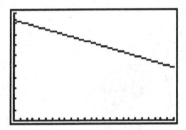

The minimum interest earned, $1008, occurs when the entire $24,000 is invested in the low-risk account. The maximum interest earned, $1872, occurs when the $0 is invested in the low-risk account (and therefore $24,000 is invested in the high-risk account.)
Using TABLE, an investment of $18,666.67 in the low risk and $5333.33 in the high risk earns $1200 interest.

17. number of minutes for GHJ27: x
    number of minutes for LGP39: $x + 5$

    $y = 9(x + 5) + 4x$
    $y = 9x + 45 + 4x$
    $y = 13x + 45$

Using TABLE, $y = 142$ when $x = 7.5$.

$x + 5 = 7.5 + 5 = 12.5$

It will take 12.5 minutes to complete the job.

19. number of $850 computers: x
    number of $600 computers: $58 - x$
    amount collected: y

    $y = 850x + 600(58 - x)$
    $y = 850x + 34800 - 600x$
    $y = 250x + 34800$

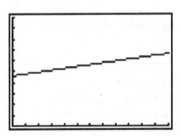

They sold 22 $850-computers.

21. Arthur's hours: x

    Lewis' hours: $x + \dfrac{1}{2}$

    $y = 300x + 200(x + \frac{1}{2})$
    $y = 300x + 200x + 100$
    $y = 500x + 100$

    Using TABLE, $y = 3000$ when $x = 5.8$.

    $x + \frac{1}{2} = 5.8 + \frac{1}{2} = 6.3$.

    It will take 6.3 hours.

23. number of AM radios: x
number of AM/FM radios: $24 - x$
total cost: y

$$y = 70 + 35x + 50(24 - x)$$
$$y = 70 + 35x + 1200 - 50x$$
$$y = 1270 - 15x$$

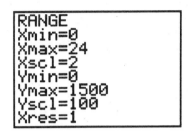

```
RANGE
Xmin=0
Xmax=24
Xscl=2
Ymin=0
Ymax=1500
Yscl=100
Xres=1
```

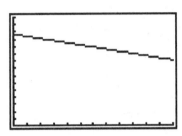

$y = 1000$ when $x = 18$.
He purchased 18 AM radios and
$24 - 18 = 6$ AM/FM radios.

3.2 Exercises

1. $x - (x - 3) = 4$
$\quad x - x + 3 = 4$
$\qquad\qquad 3 = 4$
Contradiction

3. $4(w - 2) = 4w - 8$
$\quad 4w - 8 = 4w - 8$
Identity

5. $2(y + 1) - (y - 5) = 4(y - 2) - 3(y - 5)$
$\quad 2y + 2 - y + 5 = 4y - 8 - 3y + 15$
$\qquad\qquad y + 7 = y + 7$
Identity

7. $6(2y + 1) - 4(3y - 1) = y - (y + 10)$
$\quad 12y + 6 - 12y + 4 = y - y - 10$
$\qquad\qquad\qquad 10 = -10$
Contradiction

9. $5(x - 3) - (x + 2) = 8 - x$

$x = 0$: $\quad 5(0 - 3) - (0 + 2) = 8 - 0$
$\qquad\qquad\qquad 5(-3) - 2 = 8$
$\qquad\qquad\qquad\quad -15 - 2 = 8$
$\qquad\qquad\qquad\qquad\quad -17 = 8$
does not satisfy

$x = 5$: $\quad 5(5 - 3) - (5 + 2) = 8 - 5$
$\qquad\qquad\qquad\quad 5(2) - 7 = 3$
$\qquad\qquad\qquad\quad 10 - 7 = 3$
$\qquad\qquad\qquad\qquad\quad 3 = 3$
satisfies

$x = 0.8$: $\quad 5(-0.8 - 3) - (-0.8 + 2) = 8 - (-0.8)$
$\qquad\qquad\qquad 5(-3.8) - 1.2 = 8.8$
$\qquad\qquad\qquad\quad -19 - 1.2 = 8.8$
$\qquad\qquad\qquad\qquad -20.2 = 8.8$
does not satisfy

11. $x^2 - 5x = 5 - x$

$x = -\dfrac{1}{2}$: $\left(-\dfrac{1}{2}\right)^2 - 5\left(-\dfrac{1}{2}\right) = 5 - \left(-\dfrac{1}{2}\right)$

$$\dfrac{1}{4} + \dfrac{5}{2} = \dfrac{11}{2}$$

$$\dfrac{11}{4} = \dfrac{11}{2}$$

does not satisfy

$x = -5$: $\quad (-5)^2 - 5(-5) = 5 - (-5)$
$\qquad\qquad\qquad 25 + 25 = 10$
$\qquad\qquad\qquad\qquad 50 = 10$
does not satisfy

$x = 5$: $\quad 5^2 - 5(5) = 5 - 5$
$\qquad\qquad 25 - 25 = 0$
$\qquad\qquad\qquad\quad 0 = 0$
satisfies

13. $3x - 12 = 12 - 3x$
$\quad 6x - 12 = 12$
$\qquad\quad 6x = 24$
$\qquad\qquad x = 4$

15. $3x - 12 = -12 + 3x$
$\qquad -12 = -12$
Identity, true for all real numbers

17. $2t + 1 = 7 + t$

   $\quad t + 1 = 7$

   $\quad\quad t = 6$

19. $2(x - 1) = 2(x + 1)$

   $\quad 2x - 2 = 2x + 2$

   $\quad\quad -2 = 2$

   Contradiction, no solution

21. $3(x + 1) - x = 2(x + 3)$

   $\quad 3x + 3 - x = 2x + 6$

   $\quad\quad 2x + 3 = 2x + 6$

   $\quad\quad\quad\quad 3 = 6$

   Contradiction, no solution

23. $5(2t - 1) - 3t = 5 - 7t$

   $\quad 10t - 5 - 3t = 5 - 7t$

   $\quad\quad 7t - 5 = 5 - 7t$

   $\quad\quad 14t - 5 = 5$

   $\quad\quad\quad 14t = 10$

   $\quad\quad\quad\quad t = \dfrac{5}{7}$

25. $6(3 - z) + 2(4z - 5) = z - (3 - z)$

   $\quad 18 - 6z + 8z - 10 = z - 3 + z$

   $\quad\quad\quad 2z + 8 = 2z - 3$

   $\quad\quad\quad\quad\quad 8 = -3$

   Contradiction, no solution

27. $4(3x - 1) - 5(3x - 2) = 2(x + 3) - 5x$

   $\quad 12x - 4 - 15x + 10 = 2x + 6 - 5x$

   $\quad\quad\quad -3x + 6 = -3x + 6$

   Identity, true for all numbers

29. $x(x - 2) - 15 = x(x + 5) - 3(x + 5)$

   $\quad x^2 - 2x - 15 = x^2 + 5x - 3x - 15$

   $\quad x^2 - 2x - 15 = x^2 + 2x - 15$

   Identity, true for all numbers

31. $6x - [2 - (x - 1)] = x - 5(x + 1)$

   $\quad 6x - (2 - x + 1) = x - 5x - 5$

   $\quad\quad 6x - (3 - x) = -4x - 5$

   $\quad\quad 6x - 3 + x = -4x - 5$

   $\quad\quad\quad 7x - 3 = -4x - 5$

   $\quad\quad\quad 11x - 3 = -5$

   $\quad\quad\quad\quad 11x = -2$

   $\quad\quad\quad\quad\quad x = -\dfrac{2}{11}$

33. $3 - [4t + 5(2t - 1) - 3t] = 3(4 - t)$

   $\quad 3 - (4t + 10t - 5 - 3t) = 12 - 3t$

   $\quad\quad 3 - (11t - 5) = 12 - 3t$

   $\quad\quad 3 - 11t + 5 = 12 - 3t$

   $\quad\quad\quad -11t + 8 = 12 - 3t$

   $\quad\quad\quad\quad\quad 8 = 12 + 8t$

   $\quad\quad\quad\quad -4 = 8t$

   $\quad\quad\quad -\dfrac{1}{2} = t$

35. $\quad\quad \dfrac{x}{2} - \dfrac{3x}{5} = 4$

   $\quad 10\left(\dfrac{x}{2} - \dfrac{3x}{5}\right) = 10(4)$

   $\quad \dfrac{10}{1} \cdot \dfrac{x}{2} - \dfrac{10}{1} \cdot \dfrac{3x}{5} = 40$

   $\quad\quad 5x - 6x = 40$

   $\quad\quad\quad -x = 40$

   $\quad\quad\quad\quad x = -40$

37. $\quad\quad \dfrac{y}{2} - \dfrac{4 - y}{3} = 4$

   $\quad 6\left(\dfrac{y}{2} - \dfrac{4 - y}{3}\right) = 6(4)$

   $\quad \dfrac{6}{1} \cdot \dfrac{y}{2} - \dfrac{6}{1} \cdot \dfrac{4 - y}{3} = 24$

   $\quad\quad 3y - 2(4 - y) = 24$

   $\quad\quad 3y - 8 + 2y = 24$

   $\quad\quad\quad 5y - 8 = 24$

   $\quad\quad\quad\quad 5y = 32$

   $\quad\quad\quad\quad y = \dfrac{32}{5}$

39. $\quad\quad \dfrac{2y - 1}{4} = y - 3$

   $\quad 4\left(\dfrac{2y - 1}{4}\right) = 4(y - 3)$

   $\quad\quad 2y - 1 = 4y - 12$

   $\quad\quad\quad -1 = 2y - 12$

   $\quad\quad\quad 11 = 2y$

   $\quad\quad\quad \dfrac{11}{2} = y$

41.

$$\frac{3a}{2} - \frac{1}{3} = \frac{a+5}{4}$$

$$12\left(\frac{3a}{2} - \frac{1}{3}\right) = 12\left(\frac{a+5}{4}\right)$$

$$\frac{12}{1} \cdot \frac{3a}{2} - \frac{12}{1} \cdot \frac{1}{3} = 3(a+5)$$

$$18a - 4 = 3a + 15$$
$$15a - 4 = 15$$
$$15a = 19$$
$$a = \frac{19}{15}$$

43. 1st number: x
2nd number: $3 + 4x$

$$x + 3 + 4x = 43$$
$$5x + 3 = 43$$
$$5x = 40$$
$$x = 8$$
$$3 + 4x = 3 + 4(8) = 35$$

The numbers are 8 and 35.

45. 1st number: x
2nd number: $5x - 8$

$$x + 5x - 8 = -20$$
$$6x - 8 = -20$$
$$6x = -12$$
$$x = -2$$
$$5x - 8 = 5(-2) - 8 = -18$$

The numbers are −2 and −18.

47. 1st integer: x
2nd integer: $x + 1$
3rd integer: $x + 2$

$$x + x + 1 + x + 2 = 66$$
$$3x + 3 = 66$$
$$3x = 63$$
$$x = 21$$
$$x + 1 = 21 + 1 = 22$$
$$x + 2 = 21 + 2 = 23$$

The integers are 21, 22, and 23.

49. 1st integer: x
2nd integer: $x + 2$
3rd integer: $x + 4$
4th integer: $x + 6$

$$x + x + 2 + x + 4 + x + 6 = 56$$
$$4x + 12 = 56$$
$$4x = 44$$
$$x = 11$$
$$x + 2 = 11 + 2 = 13$$
$$x + 4 = 11 + 4 = 15$$
$$x + 6 = 11 + 6 = 17$$

The integers are 11, 13, 15, and 17.

51. width: x
length: $2x$

$$P = 2w + 2l$$
$$42 = 2x + 2(2x)$$
$$42 = 2x + 4x$$
$$42 = 6x$$
$$7 = x$$
$$2x = 2(7) = 14$$

The rectangle is 7 meters by 14 meters.

53. 1st side: $x - 5$
2nd side: x
3rd side: $2(x - 5)$

$$P = \text{sum of three side lengths}$$
$$33 = x - 5 + x + 2(x - 5)$$
$$33 = x - 5 + x + 2x - 10$$
$$33 = 4x - 15$$
$$48 = 4x$$
$$12 = x$$
$$x - 5 = 12 - 5 = 7$$
$$2(x - 5) = 2(12 - 5) = 14$$

The sides have lengths 7 cm, 12 cm, and 14 cm.

55. <u>original rectangle</u>
width: x
length: $1 + 3x$

<u>new rectangle</u>
width: $x + 2$
length: $2(1 + 3x)$

$$\begin{aligned}\text{original perimeter} &= 2(x) + 2(1 + 3x)\\ &= 2x + 2 + 6x\\ &= 8x + 2\end{aligned}$$

new perimeter $= 2(x + 2) + 2[2(1 + 3x)]$
$$= 2x + 4 + 4 + 12x$$
$$= 14x + 8$$

The new perimeter is 3 less than 5 times the original length.

$$14x + 8 = 5(1 + 3x) - 3$$
$$14x + 8 = 5 + 15x - 3$$
$$14x + 8 = 15x + 2$$
$$8 = x + 2$$
$$6 = x$$
$$1 + 3x = 1 + 3(6) = 19$$

The original dimensions were 6 by 19.

57. gross sales: x

$$0.08x = 500$$
$$x = 6250$$

The gross sales need to be $6250.

59. final exam score: x

$$0.45(x) + 0.55(72) = 80$$
$$0.45x + 39.6 = 80$$
$$0.45x = 40.4$$
$$x = 89.8$$

The minimum score is 89.8.

61. number of $5 bills: x
number of $10 bills: $x + 1$
number of $1 bills: $25 - (x + x + 1) = 24 - 2x$

$$5(x) + 10(x + 1) + 1(24 - 2x) = 164$$
$$5x + 10x + 10 + 24 - 2x = 164$$
$$13x + 34 = 164$$
$$13x = 130$$
$$x = 10$$
$$x + 1 = 10 + 1 = 11$$
$$24 - 2x = 24 - 2(10) = 4$$

He has 10 $5-bills, 11 $10-bills, and 4 $1-bills.

63. number of 20-lb packages: x
number of 25-lb packages: $50 - x$

$$20(x) + 25(50 - x) = 1075$$
$$20x + 1250 - 25x = 1075$$
$$-5x + 1250 = 1075$$
$$-5x = -175$$
$$x = 35$$
$$50 - x = 50 - 35 = 15$$

There are 35 20-lb packages and 15 25-lb packages.

65. pairs of shoes: x
pairs of boots: $30 - x$

$$187 = 140 + 2(30 - x) + 1(x)$$
$$187 = 140 + 60 - 2x + x$$
$$187 = 200 - x$$
$$-13 = -x$$
$$13 = x$$

She sold 13 pairs of shoes.

67. lbs of $2/lb coffee: x

$$2(x) + 3(30) = 2.60(x + 30)$$
$$2x + 90 = 2.60x + 78$$
$$90 = 0.60x + 78$$
$$12 = 0.60x$$
$$20 = x$$

20 lb of the $2-per-pound coffee should be used.

69. number of orchestra seats: x
number of balcony seats: $56 - x$

$$48(x) + 28(56 - x) = 2328$$
$$48x + 1568 - 28x = 2328$$
$$20x + 1568 = 2328$$
$$20x = 760$$
$$x = 38$$

38 orchestra seats were purchased.

71. plumber's hours: x
assistant's hours: $x + 2$

$$22x + 13(x + 2) = 236$$
$$22x + 13x + 26 = 236$$
$$35x + 26 = 236$$
$$35x = 210$$
$$x = 6$$

The plumber worked 6 hours.

73. 1st car's hours: x
2nd car's hours: x

$55x + 60x = 345$
$115x = 345$
$x = 3$

They will meet in 3 hours which will be 6:00 P.M.

75. 1st car's hours: x
2nd car's hours: x

$35x + 50x = 595$
$85x = 595$
$x = 7$

It will take 7 hours.

77. 1st person's hours: x
2nd person's hours: $x + 3$

$17x = 7(x + 3)$
$17x = 7x + 21$
$10x = 21$
$x = 2.1$

It would take 2.1 hours.

79. number of hours going: x
number of hours returning: $17 - x$

$48x = 54(17 - x)$
$48x = 918 - 54x$
$102x = 918$
$x = 9$

Distance $= 48(9) = 432$

The convention is 432 km away.

81. 1 hr 50 min = 110 min

number of minutes for old copier: x
number of minutes for new copier: $110 - x$

$35x + 50(110 - x) = 5125$
$35x + 5500 - 50x = 5125$
$-15x + 5500 = 5125$
$-15x = -375$
$x = 25$

old copier: $35(25) = 875$

The older copier made 875 copies.

83. experienced worker's hours: x
new worker's hours: x

$60x + 30x = 6750$
$90x = 6750$
$x = 75$

It will take 75 hours.

85. new worker's hours: x
experienced worker's hours: $x + 3$

$$60(x + 3) + 30x = 6750$$
$$60x + 180 + 30x = 6750$$
$$90x + 180 = 6750$$
$$90x = 6570$$
$$x = 73$$
$$x + 3 = 73 + 3 = 76$$

It will take the new worker 73 hours and the experienced worker 76 hours.

87. experienced worker's hours: x
trainee's hours: $x + 3$

$$80x + 48(x + 3) = 656$$
$$80x + 48x + 144 = 656$$
$$128x + 144 = 656$$
$$128x = 512$$
$$x = 4$$
$$x + 3 = 4 + 3 = 7$$

The trainee works 7 hours, hence they will finish the job 7 hours after 10:00 a.m. which is 5:00 p.m.

3.3 Exercises

1. $x + 1 > x - 1$
$\quad 1 > -1$
Identity

3. $2x - (x - 3) \geq x + 5$
$\quad 2x - x + 3 \geq x + 5$
$\quad\quad x + 3 \geq x + 5$
$\quad\quad\quad 3 \geq 5$
Contradiction

5. $-6 \leq r - (r + 6) < 3$
$\quad -6 \leq r - r - 6 < 3$
$\quad\quad -6 \leq -6 < 3$
Identity

7. $0 < x - (x - 3) \leq 5$
$\quad 0 < x - x + 3 \leq 5$
$\quad\quad 0 < 3 \leq 5$
Identity

9. $4 + 3u > 10$

$u = -3.4$: $\quad 4 + 3(-3.4) > 10$
$\quad\quad\quad\quad\quad\quad -6.2 > 10$
$\quad\quad\quad\quad$ does not satisfy

$u = 2$: $\quad\quad 4 + 3(2) > 10$
$\quad\quad\quad\quad\quad\quad 10 > 10$
$\quad\quad\quad\quad$ does not satisfy

11. $6 - 4y \leq 8$

$y = -2.5$: $\quad 6 - 4(-2.5) \leq 8$
$\quad\quad\quad\quad\quad\quad\quad 16 \leq 8$
$\quad\quad\quad\quad$ does not satisfy

$y = -1$: $\quad\quad 6 - 4(-1) \leq 8$
$\quad\quad\quad\quad\quad\quad\quad 10 \leq 8$
$\quad\quad\quad\quad$ does not satisfy

13. $\quad\quad -3 \leq 3 - (4 - z) < 3$
$\quad\quad\quad -3 \leq 3 - 4 + z < 3$
$\quad\quad\quad\quad -3 \leq z - 1 < 3$

$z = -2$: $\quad -3 \leq -2 - 1 < 3$
$\quad\quad\quad\quad -3 \leq -3 < 3$
$\quad\quad\quad$ satisfies

$z = 4.6$: $\quad -3 \leq 4.6 - 1 < 3$
$\quad\quad\quad\quad -3 \leq 3.6 < 3$
$\quad\quad\quad$ does not satisfy

15. $5 < x < 8$
makes sense

17. $-6 > w \geq -8$
makes sense

$\quad -8 \leq w < -6$

19. $-3 < a < -2$
makes sense

21. $2 > x > -3$
makes sense

$-3 < x < 2$

23. $5y \geq 2 - 3y$
$8y \geq 2$
$y \geq \dfrac{1}{4}$

25. $\dfrac{2}{3}x - 4 \leq 3x + \dfrac{1}{2}$

$6\left(\dfrac{2}{3}x - 4\right) \leq 6\left(3x + \dfrac{1}{2}\right)$

$\dfrac{6}{1} \cdot \dfrac{2}{3}x - 6(4) \leq 6(3x) + \dfrac{6}{1} \cdot \dfrac{1}{2}$

$4x - 24 \leq 18x + 3$
$-24 \leq 14x + 3$
$-27 \leq 14x$
$\dfrac{-27}{14} \leq x$

or

$x \geq -\dfrac{27}{14}$

27. $3x + 4 \geq 2x - 5$
$x + 4 \geq -5$
$x \geq -9$

29. $11 - 3y < 38$
$-3y < 27$
$y > -9$

31. $3a - 8 < 8 - 3a$
$6a - 8 < 8$
$6a < 16$
$a < \dfrac{8}{3}$

33. $3t - 5 > 5t - 13$
$-2t - 5 > -13$
$-2t > -8$
$t < 4$

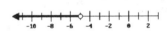

35. $2y - 3 \leq 5y + 7$
$-3y - 3 \leq 7$
$-3y \leq 10$
$y \geq -\dfrac{10}{3}$

37. $5(a - 2) - 7a > 2a + 10$
$5a - 10 - 7a > 2a + 10$
$-2a - 10 > 2a + 10$
$-4a - 10 > 10$
$-4a > 20$
$a < -5$

39. $\dfrac{3x + 2}{5} < x + 4$

$5\left(\dfrac{3x + 2}{5}\right) < 5(x + 4)$

$3x + 2 < 5x + 20$
$-2x + 2 < 20$
$-2x < 18$
$x > -9$

41. $7 - 5(t - 2) \le 2$
$7 - 5t + 10 \le 2$
$-5t + 17 \le 2$
$-5t \le -15$
$t \ge 3$

43. $2(y + 3) + 3(y - 4) < 3(y + 2) + 2(y + 1)$
$2y + 6 + 3y - 12 < 3y + 6 + 2y + 2$
$5y - 6 < 5y + 8$
$-6 < 8$

Identity, true for all real numbers

45. $1 \le c + 3 < 5$
$-2 \le c < 2$

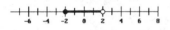

47. $6 < 2k - 1 \le 11$
$7 < 2k \le 12$
$\frac{7}{2} < k \le 6$

49. $-1 < 4 - t < 3$
$-5 < -t < -1$
$5 > t > 1$
or $1 < t < 5$

51. $1 \le 8 - 3t \le 12$
$-7 \le -3t \le 4$

$\frac{7}{3} \ge t \ge -\frac{4}{3}$

or $-\frac{4}{3} \le t \le \frac{7}{3}$

53. $2 < 5 - 3(x + 1) < 17$
$2 < 5 - 3x - 3 < 17$
$2 < 2 - 3x < 17$
$0 < -3x < 15$
$0 > x > -5$
or $-5 < x < 0$

55. $-2 < x - 5 \le 1$
$3 < x \le 6$

57. $-2 \le 5 - x < 1$
$-7 \le -x < -4$
$7 \ge x > 4$
or $4 < x \le 7$

59. $1 \le 1 - (5z - 2) \le 11$
$1 \le 1 - 5z + 2 \le 11$
$1 \le 3 - 5z \le 11$
$-2 \le -5z \le 8$

$\frac{2}{5} \ge z \ge -\frac{8}{5}$

or $-\frac{8}{5} \le z \le \frac{2}{5}$

61. $0.39 \le 0.72x - 1.5 < 8.1$
$1.89 \le 0.72x < 9.6$
$2.625 \le x < 13.\overline{3}$

63. number: x

$3x - 4 < 17$
$3x < 21$
$x < 7$

All numbers less than 7.

65. number: x

$$6x + 12 > 3x$$
$$12 > -3x$$
$$-4 < x$$

The numbers must be greater than -4.

67. side length: x

$$P \leq 72$$
$$4x \leq 72$$
$$x \leq 18$$

The maximum length is 18 feet.

69. length: x

$$P \geq 80$$
$$2w + 2l \geq 80$$
$$2(8) + 2(x) \geq 80$$
$$16 + 2x \geq 80$$
$$2x \geq 64$$
$$x \geq 32$$

The length must be at least 32 cm.

71. width: x

$$50 \leq P \leq 70$$
$$50 \leq 2w + 2l \leq 70$$
$$50 \leq 2(x) + 2(18) \leq 70$$
$$50 \leq 2x + 36 \leq 70$$
$$14 \leq 2x \leq 34$$
$$7 \leq x \leq 17$$

The width has a range 7 in. to 17 in., inclusively.

73. width: x
length: $3x + 5$

$$P = 2w + 2l$$
$$P = 2(x) + 2(3x + 5)$$
$$P = 2x + 6x + 10$$
$$P = 8x + 10$$
$$P - 10 = 8x$$

$$\frac{P - 10}{8} = x$$

$$12 \leq x \leq 16$$
$$12 \leq \frac{P - 10}{8} \leq 16$$

$$8(12) \leq 8\left(\frac{P - 10}{8}\right) \leq 8(16)$$

$$96 \leq P - 10 \leq 128$$
$$106 \leq P \leq 138$$

The range of values for the perimeter is 106 in. to 138 in.

75. shortest side: x
medium side: $x + 2$
longest side: $2x$

$$30 \leq P \leq 50$$
$$30 \leq x + x + 2 + 2x \leq 50$$
$$30 \leq 4x + 2 \leq 50$$
$$28 \leq 4x \leq 48$$
$$7 \leq x \leq 12$$

The shortest side has a range of values, 7 cm to 12 cm.

77. reserved ticket price: x
at the door ticket price: $x - 2$

$$C \geq 3750$$
$$300x + 150(x - 2) \geq 3750$$
$$300x + 150x - 300 \geq 3750$$
$$450x - 300 \geq 3750$$
$$450x \geq 4050$$
$$x \geq 9$$

The minimum price is $9.

79. number of dimes: x
number of nickels: $40 - x$

$$10(x) + 5(40 - x) \leq 285$$
$$10x + 200 - 5x \leq 285$$
$$5x + 200 \leq 285$$
$$5x \leq 85$$
$$x \leq 17$$

He can have at most 17 dimes.

81. number of elephants: x
    number of pandas: $24 - x$

    $$20(24 - x) + 25(x) \geq 575$$
    $$480 - 20x + 25x \geq 575$$
    $$480 + 5x \geq 575$$
    $$5x \geq 95$$
    $$x \geq 19$$

    The minimum number of elephants is 19.

83. hours tutoring: x
    hours at restaurant: $30 - x$

    $$8x + 6(30 - x) \geq 250$$
    $$8x + 180 - 6x \geq 250$$
    $$2x + 180 \geq 250$$
    $$2x \geq 70$$
    $$x \geq 35$$

    It is impossible to make \$250 with just 30 hours.

85. number of super-dogs: x
    number of regular-dogs: $100 - x$

    $$25(100 - x) + 45(x) \geq 3860$$
    $$2500 - 25x + 45x \geq 3860$$
    $$20x \geq 1360$$
    $$x \geq 68$$

    He must sell at least 68 super-dogs.

87. number of shares of Stock A: $1000 - x$
    number of shares of Stock B: x

    $$2(1000 - x) + 3(x) \geq 2400$$
    $$2000 - 2x + 3x \geq 2400$$
    $$x + 2000 \geq 2400$$
    $$x \geq 400$$

    The minimum number of shares of Stock B is 400.

3.4 Exercises

1. $|x| = 4$
   $x = 4$ or $x = -4$

3. $|x| < 4$
   $-4 < x < 4$

5. $|x| > 4$
   $x > 4$ or $x < -4$

7. $|x| \leq 4$
   $-4 \leq x \leq 4$

9. $|x| \geq 4$
   $x \geq 4$ or $x \leq -4$

11. No solution, absolute value is nonnegative.

13. $|x| > -4$
    True for all real numbers, since absolute value is nonnegative. It is greater than -4.

15. $|x| \leq -4$
    No solution, absolute value is nonnegative.

17. $|t - 3| = 2$
    $t - 3 = 2$ or $t - 3 = -2$
    $t = 5$ or $\qquad t = 1$

19. $|t| - 3 = 2$
    $|t| = 5$
    $t = 5$ or $t = -5$

21. $|5 - n| = 1$
    $5 - n = 1$ or $5 - n = -1$
    $-n = -4 \qquad -n = -6$
    $n = 4$ or $\qquad n = 6$

23. $|a - 5| < 3$
$-3 < a - 5 < 3$
$2 < a < 8$

25. $|a - 1| \geq 2$
$a - 1 \geq 2$ or $a - 1 \leq -2$
$a \geq 3$ or $a \leq -1$

27. $|2a - 5| < -1$
No solution, absolute value is nonnegative.

29. $|2a| - 5 < -1$
$|2a| < 4$
$-4 < 2a < 4$
$-2 < a < 2$

31. $|3x - 2| - 3 = 1$
$|3x - 2| = 4$
$3x - 2 = 4$ or $3x - 2 = -4$
$3x = 6$ $3x = -2$
$x = 2$ or $x = -\dfrac{2}{3}$

33. $|3x - 2| + 3 = 1$
$|3x - 2| = -2$
No solution, absolute value is nonnegative.

35. $|2(x - 1) + 7| = 5$
$|2x - 2 + 7| = 5$
$|2x + 5| = 5$
$2x + 5 = 5$ or $2x + 5 = -5$
$2x = 0$ $2x = -10$
$x = 0$ or $x = -5$

37. $|3 - a| \leq 2$
$-2 \leq 3 - a \leq 2$
$-5 \leq -a \leq -1$
$5 \geq a \geq 1$
or $1 \leq a \leq 5$

39. $|5 - 2a| > 1$
$5 - 2a > 1$ or $5 - 2a < -1$
$-2a > -4$ $-2a < -6$
$a < 2$ or $a > 3$

41. $|4(x - 1) - 5x| < 4$
$|4x - 4 - 5x| < 4$
$|-x - 4| < 4$
$-4 < -x - 4 < 4$
$0 < -x < 8$
$0 > x > -8$
or $-8 < x < 0$

43. $|x - 1| = 5$
$x - 1 = 5$ or $x - 1 = -5$
$x = 6$ or $x = -4$

45. $|3 - x| \leq 2$
$-2 \leq 3 - x \leq 2$
$-5 \leq -x \leq -1$
$5 \geq x \geq 1$
or $1 \leq x \leq 5$

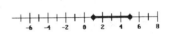

47. $|2x + 7| > 1$
$2x + 7 < -1$ or $2x + 7 > 1$
$2x < -8$ $2x > -6$
$x < -4$ or $x > -3$

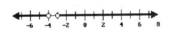

49. $|4x - 5| < 3$
$-3 < 4x - 5 < 3$
$2 < 4x < 8$
$\dfrac{1}{2} < x < 2$

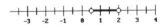

51. $|3 - 4x| < 1$

$$-1 < 3 - 4x < 1$$
$$-4 < -4x < -2$$
$$1 > x > \frac{1}{2}$$

or $\quad \frac{1}{2} < x < 1$

53. $|5t - 1| = |4t + 3|$

$5t - 1 = 4t + 3 \quad$ or $\quad 5t - 1 = -(4t + 3)$
$\quad t - 1 = 3 \qquad\qquad 5t - 1 = -4t - 3$
$\qquad\quad t = 4 \qquad\qquad 9t - 1 = -3$
$\qquad\qquad\qquad\qquad\qquad\quad 9t = -2$
$\qquad t = 4 \qquad$ or $\qquad t = -\frac{2}{9}$

55. $|4r - 3| = |2r + 9|$

$4r - 3 = 2r + 9 \quad$ or $\quad 4r - 3 = -(2r + 9)$
$2r - 3 = 9 \qquad\qquad\quad 4r - 3 = -2r - 9$
$\quad 2r = 12 \qquad\qquad\quad 6r - 3 = -9$
$\qquad r = 6 \qquad\qquad\qquad\quad 6r = -6$
$\qquad r = 6 \qquad$ or $\qquad\quad r = -1$

57. $|a - 5| = |2 - a|$

$a - 5 = 2 - a \quad$ or $\quad a - 5 = -(2 - a)$
$2a - 5 = 2 \qquad\qquad\quad a - 5 = -2 + a$
$\quad 2a = 7 \qquad\qquad\qquad -5 = -2$
$\qquad a = \frac{7}{2} \qquad\qquad$ No solution

The only solution is $a = \frac{7}{2}$.

59. $|3x - 4| = |4x - 3|$

$3x - 4 = 4x - 3 \quad$ or $\quad 3x - 4 = -(4x - 3)$
$-x - 4 = -3 \qquad\qquad\quad 3x - 4 = -4x + 3$
$\quad -x = 1 \qquad\qquad\qquad 7x - 4 = 3$
$\qquad x = -1 \qquad\qquad\qquad 7x = 7$
$\qquad x = -1 \qquad$ or $\qquad\quad x = 1$

61. $|x + 1| = |x - 1|$

$x + 1 = x - 1 \quad$ or $\quad x + 1 = -(x - 1)$
$\quad 1 = -1 \qquad\qquad\qquad x + 1 = -x + 1$
No solution $\qquad\qquad 2x + 1 = 1$
$\qquad\qquad\qquad\qquad\qquad 2x = 0$
$\qquad\qquad\qquad\qquad\qquad\quad x = 0$

The only solution is $x = 0$.

63. reading: x
For 40° F:

$$|x - 40| \le 0.75$$
$$-0.75 \le x - 40 \le 0.75$$
$$39.25 \le x \le 40.75$$
39.25° F to 40.75° F

For 50° F:

$$|x - 50| \le 0.75$$
$$-0.75 \le x - 50 \le 0.75$$
$$49.25 \le x \le 50.75$$
49.25° F to 50.75° F

For 80° F:

$$|x - 80| \le 0.75$$
$$-0.75 \le x - 80 \le 0.75$$
$$79.25 \le x \le 80.75$$
79.25° F to 80.75° F

65. percentage of votes: x

$$|x - 64| \le 6$$
$$-6 \le x - 64 \le 6$$
$$58 \le x \le 70$$
58% to 70%

3.5 Exercises

1. $x = -3, 0, 4$

3. $x = 2$

5. $3x^2 - x - 2 = 0$

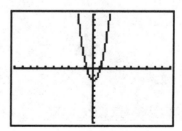

$x = -7, \quad x = 1$

7. $2x^2 + 15x = 5$
   $2x^2 + 15x - 5 = 0$

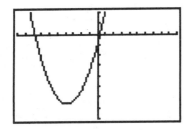

$x = -7.8, \quad x = 0.3$

9. $x^3 - x^2 - x + 2 = 0$

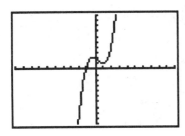

$x = -1.2$

11. $2x^4 - 5x^3 = 2x^2 - 6$
    $2x^4 - 5x^3 - 2x^2 + 6 = 0$

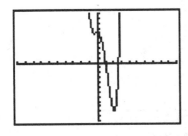

$x = 1.1, \quad x = 2.7$

CHAPTER 3 REVIEW EXERCISES

1. number of single beds: x
   number of larger beds: $24 - x$
   money collected: y

$y = 15(x) + 20(24 - x)$
$y = 15x + 480 - 20x$
$y = 480 - 5x$

Using TABLE, $y = 415$ when $x = 13$.

$\quad x = 13$
$24 - x = 24 - 13 = 11$

They delivered 13 single beds and 11 larger beds.

3. width: x
   length: $4x - 2$

$P = 2w + 2l$
$P = 2(x) + 2(4x - 2)$
$P = 2x + 8x - 4$
$P = 10x - 4$

Using TABLE, $P = 100$ when $x = 10.4$

$\quad x = 10.4$
$4x - 2 = 4(10.4) - 2 = 39.6$

The dimensions are 10.4 in. by 39.6 in.

5. $3x - 4(x - 2) = 2x - 5$
   $3x - 4x + 8 = 2x - 5$
   $-x + 8 = 2x - 5$
   $-3x + 8 = -5$
   $-3x = -13$
   $x = \dfrac{13}{3}$

7. $\dfrac{x}{2} - \dfrac{3}{2} = \dfrac{x}{5} + 1$

$10\left(\dfrac{x}{2} - \dfrac{3}{2}\right) = 10\left(\dfrac{x}{5} + 1\right)$

$\dfrac{10}{1} \cdot \dfrac{x}{2} - \dfrac{10}{1} \cdot \dfrac{3}{2} = \dfrac{10}{1} \cdot \dfrac{x}{5} + 10(1)$

$5x - 15 = 2x + 10$
$3x - 15 = 10$
$3x = 25$
$x = \dfrac{25}{3}$

9. $3x - 5x = 7x - 4x$
   $-2x = 3x$
   $0 = 5x$
   $0 = x$

11. $6(8 - a) = 3(a - 4) + 2(a - 1)$
$48 - 6a = 3a - 12 + 2a - 2$
$48 - 6a = 5a - 14$
$\qquad 48 = 11a - 14$
$\qquad 62 = 11a$
$\qquad \dfrac{62}{11} = a$

13. $7[x - 3(x - 3)] = 4x - 2$
$7(x - 3x + 9) = 4x - 2$
$\quad 7(-2x + 9) = 4x - 2$
$\quad -14x + 63 = 4x - 2$
$\qquad\qquad 63 = 18x - 2$
$\qquad\qquad 65 = 18x$
$\qquad\qquad \dfrac{65}{18} = x$

15. $5\{x - [2 - (x - 3)]\} = x - 2$
$5[x - (2 - x + 3)] = x - 2$
$\quad 5[x - (5 - x)] = x - 2$
$\quad 5(x - 5 + x) = x - 2$
$\qquad 5(2x - 5) = x - 2$
$\qquad 10x - 25 = x - 2$
$\qquad\quad 9x - 25 = -2$
$\qquad\quad 9x - 25 = 23$
$\qquad\qquad\quad x = \dfrac{23}{9}$

17. $|x| = 3$
$x = 3 \ \text{ or } \ x = -3$

19. No solution, absolute value is nonnegative.

21. $|2x| = 8$
$2x = 8 \ \text{ or } \ 2x = -8$
$\ x = 4 \ \text{ or } \ \ \ x = -4$

23. $|a + 1| = 4$
$a + 1 = 4 \ \text{ or } \ a + 1 = -4$
$\quad\ a = 3 \ \text{ or } \qquad a = -5$

25. $|4z + 5| = 0$
$4z + 5 = 0$
$\quad 4z = -5$
$\quad\ z = -\dfrac{5}{4}$

27. $|2y - 5| - 8 = -3$
$\quad\ |2y - 5| = 5$
$2y - 5 = 5 \ \text{ or } \ 2y - 5 = -5$
$\quad 2y = 10 \qquad\quad 2y = 0$
$\quad\ y = 5 \ \text{ or } \qquad\ y = 0$

29. $|x - 5| = |x - 1|$
$x - 5 = x - 1 \ \text{ or } \ x - 5 = -(x - 1)$
$\quad -5 = -1 \qquad\qquad x - 5 = -x + 1$
No solution $\qquad\qquad\ 2x - 5 = 1$
$\qquad\qquad\qquad\qquad\quad 2x = 6$
$\qquad\qquad\qquad\qquad\quad\ x = 3$

The only solution is $x = 3$.

31. $|2t - 4| = |t - 2|$
$2t - 4 = t - 2 \ \text{ or } \ 2t - 4 = -(t - 2)$
$\quad t - 4 = -2 \qquad\qquad 2t - 4 = -t + 2$
$\qquad\quad t = 2 \qquad\qquad\quad 3t - 4 = 2$
$\qquad\qquad\qquad\qquad\qquad 3t = 6$
$\qquad\qquad\qquad\qquad\qquad\ t = 2$

33. $3x - 4 \le 5$
$\quad 3x \le 9$
$\qquad x \le 3$

35. $5x - 4 \le 2x$
$\quad -4 \le -3x$
$\quad \dfrac{4}{3} \ge x$
or $\ x \le \dfrac{4}{3}$

37. $5z + 4 > 2z - 1$
$3z + 4 > -1$
$\quad 3z > -5$
$\qquad z > -\dfrac{5}{3}$

39. $5(s - 4) < 3(s - 4)$
$5s - 20 < 3s - 12$
$2s - 20 < -12$
$\quad\ 2s < 8$
$\qquad s < 4$

41. $3(r - 2) - 5(r - 1) \ge -2r - 1$
$3r - 6 - 5r + 5 \ge -2r - 1$
$\qquad -2r - 1 \ge -2r - 1$

Identity, true for all real numbers

43. $-3 \le 2x - 1 \le 5$
$-2 \le 2x \le 6$
$-1 \le x \le 3$

45.
$$-5 \le 6 - 2x \le 7$$
$$-11 \le -2x \le 1$$

$$\frac{11}{2} \ge x \ge -\frac{1}{2}$$

or $\quad -\frac{1}{2} \le x \le \frac{11}{2}$

47.
$$-3 \le 3(x - 4) \le 6$$
$$-3 \le 3x - 12 \le 6$$
$$9 \le 3x \le 18$$
$$3 \le x \le 6$$

49.
$$3[a - 3(a + 2)] > 0$$
$$3(a - 3a - 6) > 0$$
$$3(-2a - 6) > 0$$
$$-6a - 18 > 0$$
$$-6a > 18$$
$$a < -3$$

51.
$$5 - [2 - 3(2 - x)] < -3x + 1$$
$$5 - (2 - 6 + 3x) < -3x + 1$$
$$5 - (-4 + 3x) < -3x + 1$$
$$5 + 4 - 3x < -3x + 1$$
$$9 - 3x < -3x + 1$$
$$9 < 1$$

Contradiction, no solution

53.
$$3 - \{2 - 5[q - (2 - q]\} \le 3[q - (q - 5)]$$
$$3 - [2 - 5(q - 2 + q)] \le 3(q - q + 5)$$
$$3 - [2 - 5(2q - 2)] \le 3(5)$$
$$3 - (2 - 10q + 10) \le 15$$
$$3 - (12 - 10q) \le 15$$
$$3 - 12 + 10q \le 15$$
$$-9 + 10q \le 15$$
$$10q \le 24$$
$$q \le \frac{12}{5}$$

55. $|x| < 4$
$$-4 < x < 4$$

57. $|s| \ge 5$
$$s \le -5 \quad \text{or} \quad s \ge 5$$

59. $|t| < 0$
Contradiction, absolute value is nonnegative.

61. $|t - 1| < 2$
$$-2 < t - 1 < 2$$
$$-1 < t < 3$$

63. $|a - 6| \ge 3$
$$a - 6 \ge 3 \quad \text{or} \quad a - 6 \le -3$$
$$a \le 3 \quad \text{or} \qquad a \ge 9$$

65. $|r + 9| \le 4$
$$-4 \le r + 9 \le 4$$
$$-13 \le r \le -5$$

67. $|2x - 1| \geq 2$

$2x - 1 \leq -2 \quad$ or $\quad 2x - 1 \geq 2$

$\qquad 2x \leq -1 \qquad\qquad 2x \geq 3$

$\qquad x \leq -\dfrac{1}{2} \quad$ or $\quad x \geq \dfrac{3}{2}$

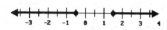

69. $|3x - 2| < 4$

$-4 < 3x - 2 < 4$

$-2 < 3x < 6$

$-\dfrac{2}{3} < x < 2$

71. $|2x - 3| \leq 5$

$-5 \leq 2x - 3 \leq 5$

$-2 \leq 2x \leq 8$

$-1 \leq x \leq 4$

73. $|3x + 5| + 2 > 7$

$\qquad |3x + 5| > 5$

$3x + 5 > 5 \quad$ or $\quad 3x + 5 < -5$

$\qquad 3x > 0 \qquad\qquad 3x < -10$

$\qquad x > 0 \quad$ or $\quad x < -\dfrac{10}{3}$

75. $|3 - 2x| + 8 \leq 4$

$\qquad |3 - 2x| \leq -4$

No solution, absolute value is nonnegative.

77. the number: x

$3x = 4x - 4$

$-x = -4$

$x = 4$

The number is 4.

79. the number: x

$5(x + 6) = x - 2$

$5x + 30 = x - 2$

$4x + 30 = -2$

$\qquad 4x = -32$

$\qquad x = -8$

The number is -8.

81. gross sales: x

$0.09x = 650$

$\qquad x = 7222.22$

His gross sales were $7222.22.

83. final exam score: x

$0.40x + 0.60(74) = 80$

$0.40x + 44.4 = 80$

$\qquad 0.40x = 35.6$

$\qquad\qquad x = 89$

You need an 89 on the final exam.

85. number of 8-lb packages: x
number of 5-lb packages: $30 - x$

$8(x) + 5(30 - x) = 186$

$8x + 150 - 5x = 186$

$3x + 150 = 186$

$3x = 36$

$x = 12$

$30 - x = 30 - 12 = 18$

There were 12 8-lb packages and 18 5-lb packages.

87. number of single beds: x
number of larger beds: $23 - x$

$295 = 10(x) + 15(23 - x)$

$295 = 10x + 345 - 15x$

$295 = 345 - 5x$

$-50 = -5x$

$10 = x$

$23 - x = 23 - 10 = 13$

They delivered 10 single beds and 13 larger beds.

89. hours editing: $30 - x$
hours tutoring: x

$12(30 - x) + 20x \geq 456$

$360 - 12x + 20x \geq 456$

$360 + 8x \geq 456$

$8x \geq 96$

$8 \geq 12$

He must tutor at least 12 hours.

91. $2x - 3 = 0$

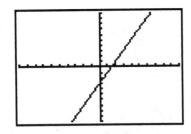

$x = 1.5$

93. $x^2 - 6x + 5 = 0$

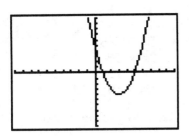

$x = 1, \quad x = 5$

95. $\qquad 6x^2 + x = 2$
$6x^2 + x - 2 = 0$

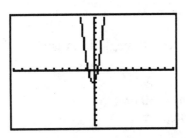

$x = -0.7, \quad x = 0.5$

97. $\qquad x^3 - x = -1$
$x^3 - x + 1 = 0$

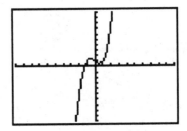

$x = -1.3$

1. amount in savings account: x
   amount in certificate: $5000 - x$

   $$I = 0.035(x) + 0.06(5000 - x)$$
   $$I = 0.035x + 300 - 0.06x$$
   $$I = -0.025x + 300$$

   $$220 = -0.025x + 300$$
   $$-80 = -0.025x$$
   $$3200 = x$$

   There is $3200 in the savings account.

2. $7x - 9 = 5x + 2$
   $2x - 9 = 2$
   $\quad\ 2x = 11$
   $\qquad x = \dfrac{11}{2}$

3. $3x - 4(x - 5) = 7(x - 2) - 8(x - 1)$
   $3x - 4x + 20 = 7x - 14 - 8x + 8$
   $\quad -x + 20 = -x - 6$
   $\qquad\quad 20 = -6$
   Contradiction, no solution

4. $6a - [2 - 3(a - 4)] = 5(a - 10)$
   $6a - (2 - 3a + 12) = 5a - 50$
   $\quad 6a - (14 - 3a) = 5a - 50$
   $\quad 6a - 14 + 3a = 5a - 50$
   $\qquad\ 9a - 14 = 5a - 50$
   $\qquad\ 4a - 14 = -50$
   $\qquad\qquad 4a = -36$
   $\qquad\qquad\ a = -9$

5. $3x + 5 \le 5x - 3$
   $-2x + 5 \le -3$
   $\quad -2x \le -8$
   $\qquad x \ge 4$

6. $3x - 2[x - 3(1 - x)] > 5$
   $3x - 2(x - 3 + 3x) > 5$
   $\quad 3x - 2(4x - 3) > 5$
   $\quad 3x - 8x + 6 > 5$
   $\qquad -5x + 6 > 5$
   $\qquad\quad -5x > -1$
   $\qquad\qquad x < \dfrac{1}{5}$

7. $-3 < 5 - 2x < 4$
   $-8 < -2x < -1$
   $4 > x > \dfrac{1}{2}$ or $\dfrac{1}{2} < x < 4$

8. $|2x - 7| + 2 = 9$
   $\quad |2x - 7| = 7$
   $2x - 7 = 7$ or $2x - 7 = -7$
   $\quad 2x = 14 \qquad\quad 2x = 0$
   $\qquad x = 7$ or $\qquad\ x = 0$

9. $|3x - 4| < 5$
   $-5 < 3x - 4 < 5$
   $-1 < 3x < 9$
   $-\dfrac{1}{3} < x < 3$

10. $|5x - 7| \ge 2$
    $5x - 7 \le -2$ or $5x - 7 \ge 2$
    $\quad 5x \le 5 \qquad\qquad 5x \ge 9$
    $\qquad x \le 1$ or $\qquad x \ge \dfrac{9}{5}$

11. number of 35-kg boxes: x
    number of 45-kg boxes: $93 - x$

    $$35(x) + 45(93 - x) = 3465$$
    $$35x + 4185 - 45x = 3465$$
    $$-10x + 4185 = 3465$$
    $$-10x = -720$$
    $$x = 72$$
    $$93 - x = 93 - 72 = 21$$

    There are 72 35-kg boxes and 21 45-kg boxes.

12. hours jogging: x
    hours walking: $x + 2$

$$8x = 3(x + 2)$$
$$8x = 3x + 6$$
$$5x = 6$$
$$x = \frac{6}{5}$$

It would take $\frac{6}{5}$ hr.

13. number of dimes: x
    number of nickels: $32 - x$

$$5(32 - x) + 10x \geq 265$$
$$160 - 5x + 10x \geq 265$$
$$160 + 5x \geq 265$$
$$5x \geq 105$$
$$x \geq 21$$

He has at least 21 dimes.

14. $3x - 4 = 0$

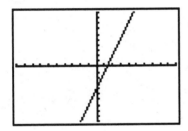

$x = 1.3$

15. $x^2 + 7x + 10 = 0$

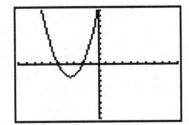

$x = -5, \quad x = -2$

1. $A = \{11, 13, 17, 19, 23, 29, 31, 37\}$

3. $A = \{11, 13, 17, 19, 23, 29, 31, 37\}$
   $C = \{10, 15, 20, 25, 30, 35\}$
   $A \cap c = \emptyset$

5. True

7. $\{x \mid -2 \le x < 8\}$

9. Closure property for addition

11. $-2 + 3 - 4 - 8 + 5 = 1 - 4 - 8 + 5$
    $= -3 - 8 + 5$
    $= -11 + 5$
    $= -6$

13. $(-9 + 3)(-4 - 2) = (-6)(-6)$
    $= 36$

15. $|y - x| - (y - x)$

    $= |2 - (-4)| - [2 - (-4)]$

    $= |6| - (6)$

    $= 6 - 6$
    $= 0$

17. $-3x(x^2 - 2x + 3)$
    $= -3x(x^2) - 3x(-2x) - 3x(3)$
    $= -3x^3 + 6x^2 - 9x$

19. number: x
    $3 + x \cdot 8$
    or $3 + 8x$

21. width: x
    length: $4x - 3$

    $A = w \cdot l$
    $A = x(4x - 3)$
    $A = 4x^2 - 3x$

$P = 2w + 2l$
$P = 2(x) + 2(4x - 3)$
$P = 2x + 8x - 6$
$P = 10x - 6$

23. $4x - 3 = 5x + 8$
    $-3 = x + 8$
    $-11 = x$

25. $2x + 3 - (x - 2) = 5 - (x - 4)$
    $2x + 3 - x + 2 = 5 - x + 4$
    $x + 5 = -x + 9$
    $2x + 5 = 9$
    $2x = 4$
    $x = 2$

27. $3(x - 4) + 2(3 - x) = 5[x - (2 - 3x)]$
    $3x - 12 + 6 - 2x = 5(x - 2 + 3x)$
    $x - 6 = 5(4x - 2)$
    $x - 6 = 20x - 10$
    $-6 = 19x - 10$
    $4 = 19x$
    $\dfrac{4}{19} = x$

29. $|2x + 1| = 9$
    $2x + 1 = 9$ or $2x + 1 = -9$
    $2x = 8$ $\qquad 2x = -10$
    $x = 4$ or $\qquad x = -5$

31. $3x - 5 \le x - 8$
    $2x - 5 \le -8$
    $2x \le -3$
    $x \le -\dfrac{3}{2}$

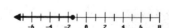

33. $6(x - 2) + (2x - 3) < 2(4x + 1)$
    $6x - 12 + 2x - 3 < 8x + 2$
    $8x - 15 < 8x + 2$
    $-15 < 2$
    Identity, true for all real numbers

35. $|x - 5| < 4$
$-4 < x - 5 < 4$
$\quad 1 < x < 9$

37. $|5 - 4x| > 7$
$5 - 4x > 7 \quad$ or $\quad 5 - 4x < -7$
$\quad -4x > 2 \qquad\qquad -4x < -12$
$\quad\quad x < -\dfrac{1}{2} \quad$ or $\qquad x > 3$

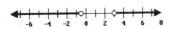

39. the number: x

$4x = 2x - 3$
$2x = -3$
$\quad x = -\dfrac{3}{2}$

The number is $-\dfrac{3}{2}$.

41. number of 300-plate packages: x
number of 100-plate packages: $28 - x$

$2(x) + 1(28 - x) = 33$
$\quad 2x + 28 - x = 33$
$\qquad\quad x + 28 = 33$
$\qquad\qquad\quad x = 5$
$\quad 28 - x = 28 - 5 = 23$

5 300-plate packages yields
$5(300) = 1500$ plates

23 100-plate packages yields
$23(100) = 2300$ plates

1500 plates + 2300 plates = 3800 plates
He bought 3800 plates.

43. 30 minutes = 30(60) = 1800 seconds
In 1800 seconds the GL-70 can print
$(1800)(80) = 144{,}000$ characters

$\dfrac{144000 \text{ characters}}{120 \text{ characters/sec}} = 1200 \text{ sec} = \dfrac{1200}{60} = 20$
minutes

It would take the VF-44 20 minutes to print
the document.

45. width: x
length: $3x + 2$

$P = 2w + 2l$
$P = 2(x) + 2(3x + 2)$
$P = 2x + 6x + 4$
$P = 8x + 4$
$P - 4 = 8x$

$\dfrac{P - 4}{8} = x$

$5 \le x \le 12$

$5 \le \dfrac{P - 4}{8} \le 12$

$40 \le P - 4 \le 96$
$44 \le P \le 100$

The perimeter varies from 44 feet to 100 feet.

47. domain: $\{2, 4, 7\}$
range: $\{3, -1, 5\}$
function

49. domain: $\{4, 3\}$
range: $\{2, 9, 7\}$
Not a function, since the domain element
4 is used twice.

51. $\{x \mid x$ is a real number$\}$

53. $5x - 4 \ne 0$
$\quad 5x \ne 4$
$\qquad x \ne \dfrac{4}{5}$

$\left\{ x \mid x \ne \dfrac{4}{5} \right\}$

55. function, passes the Vertical Line Test

57. function, passes the Vertical Line Test

59. $f(-3) = -3 + \dfrac{1}{-3}$

$= \dfrac{-10}{3}$

61. $h(x^2) = 3x^2 + 1$

63. $f(a + 3) = a + 3 + \dfrac{1}{a + 3}$

65. $f(3a) = 3a + \dfrac{1}{3a}$

67. $\qquad f(1) = 1 + \dfrac{1}{1} = 2$

$g(-1) = \dfrac{1}{(-1)^2} = \dfrac{1}{1} = 1$

$f(1) + g(-1) = 2 + 1 = 3$

69. $\quad h(2) = 3(2) + 1$
$\qquad\quad = 7$

$f[h(2)] = f[7]$

$= 7 + \dfrac{1}{7}$

$= \dfrac{50}{7}$

71. $5x - 4y = 10$

$x = 0: \ 5(0) - 4y = 10$
$\qquad\qquad -4y = 10$
$\qquad\qquad\quad y = -\dfrac{5}{2}$
$\qquad\qquad \left(0, -\dfrac{5}{2}\right)$

$y = 0; \ 5x - 4(0) = 10$
$\qquad\qquad 5x = 10$
$\qquad\qquad\ x = 2$
$\qquad\qquad (2, 0)$

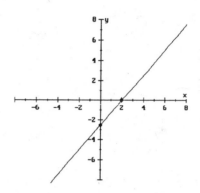

73. $5y = 3x - 10$

$x = 0: \ 5y = 3(0) - 10$
$\qquad\qquad 5y = -10$
$\qquad\qquad\ y = -2$
$\qquad\qquad (0, -2)$

$y = 0: \ 5(0) = 3x - 10$
$\qquad\qquad\ 0 = 3x - 10$
$\qquad\quad 10 = 3x$
$\qquad \dfrac{10}{3} = x$

$\left(\dfrac{10}{3}, 0\right)$

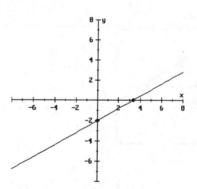

75. $3x - 2 = 8$
$\quad\ 3x = 10$
$\qquad x = \dfrac{10}{3}$

This is a vertical line passing through $\left(\dfrac{10}{3}, 0\right)$.

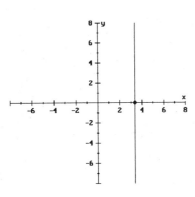

77. (a) Approximately 40,000
    (b) Approximately 4.7 hours
    (c) Between 3.5 hours and 6 hours
        (graph steepest)
    (d) After approximately 7 hours

79. Charges = flat fee + hourly fee · number of
                          hours
    $C = 350 + 75h$

81. domain: $\{x \mid x \text{ is a real number}\}$
    range: $\{y \mid y \geq -3\}$

83. $y = 7 - 2x$

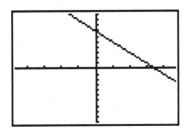

    range: $\{y \mid -2.4 \leq y \leq 16.4\}$
    no turning points

85. $y = 9x - x^3$

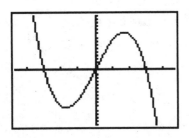

range: $\{y \mid -61.5 \leq y \leq 61.5\}$
turning points: $(-1.7, -10.4)$, $(1.7, 10.4)$

87. $s(t) = 75 + 50t - 16t^2$

    (a)

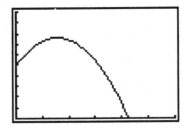

    domain: $\{t \mid 0 \leq t \leq 4.2\}$

    (b) 114.1

    (c) 114.1 feet; the maximum value of the
         function is the maximum height.

    (d) 1.6 sec (1st coordinate of maximum
         point)

CHAPTERS 1 - 3 CUMULATIVE PRACTICE TEST

1. $A = \{4, 8, 12, 16, 20\}$
   $B = \{2, 4, 6, 8, 10, 12, 14, 16, 18, 20, 22\}$

   (a) $A \cap B = \{4, 8, 12, 16, 20\} = A$

   (b) $A \cup B = \{2, 4, 6, 8, 10, 12, 14, 16, 18, 20, 22\} = B$

3. (a) $(-2)(-2) - (-2)^2(2) = (-2)(-2) - (4)(2)$
   $$= 4 - 8$$
   $$= -4$$

   (b) $5 - \{6 + [2 - 3(4 - 9)]\} = 5 - \{6 + [2 - 3(-5)]\}$
   $$= 5 - [6 + (2 + 15)]$$
   $$= 5 - (6 + 17)$$
   $$= 5 - (23)$$
   $$= -18$$

5. (a) $3x - 2 = 5x + 4$
   $$-2x - 2 = 4$$
   $$-2x = 6$$
   $$x = -3$$

   (b) $3(x - 5) - 2(x - 5) = 3 - (5 - x)$
   $$3x - 15 - 2x + 10 = 3 - 5 + x$$
   $$x - 5 = -2 + x$$
   $$-5 = -2$$
   Contradiction, no solution

   (c) $|x - 3| = 8$
   $$x - 3 = 8 \quad \text{or} \quad x - 3 = -8$$
   $$x = 11 \quad \text{or} \quad x = -5$$

7. number of 10-lb packages: x
   number of 30-lb packages: $170 - x$

   $$10(x) + 30(170 - x) = 3140$$
   $$10x + 5100 - 30x = 3140$$
   $$5100 - 20x = 3140$$
   $$-20x = -1960$$
   $$x = 98$$
   $$170 - x = 170 - 98 = 72$$

   There are 98 10-lb packages and 72 30-lb packages.

9.   Evan starts his trip by travelling 50 miles away from his home during the first hour of the day; his average rate of speed for the first hour was 50 mph. During the next 2 hours Evan did not travel. Then Evan drove closer to his home between hours 3 and 4. He was travelling an average of 40 mph during this time, and ended up only 10 miles from his home. Evan drove another 60 miles away from his home travelling at 60 mph between the 4th and 5th hour, and stopped travelling for an hour. Finally, during the last hour and a half, Evan drove home at a rate of 47 mph.

11.   (a) $x - 9 \neq 0$

$x \neq 9$

$\{x \mid x \neq 9\}$

(b) $2x - 3 \geq 0$

$2x \geq 3$

$x \geq \dfrac{3}{2}$

$\left\{x \mid x \geq \dfrac{3}{2}\right\}$

13.   (a) $f(-6) = 2$

(b) $f(3) = -2$

(c) $f(0) = 0$

(d) $f(-4) = 3$

(e) $(-4, 3)$; $(3, -2)$

(f) $f(-3)$

(g) $x = 0$, $x = 5$

(h)   domain: $\{x \mid -6 \leq x \leq 7\}$
range: $\{y \mid -2 \leq y \leq 3\}$

4.1 Exercises

1. (1, −2) and (−3, 1)

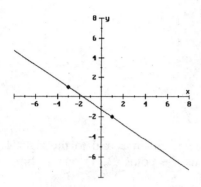

$$m = \frac{-2 - 1}{1 - (-3)}$$

$$= \frac{-3}{4}$$

3. (0, 2) and (2, 0)

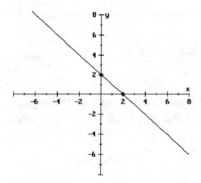

$$m = \frac{2 - 0}{0 - 2}$$

$$= \frac{2}{-2}$$

$$= -1$$

5. (−3, −4) and (−2, −5)

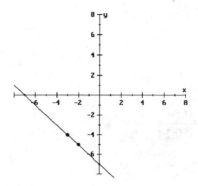

$$m = \frac{-4 - (-5)}{-3 - (-2)}$$

$$= \frac{1}{-1}$$

$$= -1$$

7. $m = \dfrac{4 - 4}{-3 - 2}$

$$= \frac{0}{-5}$$

$$= 0$$

9. $m = \dfrac{-3 - 2}{4 - 4}$

$$= \frac{-5}{0}$$

Undefined

11. $m = \dfrac{a - b}{a - b}$

$$= 1$$

13. $m = \dfrac{a^2 - b^2}{a - b}$

$$= \frac{(a - b)(a + b)}{a - b}$$

$$= a + b$$

15. $m = \dfrac{-0.2 - 0.14}{0.7 - 0.06}$

$= \dfrac{-0.34}{0.64}$

$= -\dfrac{17}{32}$

17. $m = \dfrac{7 - 0.06}{-0.2 - 0.14}$

$= \dfrac{6.94}{-0.34}$

$= -\dfrac{347}{17}$

19. To get from (0, 0) to (2, 3) (which appear to be points on the line), you go up 3 and to the right 2.

$m = \dfrac{3}{2}$

21. To get from (0, −7) to (2, 0) (which appear to be points on the line), you go up 7 and to the right 2.

$m = \dfrac{7}{2}$

23. (5, 0) and (0, −3)

$m = \dfrac{0 - (-3)}{5 - 0}$

$= \dfrac{3}{5}$

25. $m = 2 = \dfrac{2}{1}$

From (1, 3), go up 2 and to the right 1.
This gives us the point (2, 5) on the line.

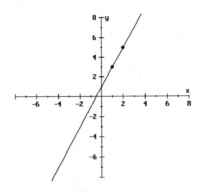

27. $m = -2 = \dfrac{-2}{1}$

From (1, 3), go down 2 and to the right 1.
This gives us the point (2, 1) on the line.

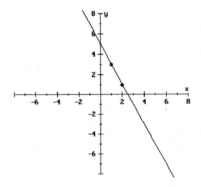

29. $m = -\dfrac{1}{4} = \dfrac{-1}{4}$

From (0, 3), go down 1 and to the right 4.
This gives us the point (4, 2) on the line.

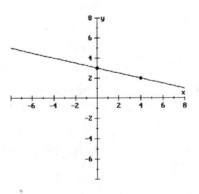

31. $m = \dfrac{2}{5}$

From (4, 0), go up 2 and to the right 5.
This gives us the point (9, 2) on the line.

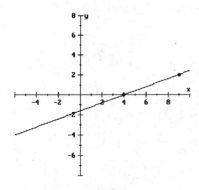

33. $m = 0$: horizontal line through (2, 5)

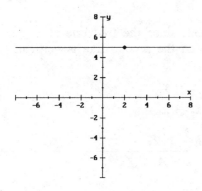

35. undefined slope: vertical line through (2, 5)

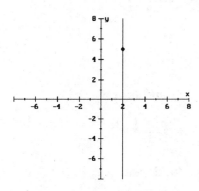

37. $y = 2x - 7$

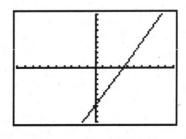

(0, −7) and (1, −5)

$m = \dfrac{-7 - (-5)}{0 - 1}$

$= 2$

39. $y = -0.4x + 5$

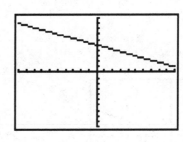

(0, 5) and (1, 4.6)

$m = \dfrac{5 - 4.6}{0 - 1}$

$= \dfrac{0.4}{-1}$

$= -0.4$

41. $2y - 3x = 8$

$2y = 3x + 8$

$y = \dfrac{3}{2}x + 4$

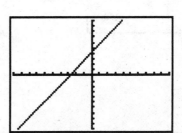

$(0, 4)$ and $(2, 7)$

$$m = \frac{7 - 4}{2 - 0}$$

$$= \frac{3}{2}$$

43.

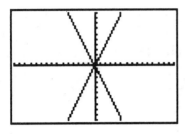

Neither

45.

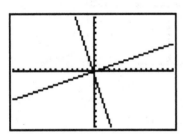

Perpendicular

47.

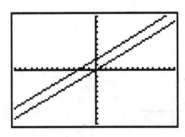

Parallel

49. $m_{P_1 P_2} = \dfrac{4 - 2}{3 - 1} = \dfrac{2}{2} = 1$

$m_{P_3 P_4} = \dfrac{-4 - (-2)}{-3 - (-1)} = \dfrac{-2}{-2} = 1$

slopes are equal: Parallel

51. $m_{P_1 P_2} = \dfrac{2 - 4}{-1 - 0} = \dfrac{-2}{-1} = 2$

$m_{P_3 P_4} = \dfrac{7 - 5}{1 - (-3)} = \dfrac{2}{4} = \dfrac{1}{2}$

Neither

53. $m_{P_1 P_2} = \dfrac{5 - 5}{-2 - 3} = \dfrac{0}{-5} = 0$

$m_{P_3 P_4} = \dfrac{-2 - 4}{1 - 1} = \dfrac{-6}{0}$ undefined

Perpendicular, since the first line is horizontal and the second line is vertical.

55. $m = \dfrac{y_2 - y_1}{x_2 - x_1}$

$-5 = \dfrac{h - (-2)}{1 - 4}$

$-5 = \dfrac{h + 2}{-3}$

$-3(-5) = (-3)\left(\dfrac{h + 2}{-3}\right)$

$15 = h + 2$

$13 = h$

57. firstline: $(0, 1)$ and (c, c)

$m_1 = \dfrac{c - 1}{c - 0} = \dfrac{c - 1}{c}$

second line: $(0, 2)$ and $(-c, c)$

$m_2 = \dfrac{c - 2}{-c - 0} = \dfrac{c - 2}{-c}$

or $\dfrac{2 - c}{c}$

To be perpendicular:

$$m_1 = -\frac{1}{m_2}$$

$$\frac{c-1}{c} = \frac{-1}{\dfrac{2-c}{c}}$$

$$\frac{c-1}{c} = \frac{-c}{2-c}$$

$$(c-1)(2-c) = (c)(-c)$$
$$2c - c^2 - 2 + c = -c^2$$
$$-c^2 + 3c - 2 = -c^2$$
$$3c - 2 = 0$$
$$3c = 2$$
$$c = \frac{2}{3}$$

59. $A(0, 0)$; $B(2, 1)$; $C(-2, 5)$; $D(0, 6)$

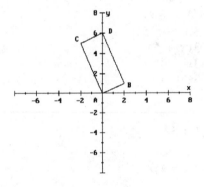

$$M_{AB} = \frac{1-0}{2-0} = \frac{1}{2}$$

$$M_{AC} = \frac{5-0}{-2-0} = -\frac{5}{2}$$

$$M_{CD} = \frac{6-5}{0-(-2)} = \frac{1}{2}$$

$$M_{BD} = \frac{6-1}{0-2} = -\frac{5}{2}$$

Since opposite sides have equal slopes, they are parallel. Hence this is a parallelogram.

61. $A(-3, 2)$; $B(-1, 6)$; $C(3, 4)$

$$M_{AB} = \frac{6-2}{-1-(-3)} = \frac{4}{2} = 2$$

$$M_{BC} = \frac{6-4}{-1-3} = \frac{2}{-4} -\frac{1}{2}$$

$$M_{AC} = \frac{4-2}{3-(-3)} = \frac{2}{6} = \frac{1}{3}$$

AB is perpendicular to BC since their slopes are negative reciprocals. Thus it is a right triangle.

63.

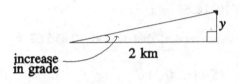

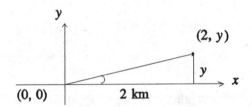

upgrade: $8\% = 0.08 = m$

$$0.08 = \frac{y-0}{2-0}$$

$$0.08 = \frac{y}{2}$$

$$0.16 = y$$

The change in elevation is 0.16 km.

65.

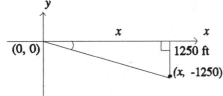

downgrade: 12% = 0.12

$$m = -0.12$$

$$\frac{-1250 - 0}{x - 0} = -0.12$$

$$\frac{-1250}{x} = -0.12$$

$$-1250 = -0.12x$$
$$10416.67 = x$$

The horizontal distance is 10,416.67 ft.

67. (2, 15) and (8, 80)

$$m = \frac{80 - 15}{8 - 2}$$

$$= \frac{65}{6}$$

$$= 10.83$$

The sprinter averages approximately 10.83 meters per second during the stretch from 2 seconds to 8 seconds.

69. (1, 56) and (5, 42)

$$m = \frac{56 - 42}{1 - 5}$$

$$= \frac{14}{-4}$$

$$= -3.5$$

The temperature falls an average of approximately 3.5° during the 4 hour period.

77. $2xy(3x - 5y) - 6y(2x^2 + xy)$
$= 6x^2y - 10xy^2 - 12x^2y - 6xy^2$
$= -6x^2y - 16xy^2$

79.

$$\frac{t}{3} - 2 = 5$$

$$3\left(\frac{t}{3} - 2\right) = 3(5)$$

$$\frac{3}{1} \cdot \frac{t}{3} - 3(2) = 3(5)$$

$$t - 6 = 15$$
$$t = 21$$

4.2 Exercises

1. $m = 5$; (1, -3)

$$y - y_1 = m(x - x_1)$$
$$y - (-3) = 5(x - 1)$$
$$y + 3 = 5x - 5$$
$$y = 5x - 8$$

3. $m = -3$; (-5, 2)

$$y - y_1 = m(x - x_1)$$
$$y - 2 = -3[x - (-5)]$$
$$y - 2 = -3(x + 5)$$
$$y - 2 = -3x - 15$$
$$y = -3x - 13$$

5. $m = \frac{2}{3}$; (6, 1)

$$y - y_1 = m(x - x_1)$$

$$y - 1 = \frac{2}{3}(x - 6)$$

$$y - 1 = \frac{2}{3}x - 4$$

$$y = \frac{2}{3}x - 3$$

7. $m = -\frac{1}{2}$; (4, 0)

$$y - y_1 = m(x - x_1)$$

$$y - 0 = -\frac{1}{2}(x - 4)$$

$$y = -\frac{1}{2}x + 2$$

9. $m = \dfrac{3}{4}$; $(0, 5)$

$y = mx + b$

$y = \dfrac{3}{4}x + 5$

11. $m = 0$; horizontal line through $(-3, -4)$

$y = -4$

13. m undefined: vertical line through $(-4, 7)$

$x = -4$

15. $(2, 3)$ and $(5, 7)$

$m = \dfrac{7 - 3}{5 - 2} = \dfrac{4}{3}$

$y - y_1 = m(x - x_1)$

$y - 3 = \dfrac{4}{3}(x - 2)$

$y - 3 = \dfrac{4}{3}x - \dfrac{8}{3}$

$y = \dfrac{4}{3}x + \dfrac{1}{3}$

17. $(-2, -1)$ and $(-3, -5)$

$m = \dfrac{-5 - (-1)}{-3 - (-2)} = \dfrac{-4}{-1} = 4$

$y - y_1 = m(x - x_1)$
$y - (-1) = 4[x - (-2)]$
$y + 1 = 4(x + 2)$
$y + 1 = 4x + 8$
$y = 4x + 7$

19. $(2, 3)$ and $(0, 5)$

$m = \dfrac{5 - 3}{0 - 2} = \dfrac{2}{-2} = -1$

$y = mx + b$
$y = -x + 5$

21. $m = 4$; $(0, 6)$

$y = mx + b$
$y = 4x + 6$

23. $m = -2$; $(-3, 0)$

$y - y_1 = m(x - x_1)$
$y - 0 = -2[x - (-3)]$
$y = -2(x + 3)$
$y = -2x - 6$

25. $(2, 3)$ and $(6, 3)$

$m = \dfrac{3 - 3}{6 - 2} = \dfrac{0}{4} = 0$

Horizontal line: $y = 3$

27. vertical line through $(-2, -4)$
$x = -2$

29. $(-3, 0)$ and $(0, 2)$

$m = \dfrac{2 - 0}{0 - (-3)} = \dfrac{2}{3}$

$y = mx + b$
$y = \dfrac{2}{3}x + 2$

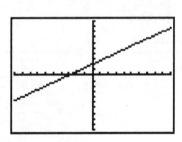

31. $y = 3x + 7$
$m = 3$
slope of parallel line: 3

$y - y_1 = m(x - x_1)$
$y - 2 = 3(x - 2)$
$y - 2 = 3x - 6$
$y = 3x - 4$

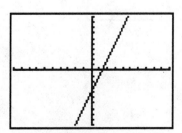

33. $\quad y = -\dfrac{2}{3}x - 1$

$$m = -\dfrac{2}{3}$$

$$m_\perp = \dfrac{3}{2}$$

$$y - (-3) = \dfrac{3}{2}[x - (-3)]$$

$$y + 3 = \dfrac{3}{2}(x + 3)$$

$$y + 3 = \dfrac{3}{2}x + \dfrac{9}{2}$$

$$y = \dfrac{3}{2}x + \dfrac{3}{2}$$

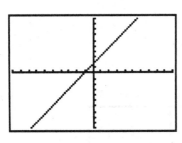

35. $\quad y = x$
$\quad m = 1$
$\quad m_\perp = -1$

$$y = mx + b$$
$$y = -1x + 0$$
$$y = -x$$

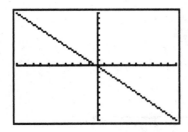

37. $\quad 2y - 3x = 12$
$\qquad 2y = 3x + 12$

$$y = \dfrac{3}{2}x + 6$$

$$m = \dfrac{3}{2}$$

parallel line: sameslope

$$y - y_1 = m(x - x_1)$$

$$y - (-2) = \dfrac{3}{2}[x - (-1)]$$

$$y + 2 = \dfrac{3}{2}(x + 1)$$

$$y + 2 = \dfrac{3}{2}x + \dfrac{3}{2}$$

$$y = \dfrac{3}{2}x - \dfrac{1}{2}$$

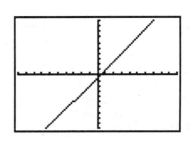

39. $\quad 4x - 3y = 9$
$\qquad -3y = -4x + 9$

$$y = \dfrac{4}{3}x - 3$$

$$m = \dfrac{4}{3}$$

$$m_\perp = -\dfrac{3}{4}$$

$$y - y_1 = m(x - x_1)$$

$$y - (-3) = -\frac{3}{4}(x - 1)$$

$$y + 3 = -\frac{3}{4}x + \frac{3}{4}$$

$$y = -\frac{3}{4}x - \frac{9}{4}$$

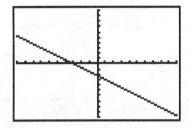

$$y = mx + b$$
$$y = -2x + 4$$

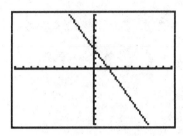

45. horizontal: $y = 3$

47. vertical: $x = -1$

41. $8x - 5y = 20$
$$-5y = -8x + 20$$

$$y = \frac{8}{5}x - 4$$

$$b = -4$$

$$m = \frac{8}{5}$$

$$m_\perp = -\frac{5}{8}$$

$$y = mx + b$$

$$y = -\frac{5}{8}x - 4$$

49. $3x - 2y = 5$
$$-2y = -3x + 5$$

$$y = \frac{3}{2}x - \frac{5}{2}$$

$$m_1 = \frac{3}{2}$$

$$3x - 2y = 6$$

$$-2y = -3x + 6$$

$$y = \frac{3}{2}x - 3$$

$$m_2 = \frac{3}{2}$$

parallel, $m_1 = m_2$

51. $2x = 3y - 4$
$$2x + 4 = 3y$$

$$\frac{2}{3}x + \frac{4}{3} = y$$

$$m_1 = \frac{2}{3}$$

$$2x + 3y = 4$$
$$3y = -2x + 4$$

$$y = -\frac{2}{3}x + \frac{4}{3}$$

$$m_2 = -\frac{2}{3}$$

neither

43. slope of line through $(3, -6)$ and $(-1, 2)$

$$m = \frac{-6 - 2}{3 - (-1)} = \frac{-8}{4} = -2$$

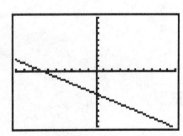

parallel line: same slope

53. $5x + y = 2$

$$y = -5x + 2$$

$$m_1 = -5$$

$$5y = x + 3$$

$$y = \frac{1}{5}x + \frac{3}{5}$$

$$m_2 = \frac{1}{5}$$

perpendicular, $m_1 = -\dfrac{1}{m_2}$

55. $3x - 7y = 1$
$$-7y = -3x + 1$$

$$y = \frac{3}{7}x - \frac{1}{7}$$

$$m_1 = \frac{3}{7}$$

$$6x = 14y + 5$$
$$6x - 5 = 14y$$

$$\frac{3}{7}x - \frac{5}{14} = y$$

$$m_2 = \frac{3}{7}$$

parallel, $m_1 = m_2$

57. Cost = flat fee + (cost per mile)(number of miles)
$$C = 29 + 0.12n$$
$$m = 0.12$$

59. Cost = flat fee + (cost per call)(number of calls)
$$B = 23 + 0.825n$$
$$m = 0.825$$

61. (x, P): $(18,200)$ and $(100, 2660)$

$$m = \frac{2660 - 200}{100 - 18} = \frac{2460}{82} = 30$$

$$P - 200 = 30(x - 18)$$
$$P - 200 = 30x - 540$$
$$P = 30x - 340$$

Find P when $x = 200$:
$$P = 30(200) - 340$$
$$= 5660$$

The profit is $5660.

63. (A, B): $(35, 75)$ and $(15, 35)$

$$m = \frac{75 - 35}{35 - 15} = \frac{40}{20} = 2$$

$$B - 75 = 2(A - 35)$$
$$B - 75 = 2A - 70$$
$$B = 2A + 5$$

Find B when $A = 40$:
$$B = 2(40) + 5$$
$$= 85$$

The score on test B would be 85.

65. (V, E): $(35000, 70)$ and $(20000, 85)$

$$m = \frac{85 - 70}{20000 - 35000} = \frac{15}{-15000} = -0.001$$

$$E - 70 = -0.001(V - 35000)$$
$$E - 70 = -0.001V + 35$$
$$E = -0.001V + 105$$

Find E when $V = 15000$:

$$E = -0.001(15000) + 105$$
$$= 90$$

The expected score is 90.

69. $\dfrac{-4(-2)(-6)}{-4 - 2(-6)} = \dfrac{8(-6)}{-4 + 12}$

$$= \frac{-48}{8}$$

$$= -6$$

71. $4x - 5 \neq 0$
$$4x \neq 5$$

$$x \neq \frac{5}{4}$$

$$\left\{ x \mid x \neq \frac{5}{4} \right\}$$

4.3 Exercises

1. $\begin{cases} 2x + y = 12 \\ 3x - y = 13 \end{cases}$

$$\begin{array}{r} 2x + y = 12 \\ \underline{3x - y = 13} \\ 5x \qquad = 25 \\ x = 5 \end{array}$$

$$2(5) + y = 12$$
$$10 + y = 12$$
$$y = 2$$

$x = 5$, $y = 2$; independent

3. $\begin{cases} -x + 5y = 11 \\ x - 2y = -2 \end{cases}$

$$\begin{aligned} -x + 5y &= 11 \\ \underline{x - 2y} &= \underline{-2} \\ 3y &= 9 \\ y &= 3 \end{aligned}$$

$$\begin{aligned} x - 2(3) &= -2 \\ x - 6 &= -2 \\ x &= 4 \end{aligned}$$

$x = 4$, $y = 3$; independent

5. $\begin{cases} 3x - y = 0 \\ 2x + 3y = 11 \end{cases}$

1st equation: $y = 3x$
Substitute into 2nd equation:

$$\begin{aligned} 2x + 3(3x) &= 11 \\ 2x + 9x &= 11 \\ 11x &= 11 \\ x &= 1 \\ y &= 3(1) = 3 \end{aligned}$$

$x = 1$, $y = 3$; independent

7. $\begin{cases} x + 7y = 20 \\ 5x + 2y = 34 \end{cases}$

1st equation: $x = 20 - 7y$
Substitute into 2nd equation:

$$\begin{aligned} 5(20 - 7y) + 2y &= 34 \\ 100 - 35y + 2y &= 34 \\ 100 - 33y &= 34 \\ -33y &= -66 \\ y &= 2 \\ x &= 20 - 7(2) = 6 \end{aligned}$$

$x = 6$, $y = 2$; independent

9. $\begin{cases} 4x + 5y = 0 \\ 2x + 3y = -2 \end{cases}$

Multiply 2nd equation by -2:

$$-2(2x + 3y) = -2(-2)$$
$$-4x - 6y = 4$$

Add to 1st equation:

$$\begin{aligned} 4x + 5y &= 0 \\ \underline{-4x - 6y} &= \underline{4} \\ -y &= 4 \\ y &= -4 \end{aligned}$$

$$\begin{aligned} 2x + 3(-4) &= -2 \\ 2x - 12 &= -2 \\ 2x &= 10 \\ x &= 5 \end{aligned}$$

$x = 5$, $y = -4$; independent

11. $\begin{cases} 2x + 3y = 7 \\ 4x + 6y = 14 \end{cases}$

Multiply 1st equation by -2:

$$-2(2x + 3y) = -2(7)$$
$$-4x - 6y = -14$$

Add to 2nd equation:

$$\begin{aligned} -4x - 6y &= -14 \\ \underline{4x + 6y} &= \underline{14} \\ 0 &= 0 \end{aligned}$$

Dependent

13. $\begin{cases} 5x + 6y = 3 \\ 10x - 12y = 5 \end{cases}$

Multiply 1st equation by -2:

$$-2(5x - 6y) = -2(3)$$
$$-10x + 12y = -6$$

Add to 2nd equation:

$$\begin{aligned} -10x + 12y &= -6 \\ \underline{10x - 12y} &= \underline{5} \\ 0 &= -1 \end{aligned}$$

Inconsistent

15. $\begin{cases} 2x - 3y = 10 \\ 3x - 2y = 15 \end{cases}$

Multiply 1st equation by 3:

$6x - 9y = 30$

Multiply 2nd equation by -2:

$-6x - 4y = -30$

Adding:

$\begin{array}{r} 6x - 9y = 30 \\ \underline{-6x + 4y = -30} \\ -5y = 0 \\ y = 0 \end{array}$

$2x - 3(0) = 10$
$\qquad 2x = 10$
$\qquad\ \ x = 5$

$x = 5, \ y = 0; \quad$ independent

17. $\begin{cases} \quad\ y = 2x + 3 \\ 2x + y = -1 \end{cases}$

Substitute 1st equation into 2nd equation:

$2x + (2x + 3) = -1$
$\qquad 4x + 3 = -1$
$\qquad\quad\ 4x = -4$
$\qquad\qquad x = -1$
$\qquad\qquad y = 2(-1) + 3 = 1$

$x = -1, \ y = 1; \quad$ independent

19. $\begin{cases} 6a - 3b = 1 \\ 8a + 5b = 7 \end{cases}$

Multiply 1st equation by 5:
$30a - 15b = 5$

Multiply 2nd equation by 3:
$24a + 15b = 21$

Adding:

$\begin{array}{r} 30a - 15b = 5 \\ \underline{24a + 15b = 21} \\ 54a \qquad\quad = 26 \\ a = \dfrac{13}{27} \end{array}$

$6\left(\dfrac{13}{27}\right) - 3b = 1$

$\dfrac{26}{9} - 3b = 1$

$\qquad -3b = -\dfrac{17}{9}$

$\qquad\quad b = \dfrac{17}{27}$

$a = \dfrac{13}{27}, \ b = \dfrac{17}{27}; \quad$ independent

21. $\begin{cases} s = 3t - 5 \\ t = 3s - 5 \end{cases}$

Substitute 1st equation into 2nd equation:

$t = 3(3t - 5) - 5$
$t = 9t - 15 - 5$
$t = 9t - 20$
$-8t = -20$

$t = \dfrac{5}{2}$

$s = 3\left(\dfrac{5}{2}\right) - 5 = \dfrac{5}{2}$

$s = \dfrac{5}{2}, \ t = \dfrac{5}{2}; \quad$ independent

23. $\begin{cases} 3m - 2n = 8 \\ \qquad 3n = m - 8 \end{cases}$

2nd equation: $m = 3n + 8$

Substitute into 1st equation:

$3(3n + 8) - 2n = 8$
$\ 9n + 24 - 2n = 8$
$\qquad 7n + 24 = 8$
$\qquad\qquad 7n = -16$

$n = \dfrac{16}{7}$

$m = 3\left(-\dfrac{16}{7}\right) + 8 = \dfrac{8}{7}$

$m = \dfrac{8}{7}, \ n = -\dfrac{16}{7}; \quad$ independent

25. $\begin{cases} 3p - 4q = 5 \\ 3q - 4p = -9 \end{cases}$

$\begin{cases} 3p - 4q = 5 \\ -4p + 3q = -9 \end{cases}$

Multiply 1st equation by 4:
$12p - 16q = 20$

Multiply 2nd equation by 3:
$-12p + 9q = -27$

Adding:

$$\begin{array}{r} 12p - 16q = 20 \\ -12p + 9q = -27 \\ \hline -7q = -7 \\ q = 1 \end{array}$$

$$\begin{array}{r} 3p - 4(1) = 5 \\ 3p - 4 = 5 \\ 3p = 9 \\ p = 3 \end{array}$$

$p = 3$, $q = 1$; independent

27. $\begin{cases} \dfrac{u}{3} - v = 1 \\ u - \dfrac{v}{2} = 5 \end{cases}$

Multiply 1st equation by 3 and 2nd equation by 2:

$\begin{cases} u - 3v = 3 \\ 2u - v = 10 \end{cases}$

Multiply 1st equation by -2:

$-2u + 6v = -6$

Adding:

$$\begin{array}{r} -2u + 6v = -6 \\ 2u - v = 10 \\ \hline 5v = 4 \\ v = \dfrac{4}{5} \end{array}$$

$2u - \dfrac{4}{5} = 10$

$5\left(2u - \dfrac{4}{5}\right) = 5(10)$

$10u - 4 = 50$
$10u = 54$

$u = \dfrac{27}{5}$

$u = \dfrac{27}{5}$, $v = \dfrac{4}{5}$; independent

29. $\begin{cases} \dfrac{10}{4} + \dfrac{z}{6} = 4 \\ \dfrac{w}{2} - \dfrac{z}{3} = 4 \end{cases}$

Multiply 1st equation by 12 and 2nd equation by 6:

$\begin{cases} 3w + 2z = 48 \\ 3w - 2z = 24 \end{cases}$

Adding:

$$\begin{array}{r} 3w + 2z = 48 \\ 3w - 2z = 24 \\ \hline 6w = 72 \\ w = 12 \end{array}$$

$3(12) + 2z = 48$
$36 + 2z = 48$
$2z = 12$
$z = 6$

$w = 12$, $z = 6$; independent

31. $\begin{cases} \dfrac{x}{6} + \dfrac{y}{8} = \dfrac{3}{4} \\ \dfrac{x}{4} + \dfrac{y}{3} = \dfrac{17}{12} \end{cases}$

Multiply 1st equation by 24 and 2nd equation by 12:

$\begin{cases} 4x + 3y = 18 \\ 3x + 4y = 17 \end{cases}$

Multiply 1st equation by 3:

$12x + 9y = 54$

Multiply 2nd equation by -4:

$-12x - 16y = -68$

Adding:

$$12x + 9y = 54$$
$$\underline{-12x - 16y = -68}$$
$$-7y = -14$$
$$y = 2$$

$$4x + 3(2) = 18$$

$$4x + 6 = 18$$
$$4x = 12$$
$$x = 3$$

$x = 3$, $y = 2$; independent

33. $\begin{cases} \dfrac{x + 3}{2} + \dfrac{y - 4}{3} = \dfrac{19}{6} \\[2mm] \dfrac{x - 2}{3} + \dfrac{y - 2}{2} = 2 \end{cases}$

Multiply 1st equation by 6:

$$3(x + 3) + 2(y - 4) = 19$$
$$3x + 9 + 2y - 8 = 19$$
$$3x + 2y + 1 = 19$$
$$3x + 2y = 18$$

Multiply 2nd equation by 6:

$$2(x - 2) + 3(y - 2) = 12$$
$$2x - 4 + 3y - 6 = 12$$
$$2x + 3y - 10 = 12$$
$$2x + 3y = 22$$

$\begin{cases} 3x + 2y = 18 \\ 2x + 3y = 22 \end{cases}$

Multiply 1st equation by -2:
$$-6x + 4y = -36$$

Multiply 2nd equation by 3:
$$6x + 9y = 66$$

Adding:

$$-6x - 4y = -36$$
$$\underline{6x + 9y = 66}$$
$$5y = 30$$
$$y = 6$$

$$3x + 2(6) = 18$$
$$3x + 12 = 18$$
$$3x = 6$$
$$x = 2$$
$x = 2$, $y = 6$; independent

35. $\begin{cases} 0.1x + 0.01y = 0.37 \\ 0.02x + 0.05y = 0.41 \end{cases}$

Multiply both equations by 100:

$\begin{cases} 10x + y = 37 \\ 2x + 5y = 41 \end{cases}$

Multiply 2nd equation by -5:
$$-10x - 25y = -205$$

Adding:

$$10x + y = 37$$
$$\underline{-10x - 25y = -205}$$
$$-24y = -168$$
$$y = 7$$

$$2x + 5(7) = 41$$
$$2x + 35 = 41$$
$$2x = 6$$
$$x = 3$$

$x = 3$, $y = 7$: independent

37. $\begin{cases} \dfrac{x}{2} + 0.05y = 0.35 \\[3mm] 0.3x + \dfrac{y}{4} = 0.65 \end{cases}$

Multiply both equations by 100:

$\begin{cases} 50x + y = 35 \\ 30x + 25y = 65 \end{cases}$

Multiply 2nd equation by $-\dfrac{1}{5}$:

$$-6x - 5y = -13$$

Adding:

$$50x + 5y = 35$$
$$\underline{-6x - 5y = -13}$$
$$44x = -22$$
$$x = \frac{1}{2}$$

$$50\left(\frac{1}{2}\right) + 5y = 35$$

$$25 + 5y = 35$$
$$5y = 10$$
$$y = 2$$

$x = \dfrac{1}{2}$, $y = 2$: independent

39. $\begin{cases} 5x - 2y = 12 \\ 4x + 3y = 5 \end{cases}$

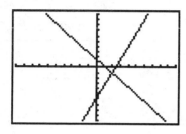

$x = 2, \; y = -1$

41. $\begin{cases} 2x + 7y = 4 \\ 9x - 8y = 15 \end{cases}$

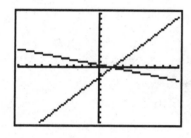

$x = 1.7, \; y = 0.1$

43. $\begin{cases} 0.3x + 0.5y = 1 \\ 0.1x - 0.4y = 2 \end{cases}$

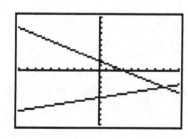

$x = 8.2, \; y = -2.9$

45. amount at 5%: x
    amount at 8%: y

$$\begin{cases} x + y = 14000 \\ 0.05x + 0.08y = 835 \end{cases}$$

Multiply 2nd equation by 100:

$$\begin{cases} x + y = 14000 \\ 5x + 8y = 83500 \end{cases}$$

Multiply 1st equation by −5 and add to 2nd equation:

$$\begin{aligned} -5x - 5y &= -70000 \\ \underline{5x + 8y} &= \underline{83500} \\ 3y &= 13500 \\ y &= 4500 \end{aligned}$$

$$\begin{aligned} x + 4500 &= 14000 \\ x &= 9500 \end{aligned}$$

She should invest \$9500 at 5% and \$4500 at 8%.

47. amount at 10%: x
    amount at 8%: y

$$\begin{cases} x + y = 10000 \\ 0.10x = 0.08y \end{cases}$$

Multiply 2nd equation by 100:

$$\begin{cases} x + y = 10000 \\ 10x = 8y \end{cases}$$

1st equation:
$x = 10000 - y$

Subsitute into 2nd equation:

$$\begin{aligned} 10(10000 - y) &= 8y \\ 100000 - 10y &= 8y \\ 100000 &= 18y \\ 5555.56 &= y \end{aligned}$$

$$\begin{aligned} x + 5555.56 &= 10000 \\ x &= 4444.44 \end{aligned}$$

Invest \$4444.44 at 10% and \$5555.56 at 8%.

49. width: x
    length: y

$$\begin{cases} 2x + 2y = 36 \\ y = 2 + x \end{cases}$$

Substitute 2nd equation into 1st:

$$2x + 2(2 + x) = 36$$
$$2x + 4 + 2x = 36$$
$$4x + 4 = 36$$
$$4x = 32$$
$$x = 8$$
$$y = 2 + 8 = 10$$

The dimensions are 8 cm by 10 cm.

51. cost per roll of 35-mm film: x
    cost per roll of movie film: y

$$\begin{cases} 5x + 3y = 35.60 \\ 3x + 5y = 43.60 \end{cases}$$

Multiply 1st equation by −3 and 2nd equation by 5, then add:

$$-15x - 9y = -106.80$$
$$\underline{15x + 25y = \quad 218}$$
$$16y = 111.2$$
$$y = 6.95$$

$$5x + 3(6.95) = 35.60$$
$$5x + 20.85 = 35.60$$
$$5x = 14.75$$
$$x = 2.95$$

35-mm film costs $2.95 per roll and movie film costs $6.95 per roll.

53. cost of jelly donut: x
    cost of cream-filled donut: y

$$\begin{cases} 5x + 7y = 3.16 \\ 8x + 4y = 3.04 \end{cases}$$

Multiply 1st equation by −4 and 2nd equation by 7, then add:

$$-20x - 28y = -12.64$$
$$\underline{56x + 28y = \quad 21.28}$$
$$36x \qquad = 8.64$$
$$x = 0.24$$

$$5(0.24) + 7y = 3.16$$
$$1.2 + 7y = 3.16$$
$$7y = 1.96$$
$$y = 0.28$$

A jelly donut costs $0.24 and a cream-filled donut costs $0.28.

55. number of expensive models: x
    number of less expensive models: y

$$\begin{cases} 6x + 5y = 730 \\ 3x + 2y = 340 \end{cases}$$

Multiply 2nd equation by −2 and add to 1st:

$$6x + 5y = 730$$
$$\underline{-6x - 4y = 680}$$
$$y = 50$$

$$3x + 2(50) = 340$$
$$3x + 100 = 340$$
$$3x = 240$$
$$x = 80$$

They can produce 80 expensive models and 50 less expensive models.

57. annual medical expenses: x
    annual cost: y

husband's plan:

cost = $140 · 12 + 10% of
      per month   months   medical expenses

$$y = (140)(12) + 0.10x$$
$$y = 1680 + 0.10x$$

wife's plan:

cost = $100 · 12 + 25% of
      per month   months   medical expenses

$$y = 1680 + 0.10x$$
$$y = 1200 + 0.25x$$

Substituting 1st equation into 2nd:

$$1680 + 0.10x = 1200 + 0.25x$$
$$480 + 0.10x = 0.25x$$
$$480 = 0.15x$$
$$3200 = x$$

The two plans are equivalent when the annual medical expenses are $3200.

59. number of miles: x
    cost: y

Cheapo
$$y = 29 + 0.12x$$

Cut-Rate
$$y = 22 + 0.15x$$

$$\begin{cases} y = 29 + 0.12x \\ y = 22 + 0.15x \end{cases}$$

Substitute 1st equation into 2nd:

$$29 + 0.12x = 22 + 0.15x$$
$$7 + 0.12x = 0.15x$$
$$7 = 0.03x$$
$$233\frac{1}{3} = x$$

$233\frac{1}{3}$ miles must be driven.

61.
$$C = R$$
$$1.6x + 7200 = 2.1x$$
$$7200 = 0.5x$$
$$14400 = x$$

They must sell 14,400 units.

63. number of calls: x
cost: y

1st plan:
$$y = 18 + 0.11x$$

2nd plan:
$$y = 24 + 50(0) + 0.09(x - 50)$$
$$y = 24 + 0.09x - 4.5$$
$$y = 0.09x + 19.5$$

1st plan = 2nd plan
$$18 + 0.11x = 0.09x + 19.5$$
$$0.11x = 0.09x + 1.5$$
$$0.02x = 1.5$$
$$x = 75$$

75 calls per month make the plans equivalent.

65. number of $5-bills: x
number of $10-bills: y

$$\begin{cases} x + y = 43 \\ 5x + 10y = 340 \end{cases}$$

Multiply 1st equation by -5, then add to 2nd:

$$\begin{array}{r} -5x - 5y = -215 \\ 5x + 10y = 340 \\ \hline 5y = 125 \\ y = 25 \end{array}$$

$$x + 25 = 43$$
$$x = 18$$

There are 18 $5-bills and 25 $10-bills.

67. charge per mile: x
flat fee: y

$$\begin{cases} 85x + y = 44.30 \\ 125x + y = 51.50 \end{cases}$$

$$y = 44.30 - 85x$$

Substituting:

$$125x + 44.30 - 85x = 51.50$$
$$40x + 44.30 = 51.50$$
$$40x = 7.2$$
$$x = 0.18$$

$$85(0.18) + y = 44.30$$
$$15.30 + y = 44.30$$
$$y = 29$$

The charge per mile is $0.18 and the flat fee is $29.

69. speed of plane: x
speed of wind: y

rate with tailwind: $x + y$
rate with headwind: $x - y$

$$\begin{cases} 6(x + y) = 2310 \\ 6(x - y) = 1530 \end{cases}$$

Divide both equations by 6, then add:

$$\begin{array}{r} x + y = 385 \\ x - y = 255 \\ \hline 2x = 640 \\ x = 320 \end{array}$$

$$320 + y = 385$$
$$y = 65$$

The plane's speed is 320 km/hr and the winds speed is 65 km/hr.

Chapter 4   89   4.3 Exercises

71. $\begin{cases} \dfrac{y - 3}{x - 2} = -1 \\ \dfrac{y - (-2)}{x - 1} = 2 \end{cases}$

$\begin{cases} y - 3 = -x + 2 \\ y + 2 = 2x - 2 \end{cases}$

$\begin{cases} y = -x + 5 \\ y = 2x - 4 \end{cases}$

Substituting:

$-x + 5 = 2x - 4$

$5 = 3x - 4$

$9 = 3x$

$3 = x$

$y = -3 + 5 = 2$

$(3, 2)$

77. $\dfrac{x}{3} - \dfrac{x}{5} = \dfrac{16}{5}$

$15\left(\dfrac{x}{3} - \dfrac{x}{5}\right) = 15\left(\dfrac{16}{5}\right)$

$\dfrac{15}{1} \cdot \dfrac{x}{3} - \dfrac{15}{1} \cdot \dfrac{x}{5} = \dfrac{15}{1} \cdot \dfrac{16}{5}$

$5x - 3x = 48$

$2x = 48$

$x = 24$

79. $x^3 - 2x = 3$

$x^3 - 2x - 3 = 0$

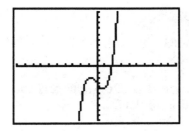

$x = 1.9$

1. $y \leq x + 3$ (solid line)

    $y = x + 3$

x	y
0	3
-3	0

Test point: (0, 0)

$0 \leq 0 + 3$
$0 \leq 3$
True

Shade the half-plane containing (0, 0).

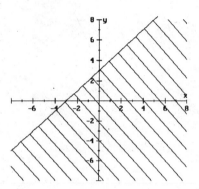

3. $x > y - 2$ (dashed line)

x	y
0	2
-2	0

Test point: (0, 0)
$0 > 0 - 2$
$0 > -2$
True

Shade the half-plane containing (0, 0).

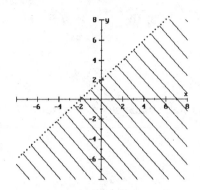

5. $x + y < 3$ (dashed line)

    $x + y = 3$

x	y
0	3
3	0

Test point: (0, 0)

$0 + 0 < 3$
$0 < 3$
True

Shade the half-plane containing (0, 0).

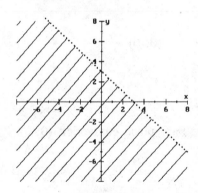

7. $x + y \geq 3$ (solid line)

    $x + y = 3$

x	y
0	3
3	0

Test point: (0, 0)

0 + 0 ≥ 3
    0 ≥ 3
    False

Shade the half-plane <u>not</u> containing (0, 0).

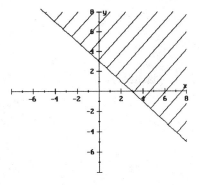

9. $2x + y \leq 6$ (solid line)

    $2x + y = 6$

x	y
0	6
3	0

Test point: (0, 0)

$2(0) + 0 \leq 6$
        $0 \leq 6$
        True

Shade the half-plane containing (0, 0).

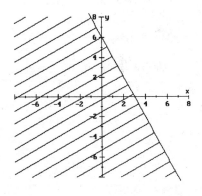

11. $x + 2y > 6$ (dashed line)

     $x + 2y = 6$

x	y
0	3
6	0

Test point: (0, 0)

$0 + 2(0) > 6$
        $0 > 6$
        False

Shade the half-plane <u>not</u> containing (0, 0).

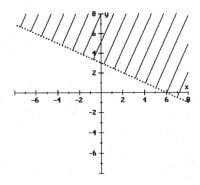

13. $3x + 2y \geq 12$ (solid line)

     $3x + 2y = 12$

x	y
0	6
4	0

Test point: (0, 0)

$3(0) + 2(0) \geq 12$
            $0 \geq 12$
            False

Shade the half-plane <u>not</u> containing (0, 0).

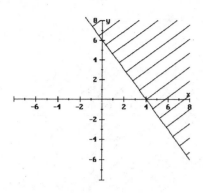

$$2(0) + 5(0) > 10$$
$$0 > 10$$
False

Shade the half-plane <u>not</u> containing (0, 0).

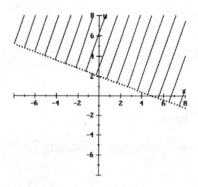

15. $2x + 5y < 10$ (dashed line)

$2x + 5y = 10$

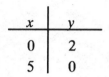

x	y
0	2
5	0

Test point: (0, 0)

$$2(0) + 5(0) < 10$$
$$0 < 10$$
True

Shade the half-plane containing (0, 0).

19. $2x + 5y \geq 10$ (solid line)

$2x + 5y = 10$

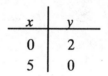

x	y
0	2
5	0

Test point: (0, 0)

$$2(0) + 5(0) \geq 10$$
$$0 \geq 10$$
False

Shade the half-plane <u>not</u> containing (0, 0).

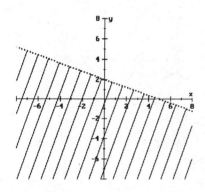

17. $2x + 5y > 10$ (dashed line)

$2x + 5y = 10$

x	y
0	2
5	0

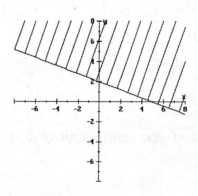

21. $2x - 5y \geq 10$ (solid line)

$2x - 5y = 10$

x	y
0	-2
5	0

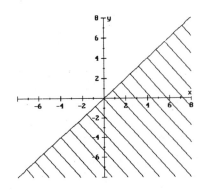

Test point: (0, 0)

$2(0) - 5(0) \geq 10$
$0 \geq 10$
False

Shade the half-plane <u>not</u> containing (0, 0).

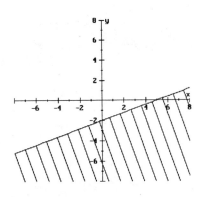

23. $y \leq x$ (solid line)

$y = x$

x	y
0	0
1	1

Test point: (0, 1)

$1 \leq 0$
False

Shade the half-plane <u>not</u> containing (0, 1).

25. $2x - y < 4$ (dashed line)

$2x - y = 4$

x	y
0	-4
2	0

Test point: (0, 0)

$2(0) - 0 < 4$
$0 < 4$
True

Shade the half-plane containing the point (0, 0).

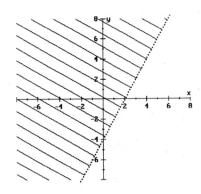

27. $4x - y \geq 8$ (solid line)

$4x - y = 8$

x	y
0	-8
2	0

Test point: (0, 0)

$4(0) - 0 \geq 8$

$\qquad 0 \geq 8$

$\qquad$ False

Shade the half-plane <u>not</u> containing the point (0, 0).

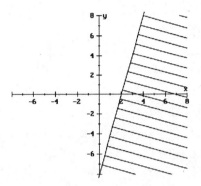

29. $3x - 4y > 12$ (dashed line)

$3x - 4y = 12$

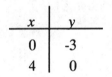

x	y
0	-3
4	0

Test point: (0, 0)

$3(0) - 4(0) > 12$

$\qquad 0 > 12$

$\qquad$ False

Shade the half-plane <u>not</u> containing the point (0, 0).

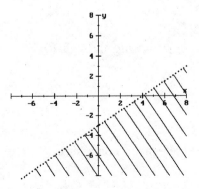

31. $7x - 3y < 15$ (dashed line)

$7x - 3y = 15$

x	y
0	-5
15/7	0

Test point: (0, 0)

$7(0) - 3(0) < 15$

$\qquad 0 < 15$

$\qquad$ True

Shade the half-plane containing (0, 0).

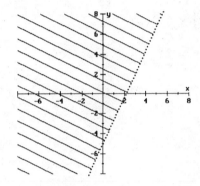

33. $7x + 3y > 15$ (dashed line)

$7x + 3y = 15$

x	y
0	5
15/7	0

Test point: (0, 0)

$7(0) + 3(0) > 15$

$\qquad 0 > 15$

$\qquad$ False

Shade the half-plane <u>not</u> containing the point (0, 0).

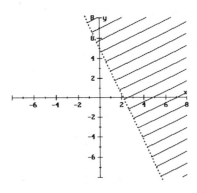

35. $\dfrac{x}{2} + \dfrac{y}{3} < 4$ (dashed line)

$\dfrac{x}{2} + \dfrac{y}{3} = 4$

$6\left(\dfrac{x}{2} + \dfrac{y}{3}\right) = 6(4)$

$3x + 2y = 24$

x	y
0	12
8	0

Test point: (0, 0)

$\dfrac{0}{2} + \dfrac{0}{3} < 4$

$0 < 4$
True

Shade the half-plane containing (0, 0).

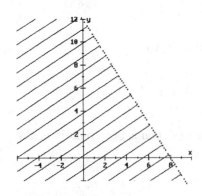

37. $\dfrac{x}{3} - \dfrac{y}{4} \geq 3$ (solid line)

$\dfrac{x}{3} - \dfrac{y}{4} = 3$

$12\left(\dfrac{x}{3} - \dfrac{y}{4}\right) = 12(3)$

$4x - 3y = 36$

x	y
0	-12
9	0

Test point: (0, 0)

$\dfrac{0}{3} - \dfrac{0}{4} \geq 3$

$0 \geq 3$
False

Shade the half-plane <u>not</u> containing (0, 0).

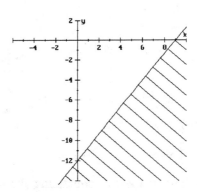

39. $y < 3$ (dashed line)

$y = 3$

Horizontal line passing through (0, 3).

Test point: (0, 0)

$0 < 3$
True

Shade the half-plane containing (0, 0).

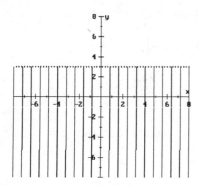

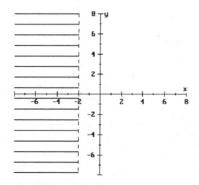

41. $x \geq -2$ (solid line)

$x = -2$

Vertical line passing through the point (-2, 0).

Test point: (0, 0)

$0 \geq -2$
 True

Shade the half-plane containing the point (0, 0).

45. $y < 0$ (dashed line)

$y = 0$

Horizontal line passing through the point (0, 0).

Test point: (1, 1)

$1 < 0$
 False

Shade the half-plane not containing the point (1, 1).

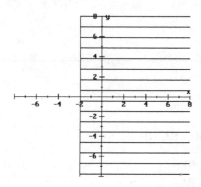

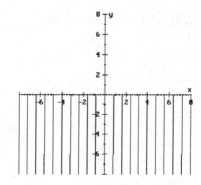

43. $x < -2$ (dashed line)

$x = -2$

Vertical line passing through the point (-2, 0).

Test point: (0, 0)

$0 < -2$
 False

Shade the half-plane not containing the point (0, 0).

47. $x \leq 0$ (solid line)

$x = 0$

Vertical line containing the point (0, 0).

Test point: (1, 1)

$1 < 0$
 False

Shade the half-plane not containing the point (1, 1).

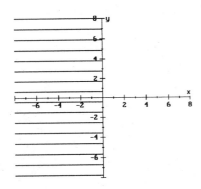

Horizontal line containing the point (0, 0).

Test point: (1, 1)

$$\frac{1}{2} > 0$$
True

Shade the half-plane containing the point (1, 1).

49. $x < \dfrac{1}{2}$ (dashed line)

$x = \dfrac{1}{2}$

Vertical line containing the point $\left(\dfrac{1}{2}, 0\right)$.

Test point: (0, 0)

$$0 < \frac{1}{2}$$
True

Shade the half-plane containing the point (0, 0).

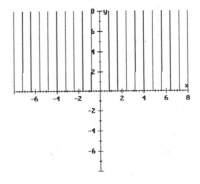

53. $P \le 80$ (solid line)
    $2x + 2y \le 80$

    $2x + 2y = 80$

x	y
0	40
40	0

Test point: (0, 0)

$$0 + 0 \le 40$$
$$0 \le 40$$
True

Shade the half-plane containing the point (0, 0).

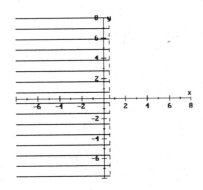

51. $\dfrac{y}{2} > 0$ (dashed line)

    $\dfrac{y}{2} = 0$

    $2\left(\dfrac{y}{2}\right) = 2(0)$

    $y = 0$

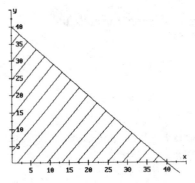

Three possible feasible points: (1, 1), (1, 2), (10, 10)

55. $10s + 18p \le 120$ (solid line)

$10s + 18p = 120$

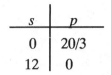

s	p
0	20/3
12	0

Test point: (0, 0)

$10(0) + 18(0) \le 120$

$0 \le 120$

True

Shade the half-plane containing the point (0, 0).

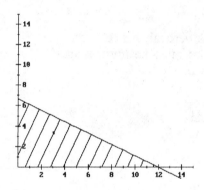

Three possible feasible points: $(1, 1)$, $(2, 2)$, $(1, 2)$

57. $12r + 8a \le 2500$ (solid line)

$12r + 8a = 2500$

r	a
0	312.5
208.3	0

Test point: (0, 0)

$12(0) + 8(0) \le 2500$

$0 \le 2500$

True

Shade the half-plane containing the point (0, 0).

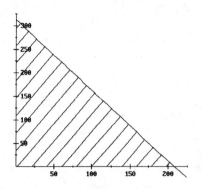

Three possible feasible points: (10, 10), (20, 10), (10, 20)

59. number of regular clerks: r
number of special clerks: s

$15r + 10s \ge 90$ (solid line)

$15r + 10s = 90$

r	s
0	9
6	0

Test point: (0, 0)

$15(0) + 10(0) \ge 90$

$0 \ge 90$

False

Shade the half-plane <u>not</u> containing the point (0, 0).

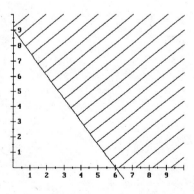

Three possible feasible points:
(10, 1), (1, 10), (20, 20)

61. $|-6 - 3| - |6| - |3|$

$= |-9| - |6| - |3|$

$= 9 - 6 - 3$
$= 0$

63. Distributive property

65. Bobby's hours: x
Linda's hours: $x + 1$

$6x = 5(x + 1)$
$6x = 5x + 5$
$x = 5$

It will take 5 hours.

CHAPTER 4 REVIEW EXERCISES

1. $m = \dfrac{y_2 - y_1}{x_2 - x_1}$

$= \dfrac{0 - (-2)}{-1 - 3}$

$= \dfrac{2}{-4}$

$= -\dfrac{1}{2}$

3. $y = 3x - 5$
$m = 3$

5. $4y - 3x = 1$
$\quad 4y = 3x + 1$

$\quad y = \dfrac{3}{4}x + \dfrac{1}{4}$

$\quad m = \dfrac{3}{4}$

7. $m = \dfrac{y_2 - y_1}{x_2 - x_1}$

$= \dfrac{5 - 4}{3 - 1}$

$= \dfrac{1}{2}$

parallel lines have equal slopes

9. $m = \dfrac{y_2 - y_1}{x_2 - x_1}$

$= \dfrac{9 - 7}{4 - 4}$

$= \dfrac{2}{0}$
undefined

This line is vertical, hence a line perpendicular must be horizontaland its slope is 0.

11. $m = \dfrac{y_2 - y_1}{x_2 - x_1}$

$= \dfrac{6 - 6}{-7 - 2}$

$= \dfrac{0}{-9}$

$= 0$

13. $y = 3x - 7$
$m = 3$

parallel lines have equal slopes.

15. $3y - 5x + 6 = 0$
$\quad 3y = 5x - 6$

$\quad y = \dfrac{5}{3}x - 2$

$\quad m = \dfrac{5}{3}$

Perpendicular lines have slopes that are negative reciprocals.
$m_\perp = -\dfrac{3}{5}$

17. $x = 3$ is a vertical line, hence a line parallel will also be vertical and its slope is undefined.

19. $m = \dfrac{y_2 - y_1}{x_2 - x_1}$

$\quad 4 = \dfrac{a - 2}{4 - 1}$

$\quad 4 = \dfrac{a - 2}{3}$

$\quad 12 = a - 2$
$\quad 14 = a$

21. $(-3, a)$ and $(0, 3)$

$\quad m_1 = \dfrac{a - 3}{-3 - 0}$

$\quad m_1 = \dfrac{a - 3}{-3}$

$\quad (7, a)$ and $(0, 0)$

$\quad m_2 = \dfrac{a - 0}{7 - 0}$

$\quad m_2 = \dfrac{a}{7}$

parallel lines have equal slopes.

$\qquad m_1 = m_2$

$\quad 21\left(\dfrac{a - 3}{-3}\right) = 21\left(\dfrac{a}{7}\right)$

$\quad -7(a - 3) = 3a$
$\quad -7a + 21 = 3a$
$\qquad\quad 21 = 10a$

$\qquad \dfrac{21}{10} = a$

23. $m = \dfrac{-4 - 3}{1 - (-2)}$

$\quad\ = -\dfrac{7}{3}$

$y - y_1 = m(x - x_1)$

$y - 3 = -\dfrac{7}{3}[x - (-2)]$

$y - 3 = -\dfrac{7}{3}(x + 2)$

$y - 3 = -\dfrac{7}{3}x - \dfrac{14}{3}$

$\quad\ y = -\dfrac{7}{3}x - \dfrac{5}{3}$

25. $\quad m = \dfrac{5 - (-5)}{3 - (-3)}$

$\qquad\ = \dfrac{10}{6}$

$\qquad\ = \dfrac{5}{3}$

$y - y_1 = m(x - x_1)$

$y - 5 = \dfrac{5}{3}(x - 3)$

$y - 5 = \dfrac{5}{3}x - 5$

$\quad\ y = \dfrac{5}{3}x$

27. $y - y_1 = m(x - x_1)$

$y - 5 = \dfrac{2}{5}(x - 2)$

$y - 5 = \dfrac{2}{5}x - \dfrac{4}{5}$

$\quad\ y = \dfrac{2}{5}x + \dfrac{21}{5}$

29. $y - y_1 = m(x - x_1)$
$\ \ y - 7 = 5(x - 4)$
$\ \ y - 7 = 5x - 20$
$\qquad\ y = 5x - 13$

31. $y = mx + b$
$\ \ y = 5x + 3$

33. $y = 3$

35. $y = \dfrac{3}{2}x - 1$

$m = \dfrac{3}{2}$

parallel lines have equal slopes

$m = \dfrac{3}{2}$; $\;(0, 0)$

$y = mx + b$

$y = \dfrac{3}{2}x + 0$

$y = \dfrac{3}{2}x$

37. $2y - 5x = 1$
$\quad\; 2y = 5x + 1$

$y = \dfrac{5}{2}x + \dfrac{1}{2}$

$m = \dfrac{5}{2}$

$m_{\perp} = -\dfrac{2}{5}$

$m_{\perp} = -\dfrac{2}{5}$; $\;(0, 6)$

$y = mx + b$
$y = -\dfrac{2}{5}x + 6$

39. $\quad 3x = -5y$

$-\dfrac{3}{5}x = y$

$m = -\dfrac{3}{5}$

$m_{\perp} = \dfrac{5}{3}$

$m_{\perp} = \dfrac{5}{3}$; $\;(0, 0)$

$y = mx + b$

$y = \dfrac{5}{3}x + 0$

$y = \dfrac{5}{3}x$

41. $(3, 0)$ and $(0, -5)$

$m = \dfrac{-5 - 0}{0 - 3} = \dfrac{5}{3}$

$y = mx + b$

$y = \dfrac{5}{3}x - 5$

43. $3x - 2y = 5$
$\qquad -2y = -3x + 5$

$y = \dfrac{3}{2}x - \dfrac{5}{2}$

$m = \dfrac{3}{2}$

$5y = x + 3$

$y = \dfrac{1}{5}x + \dfrac{3}{5}$

$b = \dfrac{3}{5}$

$m = \dfrac{3}{2}$; $\;\left(0, \dfrac{3}{5}\right)$

$y = mx + b$
$y = \dfrac{3}{2}x + \dfrac{3}{5}$

45. (x, P); $(250, 12000)$ and $(300, 20000)$

$m = \dfrac{20000 - 12000}{300 - 250}$

$= \dfrac{8000}{50}$

$= 160$

$$P - 12000 = 160(x - 250)$$
$$P - 12000 = 160x - 40000$$
$$P = 160x - 28000$$

If $x = 400$:
$$P = 160(400) - 28000$$
$$= 36000$$

He would make $36,000.

47. $\begin{cases} x - y = 4 \\ 2x - 3y = 7 \end{cases}$

$\begin{cases} -2(x - y) = -2(4) \\ 2x - 3y = 7 \end{cases}$

$$\begin{array}{r} -2x + 2y = -8 \\ \underline{2x - 3y = 7} \\ -y = -1 \\ y = 1 \end{array}$$

$$x - 1 = 4$$
$$x = 5$$

$$x = 5, \quad y = 1$$

49. $\begin{cases} \dfrac{x}{6} - \dfrac{y}{4} = \dfrac{4}{3} \\ \dfrac{x}{5} - \dfrac{y}{2} = \dfrac{8}{5} \end{cases}$

$\begin{cases} 12\left(\dfrac{x}{6} - \dfrac{y}{4}\right) = 12\left(\dfrac{4}{3}\right) \\ -10x\left(\dfrac{x}{5} - \dfrac{4}{2}\right) = -10\left(\dfrac{8}{5}\right) \end{cases}$

$$\begin{array}{r} 2x - 3y = 16 \\ \underline{-2x + 5y = -16} \\ 2y = 0 \\ y = 0 \end{array}$$

$$2x - 3(0) = 16$$
$$2x = 16$$
$$x = 8$$

$$x = 8, \quad y = 0$$

51. $\begin{cases} 3x - \dfrac{y}{4} = 2 \\ 6x - \dfrac{y}{2} = 4 \end{cases}$

$\begin{cases} 4\left(3x - \dfrac{y}{4}\right) = 4(2) \\ 2\left(6x - \dfrac{y}{2}\right) = 2(4) \end{cases}$

$\begin{cases} 12x - y = 8 \\ 12x - y = 8 \end{cases}$

Dependent

53. $\begin{cases} x = 2y - 3 \\ y = 3x + 2 \end{cases}$

Substituting:
$$y = 3(2y - 3) + 2$$
$$y = 6y - 9 + 2$$
$$y = 6y - 7$$
$$-5y = -7$$

$$y = \frac{7}{5}$$

$$x = 2\left(\frac{7}{5}\right) - 3 = -\frac{1}{5}$$

$$x = -\frac{1}{5}, \quad y = \frac{7}{5}$$

55. $\begin{cases} 2x - 5y = 12 \\ 3x + 4y = 5 \end{cases}$

$\begin{cases} y = \dfrac{2}{5}x - \dfrac{12}{5} \\ y = -\dfrac{3}{4}x + \dfrac{5}{4} \end{cases}$

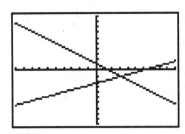

$x = 3.2, \quad y = -1.1$

57. $\begin{cases} 0.8x - y = 4 \\ 2x - 0.4y = 3.2 \end{cases}$

$\begin{cases} y = 0.8x - 4 \\ y = 5x - 8 \end{cases}$

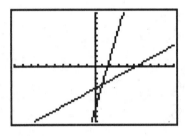

$x = 1.0, \quad y = -3.2$

59. amount in CD: x
amount in bond: y

$\begin{cases} x + y = 8500 \\ 0.0475x + 0.0665y = 512.05 \end{cases}$

1st equation: $x = 8500 - y$
Substituting:

$0.0475(8500 - y) + 0.0665y = 512.05$
$403.75 - 0.0475y + 0.0665y = 512.05$
$403.75 + 0.019y = 512.05$
$0.019y = 108.3$
$y = 5700$

$x + 5700 = 8500$
$x = 2800$

$2800 is invested in the CD and $5700
is invested in the bond.

61. $y - 2x < 4$ \qquad (dashed line)

$y - 2x = 4$

x	y
0	4
-2	0

Test point: $(0, 0)$

$0 - 2(0) < 4$
$0 < 4$
True

Shade half-plane containing $(0, 0)$.

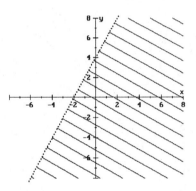

63. $2y - 3x > 6$ \qquad (dashed line)

$2y - 3x = 6$

x	y
0	3
-2	0

Test point: $(0, 0)$

$2(0) - 3(0) > 6$
$0 > 6$
False

Shade half-plane <u>not</u> containing $(0, 0)$.

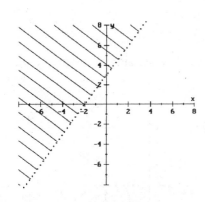

x	y
0	18
12	0

Test point: (0, 0)

$$3(0) + 2(0) \geq 36$$
$$0 \geq 36$$
False

Shade half-plane <u>not</u> containing (0, 0).

65. $5y - 8x < 20$ (solid line)

$$5y - 8x = 20$$

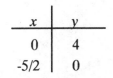

x	y
0	4
-5/2	0

Test point: (0, 0)

$$5(0) - 8(0) \leq 20$$
$$0 \leq 20$$
True

Shade half-plane containing (0, 0).

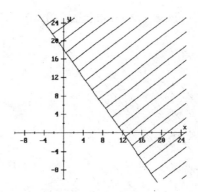

69. $y < 5$ (dashed line)

$$y = 5$$

Horizontal line passing through (0, 5).

Test point: (0, 0)

$$0 < 5$$
True

Shade half-plane containing (0, 0).

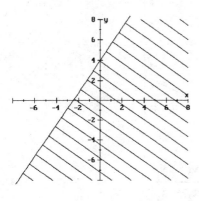

67. $\dfrac{x}{2} + \dfrac{y}{3} \geq 6$ (solid line)

$$6\left(\dfrac{x}{2} + \dfrac{y}{3}\right) \geq 6(6)$$

$$3x + 2y \geq 36$$

$$3x + 2y = 36$$

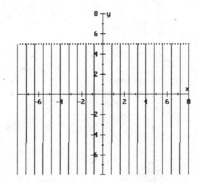

71. $2x + 2y > 100$ (dashed line)

$2x + 2y = 100$

x	y
0	50
50	0

Test point: (0, 0)

$2(0) + 2(0) > 100$

$0 > 100$

False

Shade half-plane <u>not</u> containing (0, 0).

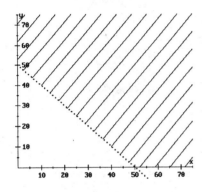

Three feasible points: $(60, 60)$, $(60, 70)$, $(70, 60)$

CHAPTER 4 PRACTICE TEST

1. (a) $4x + 3y - 24 = 0$

$4x + 3y = 24$

x	y
0	8
6	0

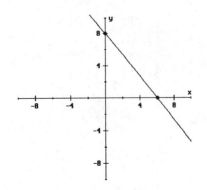

(b) $2x = 8$

$x = 4$

Vertical line passing through (4, 0).

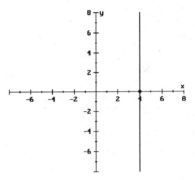

(c) $y = \dfrac{1}{2}x - 6$

x	y
0	-6
12	0

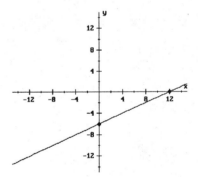

2. (a) $m = \dfrac{y_2 - y_1}{x_2 - x_1}$

$= \dfrac{9 - 4}{3 - (-2)}$

$= \dfrac{5}{5}$

$= 1$

(b) $3x - 2y = 8$

$-2y = -3x + 8$

$y = \dfrac{3}{2}x - 4$

$m = \dfrac{3}{2}$

3.　　　$m = \dfrac{y_2 - y_1}{x_2 - x_1}$

$2 = \dfrac{5 - 2}{2 - a}$

$2 = \dfrac{3}{2 - a}$

$2(2 - a) = \left(\dfrac{3}{2 - a}\right)(2 - a)$

$4 - 2a = 3$

$-2a = -1$

$a = \dfrac{1}{2}$

4. (a)　　$m = \dfrac{-3 - 5}{2 - 3}$

$= 8$

$y - y_1 = m(x - x_1)$
$y - 5 = 8(x - 3)$
$y - 5 = 8x - 24$
$y = 8x - 19$

(b) $y - y_1 = m(x - x_1)$
$y - 0 = -4(x - 1)$
$y = -4x + 4$

(c) $(2, 5)$ and $(0, 3)$

$m = \dfrac{5 - 3}{2 - 0}$

$= 1$
$y = mx + b$
$y = x + 3$

(d) $x + 3y = 8$
$3y = -x + 8$

$y = -\dfrac{1}{3}x + \dfrac{8}{3}$

$m = -\dfrac{1}{3}$

$(2, -3)$

$y - y_1 = m(x - x_1)$

$y - (-3) = -\dfrac{1}{3}(x - 2)$

$y + 3 = -\dfrac{1}{3}x + \dfrac{2}{3}$

$y = -\dfrac{1}{3}x - \dfrac{7}{3}$

(e) $x + 3y = 8$
$3y = -x + 8$

$y = -\dfrac{1}{3}x + \dfrac{8}{3}$

$m = -\dfrac{1}{3}$

$m_\perp = 3$

$y - y_1 = m(x - x_1)$
$y - (-3) = 3(x - 2)$
$y + 3 = 3x - 6$
$y = 3x - 9$

(f) $y = -1$

5. (A, B): $(60, 90)$; $(80, 150)$

$m = \dfrac{150 - 90}{80 - 60}$

$= \dfrac{60}{20}$

$= 3$

$B - 90 = 3(A - 60)$
$B - 90 = 3A - 180$
$B = 3A - 90$

If $A = 85$:
$B = 3(85) - 90$
$= 165$

They should score 165 on test B.

6. (a) $\begin{cases} 2x + 3y = 1 \\ 3x + 4y = 4 \end{cases}$

$\begin{cases} -3(2x + 3y) = -3(1) \\ 2(3x + 4y) = 2(4) \end{cases}$

$\begin{array}{r} -6x - 9y = -3 \\ 6x + 8y = 8 \\ \hline -y = 5 \\ y = -5 \end{array}$

$2x + 3(-5) = 1$
$2x - 15 = 1$
$2x = 16$
$x = 8$

$x = 8, \quad y = -5$

(b) $\begin{cases} \dfrac{a}{3} + \dfrac{b}{2} = 2 \\ a = \dfrac{b}{3} - 5 \end{cases}$

$\begin{cases} 6\left(\dfrac{a}{3} + \dfrac{b}{2}\right) = 6(2) \\ 3(a) = 3\left(\dfrac{b}{3} - 5\right) \end{cases}$

$\begin{cases} 2a + 3b = 12 \\ 3a = b - 15 \end{cases}$

2nd equation: $b = 3a + 15$
Substituting into 1st equation:

$2a + 3(3a + 15) = 12$
$2a + 9a + 45 = 12$
$11a + 45 = 12$
$11a = -33$
$a = -3$

$b = 3(-3) + 15$
$= 6$

$a = -3, \quad b = 6$

7. amount at 8½%: x
   amount at 9%: y

$\begin{cases} x + y = 3500 \\ 0.085x + 0.09y = 309 \end{cases}$

1st equation: $x = 3500 - y$
Substituting into 2nd equation:

$0.085(3500 - y) + 0.09y = 309$
$297.5 - 0.085y + 0.09y = 309$
$297.5 + 0.005y = 309$
$0.005y = 11.5$
$y = 2300$

$x + 2300 = 3500$
$x = 1200$

He invested \$1200 at 8½% and \$2300 at 9%.

8. $3x - 8y > 12$ \quad (dashed line)

$3x - 8y = 12$

x	y
0	-3/2
4	0

Test point: $(0, 0)$

$3(0) - 8(0) > 12$
$0 > 12$
False

Shade half-plane <u>not</u> containing $(0, 0)$.

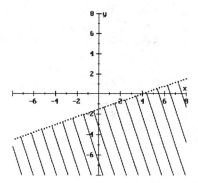

9. $315c + 425s \leq 7200$ (solid line)

$315c + 425s = 7200$

c	s
0	16.9
22.9	0

Test point: (0, 0)

$315(0) + 425(0) \leq 7200$

$\qquad\qquad\quad 0 \leq 7200$

$\qquad\qquad\qquad$ True

Shade half-plane containing (0, 0).

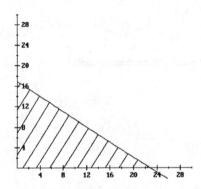

Three feasible points: (5, 5), (2, 5), (5, 2)

10. $\begin{cases} 6x - y = 1 \\ 2x + y = 3 \end{cases}$

$\begin{cases} y = 6x - 1 \\ y = -2x + 3 \end{cases}$

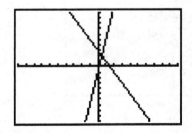

$x = 0.5, \quad y = 2$

5.1 Exercises

1.　　polynomial

　　(a)　binomial
　　(b)　degree:　2
　　(c)　1 variable
　　(d)　5;　4

3.　　polynomial

　　(a)　monomial
　　(b)　degree:　0
　　(c)　0 variables
　　(d)　59

5.　not a polynomial

7.　　polynomial

　　(a)　monomial
　　(b)　degree:　$1 + 3 + 7 = 11$
　　(c)　3 variables
　　(d)　−4

9.　not a polynomial

11.　　polynomial

　　(a)　4 terms
　　(b)　degree:　3
　　(c)　1 variable
　　(d)　4;　−1;　−2;　1

13.　(a)　Revenue　=　cost　·　number
　　　　　　　per pound　　of pounds

$$R = d(430 - d^2)$$
$$R = 430d - d^3$$
(b)

(b)

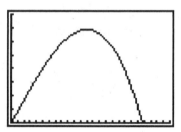

```
RANGE
Xmin=0
Xmax=25
Xscl=1
Ymin=0
Ymax=4000
Yscl=500
Xres=1
```

$d \approx \$11.97$

(c)　\$3432.03

15.　(a)　Revenue　=　number　·　price
　　　　　　　　of items　　per item

$$R = n(2 + 0.45n - 0.001n^2)$$
$$R = 2n + 0.45n^2 - 0.001n^3$$

(b)

```
RANGE
Xmin=0
Xmax=600
Xscl=100
Ymin=0
Ymax=15000
Yscl=1000
Xres=1
```

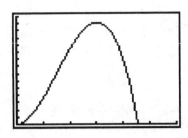

number of items producing maximum
revenue: 302.2
price: $2 + 0.45(302.2) - 0.001(302.2)^2$
$= \$46.67$

(c) $14,102.21

17. width: w
length: $3w$
height: $w - 6$

$S = 2 \cdot$ width $\cdot$ length $+ 2 \cdot$ length $\cdot$ height
$\quad + 2 \cdot$ width $\cdot$ height
$S = 2w(3w) + 2(3w)(w - 6) + 2w(w - 6)$
$S = 6w^2 + 6w^2 - 36w + 2w^2 - 12w$
$S = 14w^2 - 48w$

When $w = 20$:

$S = 14(20)^2 - 48(20)$
$\quad = 4640$ sq. in.

When $w = 26$:

$S = 14(26)^2 - 48(26)$
$\quad = 8216$ sq. in.

When $w = 42$:

$S = 14(42)^2 - 48(42)$
$\quad = 22,680$ sq. in.

19. $A = 0.28m + 0.39m^2 - 0.02m^3$

(a) Using TABLE:

M	A
0	0
5	8.7
10	21.8
15	24.5
20	1.6
25	0
30	0

(b)

```
RANGE
Xmin=0
Xmax=25
Xscl=1
Ymin=0
Ymax=30
Yscl=1
Xres=1
```

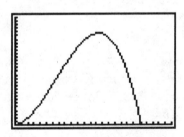

It is at a maximum in 13 minutes.

21. $-3 - (-4) + (-5) - (-8)$
$\quad = -3 + 4 + (-5) + 8$
$\quad = 1 + (-5) + 8$
$\quad = -4 + 8$
$\quad = 4$

23. 1st side: $2x$

2nd side: x

3rd side: 24

$P = 1^{st}$ side + 2nd side + 3rd side

$75 = 2x + x + 24$

$75 = 3x + 24$

$51 = 3x$

$17 = x$

$2x = 2(17) = 34$

The 1st side is 34 in. and the 2nd side is 17 in.

5.2 Exercises

1. $(3x^2 - 2x + 5) + (2x^2 - 7x + 4)$

$= 3x^2 - 2x + 5 + 2x^2 - 7x + 4$

$= 5x^2 - 9x + 9$

3. $(15x^2 - 3xy - 4y^2) - (16x^2 - 3x + 2)$

$= 15x^2 - 3xy - 4y^2 - 16x^2 + 3x - 2$

$= -x^2 - 3xy - 4y^2 + 3x - 2$

5. $(3a^2 - 2ab + 4b^2) + \left[(3a^2 - b^2) - (3ab)\right]$

$= (3a^2 - 2ab + 4b^2) + (3a^2 - b^2 - 3ab)$

$= 3a^2 - 2ab + 4b^2 + 3a^2 - b^2 - 3ab$

$= 6a^2 - 5ab + 3b^2$

7. $(3a^2 - 2ab + 4b^2) - \left[(3a^2 - b^2) - (3ab)\right]$

$= (3a^2 - 2ab + 4b^2) - (3a^2 - b^2 - 3ab)$

$= 3a^2 - 2ab + 4b^2 - 3a^2 + b^2 + 3ab$

$= ab + 5b^2$

9. $\left[(3a^2 - b^2) - (3ab)\right] - (3a^2 - 2ab + 4b^2)$

$= (3a^2 - b^2 - 3ab) - (3a^2 - 2ab + 4b^2)$

$= 3a^2 - b^2 - 3ab - 3a^2 + 2ab - 4b^2$

$= -ab - 5b^2$

11. $(3xy + 2y^2) + (-7y^2 - 3y + 4)$

$= 3xy + 2y^2 - 7y^2 - 3y + 4$

$= 3xy - 5y^2 - 3y + 4$

13. $(-8x^2 - 5xy + 9y^2) - (3x^2 - 2xy + 7y^2)$

$= -8x^2 - 5xy + 9y^2 - 3x^2 + 2xy - 7y^2$

$= -11x^2 - 3xy + 2y^2$

15. $(3a^2 - 2b^2) - \left[(2a^2 - 3ab + 4b^2) + (6a^2 + 2ab - 2b^2)\right]$

$= (3a^2 - 2b^2) - (2a^2 - 3ab + 4b^2 + 6a^2 + 2ab - 2b^2)$

$(3a^2 - 2b^2) - (8a^2 - ab + 2b^2)$

$= 3a^2 - 2b^2 - 8a^2 + ab - 2b^2$

$= -5a^2 + ab - 4b^2$

17. $x^2y(3x^2 - 2xy + 2y^2)$

$= x^2y(3x^2) + x^2y(-2xy) + x^2y(2y^2)$

$= 3x^4y - 2x^3y^2 + 2x^2y^3$

19. $(3x^2 - 2xy + 2y^2)(x^2y)$

$= 3x^2(x^2y) - 2xy(x^2y) + 2y^2(x^2y)$

$= 3x^4y - 2x^3y^2 + 2x^2y^3$

21. $(x + 30(x + 4) = x^2 + 4x + 3x + 12$

$= x^2 + 7x + 12$

23. $(3x - 1)(2x + 1) = 6x^2 + 3x - 2x - 1$

$= 6x^2 + x - 1$

25. $(3x + 1)(2x - 1) = 6x^2 - 3x + 2x - 1$

$= 6x^2 - x - 1$

27. $(3a - b)(a + b)$

$= 3a^2 + 3ab - ab - b^2$

$= 3a^2 + 2ab - b^2$

29. $(2r - s)^2 = (2r - s)(2r - s)$

$= 4r^2 - 2rs - 2rs + s^2$

$= 4r^s - 4rs + s^2$

31. $(2r + s)(2r - s)$

$= 4r^2 - 2rs + 2rs - s^2$

$= 4r^2 - s^2$

33. $(3x - 2y)(3x + 2y)$

$= 9x^2 + 6xy - 6xy - 4y^2$

$= 9x^2 - 4y^2$

35. $(3x - 2y)^2 = (3x - 2y)(3x - 2y)$

$= 9x^2 - 6xy - 6xy + 4y^2$

$= 9x^2 - 12xy + 4y^2$

37. $(x - y)(a - b) = xa - xb - ya + yb$

39. $(2r + 3s)(2a + 3b)$

$= 4ra + 6rb + 6sa + 9sb$

41. $(2y^2 - 3)(y^2 + 1) = 2y^4 + 2y^2 - 3y^2 - 3$
$$= 2y^4 - y^2 - 3$$

43. $(5x^2 - 3x + 4)(x + 3)$
$$= (5x^2 - 3x + 4)(x) + (5x^2 - 3x + 4)(3)$$
$$= 5x^3 - 3x^2 + 4x + 15x^2 - 9x + 12$$
$$= 5x^3 + 12x^2 - 5x + 12$$

45. $(9a^2 + 3a - 5)(3a + 1)$
$$= (9a^2 + 3a - 5)(3a) + (9a^2 + 3a - 5)(1)$$
$$= 27a^3 + 9a^2 - 15a + 9a^2 + 3a - 5$$
$$= 27a^3 + 18a^2 - 12a - 5$$

47. $(a^2 - ab + b^2)(a + b)$
$$= (a^2 - ab + b^2)(a) + (a^2 - ab + b^2)(b)$$
$$= a^3 - a^2b + ab^2 + a^2b - ab^2 + b^3$$
$$= a^3 + b^3$$

49. $(x - y - z)^2$
$$= (x - y \cdot z)(x - y - z)$$
$$= x(x - y - z) - y(x - y \cdot z) - z(x - y - z)$$
$$= x^2 - xy - xz - xy + y^2 + yz - xz + yz + z^2$$
$$= x^2 - 2xy - 2xz + 2yz + y^2 + z^2$$

51. $(x^2 - 2x + 4)(x + 2)$
$$= (x^2 - 2x + 4)(x) + (x^2 - 2x + 4)(2)$$
$$= x^3 - 2x^2 + 4x + 2x^2 - 4x + 8$$
$$= x^3 + 8$$

53. $(a + b + c + d)(a + b)$
$$= (a + b + c + d)(a) + (a + b + c + d)(b)$$
$$= a^2 + ab + ac + ad + ab + b^2 + bc + bd$$
$$= a^2 + 2ab + ac + ad + b^2 + bc + bd$$

55. width: w
length: $3w + 5$

Area = width $\cdot$ length
$$A = w(3w + 5)$$
$$A = 3w^2 + 5w$$

Perimeter = 2 $\cdot$ width + 2 $\cdot$ length
$$P = 2w + 2(3w + 5)$$
$$P = 2w + 6w + 10$$
$$P = 8w + 10$$

57.

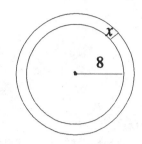

radius of inner circle: 8
area of inner circle: $\pi(8)^2 = 64\pi$
radius of outer circle: $x + 8$
area of outer circle: $\pi(x + 8)^2$

Area   =   area of   –   area of
of walkway   outer circle   inner circle

$$A = \pi(x + 8)^2 - 64\pi$$
$$A = \pi(x^2 + 16x + 64) - 64\pi$$
$$A = \pi x^2 + 16\pi x + 64\pi - 64\pi$$
$$A = \pi x^2 + 16\pi x \text{ sq. ft}$$

59.

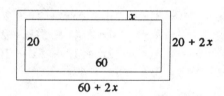

Area   =   area of   –   area of
of walkway   outer rectangle   inner rectangle

$$A = (60 + 2x)(20 + 2x) - (60)(20)$$
$$A = 1200 + 160x + 4x^2 - 1200$$
$$A = 4x^2 + 160x \text{ sq. ft}$$

61. $f(x) - g(x) = (2x - 3) - (3x^2 - 5x + 2)$
$$= 2x - 3 - 3x^2 + 5x - 2$$
$$= -3x^2 + 7x - 5$$

63. $h(x) - [f(x) - g(x)]$
$$= (x^3 - x) - [(2x - 3) - (3x^2 - 5x + 2)]$$
$$= (x^3 - x) - (2x - 3 - 3x^2 + 5x - 2)$$
$$= (x^3 - x) - (-3x^2 + 7x - 5)$$
$$= x^3 - x + 3x^2 - 7x + 5$$
$$= x^3 + 3x^2 - 8x + 5$$

65. $f(x) \cdot g(x)$
$= (2x - 3)(3x^2 - 5x + 2)$
$= 2x(3x^2 - 5x + 2) - 3(3x^2 - 5x + 2)$
$= 6x^3 - 10x^2 + 4x - 9x^2 + 15x - 6$
$= 6x^3 - 19x^2 + 19x - 6$

67. $h(x)[f(x) + g(x)]$
$= (x^3 - x)\left[(2x - 3) + (3x^2 - 5x + 2)\right]$
$= (x^3 - x)(2x - 3 + 3x^2 - 5x + 2)$
$= (x^3 - x)(3x^2 - 3x - 1)$
$= x^3(3x^2 - 3x - 1) - x(3x^2 - 3x - 1)$
$= 3x^5 - 3x^4 - x^3 - 3x^3 + 3x^2 + x$
$= 3x^5 - 3x^4 - 4x^3 + 3x^2 + x$

69. $g(x) = 3x^2 - 5x + 2$
$g(2) = 3(2)^2 - 5(2) + 2$
$\quad\; = 3(4) - 5(2) + 2$
$\quad\; = 4$

$g(x) + g(2) = 3x^2 - 5x + 2 + 4$
$\qquad\qquad = 3x^2 - 5x + 6$

71. $g(x + 2) = 3(x + 2)^2 - 5(x + 2) + 2$
$\qquad\quad = 3(x^2 + 4x + 4) - 5(x + 2) + 2$
$\qquad\quad = 3x^2 + 12x + 12 - 5x - 10 + 2$
$\qquad\quad = 3x^2 + 7x + 4$

73. $h(x) - 1 = (x^3 - x) - 1$
$\qquad\qquad = x^3 - x - 1$

77. $\dfrac{-4 - [4 - 5(2 - 8)]}{4 - 5 \cdot 2 - 8} = \dfrac{-4 - [4 - 5(-6)]}{4 - 10 - 8}$

$= \dfrac{-4 - (4 + 30)}{-14}$

$= \dfrac{-4 - 34}{-14}$

$= \dfrac{-38}{-14}$

$= \dfrac{19}{7}$

79. $|3 - 2x| \le 5$

$-5 \le 3 - 2x \le 5$
$-8 \le -2x \le 2$
$\;\; 4 \ge x \ge -1$ or $-1 \le x \le y$

1. $(x + 4)(x + 5) = x^2 + (4 + 5)x + 20$
$\qquad\qquad\qquad = x^2 + 9x + 20$

3. $(x + 3)(x - 7) = x^2 + (3 - 7)x - 21$
$\qquad\qquad\qquad = x^2 - 4x - 21$

5. $(x - 8)(x - 11) = x^2 + (-8 - 11)x + 88$
$\qquad\qquad\qquad\; = x^2 - 19x + 88$

7. $(x + 5)(x - 6) = x^2 + (-6 + 5)x - 30$
$\qquad\qquad\qquad = x^2 - x - 30$

9. $x + 5(x - 6) = x + 5x - 30$
$\qquad\qquad\;\; = 6x - 30$

11. $(x - 7) - (x - 4) = x - 7 - x + 4$
$\qquad\qquad\qquad\; = -3$

13. $(2x + 1)(x - 4) = 2x^2 + (-8 + 1)x - 4$
$\qquad\qquad\qquad\; = 2x^2 - 7x - 4$

15. $(5a - 4)(5a + 4) = (5a)^2 - (4)^2$
$\qquad\qquad\qquad\quad = 25a^2 - 16$

17. $(3z + 5)^2 = (3z)^2 + 2(3z)(5) + 5^2$
$\qquad\qquad\; = 9z^2 + 30z + 25$

19. $(3z - 5)^2 = (3z)^2 - 2(3z)(5) + 5^2$
$\qquad\qquad\; = 9z^2 - 30z + 25$

21. $(3z - 5)(3z + 5) = (3z)^2 - 5^2$
$\qquad\qquad\qquad\; = 9z^2 - 25$

23. $(5r^2 + 3s)(3r + 5s) = 15r^3 + 25r^2s + 9rs + 15s^2$

25. $(3s - 2y)^2 = (3s)^2 - 2(3s)(2y) + (2y)^2$
$\qquad\qquad\; = 9s^2 - 12sy + 4y^2$

27. $(3y + 10z)^2 = (3y)^2 + 2(3y)(10z) + (10z)^2$
$\qquad\qquad\;\; = 9y^2 + 60yz + 100z^2$

29. $(3y + 10z)(3y - 10z) = (3y)^2 - (10z)^2$
$= 9y^2 - 100z^2$

31. $(3a + 2b)(3a + 4b)$
$= 9a^2 + (12 + 6)ab + 8b^2$
$= 9a^2 + 18ab + 8b^2$

33. $(8a - 1)^2 = (8a)^2 - 2(8a)(1) + 1^2$
$= 64a^2 - 16a + 1$

35. $(5r - 2s)(5r - 2s)$
$= (5r)^2 - 2(5r)(2s) + (2s)^2$
$= 25r^2 - 20rs + 4s^2$

37. $(7x - 8)(8x - 7)$
$= 56x^2 + (-49 - 64)x + 56$
$= 56x^2 - 113x + 56$

39. $(5t - 3s^2)(5t + 3s) = 25t^2 + 15st - 15s^2t - 9s^3$

41. $(3y^3 - 4x)^2$
$= (3y^3)^2 - 2(3y^3)(4x) + (4x)^2$
$= 9y^6 - 24xy^3 + 16x^2$

43. $(3y^3 - 4x)(3y^3 + 4x) = (3y^3)^2 - (4x)^2$
$= 9y^6 - 16x^2$

45. $(2rst - 7xyz)^2$
$= (2rst)^2 - 2(2rst)(7xyz) + (7xyz)^2$
$= 4r^2s^2t^2 - 28rstxyz + 49x^2y^2z^2$

47. $(x - 4) - (3x + 1)^2$
$= x - 4 - [(3x)^2 + 2(3x)(1) + 1^2]$
$= x - 4 - (9x^2 + 6x + 1)$
$= x - 4 - 9x^2 - 6x - 1$
$= -9x^2 - 5x - 5$

49. $(a - b)^2 - (b + a)^2$
$= a^2 - 2ab + b^2 - (b^2 + 2ab + a^2)$
$= a^2 - 2ab + b^2 - b^2 - 2ab - a^2$
$= -4ab$

51. $(3a + 2)(-5a)(2a - 1)$
$= (-15a^2 - 10a)(2a - 1)$
$= -30a^3 + 15a^2 - 20a^2 + 10a$
$= -30a^3 - 5a^2 + 10a$

53. $(3a + 2) - 5a(2a - 1)$
$= 3a + 2 - 10a^2 + 5a$
$= -10a^2 + 8a + 2$

55. $3r^3 - 3r(r - s)(r + s)$
$= 3r^3 - 3r(r^2 - s^2)$
$= 3r^3 - 3r^3 + 3rs^2$
$= 3rs^2$

57. $(3y + 1)(y + 2) - (2y - 3)^2$
$= 3y^2 + (6 + 1)y + 2 - [(2y)^2 - 2(2y)(3) + 3^2]$
$= 3y^2 + 7y + 2 - (4y^2 - 12y + 9)$
$= 3y^2 + 7y + 2 - 4y^2 + 12y - 9$
$= -y^2 + 19y - 7$

59. $(5a - 3b)^3$
$= (5a - 3b)(5a - 3b)(5a - 3b)$
$= [(5a)^2 - 2(5a)(3b) + (3b)^2](5a - 3b)$
$= (25a^2 - 30ab + 9b^2)(5a - 3b)$
$= 125a^3 - 150a^2b + 45ab^2 - 75a^2b + 90ab^2 - 27b^3$
$= 125a^3 - 225a^2b + 135ab^2 - 27b^3$

61. $3x(x + 1) - 2x(x + 1)^2$
$= 3x^2 + 3x - 2x(x^2 + 2x + 1)$
$= 3x^2 + 3x - 2x^3 - 4x^2 - 2x$
$= -2x^3 - x^2 + x$

63. $[(a + b) + 1][(a + b) - 1]$
$= (a + b)^2 - 1^2$
$= a^2 + 2ab + b^2 - 1$

65. $[(a - 2b) + 5z]^2$
$= (a - 2b)^2 + 2(a - 2b)(5z) + (5z)^2$
$= a^2 - 2a(2b) + (2b)^2 + 10z(a - 2b) + 25z^2$
$= a^2 - 4ab + 4b^2 + 10az - 20bz + 25z^2$

67. $[a - 2b + 5z][a - 2b - 5z]$
$= [(a - 2b) + 5z][(a - 2b) - 5z]$
$= (a - 2b)^2 - (5z)^2$
$= a^2 - 2a(2b) + (2b)^2 - 25z^2$
$= a^2 - 4ab + 4b^2 - 25z^2$

69. $[(a + b) + (2x + 1)][(a + b) - (2x + 1)]$
$= (a + b)^2 - (2x + 1)^2$
$= a^2 + 2ab + b^2 - [(2x)^2 + 2(2x)(1) + 1^2]$
$= a^2 + 2ab + b^2 - (4x^2 + 4x + 1)$
$= a^2 + 2ab + b^2 - 4x^2 - 4x - 1$

71. $(a^n - 3)(a^n + 3) = (a^n)^2 - 3^2$
$= a^{2n} - 9$

77. $|3x - 2| > 5$
$3x - 2 > 5$ or $3x - 2 < -5$
$3x > 7$ $\qquad 3x < -3$
$x > \dfrac{7}{3}$ or $\qquad x < -1$

79. width: x
length: $5x + 4$

$P = 2w + 2l$
$41.6 = 2x + 2(5x + 4)$
$41.6 = 2x + 10x + 8$
$41.6 = 12x + 8$
$33.6 = 12x$
$2.8 = x$
$5x + 4 = 5(2.8) + 4 = 18$

The dimensions are 2.8 in. by 18 in.

5.4 Exercises

1. $4x^2 + 2x = 2x(2x + 1)$

3. $x^2 + x = x(x + 1)$

5. $3xy^2 - 6x^2y^3 = 3xy^2(1 - 2xy)$

7. $6x^4 + 9x^3 - 21x^2 = 3x^2(2x^2 + 3x - 7)$

9. $35x^4y^4z - 15x^3y^5z + 10x^2y^3z^2$
$= 5x^2y^3z(7x^2y - 3xy^2 + 2z)$

11. $24r^3s^4 - 18r^3s^5 - 6r^2s^3$
$= 6r^2s^3(4rs - 3rs^2 - 1)$

13. $35a^2b^3 - 21a^3b^2 + 7ab$
$= 7ab(5ab^2 - 3a^2b + 1)$

15. $3x(x + 2) + 5(x + 2) = (x + 2)(3x + 5)$

17. $3x(x + 2) - 5(x + 2) = (x + 2)(3x - 5)$

19. $2x(x + 3y) + 5y(x + 3y)$
$= (x + 3y)(2x + 5y)$

21. $3x(x + 4y) - 5y(x + 4y)$
$= (x + 4y)(3x - 5y)$

23. $(2r + 1)(a - 2) + 5(a - 2)$
$= (a - 2)[(2r + 1) + 5]$
$= (a - 2)(2r + 6)$
$= (a - 2)2(r + 3)$
$= 2(a - 2)(r + 3)$

25. $2x(a - 3)^2 + 2(a - 3)^2 = 2(a - 3)^2(x + 1)$

27. $16a^2(b - 4)^2 - 4a(b - 4)$
$= 4a(b - 4)[4a(b - 4) - 1]$
$= 4a(b - 4)(4ab - 16a - 1)$

29. $2x^2 - 8x + 3x - 12$
$= 2x(x - 4) + 3(x - 4)$
$= (x - 4)(2x + 3)$

31. $3x^2 - 12xy + 5xy - 20y^2$
$= 3x(x - 4y) + 5y(x - 4y)$
$= (x - 4y)(3x + 5y)$

33. $7ax - 7bx + 3ay - 3by$
$= 7x(a - b) + 3y(a - b)$
$= (a - b)(7x + 3y)$

35. $7ax + 7bx - 3ay - 3by$
$= 7x(a + b) - 3y(a + b)$
$= (a + b)(7x - 3y)$

37. $7ax - 7bx - 3ay + 3by$
$= 7x(a - b) - 3y(a - b)$
$= (a - b)(7x - 3y)$

39. $7ax - 7bx - 3ay - 3by$
$= 7x(a - b) - 3y(a + b)$
not factorable

41. $2r^2 + 2rs - sr - s^2$
$= 2r(r + s) - s(r + s)$
$= (r + s)(2r - s)$

43. $2r^2 + 2rs - sr + s^2$
$= 2r(r + s) - s(r - s)$
not factorable

45. $5a^2 - 5ab - 2ab + 2b^2$
$= 5a(a - b) - 2b(a - b)$
$= (a - b)(5a - 2b)$

47. $3a^2 - 6a - a + 2$
$= 3a(a - 2) - 1(a - 2)$
$= (a - 2)(3a - 1)$

49. $3a^2 - 6a + a - 2$
$= 3a(a - 2) + 1(a - 2)$
$= (a - 2)(3a + 1)$

51. $3a^2 + 6a - a - 2$
$= 3a(a + 2) - 1(a + 2)$
$= (a + 2)(3a - 1)$

53. $a^3 + 2a^2 + 4a + 8$
$= a^2(a + 2) + 4(a + 2)$
$= (a + 2)(a^2 + 4)$

55. $16x^2 - 9y^2 = (4x)^2 - (3y)^2$
$= (4x - 3y)(4x + 3y)$

57. $x^2 + 4xy + 4y^2 = (x)^2 + 2(x)(2y) + (2y)^2$
$= (x + 2y)^2$

59. $|9 - 3a| \leq 6$
$-6 \leq 9 - 3a \leq 6$
$-15 \leq -3a \leq -3$
$5 \geq a \geq 1$ or $1 \leq a \leq 5$

61. $m = \frac{-2}{3}$; From $(4, -1)$, go down 2 and
right 3 to the point $(7, -3)$ on the line.

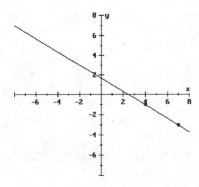

1. Factors of -45 whose sum is 4: 9, -5
$x^2 + 4x - 45 = (x + 9)(x - 5)$

3. Factors of -10 whose sum is -3: -5, 2
$y^2 - 3y - 10 = (y - 5)(y + 2)$

5. Factors of -15 whose sum is -2: -5, 3
$x^2 - 2xy - 15y^2 = (x - 5y)(x + 3y)$

7. $r^2 - 81 = r^2 - 9^2$
$= (r - 9)(r + 9)$

9. Factors of -6 whose sum is -1: -3, 2
$x^2 - xy - 6y^2 = (x - 3y)(x + 2y)$

11. Factors of 6 whose sum is -1: none
not factorable

13. Factors of 12 whose sum is -7: -3, -4
$r^2 - 7rs + 12s^2 = (r - 3s)(r - 4s)$

15. Factors of -12 whose sum is -1: -4, 3
$r^2 - rs - 12s^2 = (r - 4s)(r + 3s)$

17. $9x^2 - 49y^2 = (3x)^2 - (7y)^2$
$= (3x - 7y)(3x + 7y)$

19. Factors of $15 \cdot 4 = 60$ whose sum is 17: 5, 12

$15x^2 + 17x + 4 = 15x^2 + 5x + 12x + 4$
$= 5x(3x + 1) + 4(3x + 1)$
$= (3x + 1)(5x + 4)$

21. Factors of $15(-6) = -90$ whose sum
is -1: -10, 9

$15x^2 - xy - 6y^2$
$= 15x^2 - 10xy + 9xy - 6y^2$
$= 5x(3x - 2y) + 3y(3x - 2y)$
$= (3x - 2y)(5x + 3y)$

23. Factors of $15(-6) = -90$ whose sum
is 1: 10, -9

$15x^2 + xy - 6y^2$
$= 15x^2 + 10xy - 9xy - 6y^2$
$= 5x(3x + 2y) - 3y(3x + 2y)$
$= (3x + 2y)(5x - 3y)$

25. $10xy + 3x^2 - 25y^2 = 3x^2 + 10xy - 25y^2$

Factors of $3(-25) = -75$ whose sum is 10: 15, -5

$$3x^2 + 10xy - 25y^2$$
$$= 3x^2 + 15xy - 5xy - 25y^2$$
$$= 3x(x + 5y) - 5y(x + 5y)$$
$$= (x + 5y)(3x - 5y)$$

27. Factors of $10(-10) = -100$ whose sum is 21: 25, -4

$$10a^2 + 21ab - 10b^2$$
$$= 10a^2 + 25ab - 4ab - 10b^2$$
$$= 5a(2a + 5b) - 2b(2a + 5b)$$
$$= (2a + 5b)(5a - 2b)$$

29. $25 - 5y - 2y^2 = -(2y^2 + 5y - 25)$

Factors of $2(-25) = -50$ whose sum is 5: 10, -5

$$-(2y^2 + 5y - 25)$$
$$= -(2y^2 + 10y - 5y - 25)$$
$$= -[2y(y + 5) - 5(y + 5)]$$
$$= -(y + 5)(2y - 5)$$

31. $25 - 5y - y^2 = -(y^2 + 5y - 25)$

Factors of -25 whose sum is 5: none

Not factorable

33. $2x^3 + 4x^2 - 16x = 2x(x^2 + 2x - 8)$
$$= 2x(x + 4)(x - 2)$$

35. $x^3 + 3x^2 - 28x = x(x^2 + 3x - 28)$
$$= x(x + 7)(x - 4)$$

37. $2x(x - 3) + (x - 1)(x + 2)$
$$= 2x^2 - 6x + x^2 + x - 2$$
$$= 3x^2 - 5x - 2$$
$$= 3x^2 - 6x + x - 2$$
$$= 3x(x - 2) + 1(x - 2)$$
$$= (x - 2)(3x + 1)$$

39. $18ab - 15abx - 18abx^2$
$$= 3ab(6 - 5x - 6x^2)$$
$$= -3ab(6x^2 + 5x - 6)$$
$$= -3ab(6x^2 + 9x - 4x - 6)$$
$$= -3ab[3x(2x + 3) - 2(2x + 3)]$$
$$= -3ab(2x + 3)(3x - 2)$$

41. $90y^4 - 114y^3 + 36y^2$
$$= 6y^2(15y^2 - 19y + 6)$$
$$= 6y^2(15y^2 - 10y - 9y + 6)$$
$$= 6y^2[5y(3y - 2) - 3(3y - 2)]$$
$$= 6y^2(3y - 2)(5y - 3)$$

43. $6r^4 - r^2 - 2$
$$= 6r^4 - 4r^2 + 3r^2 - 2$$
$$= 2r^2(3r^2 - 2) + 1(3r^2 - 2)$$
$$= (3r^2 - 2)(2r^2 + 1)$$

45. $3x^2 - 27 = 3(x^2 - 9)$
$$= 3(x^2 - 3^2)$$
$$= 3(x - 3)(x + 3)$$

47. $20xy^5 + 2x^2y^3 - 8x^3y$
$$= 2xy(10y^4 + xy^2 - 4x^2)$$

49. $108x^3y + 72x^2y^3 - 15xy^5$
$$= 3xy(36x^2 + 24xy^2 - 5y^4)$$
$$= 3xy(36x^2 + 30xy^2 - 6xy^2 - 5y^4)$$
$$= 3xy[6x(6x + 5y^2) - y^2(6x + 5y^2)]$$
$$= 3xy(6x + 5y^2)(6x - y^2)$$

51. $12a^7b + a^4b - 6ab$
$$= ab(12a^6 + a^3 - 6)$$
$$= ab(12a^6 + 9a^3 - 8a^3 - 6)$$
$$= ab[3a^3(4a^3 + 3) - 2(4a^3 + 3)]$$
$$= ab(4a^3 + 3)(3a^3 - 2)$$

53. $15a^8 + 19a^4 - 10$
$$= 15a^8 + 25a^4 - 6a^4 - 10$$
$$= 5a^4(3a^4 + 5) - 2(3a^4 + 5)$$
$$= (3a^4 + 5)(5a^4 - 2)$$

55. $16x^2 - 9y^2 = (4x)^2 - (3y)^2$
$$= (4x - 3y)(4x + 3y)$$

57. $x^2 + 4xy + 4y^2 = (x)^2 + 2(x)(2y) + (2y)^2$
$$= (x + 2y)^2$$

59. $x^2 - 4xy + 4y^2 = (x)^2 - 2(x)(2y) + (2y)^2$
$$= (x - 2y)^2$$

61. Factors of -4 whose sum is -4: none
not factorable

63. $6x^2y^2 - 5xy + 1$
$= 6x^2y^2 - 3xy - 2xy + 1$
$= 3xy(2xy - 1) - 1(2xy - 1)$
$= (2xy - 1)(3xy - 1)$

65. $81r^2s^2 - 16 = (9rs)^2 - 4^2$
$= (9rs - 4)(9rs + 4)$

67. $9x^2y^2 - 4z^2 = (3xy)^2 - (2z)^2$
$= (3xy - 2z)(3xy + 2z)$

69. $25a^2 - 4a^2b^2 = a^2(25 - 4b^2)$
$= a^2[5^2 - (2b)^2]$
$= a^2(5 - 2b)(5 + 2b)$

71. $1 - 14a + 49a^2 = 1^2 - 2(1)(7a) + (7a)^2$
$= (1 - 7a)^2$

73. $8a^3 - b^3 = (2a)^3 - b^3$
$= (2a - b)[(2a)^2 + (2a)(b) + b^2]$
$= (2a - b)(4a^2 + 2ab + b^2)$

75. $x^3 + 125y^3 = x^3 + (5y)^3$
$= (x + 5y)[x^2 - x(5y) + (5y)^2]$
$= (x + 5y)(x^2 - 5xy + 25y^2)$

77. not factorable

79. $4y^2 - 20y + 10 = 2(2y^2 - 10y + 5)$

81. $12a^3c + 36a^2c^2 + 27ac^3$
$= 3ac(4a^2 + 12ac + 9c^2)$
$= 3ac[(2a)^2 + 2(2a)(3c) + (3c)^2]$
$= 3ac(2a + 3c)^2$

83. $4x^4 - 81y^4 = (2x^2)^2 - (9y^2)^2$
$= (2x^2 - 9y^2)(2x^2 + 9y^2)$

85. $25a^6 + 10a^3b + b^2$
$= (5a^3)^2 + 2(5a^3)(b) + b^2$
$= (5a^3 + b)^2$

87. $12y^6 + 27y^2 = 3y^2(4y^4 + 9)$

89. $(a + b)^2 - 4 = (a + b)^2 - 2^2$
$= (a + b - 2)(a + b + 2)$

91. $a^2 - 2ab + b^2 - 16$
$= (a - b)^2 - 16$
$= (a - b)^2 - 4^2$
$= (a - b - 4)(a - b + 4)$

93. $x^2 + 6x + 9 - r^2$
$= [x^2 + 2(x)(3) + 3^2] - r^2$
$= (x + 3)^2 - r^2$
$= (x + 3 - r)(x + 3 + r)$

95. $a^3 + a^2 - 4a - 4$
$= a^2(a + 1) - 4(a + 1)$
$= (a + 1)(a^2 - 4)$
$= (a + 1)(a^2 - 2^2)$
$= (a + 1)(a - 2)(a + 2)$

97. $x^4 + x^3 - x - 1$
$= x^3(x + 1) - 1(x + 1)$
$= (x + 1)(x^3 - 1)$
$= (x + 1)(x^3 - 1^3)$
$= (x + 1)(x - 1)(x^2 + x + 1)$

109. $x - 6y = 6$
$-6y = -x + 6$
$y = \frac{1}{6}x - 1$
$m = \frac{1}{6}$
$m_\perp = -6$

$y - y_1 = m(x - x_1)$
$y - 1 = -6(x - 5)$
$y - 1 = -6x + 30$
$y = -6x + 31$

111. (T, n): $(74, 20)$ and $(84, 40)$
$m = \dfrac{40 - 20}{84 - 74}$
$= 2$

$n - 20 = 2(T - 74)$
$n - 20 = 2T - 148$
$n = 2T - 128$

When $T = 90$:
$n = 2(90) - 128$
$= 52$
He should expect to sell 52 pairs.

1. $(x + 2)(x - 3) = 0$
   $x + 2 = 0$ or $x - 3 = 0$
   $\quad x = -2$ or $\quad\quad x = 3$

3. $0 = (2y - 1)(y - 4)$
   $2y - 1 = 0$ or $y - 4 = 0$
   $\quad 2y = 1$ or $\quad\quad y = 4$
   $\quad\quad y = \dfrac{1}{2}$ or $\quad\quad y = 4$

5. $\quad\quad\quad\quad x^2 = 25$
   $\quad\quad x^2 - 25 = 0$
   $(x - 5)(x + 5) = 0$
   $x - 5 = 0$ or $x + 5 = 0$
   $\quad x = 5$ or $\quad\quad x = -5$

7. $x(x - 4) = 0$
   $x = 0$ or $x - 4 = 0$
   $x = 0$ or $\quad\quad x = 4$

9. $12 = x(x - 4)$
   $12 = x^2 - 4x$
   $\quad 0 = x^2 - 4x - 12$
   $\quad 0 = (x - 6)(x + 2)$
   $x - 6 = 0$ or $x + 2 = 0$
   $\quad x = 6$ or $\quad\quad x = -2$

11. $5y(y - 7) = 0$
    $5y = 0$ or $y - 7 = 0$
    $\quad y = 0$ or $\quad\quad y = 7$

13. $\quad\quad\quad x^2 - 16 = 0$
    $(x - 4)(x + 4) = 0$
    $x - 4 = 0$ or $x + 4 = 0$
    $\quad x = 4$ or $\quad\quad x = -4$

15. $0 = 9c^2 - 16$
    $0 = (3c - 4)(3c + 4)$
    $3c - 4 = 0$ or $3c + 4 = 0$
    $\quad 3c = 4$ $\quad\quad\quad 3c = -4$
    $\quad\quad c = \dfrac{4}{3}$ or $\quad\quad c = -\dfrac{4}{3}$

17. $\quad\quad\quad 8a^2 - 18 = 0$
    $\quad\quad 2(4a^2 - 9) = 0$
    $\quad\quad\quad 4a^2 - 9 = 0$
    $(2a - 3)(2a + 3) = 0$

$2a - 3 = 0$ or $2a + 3 = 0$
$\quad 2a = 3$ $\quad\quad\quad 2a = -3$
$\quad\quad a = \dfrac{3}{2}$ or $\quad\quad a = -\dfrac{3}{2}$

19. $0 = x^2 - x - 6$
    $0 = (x - 3)(x + 2)$
    $x - 3 = 0$ or $x + 2 = 0$
    $\quad x = 3$ or $\quad\quad x = -2$

21. $\quad 2y^2 - 3y + 1 = 0$
    $(2y - 1)(y - 1) = 0$
    $2y - 1 = 0$ or $y - 1 = 0$
    $\quad 2y = 1$ $\quad\quad\quad y = 1$
    $\quad\quad y = \dfrac{1}{2}$ or $\quad\quad y = 1$

23. $0 = 8x^2 + 4x - 112$
    $0 = 4(2x^2 + x - 28)$
    $0 = 2x^2 + x - 28$
    $0 = (2x - 7)(x + 4)$
    $2x - 7 = 0$ or $x + 4 = 0$
    $\quad 2x = 7$ $\quad\quad\quad x = -4$
    $\quad\quad x = \dfrac{7}{2}$ or $\quad\quad x = -4$

25. $\quad x^3 + x^2 - 6x = 0$
    $\quad x(x^2 + x - 6) = 0$
    $x(x + 3)(x - 2) = 0$
    $x = 0$ or $x + 3 = 0$ or $x - 2 = 0$
    $x = 0$ or $\quad x = -3$ or $\quad\quad x = 2$

27. $\quad t^4 + 2t^3 + t^2 = 0$
    $t^2(t^2 + 2t + 1) = 0$
    $t^2(t + 1)(t + 1) = 0$
    $t^2 = 0$ or $t + 1 = 0$
    $\quad t = 0$ or $\quad\quad t = -1$

29. $\quad\quad\quad\quad\quad 20m^3 = 5m$
    $\quad\quad\quad\quad 20m^3 - 5m = 0$
    $\quad\quad\quad\quad 5m(4m^2 - 1) = 0$
    $\quad 5m(2m - 1)(2m + 1) = 0$
    $5m = 0$ or $2m - 1 = 0$ or $2m + 1 = 0$
    $\quad m = 0$ $\quad\quad 2m = 1$ $\quad\quad 2m = -1$
    $\quad m = 0$ or $\quad m = \dfrac{1}{2}$ or $\quad m = -\dfrac{1}{2}$

31. $x(x - 3) = x^2 - 10$
    $x^2 - 3x = x^2 - 10$
    $\qquad -3x = -10$
    $\qquad\quad x = \dfrac{10}{3}$

33. $(t - 4)(t + 1) = (t - 3)(t - 2)$
    $t^2 - 3t - 4 = t^2 - 5t + 6$
    $\quad -3t - 4 = -5t + 6$
    $\qquad 2t - 4 = 6$
    $\qquad\quad 2t = 10$
    $\qquad\quad t = 5$

35. $(2t - 4)(t + 1) = (t - 3)(t - 2)$
    $2t^2 - 2t - 4 = t^2 - 5t + 6$
    $\quad t^2 + 3t - 10 = 0$
    $\quad (t + 5)(t - 2) = 0$
    $t + 5 = 0 \quad$ or $\quad t - 2 = 0$
    $\qquad t = -5$ or $\qquad t = 2$

37. $\qquad\quad (x + 6)^2 = 16$
    $x^2 + 12x + 36 = 16$
    $x^2 + 12x + 20 = 0$
    $(x + 10)(x + 2) = 0$
    $x + 10 = 0 \quad$ or $\quad x + 2 = 0$
    $\qquad x = -10$ or $\qquad x = -2$

39. $\qquad\quad (3x - 4)^2 = 20 - 24x$
    $9x^2 - 24x + 16 = 20 - 24x$
    $\qquad\quad 9x^2 - 4 = 0$
    $(3x - 2)(3x + 2) = 0$
    $3x - 2 = 0 \quad$ or $\quad 3x + 2 = 0$
    $\qquad 3x = 2 \qquad\qquad 3x = -2$
    $\qquad x = \dfrac{2}{3}$ or $\qquad x = -\dfrac{2}{3}$

41. $2(x - 3)(x + 2) = 0$
    $\quad (x - 3)(x + 2) = 0$
    $x - 3 = 0 \cdot$ or $\quad x + 2 = 0$
    $\quad x = 3$ or $\qquad x = -2$

43. $2x - 3(x + 2) = 4$
    $2x - 3x - 6 = 4$
    $\quad -x - 6 = 4$
    $\qquad -x = 10$
    $\qquad\quad x = -10$

45. $\qquad\quad 5x^2 = -3x + 2$
    $5x^2 + 3x - 2 = 0$
    $(5x - 2)(x + 1) = 0$
    $5x - 2 = 0 \quad$ or $\quad x + 1 = 0$
    $\qquad 5x = 2 \qquad\qquad x = -1$
    $\qquad x = \dfrac{2}{5}$ or $\qquad x = -1$

47. $x^3 + 7x + 6 = 0$

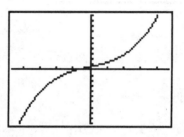

$x = -0.787$

49. $\qquad\quad 2x^3 = 6x^2$
    $2x^3 - 6x^2 = 0$
    $2x^2(x - 3) = 0$
    $2x^2 = 0 \quad$ or $\quad x - 3 = 0$
    $\quad x^2 = 0 \qquad\qquad x = 3$
    $\quad x = 0$ or $\qquad\quad x = 3$

51. $x^3 = -2x^2 - 6$
    $x^3 + 2x^2 + 6 = 0$

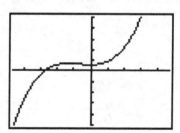

$x = -2.778$

53. $3x^2 + 2x - 5 = 2x^2 + x - 2$
   $x^2 + x - 3 = 0$

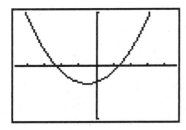

   $x = -2.303$ or $x = 1.303$

55. the number: x

   $x^2 - 5 = 1 + 5x$
   $x^2 - 5x - 6 = 0$
   $(x - 6)(x + 1) = 0$
   $x - 6 = 0$ or $x + 1 = 0$
   $x = 6$ or $x = -1$

   Since the number is positive, it is 6.

57. the number: x

   $x^2 + 8^2 = 68$
   $x^2 + 64 = 68$
   $x^2 - 4 = 0$
   $(x - 2)(x + 2) = 0$
   $x - 2 = 0$ or $x + 2 = 0$
   $x = 2$ or $x = -2$

   Since the number is positive, it is 2.

59. the number: x

   $(x + 6)^2 = 169$
   $x^2 + 12x + 36 = 169$
   $x^2 + 12x - 133 = 0$
   $(x + 19)(x - 7) = 0$
   $x + 19 = 0$ or $x - 7 = 0$
   $x = -19$ or $x = 7$

   The numbers are -19 and 7.

61. the number: x

   $$x + \frac{1}{x} = \frac{13}{6}$$

   $$6x\left(x + \frac{1}{x}\right) = 6x\left(\frac{13}{6}\right)$$

   $6x^2 + 6 = 13x$
   $6x^2 - 13x + 6 = 0$
   $(3x - 2)(2x - 3) = 0$
   $3x - 2 = 0$ or $2x - 3 = 0$
   $3x = 2$ $2x = 3$
   $x = \frac{2}{3}$ or $x = \frac{3}{2}$

   The numbers are $\frac{2}{3}$ and $\frac{3}{2}$.

63. the number: x

   $$x + 2\left(\frac{1}{x}\right) = 3$$

   $$x + \frac{2}{x} = 3$$

   $$x\left(x + \frac{2}{x}\right) = x(3)$$

   $x^2 + 2 = 3x$
   $x^2 - 3x + 2 = 0$
   $(x - 2)(x - 1) = 0$
   $x - 2 = 0$ or $x - 1 = 0$
   $x = 2$ or $x = 1$

   The numbers are 2 and 1.

65. $P = 10000(-d^2 + 12d - 35)$

   (a) When $d = 5$:
        $P = 10000\left[-5^2 + 12(5) - 35\right]$
           $= 0$

   The profit is \$0.

   (b) $10000 = 10000(-d^2 + 12d - 35)$
               $1 = -d^2 + 12d - 35$
   $d^2 - 12d + 36 = 0$
   $(d - 6)(d - 6) = 0$
        $d - 6 = 0$
           $d = 6$

   The price must be \$6.

67. $s = -16t^2 + 40$

(a) When $t = 1$:
$$s = -16(1)^2 + 40$$
$$= 24$$

He is 24 ft above the pool.

(b)
$$0 = -16t^2 + 40$$
$$16t^2 = 40$$
$$t^2 = \frac{5}{2}$$
$$t = \sqrt{\frac{5}{2}} \approx 1.6$$

He will hit the water in 1.6 seconds.

(c) When $t = 0$:
$$s = -16(0)^2 + 40$$
$$= 40$$

The board is 40 feet high.

69. width: x
length: $x + 2$

$$A = wl$$
$$80 = x(x + 2)$$
$$80 = x^2 + 2x$$
$$0 = x^2 + 2x - 80$$

$$0 = (x + 10)(x - 8)$$
$$x + 10 = 0 \quad \text{or} \quad x - 8 = 0$$
$$x = -10 \quad \text{or} \quad x = 8$$

Since x is a dimension it must be positive.

$$x = 8$$
$$x + 2 = 8 + 2 = 10$$

The dimensions are 8 feet by 10 feet.

71.

Area of $=$ area of $-$ area of
walkway   outer rectangle   inner rectangle

$$400 = (20 + 2x)(55 + 2x) - (20)(55)$$
$$400 = 1100 + 150x + 4x^2 - 1100$$
$$400 = 4x^2 + 150x$$
$$0 = 4x^2 + 150x - 400$$
$$0 = 2x^2 + 75x - 200$$
$$0 = (2x - 5)(x + 40)$$
$$2x - 5 = 0 \quad \text{or} \quad x + 40 = 0$$
$$2x = 5 \qquad\qquad x = -40$$
$$x = \frac{5}{2}$$

Since $x > 0$, the width of the walkway is 5/2 feet.

75. 1st car's hours: x
2nd car's hours: $x + 1$

$$52x = 40(x + 1)$$
$$52x = 40x + 40$$
$$12x = 40$$
$$x = \frac{10}{3}$$

It will take 10/3 hours.

77. $3x - 2y = 8$
$$-2y = -3x + 8$$
$$y = \frac{3}{2}x - 4$$
$$m = \frac{3}{2}$$

5.7 Exercises

1.
$$
\begin{array}{r}
x + 4 \\
x - 5 \overline{\smash{\big)}\ x^2 - x - 20} \\
\underline{-(x^2 - 5x)} \\
4x - 20 \\
\underline{-(4x - 20)} \\
0
\end{array}
$$

3.
$$
\begin{array}{r}
a + 2 \\
3a + 4 \overline{\smash{\big)}\ 3a^2 + 10a + 8} \\
\underline{-(3a^2 + 4a)} \\
6a + 8 \\
\underline{-(6a + 8)} \\
0
\end{array}
$$

5.
$$\begin{array}{r} 7z - 2 \\ 3z + 1 \overline{\smash{\big)}\ 21z^2 + z - 16} \\ \underline{-(21z^2 + 7z)} \\ -6z - 16 \\ \underline{-(6z - 2)} \\ -14 \end{array}$$

$$7z - 2 - \frac{14}{3z + 1}$$

7.
$$\begin{array}{r} a^2 + 2a + 3 \\ a - 1 \overline{\smash{\big)}\ a^3 + a^2 + a - 8} \\ \underline{-(a^3 - a^2)} \\ 2a^2 + a \\ \underline{-(2a^2 - 2a)} \\ 3a - 8 \\ \underline{-(3a - 3)} \\ -5 \end{array}$$

$$a^2 + 2a + 3 - \frac{5}{a - 1}$$

9.
$$\begin{array}{r} 3x^2 + x - 1 \\ x - 4 \overline{\smash{\big)}\ 3x^3 - 11x^2 - 5x + 12} \\ \underline{-(3x^3 - 12x^2)} \\ x^2 - 5x \\ \underline{-(x^2 - 4x)} \\ -x + 12 \\ \underline{-(-x + 4)} \\ 8 \end{array}$$

$$3x^2 + x - 1 + \frac{8}{x - 4}$$

11.
$$\begin{array}{r} 2z - 3 \\ 2z + 3 \overline{\smash{\big)}\ 4z^2 + 0z - 15} \\ \underline{-(4z^2 + 6z)} \\ -6z - 15 \\ \underline{-(-6z - 9)} \\ -6 \end{array}$$

$$2z - 3 - \frac{6}{2z + 3}$$

13.
$$\begin{array}{r} x^3 + 3x^2 + 2x + 1 \\ x - 2 \overline{\smash{\big)}\ x^4 + x^3 - 4x^2 - 3x - 2} \\ \underline{-(x^4 - 2x^3)} \\ 3x^3 - 4x^2 \\ \underline{-(3x^3 - 6x^2)} \\ 2x^2 - 3x \\ \underline{-(2x^2 - 4x)} \\ x - 2 \\ \underline{-(x - 2)} \\ 0 \end{array}$$

15.
$$\begin{array}{r} 3y^3 - 2y^2 + 5 \\ y - 1 \overline{\smash{\big)}\ 3y^4 - 5y^3 + 2y^2 + 5y - 10} \\ \underline{-(3y^4 - 3y^3)} \\ -2y^3 + 2y^2 \\ \underline{-(-2y^3 + 2y^2)} \\ 5y - 10 \\ \underline{-(5y - 5)} \\ -5 \end{array}$$

$$3y^3 - 2y^2 + 5 - \frac{5}{y - 1}$$

17.
$$\begin{array}{r} 2a^3 - 5 \\ a + 2 \overline{\smash{\big)}\ 2a^4 + 4a^3 - 5a + 6} \\ \underline{-(2a^4 + 4a^3)} \\ -5a + 6 \\ \underline{-(-5a - 10)} \\ 16 \end{array}$$

$$2a^3 - 5 + \frac{16}{a + 2}$$

19.
$$\begin{array}{r} y^2 - y + 1 \\ y + 1 \overline{\smash{\big)}\ y^3 + 0y^2 + 0y - 1} \\ \underline{-(y^3 + y^2)} \\ -y^2 + 0y \\ \underline{-(-y^2 - y)} \\ y - 1 \\ \underline{-(y + 1)} \\ -2 \end{array}$$

$$y^2 - y + 1 - \frac{2}{y + 1}$$

21.

$$\begin{array}{r} 4a^2 + 2a + 1 \\ 2a - 1\overline{)8a^3 + 0a^2 + 0a + 1} \\ \underline{-(8a^3 - 4a^2)} \\ 4a^2 + 0a \\ \underline{-(4a^2 - 2a)} \\ 2a + 1 \\ \underline{-(2a - 1)} \\ 2 \end{array}$$

$$4a^2 + 2a + 1 + \frac{2}{2a - 1}$$

23.

$$\begin{array}{r} 2y^3 - 3y^2 - 2y + 2 \\ y^2 + 1\overline{)2y^5 - 3y^4 + 0y^3 - y^2 + y + 4} \\ \underline{-(2y^5 \qquad + 2y^3)} \\ -3y^4 - 2y^3 - y^2 \\ \underline{-(-3y^4 \qquad - 3y^2)} \\ -2y^3 + 2y^2 + y \\ \underline{-(-2y^3 \qquad - 2y)} \\ 2y^2 + 3y + 4 \\ \underline{-(2y^2 \qquad + 2)} \\ 3y + 2 \end{array}$$

$$2y^3 - 3y^2 - 2y + 2 + \frac{3y + 2}{y^2 + 1}$$

25.

$$\begin{array}{r} 3z^3 - 6z^2 + 11z - 25 \\ z^2 + 2z - 1\overline{)3z^5 + 0z^4 - 4z^3 + 3z^2 + 12z - 10} \\ \underline{-(3z^5 + 6z^4 - 3z^3)} \\ -6z^4 - z^3 + 3z^3 \\ \underline{-(-6z^4 - 12z^3 + 6z^2)} \\ 11z^3 - 3z^2 + 12z \\ \underline{-(11z^3 + 22z^2 - 11z)} \\ -25z^2 + 23z - 10 \\ \underline{-(-25z^2 - 50z + 25)} \\ 73z - 35 \end{array}$$

$$3z^3 - 6z^2 + 11z - 25 + \frac{73z - 35}{z^2 + 2z - 1}$$

27.

$$\begin{array}{r} 7x^4 - 3 \\ x^2 - 2\overline{)7x^6 - 14x^4 - 3x^2 + x + 1} \\ \underline{-(7x^6 - 14x^4)} \\ 3x^2 + x + 1 \\ \underline{-(-3x^2 \qquad + 6)} \\ x - 5 \end{array}$$

$$7x^4 - 3 + \frac{x - 5}{x^2 - 2}$$

29. $f(x) = 5x^2 - 3x + 2$

$$\begin{aligned} f(4) &= 5(4)^2 - 3(4) + 2 \\ &= 5(16) - 3(4) + 2 \\ &= 70 \end{aligned}$$

$$\begin{aligned} f(x^2) &= 5(x^2)^2 - 3(x^2) + 2 \\ &= 5x^4 - 3x^2 + 2 \end{aligned}$$

CHAPTER 5 REVIEW EXERCISES

1. $(2x^3 - 4) - [(2x^2 - 3x + 4) + (5x^2 - 3)]$
   $= (2x^3 - 4) - (2x^2 - 3x + 4 + 5x^2 - 3)$
   $= (2x^3 - 4) - (7x^2 - 3x + 1)$
   $= 2x^3 - 4 - 7x^2 + 3x - 1$
   $= 2x^3 - 7x^2 + 3x - 5$

3. $3x^2(2xy - 3y + 1)$
   $= 6x^3y - 9x^2y + 3x^2$

5. $3ab(2a - 3b) - 2a(3ab - 4b^2)$
   $= 6a^2b - 9ab^2 - 6a^2b + 8ab^2$
   $= -ab^2$

7. $3x - 2(x - 3) - [x - 2(5 - x)]$
   $= 3x - 2x + 6 - (x - 10 + 2x)$
   $= x + 6 - (3x - 10)$
   $= x + 6 - 3x + 10$
   $= -2x + 16$

9. $(x - 3)(x + 2)$
   $= x^2 + (2 - 3)x - 6$
   $= x^2 - x - 6$

11. $(3y - 2)(2y - 1)$
   $= 6y^2 + (-3 - 4)y + 2$
   $= 6y^2 - 7y + 2$

13. $(3x - 4y)^2$
   $= (3x)^2 - 2(3x)(4y) + (4y)^2$
   $= 9x^2 - 24xy + 16y^2$

15. $(4x^2 - 5y)^2$
$= (4x^2)^2 - 2(4x^2)(5y) + (5y)^2$
$= 16x^4 - 40x^2y + 25y^2$

17. $(7x^2 - 5y^3)(7x^2 + 5y^3)$
$= (7x^2)^2 - (5y^3)^2$
$= 49x^4 - 25y^6$

19. $(5y^2 - 3y + 7)(2y - 3)$
$= 10y^3 - 6y^2 + 14y - 15y^2 + 9y - 21$
$= 10y^3 - 21y^2 + 23y - 21$

21. $(2x + 3y + 4)(2x + 3y - 5)$
$= [(2x + 3y) + 4][(2x + 3y) - 5]$
$= (2x + 3y)^2 + (-5 + 4)(2x + 3y) - 20$
$= (2x)^2 + 2(2x)(3y) + (3y)^2 - (2x + 3y) - 20$
$= 4x^2 + 12xy + 9y^2 - 2x - 3y - 20$

23. $[(x - y) + 5]^2$
$= (x - y)^2 + 2(x - y)(5) + 5^2$
$= x^2 - 2xy + y^2 + 10x - 10y + 25$

25. $(a + b - 4)^2$
$= (a + b - 4)(a + b - 4)$
$= a^2 + ab - 4a + ab + b^2 - 4b$
$\quad - 4a - 4b + 16$
$= a^2 + 2ab + b^2 - 8a - 8b + 16$

27. $6x^2y - 12xy^2 + 9xy = 3xy(2x - 4y + 3)$

29. $3x(a + b) - 2(a + b) = (a + b)(3x - 2)$

31. $2(a - b)^2 + 3(a - b)$
$= (a - b)[2(a - b) + 3]$
$= (a - b)(2a - 2b + 3)$

33. $5ax - 5a + 3bx - 3b$
$= 5a(x - 1) + 3b(x - 1)$
$= (x - 1)(5a + 3b)$

35. $5y^2 - 5y + 3y - 3 = 5y(y - 1) + 3(y - 1)$
$\qquad = (y - 1)(5y + 3)$

37. $a^2 + a(a - 10) = a^2 + a^2 - 10a$
$\qquad = 2a^2 - 10a$
$\qquad = 2a(a - 5)$

39. $2t(t + 2) - (t - 2)(t + 4)$
$= 2t^2 + 4t - (t^2 + 2t - 8)$
$= 2t^2 + 4t - t^2 - 2t + 8$
$= t^2 + 2t + 8$
not factorable

41. Factors of -35 whose sum is -2: $-7, 5$

$x^2 - 2x - 35 = (x - 7)(x + 5)$

43. Factors of -14 whose sum is 5: $-2, 7$

$a^2 + 5ab - 14b^2 = (a - 2b)(a + 7b)$

45. Factors of $35 \cdot 2 = 70$ whose sum
is 17: $7, 10$

$35a^2 + 17ab + 2b^2$
$= 35a^2 + 7ab + 10ab + 2b^2$
$= 7a(5a + b) + 2b(5a + b)$
$= (5a + b)(7a + 2b)$

47. $a^2 - 6ab^2 + 9b^4$
$= (a)^2 - 2(a)(3b^2) + (3b^2)^2$
$= (a - 3b^2)^2$

49. $3a^3 - 21a^2 + 30a = 3a(a^2 - 7a + 10)$
$\qquad = 3a(a - 5)(a - 2)$

51. $2x^3 - 50xy^2 = 2x(x^2 - 25y^2)$
$\qquad = 2x[(x)^2 - (5y)^2]$
$\qquad = 2x(x - 5y)(x + 5y)$

53. $6x^2 + 5x - 6$
$= 6x^2 + 9x - 4x - 6$
$= 3x(2x + 3) - 2(2x + 3)$
$= (2x + 3)(3x - 2)$

55. $8x^3 + 125y^3$
$= (2x)^3 + (5y)^3$
$= (2x + 5y)[(2x)^2 - (2x)(5y) + (5y)^2]$
$= (2x + 5y)(4x^2 - 10xy + 25y^2)$

57. $6a^2 - 17ab - 3b^2$
$= 6a^2 - 18ab + ab - 3b^2$
$= 6a(a - 3b) + b(a - 3b)$
$= (a - 3b)(6a + b)$

59. $21a^4 + 41a^2b^2 + 10b^4$
$= 21a^4 + 6a^2b^2 + 35a^2b^2 + 10b^4$
$= 3a^2(7a^2 + 2b^2) + 5b^2(7a^2 + 2b^2)$
$= (7a^2 + 2b^2)(3a^2 + 5b^2)$

61. $25x^4 - 40x^2y^2 + 16y^4$
$= (5x^2)^2 - 2(5x^2)(4y^2) + (4y^2)^2$
$= (5x^2 - 4y^2)^2$

63. $20x^3y - 60x^2y^2 + 45xy^3$
$= 5xy(4x^2 - 12xy + 9y^2)$
$= 5xy[(2x)^2 - 2(2x)(3y) + (3y)^2]$
$= 5xy(2x - 3y)^2$

65. $6a^4b - 8a^3b^2 - 8a^2b^3$
$= 2a^2b(3a^2 - 4ab - 4b^2)$
$= 2a^2b(3a^2 - 6ab + 2ab - 4b^2)$
$= 2a^2b[3a(a - 2b) + 2b(a - 2b)]$
$= 2a^2b(a - 2b)(3a + 2b)$

67. $6x^5 - 10x^3 - 4x$
$= 2x(3x^4 - 5x^2 - 2)$
$= 2x(3x^4 - 6x^2 + x^2 - 2)$
$= 2x[3x^2(x^2 - 2) + 1(x^2 - 2)]$
$= 2x(x^2 - 2)(3x^2 + 1)$

69. $(a - b)^2 - 4 = (a - b)^2 - 2^2$
$\qquad\qquad = (a - b - 2)(a - b + 2)$

71. $9y^2 + 30y + 25 - 9x^2$
$= [(3y)^2 + 2(3y)(5) + 5^2] - (3x)^2$
$= (3y + 5)^2 - (3x)^2$
$= (3y + 5 - 3x)(3y + 5 + 3x)$

73.
$$
\begin{array}{r}
3x^2 + 2x + 11 \\
x - 2\overline{)3x^3 - 4x^2 + 7x - 5} \\
\underline{-(3x^3 - 6x^2)} \\
2x^2 + 7x \\
\underline{-(2x^2 - 4x)} \\
11x - 5 \\
\underline{-(11x - 22)} \\
17
\end{array}
$$

$3x^2 + 2x + 11 + \dfrac{17}{x - 2}$

75.
$$
\begin{array}{r}
4a^2 + 6a + 9 \\
2a - 3\overline{)8a^3 + 0a^2 + 0a - 27} \\
\underline{-(8a^3 - 12a^2)} \\
12a^2 + 0a \\
\underline{-(12a^2 - 18a)} \\
18a - 27 \\
\underline{-(18a - 27)} \\
0
\end{array}
$$

77. $x(x + 5)(3x - 4) = 0$
$x = 0$ or $x + 5 = 0$ or $3x - 4 = 0$
$x = 0 \qquad\qquad x = -5 \qquad\qquad 3x = 4$
$x = 0$ or $\qquad x = -5$ or $\qquad x = \dfrac{4}{3}$

79. $\qquad\qquad\qquad x^4 = 25x^2$
$\qquad\qquad x^4 - 25x^2 = 0$
$\qquad\qquad x^2(x^2 - 25) = 0$
$\qquad x^2(x - 5)(x + 5) = 0$
$x^2 = 0$ or $x - 5 = 0$ or $x + 5 = 0$
$\quad x = 0$ or $\qquad x = 5$ or $\qquad x = -5$

81. $(x + 6)(x - 3) = (4x - 2)(x + 4)$
$\quad x^2 + 3x - 18 = 4x^2 + 14x - 8$
$\qquad\qquad\qquad 0 = 3x^2 + 11x + 10$
$\qquad\qquad\qquad 0 = (3x + 5)(x + 2)$
$3x + 5 = 0 \qquad$ or $\quad x + 2 = 0$
$\quad 3x = -5 \qquad\qquad\qquad x = -2$
$\quad x = -\dfrac{5}{3}$ or $\qquad x = -2$

83. $\qquad\qquad 2x^3 = 6 - 5x$
$2x^3 + 5x - 6 = 0$

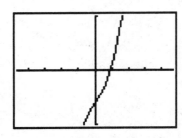

$x = 0.904$

1. $(9a^2 + 6ab + 4b^2)(3a - 2b)$
    $= 27a^3 + 18a^2b + 12ab^2 - 18a^2b$
    $\quad - 12ab^2 - 8b^3$
    $= 27a^3 - 8b^3$

2. $(x - 3)(x + 3) - (x - 3)^2$
    $= x^2 - 9 - (x^2 - 6x + 9)$
    $= x^2 - 9 - x^2 + 6x - 9$
    $= 6x - 18$

3. $(x - y - 2)(x - y + 2)$
    $= [(x - y) - 2][(x - y) + 2]$
    $= (x - y)^2 - 2^2$
    $= x^2 - 2xy + y^2 - 4$

4. $(3x + y^2)^2 = (3x)^2 + 2(3x)(y^2) + (y^2)^2$
    $\qquad\qquad = 9x^2 + 6xy^2 + y^4$

5. $10x^3y^2 - 6x^2y^3 + 2xy$
    $= 2xy(5x^2y - 3xy^2 + 1)$

6. $6ax + 15a - 2bx - 5b$
    $= 3a(2x + 5) - b(2x + 5)$
    $= (2x + 5)(3a - b)$

7. $x^2 + 11xy - 60y^2 = (x + 15y)(x - 4y)$

8. $10x^2 - 11x - 6 = 10x^2 - 15x + 4x - 6$
    $\qquad\qquad = 5x(2x - 3) + 2(2x - 3)$
    $\qquad\qquad = (2x - 3)(5x + 2)$

9. Factors of $2 \cdot 7 = 14$ whose sum is
    -3: none
    not factorable

10. $3a^3b - 3ab^3 = 3ab(a^2 - b^2)$
    $\qquad\qquad = 3ab(a - b)(a + b)$

11. $r^3s - 10r^2s^2 + 25rs^3$
    $= rs(r^2 - 10rs + 25s^2)$
    $= rs[(r)^2 - 2(r)(5s) + (5s)^2]$
    $= rs(r - 5s)^2$

12. $8a^3 - 1 = (2a)^3 - 1^3$
    $\qquad\quad = (2a - 1)[(2a)^2 + (2a)(1) + 1^2]$
    $\qquad\quad = (2a - 1)(4a^2 + 2a + 1)$

13. $(2x + y)^2 - 64$
    $= (2x + y)^2 - 8^2$
    $= [(2x + y) - 8][(2x + y) + 8]$
    $= (2x + y - 8)(2x + y + 8)$

14. $a^3 + 4a^2 - a - 4$
    $= a^2(a + 4) - 1(a + 4)$
    $= (a + 4)(a^2 - 1)$
    $= (a + 4)(a - 1)(a + 1)$

15.
$$\begin{array}{r} x^2 + 3x + 16 \\ x - 3 \overline{)\, x^3 + 0x^2 + 7x - 8} \\ \underline{-(x^3 - 3x^2)} \\ 3x^2 + 7x \\ \underline{-(3x^2 + 9x)} \\ 16x - 8 \\ \underline{-(16x - 48)} \\ 40 \end{array}$$

$$x^2 + 3x + 16 + \frac{40}{x - 3}$$

16. (a) $\quad 2x^3 + 8x^2 - 24x = 0$
    $\qquad 2x(x^2 + 4x - 12) = 0$
    $\qquad 2x(x + 6)(x - 2) = 0$
    $\qquad 2x = 0 \ \text{ or } \ x + 6 = 0 \ \text{ or } \ x - 2 = 0$
    $\qquad\ x = 0 \ \text{ or } \qquad x = -6 \ \text{ or } \qquad x = 2$

    (b) $\quad (2x - 7)(x - 4) = (4x - 19)(x - 2)$
    $\qquad 2x^2 - 15x + 28 = 4x^2 - 27x + 38$
    $\qquad\qquad\qquad 0 = 2x^2 - 12x + 10$
    $\qquad\qquad\qquad 0 = x^2 - 6x + 5$
    $\qquad\qquad\qquad 0 = (x - 5)(x - 1)$
    $\qquad x - 5 = 0 \ \text{ or } \ x - 1 = 0$
    $\qquad\quad x = 5 \ \text{ or } \qquad x = 1$

6.1 Exercises

1. $x - 2 \neq 0$
   $x \neq 2$

3. x and y can be any number

5. $5x \neq 0$
   $x \neq 0$

7. $3x - y = 0$
   $3x \neq y$

9. $\dfrac{3x^2 y}{15xy} = \dfrac{3x \cdot x \cdot y}{3 \cdot 5 \cdot x \cdot y}$

   $= \dfrac{x}{5}$

11. $\dfrac{16x^2 y^3 a}{18xy^4 a^3} = \dfrac{2 \cdot 8 \cdot x \cdot x \cdot y^3 \cdot a}{2 \cdot 9 \cdot x \cdot y^3 \cdot y \cdot a \cdot a^2}$

    $= \dfrac{8x}{9ya^2}$

13. $\dfrac{(x - 5)(x + 4)}{(x - 3)(x + 4)} = \dfrac{x - 5}{x - 3}$

15. $\dfrac{(2x - 3)(x - 5)}{(2x + 3)(x + 3)}$ cannot be reduced.

17. $\dfrac{2x - 3)(x - 5)}{(2x - 3)(5 - x)} = \dfrac{(2x - 3)(-1)(5 - x)}{(2x - 3)(5 - x)}$

    $= -1$

19. $\dfrac{3x^2 - 3x}{6x^2 + 18x} = \dfrac{3x(x - 1)}{6x(x + 3)}$

    $= \dfrac{x - 1}{2(x + 3)}$

21. $\dfrac{x^2 - 9}{x^2 - 6x + 9} = \dfrac{(x - 3)(x + 3)}{(x - 3)(x - 3)}$

    $= \dfrac{x + 3}{x - 3}$

23. $\dfrac{w^2 - 8wz + 7z^2}{w^2 + 8wz + 7z^2} = \dfrac{(w - 7z)(w - z)}{(w + 7z)(w + z)}$

    cannot be reduced

25. $\dfrac{x^2 + x - 12}{x - 3} = \dfrac{(x + 4)(x - 3)}{x - 3}$

    $= x + 4$

27. $\dfrac{x^2 + x - 12}{3 - x} = \dfrac{(x + 4)(x - 3)}{-(x - 3)}$

    $= \dfrac{x + 4}{-1}$

    $= -(x + 4)$

29. $\dfrac{3a^2 - 13a - 30}{15a^2 + 28a + 5} = \dfrac{(3a + 5)(a - 6)}{(3a + 5)(5a + 1)}$

    $= \dfrac{a - 6}{5a + 1}$

31. $\dfrac{2x^3 - 2x^2 - 12x}{3x^2 - 6x} = \dfrac{2x(x - 3)(x + 2)}{3x(x - 2)}$

    $= \dfrac{2(x - 3)(x + 2)}{3(x - 2)}$

33. $\dfrac{2x^2}{3x^2 - 6} = \dfrac{2x^2}{3(x^2 - 2)}$
    cannot be reduced

35. $\dfrac{3a^2 - 7a - 6}{3 + 5a - 2a^2} = \dfrac{3a^2 - 7a - 6}{-(2a^2 - 5a - 3)}$

    $= \dfrac{(3a + 2)(a - 3)}{-(2a + 1)(a - 3)}$

    $= -\dfrac{3a + 2}{2a + 1}$

37. $\dfrac{6r^2 - r - 2}{2 + r - 6r^2} = \dfrac{6r^2 - r - 2}{-(6r^2 - r - 2)}$

    $= \dfrac{1}{-1}$

    $= -1$

39. $\dfrac{a^3 + b^3}{(a + b)^3} = \dfrac{(a + b)(a^2 - ab + b^2)}{(a + b)^3}$

$\qquad\qquad = \dfrac{a^2 - ab + b^2}{(a + b)^2}$

41. $\dfrac{ax + bx - 2ay - 2by}{4xy - 2x^2} = \dfrac{(a + b)(x - 2y)}{-2x(-2y + x)}$

$\qquad\qquad\qquad = \dfrac{(a + b)(x - 2y)}{-2x(x - 2y)}$

$\qquad\qquad\qquad = -\dfrac{a + b}{2x}$

43. $\dfrac{(a + b)^2 - (x + y)^2}{a + b + x + y} = \dfrac{(a + b - x - y)(a + b + x + y)}{a + b + x + y}$

$\qquad\qquad\qquad = a + b - x - y$

45. $\dfrac{2x}{3y} = \dfrac{2x(3x^2)}{3y(3x^2)}$

$\qquad = \dfrac{6x^3}{9x^2y}$

47. $\dfrac{5}{3a^2b} = \dfrac{5(5a^2b)}{3a^2b(5a^2b)}$

$\qquad = \dfrac{25a^2b}{15a^4b^2}$

49. $\dfrac{3x}{x - 5} = \dfrac{3x(x + 5)}{(x - 5)(x + 5)}$

$\qquad = \dfrac{3x^2 + 15x}{(x - 5)(x + 5)}$

51. $\dfrac{a - b}{x + y} = \dfrac{?}{x^2 + 2xy + y^2}$

$\dfrac{a - b}{x + y} = \dfrac{?}{(x + y)^2}$

$\dfrac{a - b}{x + y} = \dfrac{(a - b)(x + y)}{(x + y)(x + y)}$

53. $\dfrac{y - 2}{2y - 3} = \dfrac{?}{12 - 5y - 2y^2}$

$\dfrac{y - 2}{2y - 3} = \dfrac{?}{-(2y - 3)(y + 4)}$

$\dfrac{y - 2}{2y - 3} = \dfrac{(y - 2)(-1)(y + 4)}{(2y - 3)(-1)(y + 4)}$

$\dfrac{y - 2}{2y - 3} = \dfrac{-y^2 - 2y + 8}{12 - 5y - 2y^2}$

55. $\dfrac{x - y}{a^2 - b^2} = \dfrac{?}{a^2x + a^2y - b^2x - b^2y}$

$\dfrac{x - y}{a^2 - b^2} = \dfrac{?}{(a^2 - b^2)(x + y)}$

$\dfrac{x - y}{a^2 - b^2} = \dfrac{(x - y)(x + y)}{(a^2 - b^2)(x + y)}$

59. $10(x + 24) - 6x = 400$
$\quad 10x + 240 - 6x = 400$
$\qquad\quad 4x + 240 = 400$
$\qquad\qquad\quad 4x = 160$
$\qquad\qquad\quad\; x = 40$

61. $f(x) = 2x^3 - x^2 + 3$
$f(x^2) = 2(x^2)^3 - (x^2)^2 + 3$
$\qquad\; = 2x^6 - x^4 + 3$

6.2 Exercises

1. $\dfrac{10a^2b}{3xy^3} \cdot \dfrac{9xy^4}{5a^4b^7}$

$= \dfrac{2 \cdot 5 \cdot a^2 \cdot b \cdot 3 \cdot 3 \cdot x \cdot y^3 \cdot y}{3xy^3 \cdot 5a^2 \cdot a^2 \cdot b \cdot b^6}$

$= \dfrac{2 \cdot 3 \cdot y}{a^2b^6}$

$= \dfrac{6y}{a^2b^6}$

3. $\dfrac{17a^2b^3}{18yx} \div \dfrac{34a^2}{9xy} = \dfrac{17a^2b^3}{18yx} \cdot \dfrac{9xy}{34a^2}$

$\qquad = \dfrac{17a^2b^3 \cdot 9xy}{2 \cdot 9yx \cdot 17 \cdot 2a^2}$

$\qquad = \dfrac{b^3}{2 \cdot 2}$

$\qquad = \dfrac{b^3}{4}$

$= \dfrac{32r^2s^3}{12a^2b} \cdot \dfrac{2}{9bs^2}$

$= \dfrac{4 \cdot 8r^2 \cdot s^2 \cdot s \cdot 2}{4 \cdot 3a^2 \, b \cdot 9bs^2}$

$= \dfrac{16r^2s}{27a^2b^2}$

5. $\dfrac{16x^3b}{9a} \cdot 24a^2b^3 = \dfrac{16x^3b}{9a} \cdot \dfrac{24a^2b^3}{1}$

$\qquad = \dfrac{16x^3b \cdot 3 \cdot 8 \cdot a \cdot a \cdot b^3}{3 \cdot 3a}$

$\qquad = \dfrac{16x^3b \cdot 8 \cdot ab^3}{3}$

$\qquad = \dfrac{128x^3ab^4}{3}$

11. $\dfrac{x^2 - x - 2}{x + 3} \div \dfrac{3x + 9}{2x + 2}$

$= \dfrac{x^2 - x - 2}{x + 3} \cdot \dfrac{2x + 2}{3x + 9}$

$= \dfrac{(x - 2)(x + 1)}{x + 3} \cdot \dfrac{2(x + 1)}{3(x + 3)}$

$= \dfrac{2(x + 1)^2(x - 2)}{3(x + 3)^2}$

7. $\dfrac{32r^2s^3}{12a^2b} \cdot \left(\dfrac{15ab^2}{16r} \cdot \dfrac{24rs^2}{5ab} \right)$

$= \dfrac{32r^2s^3}{12a^2b} \cdot \left(\dfrac{3 \cdot 5ab \cdot b \cdot 8 \cdot 3rs^2}{8 \cdot 2r \cdot 5ab} \right)$

$= \dfrac{32r^2s^3}{12a^2b} \cdot \dfrac{9bs^2}{2}$

$= \dfrac{4 \cdot 2 \cdot 4r^2s^3 \cdot 3 \cdot 3bs^2}{4 \cdot 3a^2b \cdot 2}$

$= \dfrac{12r^2s^5}{a^2}$

13. $\dfrac{x^2 - x - 2}{x + 3} \cdot \dfrac{3x + 9}{2x + 2}$

$= \dfrac{(x - 2)(x + 1)}{x + 3} \cdot \dfrac{3(x + 3)}{2(x + 1)}$

$= \dfrac{3(x - 2)}{2}$

15. $\dfrac{5a^3 - 5a^2b}{3a^2 + 3ab} \cdot (a + b)$

$= \dfrac{5a^2(a - b)}{3a(a + b)} \cdot \dfrac{(a + b)}{1}$

$= \dfrac{5a(a - b)}{3}$

9. $\dfrac{32r^2s^3}{12a^2b} \div \left(\dfrac{15ab^2}{16r} \cdot \dfrac{24rs^2}{5ab} \right)$

$= \dfrac{32r^2s^3}{12a^2b} \div \left(\dfrac{5 \cdot 3a \cdot b \cdot b \cdot 8 \cdot 3rs^2}{8 \cdot 2r \cdot 5ab} \right)$

$= \dfrac{32r^2s^3}{12a^2b} \div \dfrac{9bs^2}{2}$

17. $\dfrac{5a^3 - 5a^2b}{3a^2 + 3ab} \div (a + b)$

$= \dfrac{5a^3 - 5a^2b}{3a^2 + 3ab} \cdot \dfrac{1}{a + b}$

$= \dfrac{5a^2(a - b)}{3a(a + b)} \cdot \dfrac{1}{a + b}$

$= \dfrac{5a(a - b)}{3(a + b)^2}$

19. $\dfrac{2x^2 + 3x - 5}{2x + 5} \cdot \dfrac{1}{1 - x}$

$= \dfrac{(2x^2 + 5)(x - 1)}{2x + 5} \cdot \dfrac{1}{-1(x - 1)}$

$= \dfrac{2}{-1}$

$= -1$

21. $\dfrac{2a^2 - 7a + 6}{4a^2 - 9} \cdot \dfrac{4a^2 + 12a + 9}{a^2 - a - 2}$

$= \dfrac{(2a - 3)(a - 2)}{(2a - 3)(2a + 3)} \cdot \dfrac{(2a + 3)(2a + 3)}{(a - 2)(a + 1)}$

$= \dfrac{2a + 3}{a + 1}$

23. $\dfrac{x^2 - y^2}{(x + y)^3} \cdot \dfrac{(x + y)^2}{(x - y)^2}$

$= \dfrac{(x + y)(x - y)}{(x + y)^3} \cdot \dfrac{(x + y)^2}{(x - y)^2}$

$= \dfrac{(x + y)^3(x - y)}{(x + y)^3(x - y)^2}$

$= \dfrac{1}{x - y}$

25. $\dfrac{9x^2 + 3x - 2}{6x^2 - 2x} \div \dfrac{3x + 2}{6x^2}$

$= \dfrac{9x^2 + 3x - 2}{6x^2 - 2x} \cdot \dfrac{6x^2}{3x + 2}$

$= \dfrac{(3x + 2)(3x - 1)}{2x(3x - 1)} \cdot \dfrac{6x^2}{3x + 2}$

$= 3x$

27. $\dfrac{2x^2 + x - 3}{x^2 - 1} \div \dfrac{2x^2 + 5x + 3}{2 - 2x}$

$= \dfrac{2x^2 + x - 3}{x^2 - 1} \cdot \dfrac{2 - 2x}{2x^2 + 5x + 3}$

$= \dfrac{(2x + 3)(x - 1)}{(x + 1)(x - 1)} \cdot \dfrac{-2(x - 1)}{(2x + 3)(x + 1)}$

$= \dfrac{-2(x - 1)}{(x + 1)^2}$ or $\dfrac{-2x + 2}{(x + 1)^2}$

29. $\dfrac{6q^2 - q - 2}{8q^2 + 4q} \cdot \dfrac{8q^2}{6q^2 - 4q}$

$= \dfrac{(3q - 2)(2q + 1)}{4q(2q + 1)} \cdot \dfrac{8q^2}{2q(3q - 2)}$

$= \dfrac{8q^2(3q - 2)(2q + 1)}{8q^2(3q - 2)(2q + 1)}$

$= 1$

31. $\dfrac{m^2 - 10m + 25}{m^2 - 25} \div (5m^2 - 25m)$

$= \dfrac{m^2 - 10m + 25}{m^2 - 25} \cdot \dfrac{1}{5m^2 - 25m}$

$= \dfrac{(m - 5)(m - 5)}{(m - 5)(m + 5)} \cdot \dfrac{1}{5m(m - 5)}$

$= \dfrac{1}{5m(m + 5)}$

33. $\left(\dfrac{2t^2 + 3t + 1}{3t^2 - t - 2}\right) \cdot \left(\dfrac{t - 1}{2t - 1}\right) \cdot \left(\dfrac{3t + 2}{t + 1}\right)$

$= \dfrac{(2t + 1)(t + 1)}{(3t + 2)(t - 1)} \cdot \dfrac{(t - 1)}{(2t - 1)} \cdot \dfrac{(3t + 2)}{(t + 1)}$

$= \dfrac{2t + 1}{2t - 1}$

35. $\dfrac{9a^2 + 9a + 2}{3a^2 - 2a - 1} \cdot \left(\dfrac{a - 1}{3a^2 + 4a + 1} \div \dfrac{3a + 2}{a + 1}\right)$

$= \dfrac{9a^2 + 9a + 2}{3a^2 - 2a - 1} \cdot \left(\dfrac{a - 1}{3a^2 + 4a + 1} \cdot \dfrac{a + 1}{3a + 2}\right)$

$= \dfrac{(3a + 2)(3a + 1)}{(3a + 1)(a - 1)} \cdot \left(\dfrac{a - 1}{(3a + 1)(a + 1)} \cdot \dfrac{a + 1}{3a + 2}\right)$

$= \dfrac{(3a + 2)}{(a - 1)} \cdot \dfrac{(a - 1)}{(3a + 1)(3a + 2)}$

$= \dfrac{1}{3a + 1}$

37. $\dfrac{2r^2 - 5r - 3}{6r - 2} \div \left(\dfrac{r - 3}{2} \div \dfrac{3r - 1}{2r + 1} \right)$

$= \dfrac{2r^2 - 5r - 3}{6r - 2} \div \left(\dfrac{r - 3}{2} \cdot \dfrac{2r + 1}{3r - 1} \right)$

$= \dfrac{(2r + 1)(r - 3)}{2(3r - 1)} \div \left[\dfrac{(r - 3)(2r + 1)}{2(3r - 1)} \right]$

$= \dfrac{(2r + 1)(r - 3)}{2(3r - 1)} \cdot \dfrac{2(3r - 1)}{(r - 3)(2r + 1)}$

$= 1$

39. $\dfrac{x^2 - y^2}{y^3 x - x^3 y} \div (x^2 + 2xy + y^2)$

$= \dfrac{x^2 - y^2}{y^3 x - x^3 y} \cdot \dfrac{1}{x^2 + 2xy + y^2}$

$= \dfrac{(x - y)(x + y)}{-xy(x - y)(x + y)} \cdot \dfrac{1}{(x + y)^2}$

$= -\dfrac{1}{xy(x + y)^2}$

41. $f(x) \cdot g(x) = \dfrac{x + 2}{x - 3} \cdot 2x^2 - 2x - 12$

$= \dfrac{x + 2}{x - 3} \cdot \dfrac{2(x - 3)(x + 2)}{1}$

$= 2(x + 2)^2$

43. $f(x) \div g(x) = \dfrac{x + 2}{x - 3} \div 2x^2 - 2x - 12$

$= \dfrac{x + 2}{x - 3} \cdot \dfrac{1}{2x^2 - 2x - 12}$

$= \dfrac{x + 2}{x - 3} \cdot \dfrac{1}{2(x - 3)(x + 2)}$

$= \dfrac{1}{2(x - 3)^2}$

45. $h(x) \div f(x) \cdot g(x)$

$= \dfrac{1}{3x^2 + 6x} \div \dfrac{x + 2}{x - 3} \cdot 2x^2 - 2x - 12$

$= \dfrac{1}{3x^2 + 6x} \cdot \dfrac{x - 3}{x + 2} \cdot \dfrac{2x^2 - 2x - 12}{1}$

$= \dfrac{1}{3x(x + 2)} \cdot \dfrac{x - 3}{x + 2} \cdot \dfrac{2(x - 3)(x + 2)}{1}$

$= \dfrac{2(x - 3)^2}{3x(x + 2)}$

47. $\begin{cases} 6x - 2y = 7 \\ 16x + 3y = 2 \end{cases}$

Multiply the 1st equation by 3 and the 2nd equation by 2, then add:

$\begin{aligned} 18x - 6y &= 21 \\ \underline{32x + 6y} &= \underline{4} \\ 50x &= 25 \\ x &= \dfrac{1}{2} \end{aligned}$

$6\left(\dfrac{1}{2}\right) - 2y = 7$

$3 - 2y = 7$

$-2y = 4$

$y = -2$

$x = \dfrac{1}{2}, \quad y = -2$

49. degree: 3

6.3 Exercises

1. $\dfrac{x + 2}{x} - \dfrac{2 - x}{x} = \dfrac{x + 2 - (2 - x)}{x}$

$= \dfrac{x + 2 - 2 + x}{x}$

$= \dfrac{2x}{x}$

$= 2$

3. $\dfrac{10a}{a - b} - \dfrac{10b}{a - b} = \dfrac{10a - 10b}{a - b}$

$= \dfrac{10(a - b)}{a - b}$

$= 10$

5. $\dfrac{3a}{3a^2 + a - 2} - \dfrac{2}{3a^2 + a - 2}$

$= \dfrac{3a - 2}{3a^2 + a - 2}$

$= \dfrac{3a - 2}{(3a - 2)(a + 1)}$

$= \dfrac{1}{a + 1}$

7. $\dfrac{x^2}{x^2 - y^2} + \dfrac{y^2}{x^2 - y^2} - \dfrac{2xy}{x^2 - y^2}$

$= \dfrac{x^2 + y^2 - 2xy}{x^2 - y^2}$

$= \dfrac{x^2 - 2xy + y^2}{x^2 - y^2}$

$= \dfrac{(x - y)(x - y)}{(x - y)(x + y)}$

$= \dfrac{x - y}{x + y}$

9. $\dfrac{5x}{x + 3} + \dfrac{3x}{x + 3} + 4$

$= \dfrac{5x}{x + 3} + \dfrac{3x}{x + 3} + \dfrac{4(x + 3)}{x + 3}$

$= \dfrac{5x + 3x + 4(x + 3)}{x + 3}$

$= \dfrac{5x + 3x + 4x + 12}{x + 3}$

$= \dfrac{12x + 12}{x + 3}$

11. $\dfrac{3x}{2y^2} - \dfrac{4x^2}{9y} = \dfrac{3x(9)}{18y^2} - \dfrac{4x^2(2y)}{18y^2}$

$= \dfrac{3x(9) - 4x^2(2y)}{18y^2}$

$= \dfrac{27x - 8x^2y}{18y^2}$

13. $\dfrac{7y}{6x^2} - \dfrac{8x^2}{9y^2} = \dfrac{7y(3y^2)}{18x^2y^2} - \dfrac{8x^2(2x^2)}{18x^2y^2}$

$= \dfrac{7y(3y^2) - 8x^2(2x^2)}{18x^2y^2}$

$= \dfrac{21y^3 - 16x^4}{18x^2y^2}$

15. $\dfrac{7y}{6x^2} \cdot \dfrac{8x^2}{9y^2} = \dfrac{7y \cdot 2 \cdot 4x^2}{2 \cdot 3x^2 \cdot 9y \cdot y}$

$= \dfrac{28}{27y}$

17. $\dfrac{36a^2}{b^2c} + \dfrac{24}{bc^3} - \dfrac{3}{7bc}$

$= \dfrac{36a^2(7c^2)}{7b^2c^3} + \dfrac{24(7b)}{7b^2c^3} - \dfrac{3(bc^2)}{7b^2c^3}$

$= \dfrac{36a^2(7c^2) + 24(7b) - 3(bc^2)}{7b^2c^3}$

$= \dfrac{252a^2c^2 + 168b - 3bc^2}{7b^2c^3}$

19. $\dfrac{3}{x} + \dfrac{x}{x + 2}$

$= \dfrac{3(x + 2)}{x(x + 2)} + \dfrac{x(x)}{x(x + 2)}$

$= \dfrac{3(x + 2) + x(x)}{x(x + 2)}$

$= \dfrac{3x + 6 + x^2}{x(x + 2)}$

$= \dfrac{x^2 + 3x + 6}{x(x + 2)}$

21. $\dfrac{a}{a - b} - \dfrac{b}{a}$

$= \dfrac{a(a)}{a(a - b)} - \dfrac{b(a - b)}{a(a - b)}$

$= \dfrac{a(a) - b(a - b)}{a(a - b)}$

$= \dfrac{a^2 - ab + b^2}{a(a - b)}$

23. $\dfrac{5}{x+7} - \dfrac{2}{x-3}$

$= \dfrac{5(x-3)}{(x+7)(x-3)} - \dfrac{2(x+7)}{(x+7)(x-3)}$

$= \dfrac{5(x-3) - 2(x+7)}{(x+7)(x-3)}$

$= \dfrac{5x - 15 - 2x - 14}{(x+7)(x-3)}$

$= \dfrac{3x - 29}{(x+7)(x-3)}$

25. $\dfrac{2}{x-7} + \dfrac{3x+1}{x+2}$

$= \dfrac{2(x+2)}{(x-7)(x+2)} + \dfrac{(3x+1)(x-7)}{(x-7)(x+2)}$

$= \dfrac{2(x+2) + (3x+1)(x-7)}{(x-7)(x+2)}$

$= \dfrac{2x + 4 + 3x^2 - 20x - 7}{(x-7)(x+2)}$

$= \dfrac{3x^2 - 18x - 3}{(x-7)(x+2)}$

27. $\dfrac{2r+s}{r-s} - \dfrac{r-2s}{r+s}$

$= \dfrac{(2r+s)(r+s)}{(r-s)(r+s)} - \dfrac{(r-2s)(r-s)}{(r-s)(r+s)}$

$= \dfrac{(2r+s)(r+s) - (r-2s)(r-s)}{(r-s)(r+s)}$

$= \dfrac{2r^2 + 3rs + s^2 - (r^2 - 3rs + 2s^2)}{(r-s)(r+s)}$

$= \dfrac{2r^2 + 3rs + s^2 - r^2 + 3rs - 2s^2}{(r-s)(r+s)}$

$= \dfrac{r^2 + 6rs - s^2}{(r-s)(r+s)}$

29. $\dfrac{4}{x-4} + \dfrac{x}{4-x}$

$= \dfrac{4}{x-4} + \dfrac{x}{(-1)(x-4)}$

$= \dfrac{4}{x-4} + \dfrac{(-1)(x)}{x-4}$

$= \dfrac{4 + (-1)(x)}{x-4}$

$= \dfrac{4 - x}{x-4}$

$= \dfrac{-1(x-4)}{x-4}$

$= -1$

31. $\dfrac{7a+3}{2a-1} + \dfrac{5a+4}{1-2a}$

$= \dfrac{7a+3}{2a-1} + \dfrac{5a+4}{(-1)(2a-1)}$

$= \dfrac{7a+3}{2a-1} + \dfrac{(-1)(5a+4)}{2a-1}$

$= \dfrac{7a+3 + (-1)(5a+4)}{2a-1}$

$= \dfrac{7a+3 - 5a - 4}{2a-1}$

$= \dfrac{2a-1}{2a-1}$

$= 1$

33. $\dfrac{a}{a-b} - \dfrac{b}{a^2-b^2}$

$= \dfrac{a}{a-b} - \dfrac{b}{(a-b)(a+b)}$

$= \dfrac{a(a+b)}{(a-b)(a+b)} - \dfrac{b}{(a-b)(a+b)}$

$= \dfrac{a(a+b) - b}{(a-b)(a+b)}$

$= \dfrac{a^2 + ab - b}{(a-b)(a+b)}$

35. $\dfrac{a}{a-b} \div \dfrac{b}{a^2-b^2}$

$= \dfrac{a}{a-b} \cdot \dfrac{a^2-b^2}{b}$

$= \dfrac{a}{a-b} \cdot \dfrac{(a-b)(a+b)}{b}$

$= \dfrac{a(a+b)}{b}$

37. $\dfrac{2y+1}{y+2} + \dfrac{3y}{y+3} + y$

$= \dfrac{(2y+1)(y+3)}{(y+2)(y+3)} + \dfrac{3y(y+2)}{(y+2)(y+3)}$

$\quad + \dfrac{y(y+2)(y+3)}{(y+2)(y+3)}$

$= \dfrac{(2y+1)(y+3) + 3y(y+2) + y(y+2)(y+3)}{(y+2)(y+3)}$

$= \dfrac{2y^2 + 7y + 3 + 3y^2 + 6y + y^3 + 5y^2 + 6y}{(y+2)(y+3)}$

$= \dfrac{y^3 + 10y^2 + 19y + 3}{(y+2)(y+3)}$

39. $\dfrac{5x+1}{x+2} + \dfrac{2x+6}{x^2+5x+6}$

$= \dfrac{5x+1}{x+2} + \dfrac{2(x+3)}{(x+2)(x+3)}$

$= \dfrac{5x+1}{x+2} + \dfrac{2}{x+2}$

$= \dfrac{5x+1+2}{x+2}$

$= \dfrac{5x+3}{x+2}$

41. $\dfrac{5x+1}{x+2} \div \dfrac{2x+6}{x^2+5x+6}$

$= \dfrac{5x+1}{x+2} \cdot \dfrac{x^2+5x+6}{2x+6}$

$= \dfrac{5x+1}{x+2} \cdot \dfrac{(x+2)(x+3)}{2(x+3)}$

$= \dfrac{5x+1}{2}$

43. $\dfrac{x+1}{x^2-3x} + \dfrac{x-2}{x^2-6x+9}$

$= \dfrac{x+1}{x(x-3)} + \dfrac{x-2}{(x-3)^2}$

$= \dfrac{(x+1)(x-3)}{x(x-3)^2} + \dfrac{(x-2)(x)}{x(x-3)^2}$

$= \dfrac{(x+1)(x-3) + (x-2)(x)}{x(x-3)^2}$

$= \dfrac{x^2 - 2x - 3 + x^2 - 2x}{x(x-3)^2}$

$= \dfrac{2x^2 - 4x - 3}{x(x-3)^2}$

45. $\dfrac{a+3}{a-3} + \dfrac{a}{a+4} - \dfrac{3}{a^2+a-12}$

$= \dfrac{a+3}{a-3} + \dfrac{a}{a+4} - \dfrac{3}{(a+4)(a-3)}$

$= \dfrac{(a+3)(a+4)}{(a-3)(a+4)} + \dfrac{a(a-3)}{(a-3)(a+4)}$

$\quad - \dfrac{3}{(a-3)(a+4)}$

$= \dfrac{(a+3)(a+4) + a(a-3) - 3}{(a-3)(a+4)}$

$= \dfrac{a^2 + 7a + 12 + a^2 - 3a - 3}{(a-3)(a+4)}$

$= \dfrac{2a^2 + 4a + 9}{(a-3)(a+4)}$

47. $\dfrac{y + 7}{y + 5} - \dfrac{y}{y - 3} + \dfrac{16}{y^2 + 2y - 15}$

$= \dfrac{y + 7}{y + 5} - \dfrac{y}{y - 3} + \dfrac{16}{(y + 5)(y - 3)}$

$= \dfrac{(y + 7)(y - 3)}{(y + 5)(y - 3)} - \dfrac{y(y + 5)}{(y + 5)(y - 3)}$

$ + \dfrac{16}{(y + 5)(y - 3)}$

$= \dfrac{(y + 7)(y - 30 - y(y + 5) + 16}{(y + 5)(y - 3)}$

$= \dfrac{y^2 + 4y - 21 - y^2 - 5y + 16}{(y + 5)(y - 3)}$

$= \dfrac{-y - 5}{(y + 5)(y - 3)}$

$= \dfrac{-(y + 5)}{(y + 5)(y - 3)}$

$= \dfrac{-1}{y - 3}$

49. $\dfrac{3x - 4}{x - 5} + \dfrac{4x}{10 + 3x - x^2}$

$= \dfrac{3x - 4}{x - 5} + \dfrac{4x}{-(x - 5)(x + 2)}$

$= \dfrac{3x - 4}{x - 5} + \dfrac{(-1)(4x)}{(x - 5)(x + 2)}$

$= \dfrac{(3x - 4)(x + 2)}{(x - 5)(x + 2)} + \dfrac{-4x}{(x - 5)(x + 2)}$

$= \dfrac{(3x - 4)(x + 2) - 4x}{(x - 5)(x + 2)}$

$= \dfrac{3x^2 + 2x - 8 - 4x}{(x - 5)(x + 2)}$

$= \dfrac{3x^2 - 2x - 8}{(x - 5)(x + 2)}$

51. $\dfrac{2a - 1}{a^2 + a - 6} + \dfrac{a + 2}{a^2 - 2a - 15} - \dfrac{a + 1}{a^2 - 7a + 10)}$

$= \dfrac{2a - 1}{(a + 3)(a - 2)} + \dfrac{a + 2}{(a - 5)(a + 3)} - \dfrac{a + 1}{(a - 5)(a - 2)}$

$= \dfrac{(2a - 1)(a - 5)}{(a + 3)(a - 2)(a - 5)} + \dfrac{(a + 2)(a - 2)}{(a + 3)(a - 2)(a - 5)}$

$ - \dfrac{(a + 1)(a + 3)}{(a + 3)(a - 2)(a - 5)}$

$= \dfrac{2a^2 - 11a + 5 + a^2 - 4 - (a^2 + 4a + 3)}{(a + 3)(a - 2)(a - 5)}$

$= \dfrac{3a^2 - 11a + 1 - a^2 - 4a - 3}{(a + 3)(a - 2)(a - 5)}$

$= \dfrac{2a^2 - 15a - 2}{(a + 3)(a - 2)(a - 5)}$

53. $\dfrac{5a}{3a - 1} + \dfrac{2a + 1}{5a + 2} + 5a + 1$

$= \dfrac{5a(5a + 2)}{(3a - 1)(5a + 2)} + \dfrac{(2a + 1)(3a - 1)}{(3a - 1)(5a + 2)}$

$ + \dfrac{(5a + 1)(3a - 1)(5a + 2)}{(3a - 1)(5a + 2)}$

$= \dfrac{5a(5a + 2) + (2a + 1)(3a - 1) + (5a + 1)(3a - 1)(5a + 2)}{(3a - 1)(5a + 2)}$

$= \dfrac{25a^2 + 10a + 6a^2 + a - 1 + 75a^3 + 20a^2 - 9a - 2}{(3a - 1)(5a + 2)}$

$= \dfrac{75a^3 + 51a^2 + 2a - 3}{(3a - 1)(5a + 2)}$

55. $\dfrac{2r + xs}{r + s} + \dfrac{2s + xr}{r + s}$

$= \dfrac{2r + xs + 2s + xr}{r + s}$

$= \dfrac{2r + 2s + xs + xr}{r + s}$

$= \dfrac{(r + s)(2 + x)}{r + s}$

$= 2 + x$

57. $\dfrac{r^2}{r^3 - s^3} + \dfrac{rs}{r^3 - s^3} + \dfrac{s^2}{r^3 - s^3}$

$= \dfrac{r^2 + rs + s^2}{r^3 - s^3}$

$= \dfrac{r^2 + rs + s^2}{(r - s)(r^2 + rs + s^2)}$

$= \dfrac{1}{r - s}$

59. $\dfrac{2s + t}{s^3 - t^3} + \dfrac{3s}{s^2 + st + t^2}$

$\dfrac{2s + t}{(s - t)(s^2 + st + t^2)} + \dfrac{3s}{s^2 + st + t^2}$

$\dfrac{2s + t}{(s - t)(s^2 + st + t^2)} + \dfrac{3s(s - t)}{(s - t)(s^2 + st + t^2)}$

$= \dfrac{2s + t + 3s(s - t)}{(s - t)(s^2 + st + t^2)}$

$= \dfrac{2s + t + 3s^2 - 3st}{(s - t)(s^2 + st + t^2)}$

$= \dfrac{3s^2 - 3st + 2s + t}{s^3 - t^3}$

61. (a) $\dfrac{x^2 + 4x}{4x} = \dfrac{x(x + 4)}{4x}$

$= \dfrac{x + 4}{4}$

(b) $\dfrac{x^2 + 4x}{4x} = \dfrac{x^2}{4x} + \dfrac{4x}{4x}$

$= \dfrac{x}{4} + 1$

63. (a) $\dfrac{15x^3y^2 - 10x^2y^3}{5x^2y^2} = \dfrac{5x^2y^2(3x - 2y)}{5x^2y^2}$

$= 3x - 2y$

(b) $\dfrac{15x^3y^2 - 10x^2y^3}{5x^2y^2} = \dfrac{15x^3y^2}{5x^2y^2} - \dfrac{10x^2y^3}{5x^2y^2}$

$= 3x - 2y$

65. (a) $\dfrac{6m^2n - 4m^3n^2 - 9mn}{15mn^2}$

$= \dfrac{mn(6m - 4m^2n - 9)}{15mn^2}$

$= \dfrac{6m - 4m^2n - 9}{15n}$

(b) $\dfrac{6m^2n - 4m^3n^2 - 9mn}{15mn}$

$= \dfrac{6m^2n}{15mn^2} - \dfrac{4m^3n^2}{15mn^2} - \dfrac{9mn}{15mn^2}$

$= \dfrac{2m}{5n} - \dfrac{4m^2}{15} - \dfrac{3}{5n}$

67. $f(x) + g(x)$

$= \dfrac{x - 2}{x + 2} + \dfrac{2x}{x^2 - 4}$

$= \dfrac{x - 2}{x + 2} + \dfrac{2x}{(x + 2)(x - 2)}$

$= \dfrac{(x - 2)(x - 2)}{(x + 2)(x - 2)} + \dfrac{2x}{(x + 2)(x - 2)}$

$= \dfrac{(x - 2)(x - 2) + 2x}{(x + 2)(x - 2)}$

$= \dfrac{x^2 - 4x + 4 + 2x}{(x + 2)(x - 2)}$

$= \dfrac{x^2 - 2x + 4}{(x + 2)(x - 2)}$

69. $f(x) \cdot h(x) = \dfrac{x - 2}{x + 2} \cdot \dfrac{x}{x - 2}$

$= \dfrac{x}{x + 2}$

71. $g(x) + h(x)$

$$= \frac{2x}{x^2 - 4} + \frac{x}{x - 2}$$

$$= \frac{2x}{(x - 2)(x + 2)} + \frac{x}{x - 2}$$

$$= \frac{2x}{(x - 2)(x + 2)} + \frac{x(x + 2)}{(x - 2)(x + 2)}$$

$$= \frac{2x + x(x + 2)}{(x - 2)(x + 2)}$$

$$= \frac{2x + x^2 + 2x}{(x - 2)(x + 2)}$$

$$= \frac{x^2 + 4x}{(x - 2)(x + 2)}$$

73. $h(x) \div g(x) = \dfrac{x}{x - 2} \div \dfrac{2x}{x^2 - 4}$

$$= \frac{x}{x - 2} \cdot \frac{x^2 - 4}{2x}$$

$$= \frac{x}{x - 2} \cdot \frac{(x - 2)(x + 2)}{2x}$$

$$= \frac{x + 2}{2}$$

77. $m = \dfrac{y_2 - y_1}{x_2 - x_1}$

$$= \frac{2 - (-2)}{-4 - 5}$$

$$= -\frac{4}{9}$$

79. $2x - 3y \geq 12$ (solid line)

    $2x - 3y = 12$

x	y
0	4
-2	0

Test point: $(0, 0)$

$$2(0) - 3(0) \geq 12$$
$$0 \geq 12$$
$$\text{False}$$

Shade half-plane <u>not</u> containing $(0, 0)$.

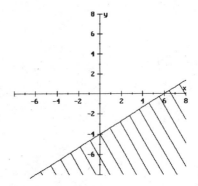

6.4 Exercises

1. $\dfrac{\dfrac{3}{xy^2}}{\dfrac{15}{x^2y}} = \dfrac{3}{xy^2} \div \dfrac{15}{x^2y}$

$$= \frac{3}{xy^2} \cdot \frac{x^2y}{15}$$

$$= \frac{x}{5y}$$

3. $\dfrac{\dfrac{3}{x - y}}{\dfrac{x - y}{3}} = \dfrac{3}{x - y} \div \dfrac{x - y}{3}$

$$= \frac{3}{x - y} \cdot \frac{3}{x - y}$$

$$= \frac{9}{(x - y)^2}$$

5. $\dfrac{a - \dfrac{1}{3}}{\dfrac{9a^2 - 1}{3a}}$

$= \dfrac{a - \dfrac{1}{3}}{\dfrac{9a^2 - 1}{3a}} \cdot \dfrac{3a}{3a}$

$= \dfrac{3a^2 - \dfrac{1}{3} \cdot \dfrac{3a}{1}}{\dfrac{9a^2 - 1}{3a} \cdot \dfrac{3a}{1}}$

$= \dfrac{3a^2 - a}{9a^2 - 1}$

$= \dfrac{a(3a - 1)}{(3a - 1)(3a + 1)}$

$= \dfrac{a}{3a + 1}$

7. $\dfrac{x + \dfrac{2}{xy^2}}{\dfrac{1}{x} + 2}$

$= \dfrac{x + \dfrac{2}{xy^2}}{\dfrac{1}{x} + 2} \cdot \dfrac{xy^2}{xy^2}$

$= \dfrac{x(xy^2) + \dfrac{2}{xy^2} \cdot \dfrac{xy^2}{1}}{\dfrac{1}{x} \cdot \dfrac{xy^2}{1} + 2xy^2}$

$= \dfrac{x^2y^2 + 2}{y^2 + 2xy^2}$

9. $\left(1 - \dfrac{4}{x^2}\right) \div \left(\dfrac{1}{x} - \dfrac{2}{x^2}\right)$

$= \dfrac{1 - \dfrac{4}{x^2}}{\dfrac{1}{x} - \dfrac{2}{x^2}} \cdot \dfrac{x^2}{x^2}$

$= \dfrac{x^2 - \dfrac{4}{x^2} \cdot \dfrac{x^2}{1}}{\dfrac{1}{x} \cdot \dfrac{x^2}{1} - \dfrac{2}{x^2} \cdot \dfrac{x^2}{1}}$

$= \dfrac{x^2 - 4}{x - 2}$

$= \dfrac{(x - 2)(x + 2)}{x - 2}$

$= x + 2$

11. $\dfrac{1 - \dfrac{5}{y}}{y + 3 - \dfrac{40}{y}}$

$= \dfrac{1 - \dfrac{5}{y}}{y + 3 - \dfrac{40}{y}} \cdot \dfrac{y}{y}$

$= \dfrac{y - \dfrac{5}{y} \cdot \dfrac{y}{1}}{y^2 + 3y - \dfrac{40}{y} \cdot \dfrac{y}{1}}$

$= \dfrac{y - 5}{y^2 + 3y - 40}$

$= \dfrac{y - 5}{(y + 8)(y - 5)}$

$= \dfrac{1}{y + 8}$

13.
$$\frac{1 - \dfrac{4}{z} + \dfrac{4}{z^2}}{\dfrac{1}{z^2} - \dfrac{2}{z^3}}$$

$$= \frac{1 - \dfrac{4}{z} + \dfrac{4}{z^2}}{\dfrac{1}{z^2} - \dfrac{2}{z^3}} \cdot \frac{z^3}{z^3}$$

$$= \frac{z^3 - \dfrac{4}{z} \cdot \dfrac{z^3}{1} + \dfrac{4}{z^2} \cdot \dfrac{z^3}{1}}{\dfrac{1}{z^2} \cdot \dfrac{z^3}{1} - \dfrac{2}{z^3} \cdot \dfrac{z^3}{1}}$$

$$= \frac{z^3 - 4z^2 + 4z}{z - 2}$$

$$= \frac{z(z - 2)^2}{z - 2}$$

$$= z(z - 2)$$

15.
$$\frac{\dfrac{4}{y^2} - \dfrac{12}{xy} + \dfrac{9}{x^2}}{\dfrac{4}{y^2} - \dfrac{9}{x^2}}$$

$$= \frac{\dfrac{4}{y^2} - \dfrac{12}{xy} + \dfrac{9}{x^2}}{\dfrac{4}{y^2} - \dfrac{9}{x^2}} \cdot \frac{x^2y^2}{x^2y^2}$$

$$= \frac{\dfrac{4}{y^2} \cdot \dfrac{x^2y^2}{1} - \dfrac{12}{xy} \cdot \dfrac{x^2y^2}{1} + \dfrac{9}{x^2} \cdot \dfrac{x^2y^2}{1}}{\dfrac{4}{y^2} \cdot \dfrac{x^2y^2}{1} - \dfrac{9}{x^2} \cdot \dfrac{x^2y^2}{1}}$$

$$= \frac{4x^2 - 12xy + 9y^2}{4x^2 - 9y^2}$$

$$= \frac{(2x - 3y)(2x - 3y)}{(2x - 3y)(2x + 3y)}$$

$$= \frac{2x - 3y}{2x + 3y}$$

17.
$$\frac{\dfrac{2x}{y} + 7 + \dfrac{5y}{x}}{3x + 2y - \dfrac{y^2}{x}}$$

$$= \frac{\dfrac{2x}{y} + 7 + \dfrac{5y}{x}}{3x + 2y - \dfrac{y^2}{x}} \cdot \frac{xy}{xy}$$

$$= \frac{\dfrac{2x}{y} \cdot \dfrac{xy}{1} + 7xy + \dfrac{5y}{x} \cdot \dfrac{xy}{1}}{3x(xy) + 2y(xy) - \dfrac{y^2}{x} \cdot \dfrac{xy}{1}}$$

$$= \frac{2x^2 + 7xy + 5y^2}{3x^2y + 2xy^2 - y^3}$$

$$= \frac{(2x + 5y)(x + y)}{y(3x - y)(x + y)}$$

$$= \frac{2x + 5y}{y(3x - y)}$$

19.
$$\frac{1 - \dfrac{3}{x} - \dfrac{10}{x^2}}{4 + \dfrac{8}{x}}$$

$$= \frac{1 - \dfrac{3}{x} - \dfrac{10}{x^2}}{4 + \dfrac{8}{x}} \cdot \frac{x^2}{x^2}$$

$$= \frac{x^2 - \dfrac{3}{x} \cdot \dfrac{x^2}{1} - \dfrac{10}{x^2} \cdot \dfrac{x^2}{1}}{4x^2 + \dfrac{8}{x} \cdot \dfrac{x^2}{1}}$$

$$= \frac{x^2 - 3x - 10}{4x^2 + 8x}$$

$$= \frac{(x - 5)(x + 2)}{4x(x + 2)}$$

$$= \frac{x - 5}{4x}$$

21. $\dfrac{9 - \dfrac{25}{t^2}}{6 + \dfrac{10}{t}}$

$= \dfrac{9 - \dfrac{25}{t^2}}{6 + \dfrac{10}{t} \cdot \dfrac{t^2}{t^2}}$

$= \dfrac{9t^2 - \dfrac{25}{t^2} \cdot \dfrac{t^2}{1}}{6t^2 + \dfrac{10}{t} \cdot \dfrac{t^2}{1}}$

$= \dfrac{9t^2 - 25}{6t^2 + 10t}$

$= \dfrac{(3t - 5)(3t + 5)}{2t(3t + 5)}$

$= \dfrac{3t - 5}{2t}$

23. $\dfrac{\dfrac{1}{x - 3} - \dfrac{1}{x}}{3}$

$= \dfrac{\dfrac{1}{x - 3} - \dfrac{1}{x}}{3} \cdot \dfrac{x(x - 3)}{x(x - 3)}$

$= \dfrac{\dfrac{1}{x - 3} \cdot \dfrac{x(x - 3)}{1} - \dfrac{1}{x} \cdot \dfrac{x(x - 3)}{1}}{3x(x - 3)}$

$= \dfrac{x - (x - 3)}{3x(x - 3)}$

$= \dfrac{x - x + 3}{3x(x - 3)}$

$= \dfrac{3}{3x(x - 3)}$

$= \dfrac{1}{x(x - 3)}$

25. $\dfrac{\dfrac{h}{h + 1} - \dfrac{h}{h - 1}}{2}$

$= \dfrac{\dfrac{h}{h + 1} - \dfrac{h}{h - 1}}{2} \cdot \dfrac{(h + 1)(h - 1)}{(h + 1)(h - 1)}$

$= \dfrac{\dfrac{h}{h + 1} \cdot \dfrac{(h + 1)(h - 1)}{1} - \dfrac{h}{h - 1} \cdot \dfrac{(h + 1)(h - 1)}{1}}{2(h + 1)(h - 1)}$

$= \dfrac{h(h - 1) - h(h + 1)}{2(h + 1)(h - 1)}$

$= \dfrac{h^2 - h - h^2 - h}{2(h + 1)(h - 1)}$

$= \dfrac{-2h}{2(h + 1)(h - 1)}$

$= \dfrac{-h}{(h + 1)(h - 1)}$

27. $\left(3 + \dfrac{1}{2x - 1}\right) \div \left(5 + \dfrac{x}{2 - 1}\right)$

$= \dfrac{3 + \dfrac{1}{2x - 1}}{5 + \dfrac{x}{2x - 1}}$

$= \dfrac{3 + \dfrac{1}{2x - 1}}{5 + \dfrac{x}{2x - 1}} \cdot \dfrac{2x - 1}{2x - 1}$

$= \dfrac{3(2x - 1) + \dfrac{1}{2x - 1} \cdot \dfrac{2x - 1}{1}}{5(2x - 1) + \dfrac{x}{2x - 1} \cdot \dfrac{2x - 1}{1}}$

$= \dfrac{6x - 3 + 1}{10x - 5 + x}$

$= \dfrac{6x - 2}{11x - 5}$

29.

$$\dfrac{\dfrac{4}{x+3} - \dfrac{4}{x}}{3}$$

$$= \dfrac{\dfrac{4}{x+3} - \dfrac{4}{x}}{3} \cdot \dfrac{x(x+3)}{x(x+3)}$$

$$= \dfrac{\dfrac{4}{x+3} \cdot \dfrac{x(x+3)}{1} - \dfrac{4}{x} \cdot \dfrac{x(x+3)}{1}}{3x(x+3)}$$

$$= \dfrac{4x - 4(x+3)}{3x(x+3)}$$

$$= \dfrac{4x - 4x - 12}{3x(x+3)}$$

$$= \dfrac{-12}{3x(x+3)}$$

$$= \dfrac{-4}{x(x+3)}$$

31.

$$f(x) = \dfrac{2x}{x-3}$$

$$f(x+3) = \dfrac{2(x+3)}{x+3-3}$$

$$= \dfrac{2x+6}{x}$$

$$\dfrac{f(x+3) - f(x)}{3}$$

$$= \dfrac{\dfrac{2x+6}{x} - \dfrac{2x}{x-3}}{3}$$

$$= \dfrac{\dfrac{2x+6}{x} - \dfrac{2x}{x-3}}{3} \cdot \dfrac{x(x-3)}{x(x-3)}$$

$$= \dfrac{\dfrac{2x+6}{x} \cdot \dfrac{x(x-3)}{1} - \dfrac{2x}{x-3} \cdot \dfrac{x(x-3)}{1}}{3x(x-3)}$$

$$= \dfrac{(2x+6)(x-3) - 2x(x)}{3x(x-3)}$$

$$= \dfrac{2x^2 - 18 - 2x^2}{3x(x-3)}$$

$$= \dfrac{-18}{3x(x-3)}$$

$$= \dfrac{-6}{x(x-3)}$$

33.

$$h(x) = \dfrac{x+2}{x}$$

$$h(x-2) = \dfrac{x-2+2}{x-2}$$

$$= \dfrac{x}{x-2}$$

$$\dfrac{(hx-2) - h(x)}{2}$$

$$= \dfrac{\dfrac{x}{x-2} - \dfrac{x+2}{x}}{2}$$

$$= \dfrac{\dfrac{x}{x-2} - \dfrac{x+2}{x}}{2} \cdot \dfrac{x(x-2)}{x(x-2)}$$

$$= \dfrac{\dfrac{x}{x-2} \cdot \dfrac{x(x-2)}{1} - \dfrac{x+2}{x} \cdot \dfrac{x(x-2)}{1}}{2x(x-2)}$$

$$= \dfrac{x^2 - (x+2)(x-2)}{2x(x-2)}$$

$$= \dfrac{x^2 - (x^2 - 4)}{2x(x-2)}$$

$$= \dfrac{x^2 - x^2 + 4}{2x(x-2)}$$

$$= \dfrac{4}{2x(x-2)}$$

$$= \dfrac{2}{x(x-2)}$$

35. $\dfrac{f(x) + g(x)}{h(x)}$

$= \dfrac{\dfrac{2x}{x-3} + \left(-\dfrac{1}{x}\right)}{\dfrac{x+2}{x}}$

$= \dfrac{\dfrac{2x}{x-3} - \dfrac{1}{x}}{\dfrac{x+2}{x}} \cdot \dfrac{x(x-3)}{x(x-3)}$

$= \dfrac{\dfrac{2x}{x-3} \cdot \dfrac{x(x-3)}{1} - \dfrac{1}{x} \cdot \dfrac{x(x-3)}{1}}{\dfrac{x+2}{x} \cdot \dfrac{x(x-3)}{1}}$

$= \dfrac{2x^2 - (x-3)}{(x+2)(x-3)}$

$= \dfrac{2x^2 - x + 3}{x^2 - x - 6}$

37. $\dfrac{g(x) + 1}{h(x)}$

$= \dfrac{-\dfrac{1}{x} + 1}{\dfrac{x+2}{x}}$

$= \dfrac{-\dfrac{1}{x} + 1}{\dfrac{x+2}{x}} \cdot \dfrac{x}{x}$

$= \dfrac{-\dfrac{1}{x} \cdot \dfrac{x}{1} + x}{\dfrac{x+2}{x} \cdot \dfrac{x}{1}}$

$= \dfrac{-1 + x}{x + 2}$

39. $\dfrac{f(x) - h(x)}{3}$

$= \dfrac{\dfrac{2x}{x-3} - \dfrac{x+2}{x}}{3}$

$= \dfrac{\dfrac{2x}{x-3} - \dfrac{x+2}{x}}{3} \cdot \dfrac{x(x-3)}{x(x-3)}$

$= \dfrac{\dfrac{2x}{x-3} \cdot \dfrac{x(x-3)}{1} - \dfrac{x+2}{x} \cdot \dfrac{x(x-3)}{1}}{3x(x-3)}$

$= \dfrac{2x^2 - (x+2)(x-3)}{3x(x-3)}$

$= \dfrac{2x^2 - (x^2 - x - 6)}{3x(x-3)}$

$= \dfrac{2x^2 - x^2 + x + 6}{3x(x-3)}$

$= \dfrac{x^2 + x + 6}{3x(x-3)}$

41. $(x - 4)(x + 1) = (x - 3)(x - 2)$

$x^2 - 3x - 4 = x^2 - 5x + 6$

$-3x - 4 = -5x + 6$

$2x - 4 = 6$

$2x = 10$

$x = 5$

43. $3x - 2y = 5$

$-2y = -3x + 5$

$y = \dfrac{3}{2}x - \dfrac{5}{2}$

$m = \dfrac{3}{2}$

Parallel lines have equal slopes.

$y - y_1 = m(x - x_1)$

$y - 5 = \dfrac{3}{2}[x - (-4)]$

$y - 5 = \dfrac{3}{2}(x + 4)$

$y - 5 = \dfrac{3}{2}x + 6$

$y = \dfrac{3}{2}x + 11$

1.
$$\frac{x}{3} - \frac{x}{2} + \frac{x}{4} = 1$$

$$12\left(\frac{x}{3} - \frac{x}{2} + \frac{x}{4}\right) = 12(1)$$

$$\frac{12}{1} \cdot \frac{x}{3} - \frac{12}{1} \cdot \frac{x}{2} + \frac{12}{1} \cdot \frac{x}{4} = 12$$

$$4x - 6x + 3x = 12$$
$$x = 12$$

3.
$$\frac{t}{3} + \frac{t}{5} < \frac{t}{6} - 11$$

$$30\left(\frac{t}{3} + \frac{t}{5}\right) < 30\left(\frac{t}{6} - 11\right)$$

$$\frac{30}{1} \cdot \frac{t}{3} + \frac{30}{1} \cdot \frac{t}{5} < \frac{30}{1} \cdot \frac{t}{6} - 30(11)$$

$$10t + 6t < 5t - 330$$
$$16t < 5t - 330$$
$$11t < -330$$
$$t < -30$$

5.
$$\frac{a - 1}{6} + \frac{a + 1}{10} = a - 3$$

$$30\left(\frac{a - 1}{6} + \frac{a + 1}{10}\right) = 30(a - 3)$$

$$\frac{30}{1} \cdot \frac{a - 1}{6} + \frac{30}{1} \cdot \frac{a + 1}{10} = 30(a - 3)$$

$$5(a - 1) + 3(a + 1) = 30(a - 3)$$
$$5a - 5 + 3a + 3 = 30a - 90$$
$$8a - 2 = 30a - 90$$
$$-2 = 22a - 90$$
$$88 = 22a$$
$$4 = a$$

7.
$$\frac{y - 5}{2} = \frac{y - 2}{5}$$

$$\frac{10}{1} \cdot \frac{y - 5}{2} = \frac{10}{1} \cdot \frac{y - 2}{5}$$

$$5(y - 5) = 2(y - 2)$$
$$5y - 25 = 2y - 4$$
$$3y - 25 = -4$$
$$3y = 21$$
$$y = 7$$

9.
$$\frac{y - 5}{2} \le \frac{y - 2}{5} + 3$$

$$10\left(\frac{y - 5}{2}\right) \le 10\left(\frac{y - 2}{5} + 3\right)$$

$$\frac{10}{1} \cdot \frac{y - 5}{2} \le \frac{10}{1} \cdot \frac{y - 2}{5} + 10(3)$$

$$5(y - 5) \le 2(y - 2) + 30$$
$$5y - 25 \le 2y - 4 + 30$$
$$5y - 25 \le 2y + 26$$
$$3y - 25 \le 26$$
$$3y \le 51$$
$$y \le 17$$

11.
$$\frac{x - 3}{4} - \frac{x - 4}{3} = 2$$

$$12\left(\frac{x - 3}{4} - \frac{x - 4}{3}\right) = 12(2)$$

$$\frac{12}{1} \cdot \frac{x - 3}{4} - \frac{12}{1} \cdot \frac{x - 4}{3} = 24$$

$$3(x - 3) - 4(x - 4) = 24$$
$$3x - 9 - 4x + 16 = 24$$
$$-x + 7 = 24$$
$$-x = 17$$
$$x = -17$$

13.
$$\frac{x - 3}{4} - \frac{x - 4}{3} \ge 2$$

$$12\left(\frac{x - 3}{4} - \frac{x - 4}{3}\right) \ge 12(2)$$

$$\frac{12}{1} \cdot \frac{x - 3}{4} - \frac{12}{1} \cdot \frac{x - 4}{3} \ge 24$$

$$3(x - 3) - 4(x - 4) \ge 24$$
$$3x - 9 - 4x + 16 \ge 24$$
$$-x + 7 \ge 24$$
$$-x \ge 17$$
$$x \le -17$$

15.
$$\frac{3x+11}{6} - \frac{2x+1}{3} = x + 5$$

$$18\left(\frac{3x+11}{6} - \frac{2x+1}{3}\right) = 18(x+5)$$

$$\frac{18}{1} \cdot \frac{3x+11}{6} - \frac{18}{1} \cdot \frac{2x+1}{3} = 18x + 90$$

$$3(3x+11) - 6(2x+1) = 18x + 90$$
$$9x + 33 - 12x - 6 = 18x + 90$$
$$-3x + 27 = 18x + 90$$
$$27 = 21x + 90$$
$$-63 = 21x$$
$$-3 = x$$

17.
$$\frac{x-3}{5} - \frac{3x+1}{4} < 8$$

$$20\left(\frac{x-3}{5} - \frac{3x+1}{4}\right) < 20(8)$$

$$\frac{20}{1} \cdot \frac{x-3}{5} - \frac{20}{1} \cdot \frac{3x+1}{4} < 160$$

$$4(x-3) - 5(3x+1) < 160$$
$$4x - 12 - 15x - 5 < 160$$
$$-11x - 17 < 160$$
$$-11x < 177$$
$$x > -\frac{177}{11}$$

19.
$$\frac{5}{x} - \frac{1}{2} = \frac{3}{x}$$

$$2x\left(\frac{5}{x} - \frac{1}{2}\right) = 2x\left(\frac{3}{x}\right)$$

$$\frac{2x}{1} \cdot \frac{5}{x} - \frac{2x}{1} \cdot \frac{1}{2} = \frac{2x}{1} \cdot \frac{3}{x}$$

$$10 - x = 6$$
$$-x = -4$$
$$x = 4$$

21. $\dfrac{4}{x} - \dfrac{1}{5} + \dfrac{7}{2x}$

$$= \frac{4(10)}{10x} - \frac{1(2x)}{10x} + \frac{7(5)}{10x}$$

$$= \frac{4(10) - 2x + 7(5)}{10x}$$

$$= \frac{40 - 2x + 35}{10x}$$

$$= \frac{-2x + 75}{10x}$$

23.
$$\frac{1}{t-3} + \frac{2}{t} = \frac{5}{3t}$$

$$3t(t-3)\left(\frac{1}{t-3} + \frac{2}{t}\right) = 3t(t-3)\left(\frac{5}{3t}\right)$$

$$\frac{3t(t-3)}{1} \cdot \frac{1}{t-3} + \frac{3t(t-3)}{1} \cdot \frac{2}{t} = \frac{3t(t-3)}{1} \cdot \frac{5}{3t}$$

$$3t + 3(t-3)(2) = 5(t-3)$$
$$3t + 6t - 18 = 5t - 15$$
$$9t - 18 = 5t - 15$$
$$4t - 18 = -15$$
$$4t = 3$$
$$t = \frac{3}{4}$$

25.
$$\frac{6}{a-3} - \frac{3}{8} = \frac{21}{4a-12}$$

$$\frac{6}{a-3} - \frac{3}{8} = \frac{21}{4(a-3)}$$

$$8(a-3)\left(\frac{6}{a-3} - \frac{3}{8}\right) = 8(a-3)\left[\frac{21}{4(a-3)}\right]$$

$$\frac{8(a-3)}{1} \cdot \frac{6}{a-3} - \frac{8(a-3)}{1} \cdot \frac{3}{8} = \frac{8(a-3)}{1} \cdot \frac{21}{4(a-3)}$$

$$48 - 3(a-3) = 2(21)$$
$$48 - 3a + 9 = 42$$
$$-3a + 57 = 42$$
$$-3a = -15$$
$$a = 5$$

27.

$$\frac{7}{x-5} + 2 = \frac{x+2}{x-5}$$

$$(x-5)\left(\frac{7}{x-5} + 2\right) = (x-5)\left(\frac{x+2}{x-5}\right)$$

$$\frac{x-5}{1} \cdot \frac{7}{x-5} + (x-5)(2) = \frac{x-5}{1} \cdot \frac{x+2}{x-5}$$

$$7 + 2x - 10 = x + 2$$
$$2x - 3 = x + 2$$
$$x - 3 = 2$$
$$x = 5$$

No solution since $x = 5$ causes a denominator to equal 0 in the original equation.

29.

$$\frac{4}{y^2 - 2y} - \frac{3}{2y} = \frac{17}{6y}$$

$$\frac{4}{y(y-2)} - \frac{3}{2y} = \frac{17}{6y}$$

$$6y(y-2)\left[\frac{4}{y(y-2)} - \frac{3}{2y}\right] = 6y(y-2)\left(\frac{17}{6y}\right)$$

$$\frac{6y(y-2)}{1} \cdot \frac{4}{y(y-2)} - \frac{6y(y-2)}{1} \cdot \frac{3}{2y} = \frac{6y(y-2)}{1} \cdot \frac{17}{6y}$$

$$24 - 9(y-2) = 17(y-2)$$
$$24 - 9y + 18 = 17y - 34$$
$$-9y + 42 = 17y - 34$$
$$42 = 26y - 34$$
$$76 = 26y$$
$$\frac{38}{13} = y$$

31.

$$\frac{5}{y^2 + 3y} - \frac{4}{3y} + \frac{1}{2}$$

$$= \frac{5}{y(y+3)} - \frac{4}{3y} + \frac{1}{2}$$

$$= \frac{5(6)}{6y(y+3)} - \frac{4(2)(y+3)}{6y(y+3)} + \frac{3y(y+3)}{6y(y+3)}$$

$$= \frac{5(6) - 4(2)(y+3) + 3y(y+3)}{6y(y+3)}$$

$$= \frac{30 - 8y - 24 + 3y^2 + 9y}{6y(y+3)}$$

$$= \frac{3y^2 + y + 6}{6y(y+3)}$$

33.

$$\frac{9}{x^2 + 4x} = \frac{6}{x^2 + 2x}$$

$$\frac{9}{x(x+4)} = \frac{6}{x(x+2)}$$

$$\frac{x(x+4)(x+2)}{1} \cdot \frac{9}{x(x+4)} = \frac{x(x+4)(x+2)}{1}$$

$$\cdot \frac{6}{x(x+2)}$$

$$9(x+2) = 6(x+4)$$
$$9x + 18 = 6x + 24$$
$$3x + 18 = 24$$
$$3x = 6$$
$$x = 2$$

35.

$$x + \frac{1}{x} = 2$$

$$x\left(x + \frac{1}{x}\right) = x(2)$$

$$x^2 + \frac{x}{1} \cdot \frac{1}{x} = 2x$$

$$x^2 + 1 = 2x$$
$$x^2 - 2x + 1 = 0$$
$$(x-1)(x-1) = 0$$
$$x - 1 = 0$$
$$x = 1$$

37.

$$\frac{1}{x^2 - x - 2} + \frac{2}{x^2 - 1} = \frac{1}{x^2 - 3x + 2}$$

$$\frac{1}{(x - 2)(x + 1)} + \frac{2}{(x - 1)(x + 1)} = \frac{1}{(x - 2)(x - 1)}$$

$$(x - 2)(x + 1)(x - 1)\left[\frac{1}{(x - 2)(x + 1)} + \frac{2}{(x - 1)(x + 1)}\right] = (x - 2)(x + 1)(x - 1)\left[\frac{1}{(x - 2)(x - 1)}\right]$$

$$\frac{(x - 2)(x + 1)(x - 1)}{1} \cdot \frac{1}{(x - 2)(x + 1)} + \frac{(x - 2)(x + 1)(x - 1)}{1} \cdot \frac{2}{(x - 1)(x + 1)} = (x - 2)(x + 1)(x - 1)\left[\frac{1}{(x - 2)(x - 1)}\right]$$

$$x - 1 + 2(x - 2) = x + 1$$
$$x - 1 + 2x - 4 = x + 1$$
$$3x - 5 = x + 1$$
$$2x - 5 = 1$$
$$2x = 6$$
$$x = 3$$

39.

$$\frac{1}{x - 4} - \frac{5}{x + 2} = \frac{6}{x^2 - 2x - 8}$$

$$\frac{1}{x - 4} - \frac{5}{x + 2} = \frac{6}{(x - 4)(x + 2)}$$

$$(x - 4)(x + 2)\left(\frac{1}{x - 4} - \frac{5}{x + 2}\right) = (x - 4)(x + 2)\left[\frac{6}{(x - 4)(x + 2)}\right]$$

$$\frac{(x - 4)(x + 2)}{1} \cdot \frac{1}{x - 4} - \frac{(x - 4)(x + 2)}{1} \cdot \frac{5}{x + 2} = \frac{(x - 4)(x + 2)}{1} \cdot \frac{6}{(x - 4)(x + 2)}$$

$$x + 2 - 5(x - 4) = 6$$
$$x + 2 - 5x + 20 = 6$$
$$-4x + 22 = 6$$
$$-4x = -16$$
$$x = 4$$

No solution since $x = 4$ causes a denominator to equal 0 in the original equation.

41. $\dfrac{n}{3n + 2} + \dfrac{6}{9n^2 - 4} - \dfrac{2}{3n - 2}$

$$= \frac{n}{3n + 2} + \frac{6}{(3n + 2)(3n - 2)} - \frac{2}{3n - 2}$$

$$= \frac{n(3n - 2)}{(3n + 2)(3n - 2)} + \frac{6}{(3n + 2)(3n - 2)} - \frac{2(3n + 2)}{(3n + 2)(3n - 2)}$$

$$= \frac{n(3n - 2) + 6 - 2(3n + 2)}{(3n + 2)(3n - 2)}$$

$$= \frac{3n^2 - 2n + 6 - 6n - 4}{(3n + 2)(3n - 2)}$$

$$= \frac{3n^2 - 8n + 2}{(3n + 2)(3n - 2)}$$

43.

$$\frac{1}{3n + 4} + \frac{8}{9n^2 - 16} = \frac{1}{3n - 4}$$

$$\frac{1}{3n + 4} + \frac{8}{(3n + 4)(3n - 4)} = \frac{1}{3n - 4}$$

$$(3n + 4)(3n - 4)\left[\frac{1}{3n + 4} + \frac{8}{(3n + 4)(3n - 4)}\right] = (3n + 4)(3n - 4)\left(\frac{1}{3n - 4}\right)$$

$$\frac{(3n + 4)(3n - 4)}{1} \cdot \frac{1}{3n - 4)} + \frac{(3n + 4)(3n - 4)}{1} \cdot \frac{8}{(3n + 40(3n - 4)} = \frac{(3n + 4)(3n - 4)}{1} \cdot \frac{1}{(3n - 4)}$$

$$3n - 4 + 8 = 3n + 4$$
$$3n + 4 = 3n + 4$$
Identity
$$\text{all reals except } n = \pm\frac{4}{3}$$

45.

$$\frac{4}{2x - 1} + \frac{2}{x + 3} = \frac{5}{2x^2 + 5x - 3}$$

$$\frac{4}{2x - 1} + \frac{2}{x + 3} = \frac{5}{(2x - 1)(x + 3)}$$

$$(2x - 1)(x + 3)\left(\frac{4}{2x - 1} + \frac{2}{x + 3}\right) = (2x - 1)(x + 3)\left[\frac{5}{(2x - 1)(x + 3)}\right]$$

$$\frac{(2x - 1)(x + 3)}{1} \cdot \frac{4}{2x - 1} + \frac{(2x - 1)(x + 3)}{1} \cdot \frac{2}{x + 3} = \frac{(2x - 1)(x + 3)}{1} \cdot \frac{5}{(2x - 1)(x + 3)}$$

$$4(x + 3) + 2(2x - 1) = 5$$
$$4x + 12 + 4x - 2 = 5$$
$$8x + 10 = 5$$
$$8x = -5$$
$$x = -\frac{5}{8}$$

47.

$$\frac{6}{x} - \frac{2}{x-1} = 1$$

$$x(x-1)\left(\frac{6}{x} - \frac{2}{x-1}\right) = x(x-1)(1)$$

$$\frac{x(x-1)}{1} \cdot \frac{6}{x} - \frac{x(x-1)}{1} \cdot \frac{2}{x-1} = x(x-1)$$

$$6(x-1) - 2x = x^2 - x$$
$$6x - 6 - 2x = x^2 - x$$
$$4x - 6 = x^2 - x$$
$$0 = x^2 - 5x + 6$$
$$0 = (x-2)(x-3)$$
$$x - 2 = 0 \quad \text{or} \quad x - 3 = 0$$
$$x = 2 \quad \text{or} \quad x = 3$$

$$5x + 10 + 9x = 4x^2 + 8x$$
$$14x + 10 = 4x^2 + 8x$$
$$0 = 4x^2 - 6x - 10$$
$$0 = 2x^2 - 3x - 5$$
$$0 = (2x-5)(x+1)$$
$$2x - 5 = 0 \quad \text{or} \quad x + 1 = 0$$
$$2x = 5 \qquad\qquad x = -1$$
$$x = \frac{5}{2} \quad \text{or} \qquad x = -1$$

49.

$$\frac{x}{x-1} = \frac{2x}{x+1}$$

$$\frac{(x-1)(x+1)}{1} \cdot \frac{x}{x-1} = \frac{(x-1)(x+1)}{1} \cdot \frac{2x}{x+1}$$

$$x(x+1) = 2x(x-1)$$
$$x^2 + x = 2x^2 - 2x$$
$$0 = x^2 - 3x$$
$$0 = x(x-3)$$
$$x = 0 \quad \text{or} \quad x - 3 = 0$$
$$x = 0 \quad \text{or} \qquad x = 3$$

51.

$$\frac{2x}{x+2} = x - 1$$

$$\frac{x+2}{1} \cdot \frac{2x}{x+2} = (x+2)(x-1)$$

$$2x = x^2 + x - 2$$
$$0 = x^2 - x - 2$$
$$0 = (x-2)(x+1)$$
$$x - 2 = 0 \quad \text{or} \quad x + 1 = 0$$
$$x = 2 \quad \text{or} \qquad x = -1$$

53.

$$\frac{5}{x} + \frac{9}{x+2} = 4$$

$$x(x+2)\left(\frac{5}{x} + \frac{9}{x+2}\right) = x(x+2)(4)$$

$$\frac{x(x+2)}{1} \cdot \frac{5}{x} + \frac{x(x+2)}{1} \cdot \frac{9}{x+2} = 4x(x+2)$$

$$5(x+2) + 9x = 4x(x+2)$$

55.

$$\frac{6}{3a + 5} - \frac{2}{a - 4} = \frac{10}{3a^2 - 7a - 20}$$

$$\frac{6}{3a + 5} - \frac{2}{a - 4} = \frac{10}{(3a + 5)(a - 4)}$$

$$(3a + 5)(a - 4)\left(\frac{6}{3a + 5} - \frac{2}{a - 4}\right) = (3a + 5)(a - 4)\left[\frac{10}{(3a + 5)(a - 4)}\right]$$

$$\frac{(3a + 5)(a - 4)}{1} \cdot \frac{6}{3a + 5} - \frac{(3a + 5)(a - 4)}{1} \cdot \frac{2}{a - 4} = \frac{(3a + 5)(a - 4)}{1} \cdot \frac{10}{(3a + 5)(a - 4)}$$

$$6(a - 4) - 2(3a + 5) = 10$$
$$6a - 24 - 6a - 10 = 10$$
$$-34 = 10$$

No solution

57.

$$\frac{3}{x^2 - x - 6} + \frac{2}{2x^2 - 5x - 3} = \frac{5}{2x^2 + 5x + 2}$$

$$\frac{3}{(x - 3)(x + 2)} + \frac{2}{(2x + 1)(x - 3)} = \frac{5}{(2x + 1)(x + 2)}$$

$$(x - 3)(x + 2)(2x + 1)\left[\frac{3}{(x - 3)(x + 2)} + \frac{2}{(2x + 1)(x - 3)}\right] = (x - 3)(x + 2)(2x + 1)\left[\frac{5}{(2x + 1)(x + 2)}\right]$$

$$\frac{(x - 3)(x + 2)(2x + 1)}{1} \cdot \frac{3}{(x - 3)(x + 2)} + \frac{(x - 3)(x + 2)(2x + 1)}{1} \cdot \frac{2}{(2x + 1)(x - 3)} = \frac{(x - 3)(x + 2)(2x + 1)}{1} \cdot \frac{5}{(2x + 1)(x + 2)}$$

$$3(2x + 1) + 2(x + 2) = 5(x - 3)$$
$$6x + 3 + 2x + 4 = 5x - 15$$
$$8x + 7 = 5x - 15$$
$$3x + 7 = -15$$
$$3x = -22$$
$$x = \frac{-22}{3}$$

59.

$$\frac{4}{4x^2-9} - \frac{5}{4x^2-8x+3} = \frac{8}{4x^2+4x-3}$$

$$\frac{4}{(2x-3)(2x+3)} - \frac{5}{(2x-1)(2x-3)} = \frac{8}{(2x+3)(2x-1)}$$

$$(2x-3)(2x+3)(2x-1)\left[\frac{4}{(2x-3)(2x+3)} - \frac{5}{(2x-1)(2x-3)}\right] =$$

$$(2x-3)(2x+3)(2x-1)\left[\frac{8}{(2x+3)(2x-1)}\right]$$

$$\frac{(2x-3)(2x+3)(2x-1)}{1} \cdot \frac{4}{(2x-3)(2x+3)} - \frac{(2x-3)(2x+3)(2x-1)}{1} \cdot \frac{5}{(2x-1)(2x-3)} = \frac{(2x-3)(2x+3)(2x-1)}{1} \cdot \frac{8}{(2x+3)(2x-1)}$$

$$4(2x - 1) - 5(2x + 3) = 8(2x - 3)$$
$$8x - 4 - 10x - 15 = 16x - 24$$
$$-2x - 19 = 16x - 24$$
$$-19 = 16x - 24$$
$$-19 = 18x - 24$$
$$5 = 18x$$
$$\frac{5}{18} = x$$

61. $3x^2y(2x^3y^2) = 3 \cdot 2x^{2+3}y^{1+2}$
    $= 6x^5y^3$

63. Answers may vary.
    One example: $f(x) = x^2$

$$f(x) = x^2$$
$$f(x + 2) = (x + 2)^2$$
$$f(2) = 2^2 = 4$$

$$f(x) + f(2) = x^2 + 4$$
$$f(x + 2) = (x + 2)^2 = x^2 + 4x + 4$$
$$f(x) + f(2) \neq f(x + 2)$$

6.6 Exercises

1. $5x + 7y = 4$
    $5x + 7y - 7y = 4 - 7y$
        $5x = 4 - 7y$

$$\frac{5x}{5} = \frac{4 - 7y}{5}$$

$$x = \frac{4 - 7y}{5}$$

3. $2x - 9y = 11$
    $2x - 9y - 2x = 11 - 2x$
        $-9y = 11 - 2x$

$$\frac{-9y}{-9} = \frac{11 - 2x}{-9}$$

$$y = \frac{11 - 2x}{-9}$$

$$y = \frac{2x - 11}{9}$$

5. $w + 4z - 1 = 2w - z + 3$
    $w + 4z - 1 - 4z + 1 - 2w = 2w - z + 3 - 4z + 1 - 2w$
            $-w = -5z + 4$

$$\frac{-w}{-1} = \frac{-5z + 4}{-1}$$

$$w = 5z - 4$$

7.
$$2(6r - 5t) > 5(2r + t)$$
$$12r - 10t > 10r + 5t$$
$$12r - 10t - 10r + 10t > 10r + 5t - 10r + 10t$$
$$2r > 15t$$

$$\frac{2r}{2} > \frac{15t}{2}$$

$$r > \frac{15t}{2}$$

9.
$$3m - 4n + 6p = 5n + 2p - 8$$
$$3m - 4n + 6p + 4n - 2p + 8 = 5n + 2p - 8 + 4n - 2p + 8$$
$$3m + 4p + 8 = 9n$$

$$\frac{3m + 4p + 8}{9} = \frac{9n}{9}$$

$$\frac{3m + 4p + 8}{9} = n$$

11.
$$\frac{a}{5} - \frac{b}{3} = \frac{a}{2} - \frac{b}{6}$$

$$30\left(\frac{a}{5} - \frac{b}{3}\right) = 30\left(\frac{a}{2} - \frac{b}{6}\right)$$

$$\frac{30}{1} \cdot \frac{a}{5} - \frac{30}{1} \cdot \frac{b}{3} = \frac{30}{1} \cdot \frac{a}{2} - \frac{30}{1} \cdot \frac{b}{6}$$

$$6a - 10b = 15a - 5b$$
$$6a - 10b - 6a + 5b = 15a - 5b - 6a + 5b$$
$$-5b = 9a$$

$$\frac{-5b}{9} = \frac{9a}{9}$$

$$\frac{-5b}{9} = a$$

13.
$$\frac{x + y}{3} - \frac{x}{2} + \frac{y}{6} = 3(x - y)$$

$$6\left(\frac{x + y}{3} - \frac{x}{2} + \frac{y}{6}\right) = 6[3(x - y)]$$

$$\frac{6}{1} \cdot \frac{x + y}{3} - \frac{6}{1} \cdot \frac{x}{2} + \frac{6}{1} \cdot \frac{y}{6} = 18(x - y)$$

$$2(x + y) - 3x + y = 18x - 18y$$
$$2x + 2y - 3x + y = 18x - 18y$$

$$-x + 3y = 18x - 18y$$
$$-x + 3y + x + 18y = 18x - 18y + x + 18y$$
$$21y = 19x$$

$$\frac{21y}{19} = \frac{19x}{19}$$

$$\frac{21y}{19} = x$$

15.
$$ax + b = cx + d$$
$$ax + b - cx - b = cx + d - cx - b$$
$$ax - cx = d - b$$
$$x(a - c) = d - b$$

$$\frac{x(a - c)}{a - c} = \frac{d - b}{a - c}$$

$$x = \frac{d - b}{a - c}$$

17.
$$3x + 2y - 5 = ax + by + 1$$
$$3x + 2y - 5 - ax - 2y + 5 = ax + by + 1 - ax - 2y + 5$$
$$3x - ax = by - 2y + 6$$
$$x(3 - a) = by - 2y + 6$$

$$\frac{x(3 - a)}{3 - a} = \frac{by - 2y + 6}{3 - a}$$

$$x = \frac{by - 2y + 6}{3 - a}$$

19. $(x + 3)(y + 7) = a$

$$\frac{(x + 3)(y + 7)}{y + 7} = \frac{a}{y + 7}$$

$$x + 3 = \frac{a}{y + 7}$$

$$x + 3 - 3 = \frac{a}{y + 7} - 3$$

$$x = \frac{a}{y + 7} - 3 = \frac{a - 3y - 21}{y + 7}$$

21.
$$y = \frac{u - 1}{u + 1}$$

$$(u + 1)(y) = (u + 1)\left(\frac{u - 1}{u + 1}\right)$$

$$uy + y = u - 1$$
$$uy + y - uy + 1 = u - 1 - uy + 1$$
$$y + 1 = u - uy$$
$$y + 1 = u(1 - y)$$

$$\frac{y + 1}{1 - y} = \frac{u(1 - y)}{1 - y}$$

$$\frac{y + 1}{1 - y} = u$$

23.
$$x = \frac{2t - 3}{3t - 2}$$

$$(3t - 2)x = (3t - 2)\left(\frac{2t - 3}{3t - 2}\right)$$

$$3tx - 2x = 2t - 3$$
$$3tx - 2x - 2t + 2x = 2t - 3 - 2t + 2x$$
$$3tx - 2t = 2x - 3$$
$$t(3x - 2) = 2x - 3$$

$$\frac{t(3x - 2)}{3x - 2} = \frac{2x - 3}{3x - 2}$$

$$t = \frac{2x - 3}{3x - 2}$$

25. $A = \frac{1}{2}bh$

$$2A = bh$$

$$\frac{2A}{h} = \frac{bh}{h}$$

$$\frac{2A}{h} = b$$

27. $A = \frac{1}{2}h(b_1 + b_2)$

$$2A = h(b_1 + b_2)$$

$$\frac{2A}{h} = \frac{h(b_1 b_2)}{h}$$

$$\frac{2A}{h} = b_1 + b_2$$

$$\frac{2A}{h} - b_2 = b_1 + b_2 - b_2$$

$$\frac{2A}{h} - b_2 = b_1$$

29.
$$A = P(1 + rt)$$
$$A = P + Prt$$
$$A - P = P + Prt - P$$
$$A - P = Prt$$

$$\frac{A - P}{Pt} = \frac{Prt}{Pt}$$

$$\frac{A - P}{Pt} = r$$

31. $C = \frac{5}{9}(F - 32)$

$$\frac{9}{5}(C) = \frac{9}{5}\left[\frac{5}{9}(F - 32)\right]$$

$$\frac{9}{5}C = F - 32$$

$$\frac{9}{5}C + 32 = F - 32 + 32$$

$$\frac{9}{5}C + 32 = F$$

33. $\frac{P_1}{V_1} = \frac{P_2}{V_2}$

$$\frac{V_2}{1} \cdot \frac{P_1}{V_1} = \frac{V_2}{1} \cdot \frac{P_2}{V_2}$$

$$\frac{P_1 V_2}{V_1} = P_2$$

35.

$$S = s_0 + v_0 t + \frac{1}{2} g t^2$$

$$S - s_0 - v_0 t = s_0 + v_0 t + \frac{1}{2} g t^2 - s_0 - v_0 t$$

$$S - s_0 - v_0 t = \frac{1}{2} g t^2$$

$$2(S - s_0 - v_0 t) = g t^2$$

$$\frac{2(S - s_0 - v_0 t)}{t^2} = \frac{g t^2}{t^2}$$

$$\frac{2(S - s_0 - v_0 t)}{t^2} = g$$

37.

$$\frac{x - \mu}{s} < 1.96 \quad , \quad (s > 0)$$

$$x - \mu < 1.96 s$$
$$x - \mu + \mu < 1.96 s + \mu$$
$$x < 1.96 s + \mu$$

39.

$$\frac{1}{f} = \frac{1}{f_1} + \frac{1}{f_2}$$

$$\frac{f f_1 f_2}{1} \cdot \frac{1}{f} = \frac{f f_1 f_2}{1} \left(\frac{1}{f_1} + \frac{1}{f_2} \right)$$

$$f_1 f_2 = \frac{f f_1 f_2}{1} \cdot \frac{1}{f_1} + \frac{f f_1 f_2}{1} \cdot \frac{1}{f_2}$$

$$f_1 f_2 = f f_2 + f f_1$$
$$f_1 f_2 - f f_1 = f f_2 + f f_1 - f f_1$$
$$f_1 f_2 - f f_1 = f f_2$$
$$f_1 (f_2 - f) = f f_2$$

$$\frac{f_1 (f_2 - f)}{f_2 - f} = \frac{f f_2}{f_2 - f}$$

$$f_1 = \frac{f f_2}{f_2 - f}$$

41. $S = 2\pi r^2 + 2\pi r h$

$$S - 2\pi r^2 = 2\pi r^2 + 2\pi r h - 2\pi r^2$$
$$S - 2\pi r^2 = 2\pi r h$$

$$\frac{S - 2\pi r^2}{2\pi r} = \frac{2\pi r h}{2\pi r}$$

$$\frac{S - 2\pi r^2}{2\pi r} = h$$

43.

$$(x - 3)^2 = 4$$
$$x^2 - 6x + 9 = 4$$
$$x^2 - 6x + 5 = 0$$
$$(x - 5)(x - 1) = 0$$
$$x - 5 = 0 \quad \text{or} \quad x - 1 = 0$$
$$x = 5 \quad \text{or} \quad x = 1$$

45. number of heavy-duty batteries: x
number of regular batteries: $18 - x$

$$60(x) + 50(18 - x) = 940$$
$$60x + 900 - 50x = 940$$
$$10x + 900 = 940$$
$$10x = 40$$
$$x = 4$$
$$18 - x = 18 - 4 = 14$$

They bought 4 heavy-duty batteries and 14 regular batteries.

6.7 Exercises

1. Let x = the number

$$\frac{3}{4}(x) = \frac{2}{5}(x) - 7$$

$$20\left(\frac{3}{4}x\right) = 20\left(\frac{2}{5}x - 7\right)$$

$$15x = 8x - 140$$
$$7x = -140$$
$$x = -20$$

The number is -20.

3. Let x = the number of men

$$\frac{7 \text{ men}}{9 \text{ women}} = \frac{x \text{ men}}{810 \text{ women}}$$

$$\frac{7}{9} = \frac{x}{810}$$

$$810\left(\frac{7}{9}\right) = 810\left(\frac{x}{810}\right)$$

$$630 = x$$

There are 630 men.

5. Let x = one number
   then $x - 21$ = the other number

   $$\frac{x - 21}{x} = \frac{5}{12}$$

   $$12x\left(\frac{x - 21}{x}\right) = 12x\left(\frac{5}{12}\right)$$

   $$12(x - 21) = 5x$$
   $$12x - 252 = 5x$$
   $$-252 = -7x$$
   $$36 = x$$
   $$x - 21 = 36 - 21 = 15$$

   The numbers are 15 and 36.

7. Let x = the number of inches in 52 cm

   $$\frac{1 \text{ inch}}{2.54 \text{ cm}} = \frac{x \text{ inches}}{52 \text{ cm}}$$

   $$\frac{1}{2.54} = \frac{x}{52}$$

   $$52\left(\frac{1}{2.54}\right) = 52\left(\frac{x}{52}\right)$$

   $$20.47 = x$$

   There are 20.47 inches in 52 cm.

9. Let x = the number of dribbles in 28 droogs
   and let y = the number of dreeps in 28 droogs

   $$\frac{5 \text{ droogs}}{4 \text{ dreeps}} = \frac{28 \text{ droogs}}{y \text{ dreeps}}$$

   $$\frac{5}{4} = \frac{28}{y}$$

   $$4y\left(\frac{5}{4}\right) = 4y\left(\frac{28}{y}\right)$$

   $$5y = 112$$
   $$y = 22.4$$

   22.4 dreeps is equivalent to 28 droogs.

   $$\frac{7 \text{ dreeps}}{25 \text{ dribbles}} = \frac{22.4 \text{ dreeps}}{x \text{ dribbles}}$$

   $$\frac{7}{25} = \frac{22.4}{x}$$

   $$25x\left(\frac{7}{25}\right) = 25x\left(\frac{22.4}{x}\right)$$

   $$7x = 560$$
   $$x = 80$$

   There are 80 dribbles in 28 droogs.

11. Let x = 2nd side's length,
    then $\frac{1}{2}x$ = 1st side's length
    and $x + 2$ = 3rd side's length

    Perimeter = sum of side lengths
    $$22 = x + \frac{1}{2}x + x + 2$$

    $$22 = \frac{5}{2}x + 2$$

    $$2(22) = 2\left(\frac{5}{2}x + 2\right)$$

    $$44 = 5x + 4$$
    $$40 = 5x$$
    $$8 = x$$

    $$\frac{1}{2}x = \frac{1}{2}(8) = 4$$
    $$x + 2 = 8 + 2 = 10$$

    The sides have lengths 4 cm, 8 cm and 10 cm.

13. Let x = the width
    then the length = $\frac{5}{2}x$

    $$P = 2 \cdot \text{width} + 2 \cdot \text{length}$$

    $$50 = 2x + 2\left(\frac{5}{2}x\right)$$

    $$50 = 2x + 5x$$
    $$50 = 7x$$

    $$\frac{50}{7} = x$$

    $$\frac{5}{2}x = \frac{5}{2}\left(\frac{50}{7}\right) = \frac{125}{7}$$

    The rectangle is $\frac{50}{7}$ cm by $\frac{125}{7}$ cm.

15.

$$\underset{\text{x = Total distance}}{\rule{0pt}{0pt}}$$

| 1/4 x = distance walking | 6 miles = ride in cab |

$$\frac{1}{4}x + 6 = x$$

$$4\left(\frac{1}{4}x + 6\right) = 4(x)$$

$$x + 24 = 4x$$
$$24 = 3x$$
$$8 = x$$

His home is 8 miles from the ballfield.

17.

$$\underset{\text{x = Total distance}}{\rule{0pt}{0pt}}$$

| 1/5x = finds nickel | 4/5x = rest of way |

| 1\5x finds nickel | 1/4(4/5x) finds dime | 2 blocks |

$$\frac{1}{5}x + \frac{1}{4}\left(\frac{4}{5}x\right) + 2 = x$$

$$\frac{1}{5}x + \frac{1}{5}x + 2 = x$$

$$5\left(\frac{1}{5}x + \frac{1}{5}x + 2\right) = 5(x)$$

$$x + x + 10 = 5x$$
$$2x + 10 = 5x$$
$$10 = 3x$$
$$\frac{10}{3} = x$$

She walked $\frac{10}{3} = 3\frac{1}{3}$ blocks.

19.

$$\frac{1}{R} = \frac{1}{R_1} + \frac{1}{R_2} + \frac{1}{R_3}$$

$$\frac{1}{1\frac{1}{4}} = \frac{1}{2} + \frac{1}{5} + \frac{1}{R_3}$$

$$\frac{1}{\frac{5}{4}} = \frac{1}{2} + \frac{1}{5} + \frac{1}{R_3}$$

$$\frac{4}{5} = \frac{1}{2} + \frac{1}{5} + \frac{1}{R_3}$$

$$10R_3\left(\frac{4}{5}\right) = 10R_3\left(\frac{1}{2} + \frac{1}{5} + \frac{1}{R_3}\right)$$

$$8R_3 = 5R_3 + 2R_3 + 10$$
$$8R_3 = 7R_3 + 10$$
$$R_3 = 10$$

The third resistance is 10 ohms.

21. amount in certificate of deposit: $x + 3000$
amount in bond: x

$$0.06(x + 3000) + 0.10(x) = 580$$
$$0.06x + 180 + 0.10x = 580$$
$$0.16x + 180 = 580$$
$$0.16x = 400$$
$$x = 2500$$
$$x + 3000 = 2500 + 3000 = 5500$$

She invested \$5500 in the certificate of deposit and \$2500 in the bond.

23. amount invested at $5\frac{1}{2}\%$: x
amount invested at 7%: $25000 - x$

$$0.055(x) + 0.07(25000 - x) = 1465$$
$$0.055x + 1750 - 0.07x = 1465$$
$$1750 - 0.015x = 1465$$
$$-0.015x = -285$$
$$x = 19000$$
$$25000 - x = 25000 - 19000 = 6000$$

He has \$19,000 in the $5\frac{1}{2}\%$ account and \$6000 in the 7% account.

25. amount in $8\frac{1}{2}$% bond: x

amount in 11% bond: $18000 - x$

$$0.085(x) + 0.11(18000 - x) = 0.10(18000)$$
$$0.085x + 1980 - 0.11x = 1800$$
$$-0.025x + 1980 = 1800$$
$$-0.025x = -180$$
$$x = 7200$$
$$18000 - x = 18000 - 7200 = 10800$$

He should invest \$7200 in the $8\frac{1}{2}$% bond and
\$10,800 in the 11% bond.

27. Let x = number of hours it takes
Carol and Bill to paint the
room working together

Portion completed + Portion completed = 1 complete
by Carol in   by Bill in   job
x hours x hours

$$\frac{x}{3} \qquad + \qquad \frac{x}{5} \qquad = \qquad 1$$

$$15\left(\frac{x}{3} + \frac{x}{5}\right) = 15(1)$$
$$5x + 3x = 15$$
$$8x = 15$$
$$x = \frac{15}{8}$$

It will take them $\frac{15}{8} = 1\frac{7}{8}$ hours to paint the room.

29. Let x = number of hours it takes
the two cleaning services working
together to clean the building

Portion completed + Portion completed = 1 complete
by QCS in   by SQCS in   job
x hours x hours

$$\frac{x}{30} \qquad + \qquad \frac{x}{20} \qquad = \qquad 1$$

$$60\left(\frac{x}{30} + \frac{x}{20}\right) = 60(1)$$
$$2x + 3x = 60$$
$$5x = 60$$
$$x = 12$$

It will take them 12 hours working together.

31. Let x = number of days for the bricklayer and assistant to complete the wall when working together

$$\begin{array}{ccccc} \text{Portion completed} & + & \text{Portion completed} & = & \text{1 complete} \\ \text{by bricklayer} & & \text{by assistant} & & \text{job} \\ \text{in } x \text{ days} & & \text{in } x \text{ days} & & \end{array}$$

$$\frac{x}{2\frac{2}{3}} \quad + \quad \frac{x}{5} \quad = \quad 1$$

$$\frac{x}{\frac{8}{3}} + \frac{x}{5} = 1$$

$$\frac{3x}{8} + \frac{x}{5} = 1$$

$$40\left(\frac{3x}{8} + \frac{x}{5}\right) = 40(1)$$

$$15x + 8x = 40$$
$$23x = 40$$
$$x = \frac{40}{23}$$

It will take them $\frac{40}{23} = 1\frac{17}{23}$ days to complete the wall.

33. Let x = number of hours SQCS works

$$\begin{array}{ccccc} \text{Portion completed} & + & \text{Portion completed} & = & \text{1 complete} \\ \text{by QCS} & & \text{by SQCS} & & \text{job} \\ \frac{10}{30} & + & \frac{x}{20} & = & 1 \end{array}$$

$$\frac{1}{3} + \frac{x}{20} = 1$$

$$60\left(\frac{1}{3} + \frac{x}{20}\right) = 60(1)$$

$$20 + 3x = 60$$
$$3x = 40$$
$$x = \frac{40}{3} = 13\frac{1}{3}$$

To complete the entire job it will take $10 + 13\frac{1}{3} = 23\frac{1}{3}$ hours.

35. Let x = number of minutes for tub to fill

$$\frac{x}{10} - \frac{x}{15} = 1$$

$$30\left(\frac{x}{10} - \frac{x}{15}\right) = 30(1)$$

$$3x - 2x = 30$$
$$x = 30$$

It will take 30 minutes.

37. Let r = rate for Bill
then $r + 10$ = rate for Jill

$$t_{\text{Bill}} = t_{\text{Jill}}$$

$$\frac{d_{\text{Bill}}}{r_{\text{Bill}}} = \frac{d_{\text{Jill}}}{r_{\text{Jill}}}$$

$$\frac{10}{r} = \frac{15}{r + 10}$$

$$r(r + 10)\left(\frac{10}{r}\right) = r(r + 10)\left(\frac{15}{r + 10}\right)$$

$$10(r + 10) = 15r$$
$$10r + 100 = 15r$$
$$100 = 5r$$
$$20 = r$$

Bill rides 20 kph.

39. hours at slower speed: x
hours at faster speed: $14 - x$

$$\begin{array}{ccccc} \text{distance} & + & \text{distance} & = & \text{total} \\ \text{at slower} & & \text{at faster} & & \text{distance} \\ \text{speed} & & \text{speed} & & \end{array}$$

$$20x + 50(14 - x) = 600$$
$$20x + 700 - 50x = 600$$
$$-30x + 700 = 600$$
$$-30x = -100$$
$$x = \frac{10}{3}$$

She drove at a slower speed for $\frac{10}{3}$ hours.

Her distance at this speed was $20\left(\frac{10}{3}\right) = 66\frac{2}{3}$ miles.

41. Let x = amount of 20% solution

Amount of alcohol + Amount of alcohol = Total amount
in the 20% solution   in the 50% solution   of alcohol in
  final solution

$$0.20(x) \quad + \quad 0.50(5) \quad = 0.30(x + 5)$$
$$0.20x + 2.5 = 0.30x + 1.5$$
$$10(0.20x + 2.5) = 10(0.30x + 1.5)$$
$$2x + 25 = 3x + 15$$
$$25 = x + 15$$
$$10 = x$$

10 oz. of the 20% solution should be used.

43. Let x = amount of 30% solution
Let y = amount of 75% solution

$$\begin{cases} x + y = 80 \\ 0.30x + 0.75y = 0.50(80) \end{cases}$$

$x + y = 80$
  $x = 80 - y$
Substitute into 2nd equation:

$$0.30(80 - y) + 0.75y = 40$$
$$24 - 0.30y + 0.75y = 40$$
$$24 + 0.45y = 40$$
$$0.45y = 16$$

$$y = 35\frac{5}{9}$$

$$x = 80 - y$$

$$= 80 - 35\frac{5}{9}$$

$$= 44\frac{4}{9}$$

$44\frac{4}{9}$ ml of the 30% solution should be mixed

with $35\frac{5}{9}$ ml of the 75% solution.

45. Let x = amount of 40% alloy
Let y = amount of 60% alloy

$$\begin{cases} x + y = 80 \\ 0.40x + 0.60y = 0.55(80) \end{cases}$$

$x = 80 - y$
Substitute into 2nd equation:

$$0.40(80 - y) + 0.60y = 44$$
$$32 - 0.40y + 0.60y = 44$$
$$32 + 0.20y = 44$$
$$0.20y = 12$$
$$y = 60$$
$$x = 80 - y$$
$$= 80 - 60$$
$$= 20$$

20 tons of the 40% alloy and 60 tons of the 60% alloy should be used.

47. Let x = amount of pure alcohol

| Amount of alcohol in 60% solution | + | Amount of alcohol in pure solution | = | Total amount of alcohol in final solution |

$$0.60(2) \quad + \quad 1(x) \quad = \quad 0.80(2 + x)$$
$$1.2 + x = 1.6 + 0.80x$$
$$10(1.2. + x) = 10(1.6 + 0.80x)$$
$$12 + 10x = 16 + 8x$$
$$12 + 2x = 16$$
$$2x = 4$$
$$x = 2$$

He must add 2 liters of pure alcohol.

49. Let x = amount drained off = water added

$$0.30(3 - x) + 0(x) = 0.20(3)$$
$$0.9 - 0.3x = 0.6$$
$$10(0.9 - 0.3x) = 10(0.6)$$
$$9 - 3x = 6$$
$$-3x = -3$$
$$x = 1$$

He drained off 1 gallon.

51. Let x = number of advance tickets
Let y = number of door tickets

$$\begin{cases} x + y = 3600 \\ 15x + 20.50y = 61700 \end{cases}$$

$$y = 3600 - x$$
Substitute into 2nd equation:

$$15x + 20.50(3600 - x) = 61700$$
$$15x + 73800 - 20.50x = 61700$$
$$-5.5x + 73800 = 61700$$
$$-5.5x = -12100$$
$$x = 2200$$

2200 advance tickets were sold.

53. Let x = number of nickels,
then $x + 5$ = number of dimes
and $2x$ = number of quarters

$$5(x) + 10(x + 5) + 25(2x) = 2000$$
$$5x + 10x + 50 + 50x = 2000$$
$$65x + 50 = 2000$$
$$65x = 1950$$
$$x = 30$$
$$x + 5 = 30 + 5 = 35$$
$$2x = 2(30) = 60$$

He has 30 nickels, 35 dimes and 60 quarters.

55. Let x = number of orchestra tickets, then $2x$ = number of general admission tickets and $900 - (x + 2x) = 900 - 3x$ = number of balcony tickets

$$25(x) + 20.50(900 - 3x) + 16(2x) = 17325$$
$$25x + 18450 - 61.5x + 32x = 17325$$
$$-4.5x + 18450 = 17325$$
$$-4.5x = -1125$$
$$x = 250$$
$$2x = 2(250) = 500$$
$$900 - 3x = 900 - 3(250) = 150$$

They sold 250 orchestra tickets, 500 general admission tickets and 150 balcony tickets.

57. Let x = score on final exam

$$0.20(85) + 0.20(65) + 0.20(72) + 0.40(x) \geq 80$$
$$17 + 13 + 14.4 + 0.40x \geq 80$$
$$44.4 + 0.40x \geq 80$$
$$0.40x \geq 35.6$$
$$x \geq 89$$

He must receive at least a grade of 89.

59. Let x = amount in high-risk bond
then $20000 - x$ = amount in savings

$$0.082(x) + 0.039(20000 - x) \geq 1000$$
$$0.082x + 780 - 0.039x \geq 1000$$
$$0.043x + 780 \geq 1000$$
$$0.043x \geq 220$$
$$x \geq 5116.28$$

She must invest at least $5116.28 in the high-risk bond.

61. Let x = final exam score

$$0.20(85) + 0.20(92) + 0.20(86) + 0.40(x) \geq 90$$
$$17 + 18.4 + 17.2 + 0.40x \geq 90$$
$$52.6 + 0.40x \geq 90$$
$$0.40x \geq 37.4$$
$$x \geq 93.5$$

He must score at least 93.5.

73. $\quad 3x^3y - x^2y^2 + 3xy^3$
$\quad = 3xy(x^2 - 2xy + y^2)$
$\quad = 3xy(x - y)^2$

75. $\qquad |2x + 8| \leq 10$
$\quad -10 \leq 2x + 8 \leq 10$
$\quad -18 \leq 2x \leq 2$
$\quad -9 \leq x \leq 1$

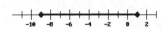

CHAPTER 6 REVIEW EXERCISES

1. $\dfrac{4x^2y^3}{16xy^5} = \dfrac{4 \cdot x \cdot x \cdot y^3}{4 \cdot 4 \cdot x \cdot y^3 \cdot y^2}$
$\qquad\qquad = \dfrac{x}{4y^2}$

3. $\dfrac{x^2 + 2x - 8}{x^2 + 3x - 10} = \dfrac{(x + 4)(x - 2)}{(x + 5)(x - 2)}$
$\qquad\qquad\qquad = \dfrac{x + 4}{x + 5}$

5. $\dfrac{x^4 - 2x^3 + 3x^2}{x^2} = \dfrac{x^2(x^2 - 2x + 3)}{x^2}$
$\qquad\qquad\qquad = x^2 - 2x + 3$

7. $\dfrac{5xa - 7a + 5xb - 7b}{3xa - 2a + 3xb - 2b} = \dfrac{(5x - 7)(a + b)}{(3x - 2)(a + b)}$
$\qquad\qquad\qquad\qquad = \dfrac{5x - 7}{3x - 2}$

9. $\dfrac{4x^2y^3z^2}{12xy^4} \cdot \dfrac{24xy^5}{16xy}$

$= \dfrac{4 \cdot 24x^3y^8z^2}{12 \cdot 16x^2y^5}$

$= \dfrac{4 \cdot 3 \cdot 8x \cdot x^2y^5 \cdot y^3z^2}{4 \cdot 3 \cdot 8 \cdot 2x^2y^5}$

$= \dfrac{xy^3z^2}{2}$

11. $\dfrac{5}{3x^2y} + \dfrac{1}{3x^2y} = \dfrac{5 + 1}{3x^2y}$

$\qquad\qquad\qquad = \dfrac{6}{3x^2y}$

$\qquad\qquad\qquad = \dfrac{2}{x^2y}$

13. $\dfrac{3x}{x - 1} + \dfrac{3}{x - 1} = \dfrac{3x + 3}{x - 1}$

15. $\dfrac{2x^2}{x^2 + x - 6} + \dfrac{2x}{x^2 + x - 6} - \dfrac{12}{x^2 + x - 6}$

$= \dfrac{2x^2 + 2x - 12}{x^2 + x - 6}$

$= \dfrac{2(x^2 + x - 6)}{x^2 + x - 6}$

$= 2$

17. $\dfrac{5}{3a^2b} - \dfrac{8}{4ab^4}$

$= \dfrac{5(4b^3)}{12a^2b^4} - \dfrac{8(3a)}{12a^2b^4}$

$= \dfrac{5(4b^3) - 8(3a)}{12a^2b^4}$

$= \dfrac{20b^3 - 24a}{12a^2b^4}$

$= \dfrac{4(5b^3 - 6a)}{12a^2b^4}$

$= \dfrac{5b^3 - 6a}{3a^2b^4}$

19. $\dfrac{3x + 1}{2x^2} - \dfrac{3x - 2}{5x}$

$= \dfrac{(3x + 1)(5)}{10x^2} - \dfrac{(3x - 2)(2x)}{10x^2}$

$= \dfrac{(3x + 1)(5) - (3x - 2)(2x)}{10x^2}$

$= \dfrac{15x + 5 - 6x^2 + 4x}{10x^2}$

$= \dfrac{-6x^2 + 19x + 5}{10x^2}$

21. $\dfrac{x - 7}{5 - x} + \dfrac{3x + 3}{x - 5}$

$= \dfrac{x - 7}{(-1)(x - 5)} + \dfrac{3x + 3}{x - 5}$

$= \dfrac{(-1)(x - 7)}{x - 5} + \dfrac{3x + 3}{x - 5}$

$= \dfrac{(-1)(x - 7) + 3x + 3}{x - 5}$

$= \dfrac{-x + 7 + 3x + 3}{x - 5}$

$= \dfrac{2x + 10}{x - 5}$

23. $\dfrac{x^2 + x - 6}{x + 4} \cdot \dfrac{2x^2 + 8x}{x^2 + x - 6}$

$= \dfrac{(x + 3)(x - 2)}{x + 4} \cdot \dfrac{2x(x + 4)}{(x + 3)(x - 2)}$

$= 2x$

25. $\dfrac{a^2 - 2ab + b^2}{a + b} \div \dfrac{(a - b)^3}{a + b}$

$= \dfrac{a^2 - 2ab + b^2}{a + b} \cdot \dfrac{a + b}{(a - b)^3}$

$= \dfrac{(a - b)^2}{a + b} \cdot \dfrac{a + b}{(a - b)^3}$

$= \dfrac{1}{a - b}$

27. $\dfrac{3x}{2x + 3} - \dfrac{5}{x - 4}$

$= \dfrac{3x(x - 4)}{(2x + 3)(x - 4)} - \dfrac{5(2x + 3)}{(2x + 3)(x - 4)}$

$= \dfrac{3x(x - 4) - 5(2x + 3)}{(2x + 3)(x - 4)}$

$= \dfrac{3x^2 - 12x - 10x - 15}{(2x + 3)(x - 4)}$

$= \dfrac{3x^2 - 22x - 15}{(2x + 3)(x - 4)}$

29. $\dfrac{3x - 2}{2x - 7} + \dfrac{5x + 2}{2x - 3}$

$= \dfrac{(3x - 2)(2x - 3)}{(2x - 7)(2x - 3)} + \dfrac{(5x + 2)(2x - 7)}{(2x - 7)(2x - 3)}$

$= \dfrac{(3x - 2)(2x - 3) + (5x + 2)(2x - 7)}{(2x - 7)(2x - 3)}$

$= \dfrac{6x^2 - 13x + 6 + 10x^2 - 31x - 14}{(2x - 7)(2x - 3)}$

$= \dfrac{16x^2 - 44x - 8}{(2x - 7)(2x - 3)}$

31. $\dfrac{5a}{a^2 - 3a} + \dfrac{2}{4a^3 + 4a^2}$

$= \dfrac{5a}{a(a - 3)} + \dfrac{2}{4a^2(a + 1)}$

$= \dfrac{5a(4a)(a + 1)}{4a^2(a - 3)(a + 1)} + \dfrac{2(a - 3)}{4a^2(a - 3)(a + 1)}$

$= \dfrac{5a(4a)(a + 1) + 2(a - 3)}{4a^2(a - 3)(a + 1)}$

$= \dfrac{20a^3 + 20a^2 + 2a - 6}{4a^2(a - 3)(a + 1)}$

$= \dfrac{2(10a^3 + 10a^2 + a - 3)}{4a^2(a - 3)(a + 1)}$

$= \dfrac{10a^3 + 10a^2 + a - 3}{2a^2(a - 3)(a + 1)}$

33. $\dfrac{5}{x^2 - 4x + 4} + \dfrac{3}{x^2 - 4}$

$= \dfrac{5}{(x - 2)(x - 2)} + \dfrac{3}{(x - 2)(x + 2)}$

$= \dfrac{5(x + 2)}{(x - 2)^2(x + 2)} + \dfrac{3(x - 2)}{(x - 2)^2(x + 2)}$

$$= \frac{5(x + 2) + 3(x - 2)}{(x - 2)^2(x + 2)}$$

$$= \frac{5x + 10 + 3x - 6}{(x - 2)^2(x + 2)}$$

$$= \frac{8x + 4}{(x - 2)^2(x + 2)}$$

39. $\left(\dfrac{2x + y}{5x^2y - xy^2}\right)\left(\dfrac{25x^2 - y^2}{10x^2 + 3xy - y^2}\right)\left(\dfrac{5x^2 - xy}{5x + y}\right)$

$$= \frac{2x + y}{xy(5x - y)} \cdot \frac{(5x - y)(5x + y)}{(5x - y)(2x + y)} \cdot \frac{x(5x - y)}{5x + y}$$

$$= \frac{1}{y}$$

35. $\dfrac{2x}{7x^2 - 14x - 21} + \dfrac{2x}{14x - 42}$

$$= \frac{2x}{7(x - 3)(x + 1)} + \frac{2x}{14(x - 3)}$$

$$= \frac{2x(2)}{14(x - 3)(x + 1)} + \frac{2x(x + 1)}{14(x - 3)(x + 1)}$$

$$= \frac{2x(2) + 2x(x + 1)}{14(x - 3)(x + 1)}$$

$$= \frac{4x + 2x^2 + 2x}{14(x - 3)(x + 1)}$$

$$= \frac{2x^2 + 6x}{14(x - 3)(x + 1)}$$

$$= \frac{2x(x + 3)}{14(x - 3)(x + 1)}$$

$$= \frac{x(x + 3)}{x(x - 3)(x + 1)}$$

41. $\dfrac{4x + 11}{x^2 + x - 6} - \dfrac{x + 2}{x^2 + 4x + 3}$

$$= \frac{4x + 11}{(x + 3)(x - 2)} - \frac{x + 2}{(x + 3)(x + 1)}$$

$$= \frac{(4x + 11)(x + 1)}{(x + 3)(x - 2)(x + 1)} - \frac{(x + 2)(x - 2)}{(x + 3)(x - 2)(x + 1)}$$

$$= \frac{(4x + 11)(x + 1) - (x + 2)(x - 2)}{(x + 3)(x - 2)(x + 1)}$$

$$= \frac{4x^2 + 15x + 11 - (x^2 - 4)}{(x + 3)(x - 2)(x + 1)}$$

$$= \frac{4x^2 + 15x + 11 - x^2 + 4}{(x + 3)(x - 2)(x + 1)}$$

$$= \frac{3x^2 + 15x + 15}{(x + 3)(x - 2)(x + 1)}$$

37. $\dfrac{5x}{x - 2} + \dfrac{3x}{x + 2} - \dfrac{2x + 3}{x^2 - 4}$

$$= \frac{5x}{x - 2} + \frac{3x}{x + 2} - \frac{2x + 3}{(x - 2)(x + 2)}$$

$$= \frac{5x(x + 2)}{(x - 2)(x + 2)} + \frac{3x(x - 2)}{(x - 2)(x + 2)} - \frac{2x + 3}{(x - 2)(x + 2)}$$

$$= \frac{5x(x + 2) + 3x(x - 2) - (2x + 3)}{(x - 2)(x + 2)}$$

$$= \frac{5x^2 + 10x + 3x^2 - 6x - 2x - 3}{(x - 2)(x + 2)}$$

$$= \frac{8x^2 + 2x - 3}{(x - 2)(x + 2)}$$

43. $\dfrac{5x}{x^2 - x - 2} + \dfrac{4x + 3}{x^3 + x^2} - \dfrac{x - 6}{x^3 - 2x^2}$

$$= \frac{5x}{(x - 2)(x + 1)} + \frac{4x + 3}{x^2(x + 1)} - \frac{x - 6}{x^2(x - 2)}$$

$$= \frac{5x(x^2)}{x^2(x - 2)(x + 1)} + \frac{(4x + 3)(x - 2)}{x^2(x - 2)(x + 1)} - \frac{(x - 6)(x + 1)}{x^2(x - 2)(x + 1)}$$

$$= \frac{5x(x^2) + (4x + 3)(x - 2) - (x - 6)(x + 1)}{x^2(x - 2)(x + 1)}$$

$$= \frac{5x^3 + 4x^2 - 5x - 6 - x^2 + 5x + 6}{x^2(x - 2)(x + 1)}$$

$$= \frac{5x^3 + 4x^2 - 5x - 6 - x^2 + 5x + 6}{x^2(x - 2)(x + 1)}$$

$$= \frac{5x^3 + 3x^2}{x^2(x - 2)(x + 1)}$$

$$= \frac{x^2(5x + 3)}{x^2(x - 2)(x + 1)}$$

$$= \frac{5x + 3}{(x - 2)(x + 1)}$$

45. $\dfrac{4x^2 + 12x + 9}{8x^3 + 27} \cdot \dfrac{12x^3 - 18x^2 + 27x}{4x^2 - 9}$

$= \dfrac{(2x + 3)(2x + 3)}{(2x + 3)(4x^2 - 6x + 9)} \cdot \dfrac{3x(4x^2 - 6x + 9)}{(2x - 3)(2x + 3)}$

$= \dfrac{3x}{2x - 3}$

47. $4x \div \left(\dfrac{8x^2 - 8xy}{2ax + bx - 2ay - by} \div \dfrac{2ax + 2bx + 3ay + 3by}{2a^2 + 3ab + b^2} \right)$

$= 4x \div \left(\dfrac{8x^2 - 8xy}{2ax + bx - 2ay - by} \cdot \dfrac{2a^2 + 3ab + b^2}{2ax + 2bx + 3ay + 3by} \right)$

$= 4x \div \left[\dfrac{8x(x - y)}{(2a + b)(x - y)} \cdot \dfrac{(2a + b)(a + b)}{(a + b)(2x + 3y)} \right]$

$= 4x \div \left(\dfrac{8x}{2x + 3y} \right)$

$= \dfrac{4x}{1} \cdot \dfrac{2x + 3y}{8x}$

$= \dfrac{2x + 3y}{2}$

49. $\left(\dfrac{x}{2} + \dfrac{3}{x} \right) \cdot \dfrac{x + 1}{x}$

$= \left(\dfrac{x(x)}{2x} + \dfrac{3(2)}{2x} \right) \cdot \dfrac{x + 1}{x}$

$= \dfrac{x(x) + 3(2)}{2x} \cdot \dfrac{x + 1}{x}$

$= \dfrac{x^2 + 6}{2x} \cdot \dfrac{x + 1}{x}$

$= \dfrac{(x^2 + 6)(x + 1)}{2x^2}$

$= \dfrac{x^3 + x^2 + 6x + 6}{2x^2}$

51. $\dfrac{\dfrac{3x^2y}{2ab}}{\dfrac{9x}{16a^2}}$

$= \dfrac{3x^2y}{2ab} \div \dfrac{9x}{16a^2}$

$= \dfrac{3x^2y}{2ab} \cdot \dfrac{16a^2}{9x}$

$= \dfrac{8xya}{3b}$

53. $\dfrac{\dfrac{3}{a} - \dfrac{2}{a}}{\dfrac{5}{a}}$

$= \dfrac{\dfrac{3}{a} - \dfrac{2}{a}}{\dfrac{5}{a}} \cdot \dfrac{a}{a}$

$= \dfrac{\dfrac{3}{a} \cdot \dfrac{a}{1} - \dfrac{2}{a} \cdot \dfrac{a}{1}}{\dfrac{5}{a} \cdot \dfrac{a}{1}}$

$= \dfrac{3 - 2}{5}$

$= \dfrac{1}{5}$

55. $\dfrac{\dfrac{3}{b + 1} + 2}{\dfrac{2}{b - 1} + b}$

$= \dfrac{\dfrac{3}{b + 1} + 2}{\dfrac{2}{b - 1} + b} \cdot \dfrac{(b + 1)(b - 1)}{(b + 1)(b - 1)}$

$= \dfrac{\dfrac{3}{b+1} \cdot \dfrac{(b+1)(b-1)}{1} + 2(b+1)(b-1)}{\dfrac{2}{b-1} \cdot \dfrac{(b+1)(b-1)}{1} + b(b+1)(b-1)}$

$$= \frac{3(b-1) + 2(b^2-1)}{2(b+1) + b(b^2-1)}$$

$$= \frac{3b - 3 + 2b^2 - 2}{2b + 2 + b^3 - b}$$

$$= \frac{2b^2 + 3b - 5}{b^3 + b + 2}$$

57.
$$\frac{x}{3} + \frac{x-1}{2} = \frac{7}{6}$$

$$6\left(\frac{x}{3} + \frac{x-1}{2}\right) = 6\left(\frac{7}{6}\right)$$

$$\frac{6}{1} \cdot \frac{x}{3} + \frac{6}{1} \cdot \frac{x-1}{2} = \frac{6}{1} \cdot \frac{7}{6}$$

$$2x + 3(x-1) = 7$$
$$2x + 3x - 3 = 7$$
$$5x - 3 = 7$$
$$5x = 10$$
$$x = 2$$

59.
$$\frac{x}{5} - \frac{x+1}{3} < \frac{1}{3}$$

$$15\left(\frac{x}{5} - \frac{x+1}{3}\right) < 15\left(\frac{1}{3}\right)$$

$$\frac{15}{1} \cdot \frac{x}{5} - \frac{15}{1} \cdot \frac{x+1}{3} < \frac{15}{1} \cdot \frac{1}{3}$$

$$3x - 5(x+1) < 5$$
$$3x - 5x - 5 < 5$$
$$-2x - 5 < 5$$
$$-2x < 10$$
$$x > -5$$

61.
$$\frac{5}{x} - \frac{1}{3} = \frac{11}{3x}$$

$$3x\left(\frac{5}{x} - \frac{1}{3}\right) = 3x\left(\frac{11}{3x}\right)$$

$$\frac{3x}{1} \cdot \frac{5}{x} - \frac{3x}{1} \cdot \frac{1}{3} = \frac{3x}{1} \cdot \frac{11}{3x}$$

$$15 - x = 11$$
$$-x = -4$$
$$x = 4$$

63.
$$\frac{x+1}{3} - \frac{x}{2} > 4$$

$$12\left(\frac{x+1}{3} - \frac{x}{2}\right) > 12(4)$$

$$\frac{12}{1} \cdot \frac{x+1}{3} - \frac{12}{1} \cdot \frac{x}{2} > 48$$

$$4(x+1) - 6x > 48$$
$$4x + 4 - 6x > 48$$
$$-2x + 4 > 48$$
$$-2x > 44$$
$$x < -22$$

65.
$$-\frac{7}{x} + 1 = -13$$

$$x\left(-\frac{7}{x} + 1\right) = x(-13)$$

$$\frac{x}{1}\left(-\frac{7}{x}\right) + x = -13x$$

$$-7 + x = -13x$$
$$-7 = -14x$$
$$\frac{1}{2} = x$$

67.
$$\frac{5}{x-2} - 1 = 0$$

$$(x-2)\left(\frac{5}{x-2} - 1\right) = (x-2)(0)$$

$$\frac{x-2}{1} \cdot \frac{5}{x-2} - (x-2)(1) = 0$$

$$5 - x + 2 = 0$$
$$-x + 7 = 0$$
$$7 = x$$

69.
$$\frac{x-2}{5} - \frac{3-x}{15} > \frac{1}{9}$$

$$45\left(\frac{x-2}{5} - \frac{3-x}{15}\right) > 45\left(\frac{1}{9}\right)$$

$$\frac{45}{1} \cdot \frac{x-2}{5} - \frac{45}{1} \cdot \frac{3-x}{15} > \frac{45}{1} \cdot \frac{1}{9}$$

$$9(x-2) - 3(3-x) > 5$$
$$9x - 18 - 9 + 3x > 5$$
$$12x - 27 > 5$$
$$12x > 32$$
$$x > \frac{8}{3}$$

71.
$$\frac{7}{x-1} + 4 = \frac{x+6}{x-1}$$

$$(x-1)\left(\frac{7}{x-1} + 4\right) = (x-1)\left(\frac{x+6}{x-1}\right)$$

$$\frac{x-1}{1} \cdot \frac{7}{x-1} + (x-1)(4) = \frac{x-1}{1} \cdot \frac{x+6}{x-1}$$

$$7 + 4x - 4 = x + 6$$
$$4x + 3 = x + 6$$
$$3x + 3 = 6$$
$$3x = 3$$
$$x = 1$$

No solution, since $x = 1$ causes a denominator to equal 0 in the original equation.

73.
$$\frac{4x+1}{x^2 - x - 6} = \frac{2}{x-3} + \frac{5}{x+2}$$

$$\frac{4x+1}{(x-3)(x+2)} = \frac{2}{x-3} + \frac{5}{x+2}$$

$$(x-3)(x+2)\left[\frac{4x+1}{(x-3)(x+2)}\right] = (x-3)(x+2)\left(\frac{2}{x-3} + \frac{5}{x+2}\right)$$

$$\frac{(x-3)(x+2)}{1} \cdot \frac{4x+1}{(x-3(x+2))} = \frac{(x-3)(x+2)}{1} \cdot \frac{2}{x-3} + \frac{(x-3)(x+2)}{1} \cdot \frac{5}{x+2}$$

$$4x + 1 = 2(x+2) + 5(x-3)$$
$$4x + 1 = 2x + 4 + 5x - 15$$
$$4x + 1 = 7x - 11$$
$$1 = 3x - 11$$
$$12 = 3x$$
$$4 = x$$

75. $5x - 3y = 2x + 7y$
$$3x - 3y = 7y$$
$$3x = 10y$$
$$x = \frac{10}{3}y$$

77. $3xy = 2xy + 4$
$$xy = 4$$
$$y = \frac{4}{x}$$

79.
$$\frac{2x+1}{y} = x$$

$$\frac{y}{1} \cdot \frac{2x+1}{y} = yx$$

$$2x + 1 = yx$$

$$\frac{2x+1}{x} = \frac{yx}{x}$$

$$\frac{2x+1}{x} = y$$

81.
$$\frac{ax + b}{cx + d} = y$$

$$\frac{cx + d}{1} \cdot \frac{ax + b}{cx + d} = (cx + d)y$$

$$ax + b = cxy + dy$$
$$ax - cxy = dy - b$$
$$x(a - cy) = dy - b$$
$$x = \frac{dy - b}{a - cy}$$

83.
$$\frac{1}{a} + \frac{1}{b} + \frac{1}{c} = \frac{1}{d}$$

$$abcd\left(\frac{1}{a} + \frac{1}{b} + \frac{1}{c}\right) = abcd\left(\frac{1}{d}\right)$$

$$\frac{abcd}{1} \cdot \frac{1}{a} + \frac{abcd}{1} \cdot \frac{1}{b} + \frac{abcd}{1} \cdot \frac{1}{c} = \frac{abcd}{1} \cdot \frac{1}{d}$$

$$bcd + acd + abd = abc$$
$$acd = abc - bcd - abd$$
$$acd = b(ac - cd - ad)$$
$$\frac{acd}{ac - cd - ad} = b$$

85. Let x = number of inches in 1 cm

$$\frac{1 \text{ inch}}{2.54 \text{ cm}} = \frac{x \text{ inches}}{1 \text{ cm}}$$

$$\frac{1}{2.54} = \frac{x}{1}$$

$$0.3937 = x$$

There is 0.3937 inches in 1 c.m.

87.

x = Total distance

$1/2\,x$ $\quad$ $1/3(1/2\,x)$ $\quad$ 1 1/2 miles

$$\frac{1}{2}x + \frac{1}{3}\left(\frac{1}{2}x\right) + 1\frac{1}{2} = x$$

$$\frac{1}{2}x + \frac{1}{6}x + \frac{3}{2} = x$$

$$6\left(\frac{1}{2}x + \frac{1}{6}x + \frac{3}{2}\right) = 6(x)$$

$$3x + x + 9 = 6x$$
$$4x + 9 = 6x$$
$$9 = 2x$$
$$\frac{9}{2} = x$$

Carol walked $\frac{9}{2} = 4\frac{1}{2}$ miles.

89. Let x = number of hours Charles and Ellen work together.

Portion of job completed by Charles	+	Portion of job completed by Ellen	=	1 whole job
$\dfrac{x}{2\frac{1}{2}}$	$+$	$\dfrac{x}{2\frac{1}{3}}$	$=$	1

$$\frac{2}{5}x + \frac{3}{7}x = 1$$

$$35\left(\frac{2}{5}x + \frac{3}{7}x\right) = 35(1)$$

$$14x + 15x = 35$$
$$29x = 35$$
$$x = \frac{35}{29}$$

It would take them $\frac{35}{29} = 1\frac{6}{29}$ days working together.

91. Let x = amount of 35% solution

$$0.35x + 0.70(5) = 0.60(x + 5)$$
$$100[0.35x + 0.70(5)] = 100[0.60(x + 5)]$$
$$35x + 350 = 60x + 300$$
$$35x + 50 = 60x$$
$$50 = 25x$$
$$2 = x$$

He should use 2 liters of the 35% solution.

93. number of children's tickets: x
number of adult tickets: $980 - x$

$$1.50(x) + 4.25(980 - x) = 3010$$
$$1.50x + 4165 - 4.25x = 3010$$
$$4165 - 2.75x = 3010$$
$$-2.75x = -1155$$
$$x = 420$$
$$980 - x = 980 - 420 = 560$$

They sold 420 children's tickets and 560 adult tickets.

CHAPTER 6 PRACTICE TEST

1. (a) $\dfrac{24x^2y^4}{64x^3y} = \dfrac{8 \cdot 3x^2 \cdot y \cdot y^3}{8 \cdot 8x^2 \cdot x \cdot y}$

$= \dfrac{3y^3}{8x}$

(b) $\dfrac{x^2 - 9}{x^2 - 6x + 9} = \dfrac{(x+3)(x-3)}{(x-3)(x-3)}$

$= \dfrac{x+3}{x-3}$

(c) $\dfrac{6x^3 - 9x^2 - 6x}{5x^3 - 10x^2}$

$= \dfrac{3x(2x+1)(x-2)}{5x^2(x-2)}$

$= \dfrac{3(2x+1)}{5x}$

2. (a) $\dfrac{4xy^3}{5ab^4} \cdot \dfrac{15}{16x^4y^5}$

$= \dfrac{4xy^3 \cdot 5 \cdot 3}{5ab^4 \cdot 4 \cdot 4x \cdot x^3y^3 \cdot y^2}$

$= \dfrac{3}{4ab^4x^3y^2}$

(b) $\dfrac{3x}{18y^2} + \dfrac{5}{8x^2y}$

$= \dfrac{x}{6y^2} + \dfrac{5}{8x^2y}$

$= \dfrac{x(4x^2)}{24x^2y^2} + \dfrac{5(3y)}{24x^2y^2}$

$= \dfrac{x(4x^2) + 5(3y)}{24x^2y^2}$

$= \dfrac{4x^3 + 15y}{24x^2y^2}$

(c) $\dfrac{r^2 - rs - 2s^2}{2s^2 + 4rs} \div \dfrac{r - 2s}{4s^2 + 8rs}$

$= \dfrac{r^2 - rs - 2s^2}{2s^2 + 4rs} \cdot \dfrac{4s^2 + 8rs}{r - 2s}$

$= \dfrac{(r - 2s)(r + s)}{2s(s + 2r)} \cdot \dfrac{4s(s + 2r)}{r - 2s}$

$= 2(r + s)$

(d) $\dfrac{9x - 2}{4x - 3} + \dfrac{x + 4}{3 - 4x}$

$= \dfrac{9x - 2}{4x - 3} + \dfrac{x + 4}{(-1)(4x - 3)}$

$= \dfrac{9x - 2}{4x - 3} + \dfrac{(-1)(x + 4)}{4x - 3}$

$= \dfrac{9x - 2 + (-1)(x + 4)}{4x - 3}$

$= \dfrac{9x - 2 - x - 4}{4x - 3}$

$= \dfrac{8x - 6}{4x - 3}$

$= \dfrac{2(4x - 3)}{4x - 3}$

$= 2$

(e) $\dfrac{3x}{x^2 - 4} + \dfrac{4}{x^2 - 5x + 6} - \dfrac{2x}{x^2 - x - 6}$

$= \dfrac{3x}{(x - 2)(x + 2)} + \dfrac{4}{(x - 2)(x - 3)} - \dfrac{2x}{(x - 3)(x + 2)}$

$= \dfrac{3x(x - 3)}{(x - 2)(x + 2)(x - 3)} + \dfrac{4(x + 2)}{(x - 2)(x + 2)(x - 3)} - \dfrac{2x(x - 2)}{(x - 2)(x + 2)(x - 3)}$

$= \dfrac{3x(x - 3) + 4(x + 2) - 2x(x - 2)}{(x - 2)(x + 2)(x - 3)}$

$= \dfrac{3x^2 - 9x + 4x + 8 - 2x^2 + 4x}{(x - 2)(x + 2)(x - 3)}$

$= \dfrac{x^2 - x + 8}{(x - 2)(x + 2)(x - 3)}$

(f) $\left(\dfrac{3}{x} - \dfrac{2}{x+1}\right) \div \dfrac{1}{x+1}$

$= \left[\dfrac{3(x+1)}{x(x+1)} - \dfrac{2(x)}{x(x+1)}\right] \div \dfrac{1}{x+1}$

$= \dfrac{3(x+1) - 2x}{x(x+1)} \div \dfrac{1}{x+1}$

$= \dfrac{3x + 3 - 2x}{x(x+1)} \div \dfrac{1}{x+1}$

$= \dfrac{x+3}{x(x+1)} \cdot \dfrac{x+1}{1}$

$= \dfrac{x+3}{x}$

3. $\dfrac{\dfrac{3}{x+1} - 2}{\dfrac{5}{x} + 1}$

$= \dfrac{\dfrac{3}{x+1} - 2}{\dfrac{5}{x} + 1} \cdot \dfrac{x(x+1)}{x(x+1)}$

$= \dfrac{\dfrac{3}{x+1} \cdot \dfrac{x(x+1)}{1} - 2x(x+1)}{\dfrac{5}{x} \cdot \dfrac{x(x+1)}{1} + 1x(x+1)}$

$= \dfrac{3x - 2x^2 - 2x}{5(x+1) + x^2 + x}$

$= \dfrac{-2x^2 + x}{5x + 5 + x^2 + x}$

$= \dfrac{-2x^2 + x}{x^2 + 6x + 5}$

4. (a) $\dfrac{2x - 4}{6} - \dfrac{5x - 2}{3} < 5$

$\dfrac{2(x-2)}{6} - \dfrac{5x-2}{3} < 5$

$\dfrac{x-2}{3} - \dfrac{5x-2}{3} < 5$

$\dfrac{x - 2 - (5x - 2)}{3} < 5$

$\dfrac{x - 2 - 5x + 2}{3} < 5$

$\dfrac{-4x}{3} < 5$

$\dfrac{3}{1}\left(-\dfrac{-4x}{3}\right) < 3(5)$

$-4x < 15$

$x > -\dfrac{15}{4}$

(b) $\dfrac{3}{x-8} = 2 - \dfrac{5-x}{x-8}$

$(x-8)\left(\dfrac{3}{x-8}\right) = (x-8)\left(2 - \dfrac{5-x}{x-8}\right)$

$\dfrac{x-8}{1} \cdot \dfrac{3}{x-8} = (x-8)(2) - \dfrac{x-8}{1} \cdot \dfrac{5-x}{x-8}$

$3 = 2x - 16 - (5 - x)$

$3 = 2x - 16 - 5 + x$

$3 = 3x - 21$

$24 = 3x$

$8 = x$

No solution, since $x = 8$ causes a denominator to equal 0 in the original equation.

(c)
$$\frac{3}{2x + 1} + \frac{4}{2x - 1} = \frac{29}{4x^2 - 1}$$

$$\frac{3}{2x + 1} + \frac{4}{2x - 1} = \frac{29}{(2x - 1)(2x + 1)}$$

$$(2x + 1)(2x - 1)\left(\frac{3}{2x + 1} + \frac{4}{2x - 1}\right) = \frac{(2x + 1)(2x - 1)}{1} \cdot \frac{29}{(2x - 1)(2x + 1)}$$

$$\frac{(2x + 1)(2x - 1)}{1} \cdot \frac{3}{2x + 1} + \frac{(2x + 1)(2x - 1)}{1} \cdot \frac{4}{2x - 1} = 29$$

$$3(2x - 1) + 4(2x + 1) = 29$$
$$6x - 3 + 8x + 4 = 29$$
$$14x + 1 = 29$$
$$14x = 28$$
$$x = 2$$

5.
$$y = \frac{x - 2}{2x + 1}$$

$$(2x + 1)(y) = \frac{2x + 1}{1} \cdot \frac{x - 2}{2x + 1}$$

$$2xy + y = x - 2$$
$$y + 2 = x - 2xy$$
$$y + 2 = x(1 - 2y)$$
$$\frac{y + 2}{1 - 2y} = x$$

6. Let x = amount of 30% solution

$$0.30(x) + 0.45(8) = 0.42(x + 8)$$
$$0.30x + 3.6 = 0.42x + 3.36$$
$$3.6 = 0.12x + 3.36$$
$$0.24 = 0.12x$$
$$2 = x$$

2 liters of the 30% solution should be used.

7. Let x = number of hours to complete the job together

Portion of job completed by Jackie	+	Portion of job completed by Eleanor	=	1 whole job
$\dfrac{x}{3\frac{1}{2}}$	+	$\dfrac{x}{2}$	=	1

$$\frac{2x}{7} + \frac{x}{2} = 1$$

$$14\left(\frac{2x}{7} + \frac{x}{2}\right) = 14(1)$$

$$2(2x) + 7x = 14$$
$$4x + 7x = 14$$
$$11x = 14$$
$$x = \frac{14}{11}$$

It will take them $\dfrac{14}{11} = 1\dfrac{3}{11}$ hours working together.

1. $3y - 5x + 9 = 0$

x	y
0	-3
9/5	0

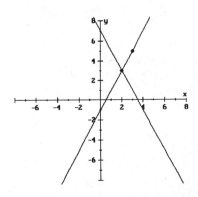

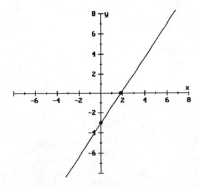

3. $y = -5$
   This is a horizontal line passing through $(0, -5)$.

7. $m = \dfrac{y_2 - y_1}{x_2 - x_1}$

   $= \dfrac{-3 - 5}{2 - 3}$

   $= \dfrac{-8}{-1}$

   $= 8$

9. $y = 5x - 8$
   $m = 5$

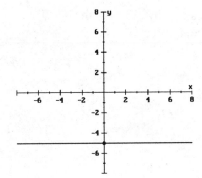

11. $m = \dfrac{y_2 - y_1}{x_2 - x_1}$

    $= \dfrac{-1 - 4}{2 - 6}$

    $= \dfrac{-5}{-4}$

    $= \dfrac{5}{4}$

Parallel lines have equal slopes.

5. $m = 2 = \dfrac{2}{1}$

   From $(2, 3)$, move up 2 then right 1 to locate $(3, 5)$ on the line.

   $m = -2 = \dfrac{-2}{1}$

   From $(2, 3)$, move down 2 then right 1 to locate $(3, 1)$ on the line.

13. $m = \dfrac{y_2 - y_1}{x_2 - x_1}$

    $3 = \dfrac{a - (-2)}{2 - a}$

    $3 = \dfrac{a + 2}{2 - a}$

$$(2 - a)(3) = \frac{(2 - a)}{1} \cdot \frac{(a + 2)}{(2 - a)}$$

$$6 - 3a = a + 2$$
$$6 = 4a + 2$$
$$4 = 4a$$
$$1 = a$$

15. $$y - y_1 = m(x - x_1)$$
$$y - 7 = 3[x - (-2)]$$
$$y - 7 = 3(x + 2)$$
$$y - 7 = 3x + 6$$
$$y = 3x + 13$$

17. $$m = \frac{7 - 1}{2 - 3} = -6$$

$$y - y_1 = m(x - x_1)$$
$$y - 1 = -6(x - 3)$$
$$y - 1 = -6x + 18$$
$$y = -6x + 19$$

19. $(0, 2)$; $m = 4$

$$y = mx + b$$
$$y = 4x + 2$$

21. $$3x + 5y = 4$$
$$5y = -3x + 4$$

$$y = -\frac{3}{5}x + \frac{4}{5}$$

$$m = -\frac{3}{5}$$

Parallel lines have equal slopes.

$$y - y_1 = m(x - x_1)$$

$$y - (-3) = -\frac{3}{5}(x - 2)$$

$$y + 3 = -\frac{3}{5}x + \frac{6}{5}$$

$$y = -\frac{3}{5}x - \frac{9}{5}$$

23. (T, P): $(86, 450)$ and $(80, 325)$

$$m = \frac{450 - 325}{86 - 80}$$

$$= \frac{125}{6}$$

$$P - 325 = \frac{125}{6}(T - 80)$$

$$P - 325 = \frac{125}{6}T - \frac{5000}{3}$$

$$P = \frac{125}{6}T - \frac{4025}{3}$$

When $T = 90$:

$$P = \frac{125}{6}(90) - \frac{4025}{3}$$

$$P = 533.33$$

The daily profit will be approximately $533.33.

25. $$\begin{cases} 4x - 3y = 2 \\ 6x - 5y = 3 \end{cases}$$

Multiply the 1st equation by -3 and the 2nd equation by 2, then add:

$$-12x + 9y = -6$$
$$\underline{12x - 10y = 6}$$
$$-y = 0$$
$$y = 0$$

$$4x - 3(0) = 2$$
$$4x = 2$$
$$x = \frac{1}{2}$$

$$x = \frac{1}{2}, \ y = 0$$

27. $$\begin{cases} -10x + 7t = -6 \\ -4s + 6t = -4 \end{cases}$$

Multiply the 1st equation by -2 and the 2nd equation by 5, then add:

$$20s - 14t = 12$$
$$\underline{-20s + 30t = -20}$$
$$16t = -8$$
$$t = -\frac{1}{2}$$

$$-4s + 6\left(-\frac{1}{2}\right) = -4$$
$$-4s - 3 = -4$$
$$-4s = -1$$
$$s = \frac{1}{4}$$

$$s = \frac{1}{4}, \quad t = -\frac{1}{2}$$

29. $\begin{cases} 5w = 4v + 7 \\ 4v = 5w - 7 \end{cases}$

$\begin{cases} 5w - 4v = 7 \\ -5w + 4z = -7 \end{cases}$

Add: $\quad 5w - 4v = 7$
$\quad\quad\quad \underline{-5w + 4z = -7}$
$\quad\quad\quad\quad\quad\quad 0 = 0$

Dependent system

31. $\begin{cases} \dfrac{x}{4} + \dfrac{5y}{6} = -\dfrac{11}{12} \\[2mm] \dfrac{5x}{3} + \dfrac{y}{2} = 4 \end{cases}$

Multiply the 1st equation by 12 and the 2nd equation by 6.

$\begin{cases} 3x + 10y = -11 \\ 10x + 3y = 24 \end{cases}$

Multiply the 1st equation by -10 and the 2nd equation by 3, then add:

$$-30x - 100y = 110$$
$$\underline{30x + 9y = 72}$$
$$-91y = 182$$
$$y = -2$$

$$10x + 3(-2) = 24$$
$$10x - 6 = 24$$
$$10x = 30$$
$$x = 3$$

$x = 3, \quad y = -2$

33. $7x - y < 14$ (dashed line)

$\quad\quad 7x - y = 14$

x	y
0	-14
2	0

Test point: $(0, 0)$

$$7(0) - 0 < 14$$
$$0 < 14$$
$$\text{True}$$

Shade the half-plane containing $(0, 0)$.

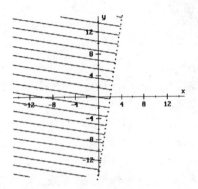

35. $2y \le 4$ (Solid line)

$\quad\quad 2y = 4$
$\quad\quad\quad y = 2$

Horizontal line passing through $(0, 2)$.

Test point: $(0, 0)$

$$2(0) \le 4$$
$$0 \le 4$$
$$\text{True}$$

Shade half-plane containing $(0, 0)$.

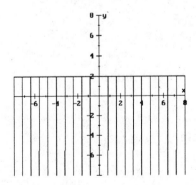

37. degree: 0

39. $-3x(2x - 5y + 4)$
  $= -6x^2 + 15xy - 12x$

41. $(3x - 2y)(x - y)$
  $= 3x^2 - 3xy - 2xy + 2y^2$
  $= 3x^2 - 5xy + 2y^2$

43. $(2a + b)(4a^2 - 2ab + b^2)$
  $= 8a^3 - 4a^2b + 2ab^2 + 4a^2b - 2ab^2 + b^3$
  $= 8a^3 + b^3$

45. $(2m + 3n)(2m - 3n)$
  $= (2m)^2 - (3n)^2$
  $= 4m^2 - 9n^2$

47. $(2m + 3n)^2$
  $= (2m)^2 + 2(2m)(3n) + (3n)^2$
  $= 4m^2 + 12mn + 9n^2$

49. $(x - 2y + 3)^2$
  $= (x - 2y + 3)(x - 2y + 3)$
  $= x^2 - 2xy + 3x - 2xy + 4y^2 - 6y + 3x - 6y + 9$
  $= x^2 - 4xy + 4y^2 + 6x - 12y + 9$

51. $y^2 - 4x^2$
  $= y^2 - (2x)^2$
  $= (y - 2x)(y + 2x)$

53. $10a^2 - 3ab - b^2$
  $= 10a^2 - 5ab + 2ab - b^2$
  $= 5a(2a - b) + b(2a - b)$
  $= (2a - b)(5a + b)$

55. $9x^2 - 25z^2$
  $= (3x)^2 - (5z)^2$
  $= (3x - 5z)(3x + 5z)$

57. Cannot be factored further

59. $25x^2 - 30xz + 9z^2$
  $= (5x)^2 - 2(5x)(3z) + (3z)^2$
  $= (5x - 3z)^2$

61. $12y^3 - 16y^2 - 3y$
  $= y(12y^2 - 16y - 3)$
  $= y(12y^2 - 18y + 2y - 3)$
  $= y[6y(2y - 3) + 1(2y - 3)]$
  $= y(2y - 3)(6y + 1)$

63. $25a^4 + 10a^2b^2 - 8b^4$
  $= 25a^4 + 20a^2b^2 - 10a^2b^2 - 8b^4$
  $= 5a^2(5a^2 + 4b^2) - 2b^2(5a^2 + 4b^2)$
  $= (5a^2 + 4b^2)(5a^2 - 2b^2)$

65. $18a^5b - 9a^3b^2 - 2ab^3$
  $= ab(18a^4 - 9a^2b - 2b^2)$
  $= ab(18a^4 - 12a^2b + 3a^2b - 2b^2)$
  $= ab[6a^2(3a^2 - 2b) + b(3a^2 - 2b)]$
  $= ab(3a^2 - 2b)(6a^2 + b)$

67. $(a - 2b)^2 - 25$
  $= (a - 2b)^2 - 5^2$
  $= [(a - 2b) - 5][(a - 2b) + 5]$
  $= (a - 2b - 5)(a - 2b + 5)$

69. $(x + y)^2 + 3(x + y) + 2$
  Let $u = x + y$:
  $u^2 + 3u + 2$
  $= (u + 2)(u + 1)$
  Replace u with $x + y$:
  $(x + y + 2)(x + y + 1)$

71. $x^3 - 25x - x^2 + 25$
  $= x^3 - x^2 - 25x + 25$
  $= x^2(x - 1) - 25(x - 1)$
  $= (x - 1)(x^2 - 25)$
  $= (x - 1)(x^2 - 5^2)$
  $= (x - 1)(x - 5)(x + 5)$

73. $2t^2 + 5 = 7t$
  $2t^2 - 7t + 5 = 0$
  $(2t - 5)(t - 1) = 0$
  $2t - 5 = 0$ or $t - 1 = 0$
  $2t = 5$ $\qquad$ $t = 1$
  $t = \dfrac{5}{2}$ or $\qquad$ $t = 1$

75. $(3c + 1)(2c - 5) = (6c - 1)(c + 2)$
  $6c^2 - 13c - 5 = 6c^2 + 11c - 2$
  $-13c - 5 = 11c - 2$
  $-5 = 24c - 2$
  $-3 = 24c$
  $-\dfrac{1}{8} = c$

77.
$$3x^2 = 5x$$
$$3x^2 - 5x = 0$$
$$x(3x - 5) = 0$$
$$x = 0 \quad \text{or} \quad 3x - 5 = 0$$
$$3x = 5$$
$$x = 0 \quad \text{or} \quad x = \frac{5}{3}$$

79.

$$\begin{array}{r} x + 3 \\ 2x + 1 \overline{\smash{\big)}\, 2x^2 + 7x - 1} \\ \underline{-(2x^2 + x)} \\ 6x - 1 \\ \underline{-(6x + 3)} \\ -4 \end{array}$$

$$x + 3 - \frac{4}{2x + 1}$$

81.

$$\begin{array}{r} x^2 - 2x + 3 \\ x^2 + 2x + 1 \overline{\smash{\big)}\, x^4 + 0x^3 + 0x^2 + 0x + 2} \\ \underline{-(x^4 + 2x^3 + x^2)} \\ -2x^3 - x^2 + 0x \\ \underline{-(2x^3 - 4x^2 - 2x)} \\ 3x^2 + 2x + 2 \\ \underline{-(3x^2 + 6x + 3)} \\ -4x - 1 \end{array}$$

$$x^2 - 2x + 3 + \frac{-4x - 1}{x^2 + 2x + 1}$$

83.
$$\frac{x^2 - 9y^2}{x^2 - 6xy + 9y^2}$$
$$= \frac{(x - 3y)(x + 3y)}{(x - 3y)(x - 3y)}$$
$$= \frac{x + 3y}{x - 3y}$$

85.
$$\frac{27x^3 - y^3}{18x^2 + 6xy + 2y^2}$$
$$= \frac{(3x - y)(9x^2 + 3xy + y^2)}{2(9x^2 + 3xy + y^2)}$$
$$= \frac{3x - y}{2}$$

87.
$$\frac{2x^2 - xy - y^2}{x + y} \cdot \frac{x^2 - y^2}{x - y}$$
$$= \frac{(2x + y)(x - y)}{x + y} \cdot \frac{(x - y)(x + y)}{x - y}$$
$$= (2x + y)(x - y)$$
$$= 2x^2 - xy - y^2$$

89.
$$\frac{6}{x - 2} - \frac{3x}{x - 2}$$
$$= \frac{6 - 3x}{x - 2}$$
$$= \frac{-3(-2 + x)}{x - 2}$$
$$= \frac{-3(x - 2)}{x - 2}$$
$$= -3$$

91.
$$\frac{9a^2 - 6ab + b^2}{3a + b} \div \frac{3a^2 - 4ab + b^2}{3a^2 - 2ab - b^2}$$
$$= \frac{9a^2 - 6ab + b^2}{3a + b} \cdot \frac{3a^2 - 2ab - b^2}{3a^2 - 4ab + b^2}$$
$$= \frac{(3a - b)(3a - b)}{3a + b} \cdot \frac{(3a + b)(a - b)}{(3a - b)(a - b)}$$
$$= 3a - b$$

93.
$$\frac{3}{2x^2 - 5xy - 3y^2} - \frac{5}{x^2 - 2xy - 3y^2}$$
$$= \frac{3}{(2x + y)(x - 3y)} - \frac{5}{(x - 3y)(x + y)}$$
$$= \frac{3(x + y)}{(2x + y)(x - 3y)(x + y)} - \frac{5(2x + y)}{(2x + y)(x - 3y)(x + y)}$$
$$= \frac{3(x + y) - 5(2x + y)}{(2x + y)(x - 3y)(x + y)}$$
$$= \frac{3x + 3y - 10x - 5y}{(2x + y)(x - 3y)(x + y)}$$
$$= \frac{-7x - 2y}{(2x + y)(x - 3y)(x + y)}$$

95.

$$\frac{x}{x-5y} - \frac{2y}{x+5y} - \frac{20y^2}{x^2-25y^2}$$

$$= \frac{x}{x-5y} - \frac{2y}{x+5y} - \frac{20y^2}{(x-5y)(x+5y)}$$

$$= \frac{x(x+5y)}{(x-5y)(x+5y)} - \frac{2y(x-5y)}{(x-5y)(x+5y)} - \frac{20y^2}{(x-5y)(x+5y)}$$

$$= \frac{x(x+5y) - 2y(x-5y) - 20y^2}{(x-5y)(x+5y)}$$

$$= \frac{x^2 + 5xy - 2xy + 10y^2 - 20y^2}{(x-5y)(x+5y)}$$

$$= \frac{x^2 + 3xy - 10y^2}{(x-5y)(x+5y)}$$

$$= \frac{(x+5y)(x-2y)}{(x-5y)(x+5y)}$$

$$= \frac{x-2y}{x-5y}$$

97.

$$\frac{\dfrac{x}{x+y} - \dfrac{4y}{x+4y}}{\dfrac{x-2y}{x+y} + 1}$$

$$= \frac{\dfrac{x}{x+y} - \dfrac{4y}{x+4y}}{\dfrac{x-2y}{x+y} + 1} \cdot \frac{(x+y)(x+4y)}{(x+y)(x+4y)}$$

$$= \frac{\dfrac{x}{x+y} \cdot \dfrac{(x+y)(x+4y)}{1} - \dfrac{4y}{x+4y} \cdot \dfrac{(x+y)(x+4y)}{2}}{\dfrac{x-2y}{x+y} \cdot \dfrac{(x+y)(x+4y)}{1} + 1(x+y)(x+4y)}$$

$$= \frac{x(x+4y) - 4y(x+y)}{(x-2y)(x+4y) + (x+y)(x+4y)}$$

$$= \frac{x^2 + 4xy - 4xy - 4y^2}{x^2 + 2xy - 8y^2 + x^2 + 5xy + 4y^2}$$

$$= \frac{x^2 - 4y^2}{2x^2 + 7xy - 4y^2}$$

99.

$$\frac{4}{x+4} - 2 = 0$$

$$(x+4)\left[\frac{4}{x+4} - 2\right] = (x+4)(0)$$

$$\frac{x+4}{1} \cdot \frac{4}{x+4} - (x+4)(2) = 0$$

$$4 - 2x - 8 = 0$$
$$-2x - 4 = 0$$
$$-4 = 2x$$
$$-2 = x$$

101.

$$\frac{3}{x-2} + 3 = \frac{x+1}{x-2}$$

$$(x-2)\left(\frac{3}{x-2} + 3\right) = (x-2)\left(\frac{x+1}{x-2}\right)$$

$$\frac{x-2}{1} \cdot \frac{3}{x-2} + (x-2)(3) = \frac{x-2}{1} \cdot \frac{x+1}{x-2}$$

$$3 + 3x - 6 = x + 1$$
$$3x - 3 = x + 1$$
$$2x - 3 = 1$$
$$2x = 4$$
$$x = 2$$

No solution, since $x = 2$ causes a denominatorto equal 0 in the original equation.

103.

$$\frac{x+3}{4} - \frac{2x+1}{3} \le \frac{1}{2}$$

$$12\left(\frac{x+3}{4} - \frac{2x+1}{3}\right) \le 12\left(\frac{1}{2}\right)$$

$$\frac{12}{1} \cdot \frac{x+3}{4} - \frac{12}{1} \cdot \frac{2x+1}{3} \le \frac{12}{1} \cdot \frac{1}{2}$$

$$3(x+3) - 4(2x+1) \le 6$$
$$3x + 9 - 8x - 4 \le 6$$
$$-5x + 5 \le 6$$
$$-5x \le 1$$
$$x \ge -\frac{1}{5}$$

105.

$$3xy - 2y = 5x + 3y$$
$$3xy - 5y = 5x$$
$$y(3x - 5) = 5x$$
$$y = \frac{5x}{3x - 5}$$

107.
$$\frac{2x + 3}{x - 2} = y$$

$$\frac{x - 2}{1} \cdot \frac{2x + 3}{x - 2} = (x - 2)y$$

$$2x + 3 = xy - 2y$$
$$3 = xy - 2y - 2x$$
$$2y + 3 = xy - 2x$$
$$2y + 3 = x(y - 2)$$
$$\frac{2y + 3}{y - 2} = x$$

109. Let x = amount of 20% solution

$$0.20(x) + 0.35(8) = 0.30(x + 8)$$
$$0.20x + 2.8 = 0.30x + 2.4$$
$$2.8 = 0.10x + 2.4$$
$$0.4 = 0.10x$$
$$4 = x$$

4 liters of the 20% solution should be used.

111. Let x = number of general admission tickets, then $505 - x$ = number of reserved seats

$$3.50(x) + 4.25(505 - x) = 1861.25$$
$$3.50x + 2146.25 - 4.25x = 1861.25$$
$$2146.25 - 0.75x = 1861.25$$
$$-0.75x = -285$$
$$x = 380$$
$$505 - x = 505 - 380 = 125$$

They sold 380 general admission tickets and 125 reserved seats.

CHAPTERS 4 - 6 CUMULATIVE PRACTICE TEST

1. (a) $(2x^2 - 3xy + 4y^2) - (5x^2 - 2xy + y^2)$
$$= 2x^2 - 3xy + 4y^2 - 5x^2 + 2xy - y^2$$
$$= -3x^2 - xy + 3y^2$$

(b) $(3a - 2b)(5a + 3b)$
$$= 15a^2 + (9 - 10)ab - 6b^2$$
$$= 15a^2 - ab - 6b^2$$

(c) $(2x^2 - y)(2x^2 + y)$
$$= (2x^2)^2 - (y)^2$$
$$= 4x^4 - y^2$$

(d) $(3y - 2z)^2$
$$= (3y)^2 - 2(3y)(2z) + (2z)^2$$
$$= 9y^2 - 12yz + 4z^2$$

(e) $(x + y - 3)^2$
$$= (x + y - 3)(x + y - 3)$$
$$= x^2 + xy - 3x + xy + y^2 - 3y - 3x - 3y + 9$$
$$= x^2 + 2xy + y^2 - 6x - 6y + 9$$

3.
$$\begin{array}{r} 2x^2 - 4x + 14 \\ x + 2\overline{)2x^3 + 0x^2 + 6x + 5} \\ \underline{-(2x^3 + 4x^2)} \\ -4x^2 + 6x \\ \underline{-(-4x^2 - 8x)} \\ 14x + 5 \\ \underline{-(14x + 28)} \\ -23 \end{array}$$

$$2x^2 - 4x + 14 - \frac{23}{x + 2}$$

5.
$$\frac{\dfrac{1}{x} - 2}{3 + \dfrac{1}{x + 1}}$$

$$= \frac{\dfrac{1}{x} - 2}{3 + \dfrac{1}{x + 1}} \cdot \frac{x(x + 1)}{x(x + 1)}$$

$$= \frac{\dfrac{1}{x} \cdot \dfrac{x(x + 1)}{1} - 2x(x + 1)}{3x(x + 1) + \dfrac{1}{x + 1} \cdot \dfrac{x(x + 1)}{1}}$$

$$= \frac{x + 1 - 2x^2 - 2x}{3x^2 + 3x + x}$$

$$= \frac{-2x^2 - x + 1}{3x^2 + 4x}$$

7.
$$y = \frac{a}{a + 1}$$

$$y(a + 1) = \frac{a}{a + 1} \cdot \frac{a + 1}{1}$$

$$ya + y = a$$
$$y = a - ya$$
$$y = a(1 - y)$$
$$\frac{y}{1 - y} = a$$

9. $y = -\dfrac{3}{4}x + 6$

x	y
0	6
8	0

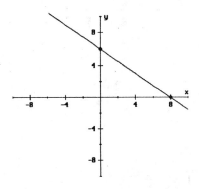

11. (a) $3y - 2x = 4$

$\qquad 3y = 2x + 4$

$\qquad\quad y = \dfrac{2}{3}x + \dfrac{4}{3}$

$\qquad\quad m = \dfrac{2}{3}$

Parallel lines have equal slopes.

$y - y_1 = m(x - x_1)$

$y - (-3) = \dfrac{2}{3}(x - 2)$

$y + 3 = \dfrac{2}{3}x - \dfrac{4}{3}$

$\qquad y = \dfrac{2}{3}x - \dfrac{13}{3}$

(b) $\qquad m_\perp = -\dfrac{3}{2}$

$y - y_1 = m(x - x_1)$

$y - (-3) = -\dfrac{3}{2}(x - 2)$

$y + 3 = -\dfrac{3}{2}x + 3$

$\qquad y = -\dfrac{3}{2}x$

13. $\begin{cases} 5x + 2y = 4 \\ 2x - 3y = 13 \end{cases}$

Multiply 1st equation by 3 and 2nd equation by 2, then add:

$\quad 15x + 6y = 12$
$\quad \underline{4x - 6y = 26}$
$\quad 19x \qquad\;\; = 38$
$\qquad\quad x = 2$

$5(2) + 2y = 4$
$\quad 10 + 2y = 4$
$\qquad\quad 2y = -6$
$\qquad\quad\; y = -3$

$x = 2, \quad y = -3$

CHAPTER 7

7.1 Exercises

1. $(x^2x^5)(x^3x) = (x^{2+5})(x^{3+1})$
$= (x^7)(x^4)$
$= x^{7+4}$
$= x^{11}$

3. $(-2a^2b^3)(3a^5b^7) = -6a^{2+5}b^{3+7}$
$= -6a^7b^{10}$

5. $(a^2)^5 = a^{2\cdot5}$
$= a^{10}$

7. $\left(\dfrac{2}{3}\right)^2 = \dfrac{2^2}{3^2}$
$= \dfrac{4}{9}$

9. $(x + y^3)^2 = x^2 + 2(x)(y^3) + (y^3)^2$
$= x^2 + 2xy^3 + y^6$

11. $(xy^3)^2 = x^2y^{3\cdot2}$
$= x^2y^6$

13. $(2^3 \cdot 3^2)^2 = 2^{3\cdot2} \cdot 3^{2\cdot2}$
$= 2^6 \cdot 3^4$
$= 64 \cdot 81$
$= 5184$

15. $(x^4y^3)^5(x^3y^2)^2$
$= x^{4\cdot5}y^{3\cdot5} \cdot x^{3\cdot2}y^{2\cdot2}$
$= x^{20}y^{15} \cdot x^6y^4$
$= x^{20+6}y^{15+4}$
$= x^{26}y^{19}$

17. $(-2a^2)^3(ab^2)^4 = (-2)^3a^{2\cdot3} \cdot a^4b^{2\cdot4}$
$= -8a^6 \cdot a^4b^8$
$= -8a^{6+4}b^8$
$= -8a^{10}b^8$

19. $(r^2st)^3(-2rs^2t)^4$
$= r^{2\cdot3}s^3t^3 \cdot (-2)^4r^4s^{2\cdot4}t^4$
$= r^6s^3t^3 \cdot 16r^4s^8t^4$
$= 16r^{6+4}s^{3+8}t^{3+4}$
$= 16r^{10}s^{11}t^7$

21. $\dfrac{x^5}{x^2} = x^{5-2}$
$= x^3$

23. $\dfrac{x^3y^2}{xy^4} = x^{3-1}y^{2-4}$
$= x^2y^{-2}$
$= \dfrac{x^2}{y^2}$

25. $\dfrac{5^4 \cdot 2^2}{25^2 \cdot 4^2} = \dfrac{5^4 \cdot 2^2}{(5^2)^2(2^2)^2}$
$= \dfrac{5^4 2^2}{5^4 2^4}$
$= 5^{4-4}2^{2-4}$
$= 5^0 2^{-2}$
$= \dfrac{1}{2^2}$
$= \dfrac{1}{4}$

27. $\dfrac{a^5b^9c}{a^4bc^5} = a^{5-4}b^{9-1}c^{1-5}$
$= ab^8c^{-4}$
$= \dfrac{ab^8}{c^4}$

29. $\dfrac{(-3)^2xy^4}{-3^2xy^5} = \dfrac{9}{-9}x^{1-1}y^{4-5}$
$= -x^0y^{-1}$
$= -\dfrac{1}{y}$

31. $\dfrac{3^2(-2)^3}{(-9^2)(-4)^2} = \dfrac{9(-8)}{-81(16)}$
$= \dfrac{1}{18}$

33. $\left(\dfrac{y^5}{y^8}\right)^3 = \dfrac{y^{5\cdot3}}{y^{8\cdot3}}$

$= \dfrac{y^{15}}{y^{24}}$

$= y^{15-24}$
$= y^{-9}$
$= \dfrac{1}{y^9}$

35. $\left(\dfrac{y^2y^7}{y^4}\right)^3 = \left(\dfrac{y^{2+7}}{y^4}\right)^3$

$= \left(\dfrac{y^9}{y^4}\right)^3$

$= (y^{9-4})^3$
$= (y^5)^3$
$= y^{5\cdot3}$
$= y^{15}$

37. $\dfrac{(3r^2s)^3(-2rs^2)^4}{(-18rs)^2}$

$= \dfrac{3^3r^{2\cdot3}s^3 \cdot (-2)^4r^4s^{2\cdot4}}{(-18)^2r^2s^2}$

$= \dfrac{27r^6s^3 \cdot 16r^4s^8}{324r^2s^2}$

$= \dfrac{4r^{6+4}s^{3+8}}{3r^2s^2}$

$= \dfrac{4r^{10}s^{11}}{3r^2s^2}$

$= \dfrac{4}{3}r^{10-2}s^{11-2}$

$= \dfrac{4}{3}r^8s^9$

39. $\left(\dfrac{2x^2y^3}{xy^4}\right)^2\left(\dfrac{3xy^2}{6}\right)^3$

$= (2x^{2-1}y^{3-4})^2\left(\dfrac{xy^2}{2}\right)^3$

$= (2xy^{-1})^2\left(\dfrac{xy^2}{2}\right)^3$

$= 2^2x^2y^{-1\cdot2} \cdot \dfrac{x^3y^{2\cdot3}}{2^3}$

$= 2^2x^2y^{-2} \cdot \dfrac{x^3y^6}{2^3}$

$= 2^{2-3}x^{2+3}y^{-2+6}$

$= 2^{-1}x^5y^4$

$= \dfrac{x^5y^4}{2}$

41. $\left(\dfrac{(6ab^2)^2}{-3ab}\right)^3$

$= \left(\dfrac{6^2a^2b^{2\cdot2}}{-3ab}\right)^3$

$= \left(\dfrac{36a^2b^4}{-3ab}\right)^3$

$= (-12a^{2-1}b^{4-1})^3$

$= (-12ab^3)^3$
$= (-12)^3a^3b^{3\cdot3}$
$= -1728a^3b^9$

43. $\left(\dfrac{-2a^2b^3}{ab}\right)^3(-3xy^2)^3$

$= (-2a^{2-1}b^{3-1})^3(-3xy^2)^3$
$= (-2ab^2)^3(-3xy^2)^3$
$= (-2)^3a^3b^{2\cdot3} \cdot (-3)^3x^3y^{2\cdot3}$
$= -8a^3b^6 \cdot (-27)x^3y^6$
$= 216a^3b^6x^3y^6$

45. $\left[(r^3s^2)^3(rs^2)^4\right]^2$
$= (r^{3\cdot3}s^{2\cdot3} \cdot r^4s^{2\cdot4})^2$
$= (r^9s^6 \cdot r^4s^8)^2$
$= (r^{9+4}s^{6+8})^2$
$= (r^{13}s^{14})^2$
$= r^{13\cdot2}s^{14\cdot2}$
$= r^{26}s^{28}$

47. $5^{-8} = 0.000003$

49. $-2 \cdot 5^{-3} + 8 = 7.984$

51. $x^{-2}x^4x^{-3} = x^{-2+4-3}$
$$= x^{-1}$$
$$= \frac{1}{x}$$

53. $(x^5y^{-4})(x^{-3}y^2x^0)$
$$= x^{5-3+0}y^{-4+2}$$
$$= x^2y^{-2}$$
$$= \frac{x^2}{y^2}$$

55. $(3^{-2})^{-3}$
$$= 3^{(-2)(-3)}$$
$$= 3^6$$

57. $(a^{-2}b^{-3})^2$
$$= a^{-2 \cdot 2}b^{-3 \cdot 2}$$
$$= a^{-4}b^{-6}$$
$$= \frac{1}{a^4b^6}$$

59. $(r^{-3}s^2)^{-4}(r^2)^{-3}$
$$= r^{(-3)(-4)}s^{2(-4)}r^{2(-3)}$$
$$= r^{12}s^{-8}r^{-6}$$
$$= r^{12-6}s^{-8}$$
$$= r^6s^{-8}$$
$$= \frac{r^6}{s^8}$$

61. $-2^{-2} = -\frac{1}{2^2}$
$$= -\frac{1}{4}$$

63. $(-3)^{-2} = \frac{1}{(-3)^2}$
$$= \frac{1}{9}$$

65. $(2^{-2})^{-3}(3^{-3})^2$
$$= 2^{(-2)(-3)}3^{(-3)(2)}$$
$$= 2^6 3^{-6}$$
$$= \frac{2^6}{3^6}$$

67. $(3^{-2}s^3)^4(9s^{-3})^{-2}$
$$= (3^{-2}s^3)^4(3^2s^{-3})^{-2}$$
$$= 3^{(-2)(4)}s^{3 \cdot 4} \cdot 3^{(2)(-2)}s^{(-3)(-2)}$$
$$= 3^{-8}s^{12} \cdot 3^{-4}s^6$$
$$= 3^{-8-4}s^{12+6}$$
$$= 3^{-12}s^{18}$$
$$= \frac{s^{18}}{3^{12}}$$

69. $\frac{x^4}{x-2} = x^{4-(-2)}$
$$= x^6$$

71. $\frac{x^{-3}y^2}{x^{-5}y^0} = x^{-3-(-5)}y^{2-0}$
$$= x^2y^2$$

73. $\left(\frac{1}{2}\right)^{-1} = \left(\frac{2}{1}\right)^1$
$$= 2$$

75. $\left(-\frac{3}{5}\right)^{-3} = \left(-\frac{5}{3}\right)^3$
$$= \frac{(-5)^3}{3^3}$$
$$= -\frac{125}{27}$$

77. $\frac{x^{-1}xy^{-2}}{x^4y^{-3}y} = \frac{x^{-1+1}y^{-2}}{x^4y^{-3+1}}$
$$= \frac{x^0y^{-2}}{x^4y^{-2}}$$
$$= x^{0-4}y^{-2-(-2)}$$
$$= x^{-4}y^0$$
$$= \frac{1}{x^4}$$

79. $\dfrac{(a^{-2}b^2)^{-3}}{ab^{-2}}$

$= \dfrac{a^{(-2)(-3)}b^{2(-3)}}{ab^{-2}}$

$= \dfrac{a^6 b^{-6}}{ab^{-2}}$

$= a^{6-1}b^{-6-(-2)}$
$= a^5 b^{-4}$
$= \dfrac{a^5}{b^4}$

81. $\left(\dfrac{x^{-1}x^{-3}}{x^{-2}}\right)^{-3}$

$= \left(\dfrac{x^{-1-3}}{x^{-2}}\right)^{-3}$

$= \left(\dfrac{x^{-4}}{x^{-2}}\right)^{-3}$

$= (x^{-4-(-2)})^{-3}$
$= (x^{-2})^{-3}$
$= (x^{-2})^{-3}$
$= x^{(-2)(-3)}$
$= x^6$

83. $\dfrac{2^{-2} \cdot 3^2}{6^{-2}} = \dfrac{2^{-2} \cdot 3^2}{(2 \cdot 3)^{-2}}$

$= \dfrac{2^{-2} \cdot 3^2}{2^{-2} \cdot 3^{-2}}$

$= 2^{-2-(-2)} \cdot 3^{2-(-2)}$

$= 2^0 \cdot 3^4$
$= 3^4 = 81$

85. $\dfrac{(3x)^{-2}(2xy^{-1})^0}{(2x^{-2}y^3)^{-2}}$

$= \dfrac{3^{-2}x^{-2} \cdot 1}{2^{-2}x^{(-2)(-2)}y^{3(-2)}}$

$= \dfrac{3^{-2}x^{-2}}{2^{-2}x^4 y^{-6}}$

$= \dfrac{2^2 y^6}{3^2 x^2 x^4}$

$= \dfrac{4y^6}{9x^{2+4}}$

$= \dfrac{4y^6}{9x^6}$

87. $\left(\dfrac{x^{-3}y^{-4}}{x^{-5}y^{-7}}\right)^{-3}$

$= (x^{-3-(-5)}y^{-4-(-7)})^{-3}$

$= (x^2 y^3)^{-3}$
$= x^{2(-3)}y^{3(-3)}$
$= x^{-6}y^{-9}$
$= \dfrac{1}{x^6 y^9}$

89. $\dfrac{(-3a^{-4}b^{-2})(-4ab^{-3})^{-1}}{(12ab^2)^{-1}}$

$= \dfrac{(-3a^{-4}b^{-2})(12ab^2)}{-4ab^{-3}}$

$= \dfrac{-36a^{-4+1}b^{-2+2}}{-4ab^{-3}}$

$= \dfrac{-36a^{-3}b^0}{-4ab^{-3}}$

$= 9a^{-3-1}b^{0-(-3)}$
$= 9a^{-4}b^3$
$= \dfrac{9b^3}{a^4}$

91. $x^2 y^{-3} = \dfrac{x^2}{y^3}$

93. $x^2 + y^{-3}$

$= x^2 + \dfrac{1}{y^3}$

$= \dfrac{x^2(y^3)}{y^3} + \dfrac{1}{y^3}$

$= \dfrac{x^2 y^3 + 1}{y^3}$

95. $(x^{-2} + y^{-2})^{-2}$

$$= \left(\frac{1}{x^2} + \frac{1}{y^2}\right)^{-2}$$

$$= \left(\frac{y^2}{x^2 y^2} + \frac{x^2}{x^2 y^2}\right)^{-2}$$

$$= \left(\frac{y^2 + x^2}{x^2 y^2}\right)^{-2}$$

$$= \left(\frac{x^2 y^2}{y^2 + x^2}\right)^{2}$$

$$= \frac{x^{2(2)} y^{2(2)}}{(y^2 + x^2)^2}$$

$$= \frac{x^4 y^4}{(y^2 + x^2)^2}$$

97. $\left(\dfrac{x^{-4} y^{-7} z^{-6}}{x^{-24} y^{-16}}\right)^0 = 1$

99. $\dfrac{x^{-1} + y^{-1}}{x y^{-1}}$

$$= \frac{\dfrac{1}{x} + \dfrac{1}{y}}{\dfrac{x}{y}}$$

$$= \frac{\dfrac{1}{x} + \dfrac{1}{y}}{\dfrac{x}{y}} \cdot \frac{xy}{xy}$$

$$= \frac{\dfrac{1}{x} \cdot \dfrac{xy}{1} + \dfrac{1}{y} \cdot \dfrac{xy}{1}}{\dfrac{x}{y} \cdot \dfrac{xy}{1}}$$

$$= \frac{y + x}{x^2}$$

101. $\dfrac{r^{-2} + s^{-1}}{r^{-1} + s^{-2}}$

$$= \frac{\dfrac{1}{r^2} + \dfrac{1}{s}}{\dfrac{1}{r} + \dfrac{1}{s^2}}$$

$$= \frac{\dfrac{1}{r^2} + \dfrac{1}{s}}{\dfrac{1}{r} + \dfrac{1}{s^2}} \cdot \frac{r^2 s^2}{r^2 s^2}$$

$$= \frac{\dfrac{1}{r^2} \cdot \dfrac{r^2 s^2}{1} + \dfrac{1}{s} \cdot \dfrac{r^2 s^2}{1}}{\dfrac{1}{r} \cdot \dfrac{r^2 s^2}{1} + \dfrac{1}{s^2} \cdot \dfrac{r^2 s^2}{1}}$$

$$= \frac{s^2 + r^2 s}{rs^2 + r^2}$$

103. $\dfrac{2a^{-1} + b^{-2}}{a^{-2} + b}$

$$= \frac{\dfrac{2}{a} + \dfrac{1}{b^2}}{\dfrac{1}{a^2} + b}$$

$$= \frac{\dfrac{2}{a} + \dfrac{1}{b^2}}{\dfrac{1}{a^2} + b} \cdot \frac{a^2 b^2}{a^2 b^2}$$

$$= \frac{\dfrac{2}{a} \cdot \dfrac{a^2 b^2}{1} + \dfrac{1}{b^2} \cdot \dfrac{a^2 b^2}{1}}{\dfrac{1}{a^2} \cdot \dfrac{a^2 b^2}{1} + b(a^2 b^2)}$$

$$= \frac{2ab^2 + a^2}{b^2 + a^2 b^3}$$

111. $-2 \le 5 - 2x < 11$

$-7 \le -2x < 6$

$\dfrac{7}{2} \ge x > -3$

or $-3 < x \le \dfrac{7}{2}$

113. $m = \dfrac{-4 - 0}{0 - 6} = \dfrac{2}{3}$

$y = mx + b$

$y = \dfrac{2}{3}x - 4$

7.2 Exercises

1. $10^{-4} \cdot 10^7 = 10^{-4+7}$
   $= 10^3$
   $= 1000$

3. $\dfrac{10^4}{10^{-5}} = 10^{4-(-5)}$
   $= 10^9$
   $= 1,000,000,000$

5. $\dfrac{10^{-4} \cdot 10^2}{10^{-3}} = \dfrac{10^{-4+2}}{10^{-3}}$
   $= \dfrac{10^{-2}}{10^{-3}}$
   $= 10^{-2-(-3)}$
   $= 10$

7. $\dfrac{10^{-4} \cdot 10^2 \cdot 10^{-3}}{10^4 \cdot 10^{-3}} = \dfrac{10^{-4+2-3}}{10^{4-3}}$
   $= \dfrac{10^{-5}}{10^1}$
   $= 10^{-5-1}$
   $= 10^{-6}$
   $= 0.000001$

9. $\dfrac{10^{-4} \cdot 10^{-5} \cdot 10^7}{10^{-6} \cdot 10^{-2} \cdot 10^0} = \dfrac{10^{-4-5+7}}{10^{-6-2+0}}$
   $= \dfrac{10^{-2}}{10^{-8}}$
   $= 10^{-2-(-8)}$
   $= 10^6$
   $= 1,000,000$

11. move 1 place right
    $1.62 \times 10^1 = 16.2$

13. move 8 places right
    $7.6 \times 10^8 = 760,000,000$

15. move 7 places left
    $8.51 \times 10^{-7} = 0.000000851$

17. move 3 places right
    $6.0 \times 10^3 = 6000$

19. $824 = 8.24 \times 10^?$
    $= 8.24 \times 10^2$

21. $5 = 5.0 \times 10^?$
    $= 5.0 \times 10^0$

23. $0.0093 = 9.3 \times 10^?$
    $= 9.3 \times 10^{-3}$

25. $827,546,000 = 8.27546 \times 10^?$
    $= 8.27546 \times 10^8$

27. $0.00000072 = 7.2 \times 10^?$
    $= 7.2 \times 10^{-7}$

29. $79.32 = 7.932 \times 10^?$
    $= 7.932 \times 10^1$

31. $\dfrac{(6000)(0.007)}{(0.021)(12,000)}$

$= \dfrac{(6 \times 10^3)(7.0 \times 10^{-3})}{(2.1 \times 10^{-2})(1.2 \times 10^4)}$

$= \dfrac{6 \cdot 7}{2.1 \cdot 1.2} \times \dfrac{10^3 \cdot 10^{-3}}{10^{-2} \cdot 10^4}$

$= 16.66\overline{6} \times 10$
$= 16.66\overline{6} \times 10^{-2}$
$= 0.16\overline{6}$

33. $\dfrac{(120)(0.005)}{(10,000)(60)}$

$$= \dfrac{(1.2 \times 10^2)(5.0 \times 10^{-3})}{(1.0 \times 10^4)(6.0 \times 10^1)}$$

$$= \dfrac{1.2 \cdot 5.0}{1.0 \cdot 6.0} \times \dfrac{10^{2-3}}{10^{4+1}}$$

$$= 1 \times 10^{-6}$$
$$= 0.000001$$

35. $t = \dfrac{D}{r}$

$$= \dfrac{3,670,000,000}{186,000}$$

$$= \dfrac{3.67 \times 10^9}{1.86 \times 10^5}$$

$$= 1.973118 \times 10^4$$
$$= 19731.18 \text{ sec}$$
$$= 5 \text{ hr } 28 \text{ min } 51 \text{ sec}$$

37. 1 day = 24 hr
$\qquad\quad$ = (24)(60) = 1440 min
$\qquad\quad$ = (1440)(60) = 86400 sec

Hydrogen used = rate $\cdot$ time in 1 day
$\qquad\qquad\qquad$ = (700,000,000)(86400)
$\qquad\qquad\qquad$ = $(70 \times 10^8)(8.64 \times 10^4)$
$\qquad\qquad\qquad$ = 60.48×10^{12}
$\qquad\qquad\qquad$ = 6.048×10^{13}

6.048×10^{13} tons are used in 1 day

Amount used = $\quad$ # of $\quad\cdot\quad$ amount used
$\quad$ in 1 yr $\qquad$ days in a yr $\qquad$ per day

$\qquad\qquad$ = $(365)(6.048 \times 10^{13})$
$\qquad\qquad$ = 2207.52×10^{13}
$\qquad\qquad$ = 2.20752×10^{16}

2.20752×10^{16} tons are used in 1 year

39. $\dfrac{1 \text{ Å}}{10^{-8} \text{ cm}} = \dfrac{x \text{ Å}}{10^{-4} \text{ cm}}$

$$\dfrac{1}{10^{-8}} = \dfrac{x}{10^{-4}}$$

$$10^{-4}\left(\dfrac{1}{10^{-8}}\right) = x$$

$$10^4 = x$$

There are 10^4 = 10,000 Å in 10^{-4} cm which is 1 micron. Hence there are 10,000 Å in 1 micron.

41. $\dfrac{10^{-8} \text{ cm}}{1 \text{ angstrom}} = \dfrac{x \text{ cm}}{0.66 \text{ angstrom}}$

$$\dfrac{10^{-8}}{1} = \dfrac{x}{0.66}$$

$$0.66(10^{-8}) = x$$
$$6.6 \times 10^{-9} = x$$

Its radius is 6.6×10^{-9} cm.

43. $\dfrac{1.6 \text{ km}}{1 \text{ mile}} = \dfrac{x \text{ km}}{5.86 \times 10^{12} \text{ miles}}$

$$\dfrac{1.6}{1} = \dfrac{x}{5.86 \times 10^{12}}$$

$$(5.86 \times 10^{12})(1.6) = x$$
$$9.376 \times 10^{12} = x$$

There are 9.376×10^{12} km in 1 light-year.

45. $\dfrac{10^{-4} \text{ cm}}{1 \text{ } \mu m} = \dfrac{x \text{ cm}}{60 \text{ } \mu m}$

$$\dfrac{10^{-4}}{1} = \dfrac{x}{60}$$

$$60 \times 10^{-4} = x$$
$$6.0 \times 10^{-3} = x$$

The diameter is 6.0×10^{-3} or 0.006 cm.

47. $\dfrac{1 \text{ atom}}{10^{-23} \text{ gm}} = \dfrac{x \text{ atoms}}{1 \text{ gm}}$

$$\dfrac{1}{10^{-23}} = \dfrac{x}{1}$$

$$x = \dfrac{1}{10^{-23}} = 10^{23}$$

There are 10^{23} atoms in 1 gram.

49. distance = number of · diameter
cheek cells   in cm
= (40000)(0.006)
= 240

They would stretch 240 cm.

51.   1 light-year = 5.86×10^{12} miles
93 million miles = 9.3×10^{7} miles

$$\frac{1 \ Au}{9.3 \times 10^{7} \ \text{miles}} = \frac{x \ Au}{5.86 \times 10^{12} \ \text{miles}}$$

$$\frac{1}{9.3 \times 10^{7}} = \frac{x}{5.86 \times 10^{12}}$$

$$(5.86 \times 10^{12})\left(\frac{1}{9.3 \times 10^{7}}\right) = x$$

$$x = 0.6301 \times 10^{5}$$
$$= 6.301 \times 10^{4}$$

There are $6.301 \times 10^{4} \ AU$ in a light-year.

55. $(3x - 2)(x - 8) - (x - 4)^2$
$= 3x^2 - 24x - 2x + 16 - (x^2 - 8x + 16)$
$= 3x^2 - 26x + 16 - x^2 + 8x - 16$
$= 2x^2 - 18x$

57. $(x - y)^3 - x^2 = [-3 - (-1)]^3 - (-3)^2$
$= (-2)^3 - (-3)^2$
$= -8 - 9$
$= -17$

7.3 Exercises

1. $8^{1/3} = 2$

3. $(-32)^{1/5} = -2$

5. $-100^{1/2} = -(100^{1/2})$
$= -(10)$
$= -10$

7. $\sqrt[3]{64} = 4$

9. $\sqrt[4]{81} = 3$

11. Not a real number

13. $-\sqrt[9]{-1} = -(-1)$
$= 1$

15. $\sqrt[3]{-343} = -7$

17. $-\sqrt[4]{1296} = -6$

19. $\sqrt[8]{256} = 2$

21. $\sqrt[7]{78,125} = 5$

23. $(-32)^{3/5} = \left[(-32)^{1/5}\right]^3$
$= (-2)^3$
$= -8$

25. $32^{-1/5} = (32^{1/5})^{-1}$
$= (2)^{-1}$
$= \dfrac{1}{2}$

27. $(-32)^{-1/5} = \left[(-32)^{1/5}\right]^{-1}$
$= (-2)^{-1}$
$= -\dfrac{1}{2}$

29. $-(81)^{-1/2} = -\left[81^{1/2}\right]^{-1}$
$= -(9)^{-1}$
$= -\dfrac{1}{9}$

31. $(-64)^{-2/3} = \left[(-64)^{1/3}\right]^{-2}$
$= (-4)^{-2}$
$= \dfrac{1}{(-4)^2}$
$= \dfrac{1}{16}$

33. $\sqrt{(-16)^2} = |-16|$
$= 16$

35. $\sqrt{-16}$ is not a real number

37. $-(\sqrt{16})^2 = -(4)^2$
$= -16$

39. $\sqrt[n]{3^{2n}} = (3^{2n})^{1/n}$
$= 3^{(2n)(1/n)}$
$= 3^2$
$= 9$

41. $\left(\dfrac{64}{27}\right)^{1/3} = \dfrac{64^{1/3}}{27^{1/3}}$

$= \dfrac{4}{3}$

43. $\left(\dfrac{81}{16}\right)^{-1/4} = \left(\dfrac{16}{81}\right)^{1/4}$

$= \dfrac{16^{1/4}}{81^{1/4}}$

$= \dfrac{2}{3}$

45. $\left(-\dfrac{1}{32}\right)^{-4/5} = (-32)^{4/5}$

$= \left[(-32)^{1/5}\right]^4$

$= (-2)^4$

$= 16$

47. $x^{1/2}x^{2/3} = x^{1/2 + 2/3}$

$= x^{7/6}$

49. $(a^{-1/2})^{-3/4} = a^{(-1/2)((-3/4)}$

$= a^{3/8}$

51. $(2^{-1} \cdot 4^{1/2})^{-2} = (2^{-1} \cdot 2)^{-2}$

$= (2^0)^{-2}$

$= 2^{(0)(-2)}$

$= 2^0$

$= 1$

53. $(r^{1/2}r^{-2/3}s^{1/2})^{-2}$

$= (r^{1/2 - 2/3}s^{1/2})^{-2}$

$= (r^{-1/6}s^{1/2})^{-2}$

$= r^{(-1/6)(-2)}s^{(1/2)(-2)}$

$= r^{1/3}s^{-1}$

$= \dfrac{r^{1/3}}{s}$

55. $(r^{-1}s^{1/2})^{-2}(r^{-1/2}s^{1/3})^2$

$= r^{(-1)(-2)}s^{(1/2)(-2)}s^{(1/2)(-2)}r^{(-1/2)(2)}s^{(1/3)(2)}$

$= r^2s^{-1}r^{-1}s^{2/3}$

$= r^{2-1}s^{-1+2/3}$

$= rs^{-1/3}$

$= \dfrac{r}{s^{1/3}}$

57. $\dfrac{x^{-1/2}}{x^{-1/3}} = x^{-1/2 - (-1/3)}$

$= x^{-1/6}$

$= \dfrac{1}{x^{1/6}}$

59. $\dfrac{a^{-1/2}b^{1/3}}{a^{1/4}b^{1/5}} = a^{-1/2 - 1/4}b^{1/3 - 1/5}$

$= a^{-3/4}b^{2/15}$

$= \dfrac{b^{2/15}}{a^{3/4}}$

61. $\left(\dfrac{x^{1/2}x^{-1}}{x^{1/3}}\right)^{-6} = \left(\dfrac{x^{1/2 - 1}}{x^{1/3}}\right)^{-6}$

$= \left(\dfrac{x^{-1/2}}{x^{1/3}}\right)^{-6}$

$= (x^{-1/2 - 1/3})^{-6}$

$= (x^{-5/6})^{-6}$

$= x^{(-5/6)(-6)}$

$= x^5$

63. $\dfrac{(4^{-1/2} \cdot 16^{3/4})^{-2}(64^{5/6})}{(-64)^{1/3}}$

$= \dfrac{\left(\dfrac{1}{4^{1/2}} \cdot 8\right)^{-2}(32)}{-4}$

$= \dfrac{\left(\dfrac{8}{2}\right)^{-2}(32)}{-4}$

$= \dfrac{4^{-2} \cdot 32}{-4}$

$= \dfrac{\dfrac{1}{4^2} \cdot 32}{-4}$

$= \dfrac{\dfrac{1}{16} \cdot 32}{-4}$

$= \dfrac{2}{-4}$

$= -\dfrac{1}{2}$

65. $\dfrac{(x^{1/2}y^{1/3})^{-2}(x^{1/3}y^{1/4})^{-12}}{xy^{1/4}}$

$= \dfrac{x^{(1/2)(-2)}y^{(1/3)(-2)}x^{(1/3)(-12)}y^{(1/4)(-12)}}{xy^{1/4}}$

$= \dfrac{x^{-1}y^{-2/3}x^{-4}y^{-3}}{xy^{1/4}}$

$= \dfrac{x^{-1-4}y^{-2/3-3}}{xy^{1/4}}$

$= \dfrac{x^{-5}y^{-11/3}}{xy^{1/4}}$

$= x^{-5-1}y^{-11/3-1/4}$

$= x^{-6}y^{-47/12}$

$= \dfrac{1}{x^{6}y^{47/12}}$

67. $(x^{1/2} + y)x^{1/2}$
$= x^{1/2} \cdot x^{1/2} + yx^{1/2}$
$= x^{1/2+1/2} + yx^{1/2}$
$= x + yx^{1/2}$

69. $(x^{1/2} - 2x^{-1/2})^2$
$= (x^{1/2})^2 - 2(x^{1/2})(2x^{-1/2}) + (2x^{-1/2})^2$
$= x^{(1/2)(2)} - 4x^{1/2-1/2} + 2^2x^{(-1/2)(2)}$
$= x - 4x^0 + 4x^{-1}$
$= x - 4 + \dfrac{4}{x} = \dfrac{x^2 - 4x + 4}{x}$

71. $\sqrt[3]{xy} = (xy)^{1/3}$

73. $\sqrt{x^2 + y^2} = (x^2 + y^2)^{1/2}$

75. $\sqrt[5]{5a^2b^3} = (5a^2b^3)^{1/5}$

77. $2\sqrt[3]{3xyz^4} = 2(3xyz^4)^{1/3}$

79. $5\sqrt[3]{(x - y)^2} = 5(x - y)^{2/3}$

81. $\sqrt[n]{x^n - y^n} = (x^n - y^n)^{1/n}$

83. $\sqrt[n]{x^{5n+1}y^{2n-1}} = (x^{5n+1}y^{2n-1})^{1/n}$

85. $x^{1/3} = \sqrt[3]{x}$

87. $mn^{1/3} = m\sqrt[3]{n}$

89. $(-a)^{2/3} = \sqrt[3]{(-a)^2}$

91. $-a^{2/3} = -\sqrt[3]{a^2}$

93. $(a^2b)^{1/3} = \sqrt[3]{a^2b}$

95. $(x^2 + y^2)^{1/2} = \sqrt{x^2 + y^2}$

97. $(x^n - y^n)^{1/2} = \sqrt{x^n - y^n}$

99. $36^{-1/2} = 0.1667$

101. $3 - 18^{-3/4} = 2.8856$

111. (a) $f(4) = -3$
     (b) $f(-3) = -6$
     (c) $f(0) = 3$
     (d) $f(2) = 0$
     (e) $x = -1, 2, 5$

113. $\dfrac{x}{3} - \dfrac{x}{2} = 4$

$6\left(\dfrac{x}{3} - \dfrac{x}{2}\right) = 6(4)$

$\dfrac{6}{1} \cdot \dfrac{x}{3} - \dfrac{6}{1} \cdot \dfrac{x}{2} = 24$

$2x - 3x = 24$
$-x = 24$
$x = -24$

7.4 Exercises

1. $\sqrt{56} = \sqrt{2^3 \cdot 7}$
$= \sqrt{2^2 \cdot 2 \cdot 7}$
$= \sqrt{2^2} \cdot \sqrt{2 \cdot 7}$
$= 2\sqrt{14}$

3. $\sqrt{48} = \sqrt{2^4 \cdot 3}$
$= \sqrt{2^4} \cdot \sqrt{3}$
$= 2^2 \cdot \sqrt{3}$
$= 4\sqrt{3}$

5. $\sqrt[5]{64} = \sqrt[5]{2^6}$
$= \sqrt[5]{2^5 \cdot 2}$
$= \sqrt[5]{2^5} \cdot \sqrt[5]{2}$
$= 2\sqrt[5]{2}$

7. $\sqrt{8}\sqrt{18} = \sqrt{8 \cdot 18}$
$= \sqrt{144}$
$= 12$

9. $\sqrt{64x^8} = \sqrt{64}\sqrt{x^8}$
$= 8x^4$

11. $\sqrt[4]{81x^{12}} = \sqrt[4]{3^4 x^{12}}$
$= \sqrt[4]{3^4} \sqrt[4]{x^{12}}$
$= 3x^3$

13. $\sqrt{128x^{60}} = \sqrt{2^7 x^{60}}$
$= \sqrt{2^6 \cdot 2x^{60}}$
$= \sqrt{2^6}\sqrt{2}\sqrt{x^{60}}$
$= 2^3 \sqrt{2}\, x^{30}$
$= 8x^{30}\sqrt{2}$

15. $\sqrt[4]{128x^{60}} = \sqrt[4]{2^7 x^{60}}$
$= \sqrt[4]{2^4 \cdot 2^3 x^{60}}$
$= \sqrt[4]{2^4} \sqrt[4]{2^3} \sqrt[4]{x^{60}}$
$= 2\sqrt[4]{8}\, x^{15}$
$= 2x^{15}\sqrt[4]{8}$

17. $\sqrt[5]{128x^{60}} = \sqrt[5]{2^7 x^{60}}$
$= \sqrt[5]{2^5 \cdot 2^2 \cdot x^{60}}$
$= \sqrt[5]{2^5} \sqrt[5]{2^2} \sqrt[5]{x^{60}}$
$= 2\sqrt[5]{4}\, x^{12}$
$= 2x^{12}\sqrt[5]{4}$

19. $\sqrt[3]{x^3 y^6} = \sqrt[3]{x^3} \sqrt[3]{y^6}$
$= xy^2$

21. $\sqrt{32a^2b^4} = \sqrt{2^5 a^2 b^4}$
$= \sqrt{2^4 \cdot 2a^2 b^4}$
$= \sqrt{2^4}\sqrt{2}\sqrt{a^2}\sqrt{b^4}$
$= 2^2\sqrt{2}\,ab^2$
$= 4ab^2\sqrt{2}$

23. $\sqrt[5]{a^{35}b^{75}} = \sqrt[5]{a^{35}} \sqrt[5]{b^{75}}$
$= a^7 b^{15}$

25. $\sqrt{x^3 y}\sqrt{xy^3} = \sqrt{x^3 y \cdot xy^3}$
$= \sqrt{x^4 y^4}$
$= \sqrt{x^4}\sqrt{y^4}$
$= x^2 y^2$

27. $\sqrt{\dfrac{1}{2}} = \dfrac{\sqrt{1}}{\sqrt{2}}$

$= \dfrac{1}{\sqrt{2}}$

$= \dfrac{1 \cdot \sqrt{2}}{\sqrt{2} \cdot \sqrt{2}}$

$= \dfrac{\sqrt{2}}{\sqrt{2^2}}$

$= \dfrac{\sqrt{2}}{2}$

29. $\dfrac{\sqrt{x}}{\sqrt{5}} = \dfrac{\sqrt{x} \cdot \sqrt{5}}{\sqrt{5} \cdot \sqrt{5}}$

$= \dfrac{\sqrt{x \cdot 5}}{\sqrt{5^2}}$

$= \dfrac{\sqrt{5x}}{5}$

31. $\sqrt{\dfrac{45}{4}} = \dfrac{\sqrt{45}}{\sqrt{4}}$

$= \dfrac{\sqrt{3^2 \cdot 5}}{2}$

$= \dfrac{3\sqrt{5}}{2}$

33. $\dfrac{1}{\sqrt{75}} = \dfrac{1}{\sqrt{3 \cdot 5^2}}$

$= \dfrac{1}{\sqrt{3} \cdot 5}$

$= \dfrac{1 \cdot \sqrt{3}}{5\sqrt{3} \cdot \sqrt{3}}$

$= \dfrac{\sqrt{3}}{5\sqrt{3^2}}$

$= \dfrac{\sqrt{3}}{5 \cdot 3}$

$= \dfrac{\sqrt{3}}{15}$

35. $\sqrt{64x^5y^8} = \sqrt{2^6x^4 \cdot xy^8}$

$= \sqrt{2^6}\sqrt{x^4}\sqrt{x}\sqrt{y^8}$

$= 2^3x^2\sqrt{x}y^4$

$= 8x^2y^4\sqrt{x}$

37. $\sqrt[3]{81x^8y^7} = \sqrt[3]{3 \cdot 3x^6 \cdot x^2y^6 \cdot y}$

$= \sqrt[3]{3^3} \cdot \sqrt[3]{3} \cdot \sqrt[3]{x^6} \cdot \sqrt[3]{x^2} \cdot \sqrt[3]{y^6} \cdot \sqrt[3]{y}$

$= 3\sqrt[3]{3}x^2\sqrt[3]{x^2}y^2\sqrt[3]{y}$

$= 3x^2y^2\sqrt[3]{3x^2y}$

39. $\sqrt[4]{54y^2}\sqrt[4]{48y^4} = \sqrt[4]{54y^2 \cdot 48y^4}$

$= \sqrt[4]{2592y^6}$

$= \sqrt[4]{2^5 \cdot 3^4y^6}$

$= \sqrt[4]{2^4 \cdot 2 \cdot 3^4 \cdot y^4 \cdot y^2}$

$= \sqrt[4]{2^4} \cdot \sqrt[4]{2} \cdot \sqrt[4]{3^4} \cdot \sqrt[4]{y^4} \cdot \sqrt[4]{y^2}$

$= 2\sqrt[4]{2} \cdot 3y \cdot \sqrt[4]{y^2}$

$= 6y\sqrt[4]{2y^2}$

41. $\sqrt[6]{(x + y^2)^6} = x + y^2$

43. Doesn't simplify further

45. $(2s\sqrt{6t})(5t\sqrt{3s}) = 10st\sqrt{6t \cdot 3s}$

$= 10st\sqrt{18ts}$

$= 10st\sqrt{2 \cdot 3^2ts}$

$= 10st\sqrt{3^2} \cdot \sqrt{2ts}$

$= 10st \cdot 3\sqrt{2ts}$

$= 30st\sqrt{2ts}$

47. $\left(3a\sqrt[3]{2b^4}\right)\left(2a^2\sqrt[3]{4b^2}\right)$

$= 6a^3\sqrt[3]{2b^4 \cdot 4b^2}$

$= 6a^3\sqrt[3]{8b^6}$

$= 6a^3\sqrt[3]{8}\sqrt[3]{b^6}$

$= 6a^3 \cdot 2 \cdot b^2$

$= 12a^3b^2$

49. $\sqrt[3]{\dfrac{x^3y^6}{8}} = \dfrac{\sqrt[3]{x^3y^6}}{\sqrt[3]{8}}$

$= \dfrac{\sqrt[3]{x^3}\sqrt[3]{y^6}}{\sqrt[3]{2^3}}$

$= \dfrac{xy^2}{2}$

51. $\sqrt[4]{\dfrac{32x^9}{y^{12}}} = \dfrac{\sqrt[4]{32x^9}}{\sqrt[4]{y^{12}}}$

$= \dfrac{\sqrt[4]{2^4 \cdot 2x^8 \cdot x}}{\sqrt[4]{y^{12}}}$

$= \dfrac{\sqrt[4]{2^4}\sqrt[4]{2}\sqrt[4]{x^8}\sqrt[4]{x}}{\sqrt[4]{y^{12}}}$

$= \dfrac{2\sqrt[4]{2}x^2\sqrt[4]{x}}{y^3}$

$= \dfrac{2x^2\sqrt[4]{2x}}{y^3}$

53. $\dfrac{\sqrt{54xy}}{\sqrt{2xy}} = \sqrt{\dfrac{54xy}{2xy}}$

$\qquad = \sqrt{27}$

$\qquad = \sqrt{3^2 \cdot 3}$

$\qquad = \sqrt{3^2}\sqrt{3}$

$\qquad = 3\sqrt{3}$

55. $\dfrac{\sqrt[4]{x^2 y^{17}}}{\sqrt[4]{x^{14}y}} = \sqrt[4]{\dfrac{x^2 y^{17}}{x^{14}y}}$

$\qquad = \sqrt[4]{\dfrac{y^{16}}{x^{12}}}$

$\qquad = \dfrac{\sqrt[4]{y^{16}}}{\sqrt[4]{x^{12}}}$

$\qquad = \dfrac{y^4}{x^3}$

57. $\sqrt{\dfrac{3xy}{5x^2 y}} = \sqrt{\dfrac{3}{5x}}$

$\qquad = \dfrac{\sqrt{3}}{\sqrt{5x}}$

$\qquad = \dfrac{\sqrt{3} \cdot \sqrt{5x}}{\sqrt{5x} \cdot \sqrt{5x}}$

$\qquad = \dfrac{\sqrt{3 \cdot 5x}}{\sqrt{(5x)^2}}$

$\qquad = \dfrac{\sqrt{15x}}{5x}$

59. $\sqrt{\dfrac{3x^2 y}{x^3 y^4}} = \sqrt{\dfrac{3}{xy^3}}$

$\qquad = \dfrac{\sqrt{3}}{\sqrt{x \cdot y^2 \cdot y}}$

$\qquad = \dfrac{\sqrt{3}}{y\sqrt{xy}}$

$\qquad = \dfrac{\sqrt{3} \cdot \sqrt{xy}}{y\sqrt{xy} \cdot \sqrt{xy}}$

$\qquad = \dfrac{\sqrt{3xy}}{y\sqrt{(xy)^2}}$

$\qquad = \dfrac{\sqrt{3xy}}{y(xy)}$

$\qquad = \dfrac{\sqrt{3xy}}{xy^2}$

61. $\sqrt[3]{\dfrac{3}{2}} = \dfrac{\sqrt[3]{3}}{\sqrt[3]{2}}$

$\qquad = \dfrac{\sqrt[3]{3} \cdot \sqrt[3]{2^2}}{\sqrt[3]{2} \cdot \sqrt[3]{2^2}}$

$\qquad = \dfrac{\sqrt[3]{3 \cdot 2^2}}{\sqrt[3]{2^3}}$

$\qquad = \dfrac{\sqrt[3]{12}}{2}$

63. $\sqrt[3]{\dfrac{9}{4}} = \dfrac{\sqrt[3]{9}}{\sqrt[3]{4}}$

$\qquad = \dfrac{\sqrt[3]{9} \cdot \sqrt[3]{2}}{\sqrt[3]{4} \cdot \sqrt[3]{2}}$

$\qquad = \dfrac{\sqrt[3]{9 \cdot 2}}{\sqrt[3]{8}}$

$\qquad = \dfrac{\sqrt[3]{18}}{2}$

65. $\sqrt[4]{\dfrac{9}{4}} = \dfrac{\sqrt[4]{9}}{\sqrt[4]{4}}$

$\qquad = \dfrac{\sqrt[4]{9} \cdot \sqrt[4]{4}}{\sqrt[4]{4} \cdot \sqrt[4]{4}}$

$\qquad = \dfrac{\sqrt[4]{9 \cdot 4}}{\sqrt[4]{16}}$

$\qquad = \dfrac{\sqrt[4]{36}}{2}$

67.
$$\sqrt[3]{\frac{81x^2y^4}{2x^3y}} = \sqrt[3]{\frac{81y^3}{2x}}$$
$$= \frac{\sqrt[3]{81y^3}}{\sqrt[3]{2x}}$$
$$= \frac{\sqrt[3]{3^3 \cdot 3y^3}}{\sqrt[3]{2x}}$$
$$= \frac{3y\sqrt[3]{3}}{\sqrt[3]{2x}}$$
$$= \frac{3y\sqrt[3]{3} \cdot \sqrt[3]{4x^2}}{\sqrt[3]{2x} \cdot \sqrt[3]{4x^2}}$$
$$= \frac{3y\sqrt[3]{3 \cdot 4x^2}}{\sqrt[3]{8x^3}}$$
$$= \frac{3y\sqrt[3]{12x^2}}{2x}$$

69.
$$\frac{3a^2\sqrt{a^2x^5}}{9a^5\sqrt{a^6x}} = \frac{3a^2}{9a^5}\sqrt{\frac{a^2x^5}{a^6x}}$$
$$= \frac{1}{3a^3}\sqrt{\frac{x^4}{a^4}}$$
$$= \frac{1}{3a^3} \cdot \frac{\sqrt{x^4}}{\sqrt{a^4}}$$
$$= \frac{1}{3a^3} \cdot \frac{x^2}{a^2}$$
$$= \frac{x^2}{3a^5}$$

71.
$$\frac{-3r^2s\sqrt{32r^2s^5}}{2r\sqrt{2r^5}}$$
$$= \frac{-3r^2s}{2r}\sqrt{\frac{32r^2s^5}{2r^5}}$$
$$= \frac{-3rs}{2}\sqrt{\frac{16s^5}{r^3}}$$

$$= \frac{-3rs}{2} \cdot \frac{\sqrt{16s^4 \cdot s}}{\sqrt{r^2 \cdot r}}$$
$$= \frac{-3rs}{2} \cdot \frac{4s^2\sqrt{s}}{r\sqrt{r}}$$
$$= \frac{-12s^3\sqrt{s}}{2\sqrt{r}}$$
$$= \frac{-12s^3\sqrt{s} \cdot \sqrt{r}}{2\sqrt{r} \cdot \sqrt{r}}$$
$$= \frac{-12s^3\sqrt{sr}}{2\sqrt{r^2}}$$
$$= -\frac{6s^3\sqrt{sr}}{r}$$

73. $\sqrt[12]{a^6} = \sqrt{a}$

75.
$$\left(\sqrt[3]{x}\right)\left(\sqrt[4]{x^3}\right)$$
$$= (x^{1/3})(x^{3/4})$$
$$= x^{1/3 + 3/4}$$
$$= x^{13/12}$$
$$= \sqrt[12]{x^{13}}$$
$$= x\sqrt[12]{x}$$

77.
$$\frac{\sqrt[3]{a^2}}{\sqrt{a}} = \frac{a^{2/3}}{a^{1/2}}$$
$$= a^{2/3 - 1/2}$$
$$= a^{1/6}$$
$$= \sqrt[6]{a}$$

79. $\sqrt[n]{x^{5n}y^{3n}} = \sqrt[n]{(x^5)^n(y^3)^n}$
$$= x^5y^3$$

83. $2x^2 + 7x - 15 = 2x^2 + 10x - 3x - 15$
$$= 2x(x + 5) - 3(x + 5)$$
$$= (x + 5)(2x - 3)$$

85. $p(x) = 2.8x^3 - 3.1x^2 + 14.7$
    $p(5.3) = 2.8(5.3)^3 - 3.1(5.3)^2 + 14.7$
    $= 344.5$

7.5 Exercises

1. $5\sqrt{3} - \sqrt{3} = (5 - 1)\sqrt{3}$
   $= 4\sqrt{3}$

3. $2\sqrt{5} - 4\sqrt{5} - \sqrt{5} = (2 - 4 - 1)\sqrt{5}$
   $= -3\sqrt{5}$

5. $8\sqrt{3} - (4\sqrt{3} - 2\sqrt{6})$
   $= 8\sqrt{3} - 4\sqrt{3} + 2\sqrt{6}$
   $= (8 - 4)\sqrt{3} + 2\sqrt{6}$
   $= 4\sqrt{3} + 2\sqrt{6}$

7. $2\sqrt{3} - 2\sqrt{5} - (\sqrt{3} - \sqrt{5})$
   $= 2\sqrt{3} - 2\sqrt{5} - \sqrt{3} + \sqrt{5}$
   $= (2 - 1)\sqrt{3} + (-2 + 1)\sqrt{5}$
   $= \sqrt{3} - \sqrt{5}$

9. $2\sqrt{x} - 5\sqrt{x} + 3\sqrt{x}$
   $= (2 - 5 + 3)\sqrt{x}$
   $= 0$

11. $5a\sqrt{b} - 3a^3\sqrt{b} + 2a\sqrt{b}$
    $= (5a - 3a^3 + 2a)\sqrt{b}$
    $= (7a - 3a^3)\sqrt{b}$

13. $3x \sqrt[3]{x^2} - 2 \sqrt[3]{x^2} + 6 \sqrt[3]{x^2}$
    $= (3x - 2 + 6) \sqrt[3]{x^2}$
    $= (3x + 4) \sqrt[3]{x^2}$

15. $\left(7 - 3 \sqrt[3]{a}\right) - \left(6 - \sqrt[3]{a}\right)$
    $= 7 - 3 \sqrt[3]{a} - 6 + \sqrt[3]{a}$
    $= 1 + (-3 + 1)\sqrt[3]{a}$
    $= 1 - 2 \sqrt[3]{a}$

17. $\sqrt{12} - \sqrt{27}$
    $= 2\sqrt{3} - 3\sqrt{3}$
    $= (2 - 3)\sqrt{3}$
    $= -\sqrt{3}$

19. $\sqrt{24} - \sqrt{27} + \sqrt{54}$
    $= 2\sqrt{6} - 3\sqrt{3} + 3\sqrt{6}$
    $= (2 + 3)\sqrt{6} - 3\sqrt{3}$
    $= 5\sqrt{6} - 3\sqrt{3}$

21. $6\sqrt{3} - 4\sqrt{81}$
    $= 6\sqrt{3} - 36$

23. $3\sqrt{24} - 5\sqrt{48} - \sqrt{6}$
    $= 6\sqrt{6} - 20\sqrt{3} - \sqrt{6}$
    $= (6 - 1)\sqrt{6} - 20\sqrt{3}$
    $= 5\sqrt{6} - 20\sqrt{3}$

25. $3 \sqrt[3]{24} - 5 \sqrt[3]{48} - \sqrt[3]{6}$
    $= 6 \sqrt[3]{3} - 10 \sqrt[3]{6} - \sqrt[3]{6}$
    $= 6 \sqrt[3]{3} + (-10 - 1)\sqrt[3]{6}$
    $= 6 \sqrt[3]{3} - 11 \sqrt[3]{6}$

27. $2a\sqrt{ab^2} - 3b\sqrt{a^2b} - ab\sqrt{ab}$
    $= 2ab\sqrt{a} - 3ab\sqrt{b} - ab\sqrt{ab}$

29. $\sqrt[3]{x^4} - x \sqrt[3]{x}$
    $= x \sqrt[3]{x} - x \sqrt[3]{x}$
    $= 0$

31. $\sqrt{20x^9y^8} + 2xy\sqrt{5x^7y^6}$
    $= 2x^4y^4\sqrt{5x} + 2x^4y^4\sqrt{5x}$
    $= (2x^4y^4 + 2x^4y^4)\sqrt{5x}$
    $= 4x^4y^4\sqrt{5x}$

33. $5 \sqrt[3]{9x^5} - 3x \sqrt[3]{x^2} + 2x \sqrt[3]{72x^2}$
    $= 5x \sqrt[3]{9x^2} - 3x \sqrt[3]{x^2} + 4x \sqrt[3]{9x^2}$
    $= (5x + 4x) \sqrt[3]{9x^2} - 3x \sqrt[3]{x^2}$
    $= 9x \sqrt[3]{9x^2} - 3x \sqrt[3]{x^2}$

35. $4 \sqrt[4]{16x} - 7 \sqrt[4]{x^5} + x \sqrt[4]{81x}$
    $= 8 \sqrt[4]{x} - 7x \sqrt[4]{x} + 3x \sqrt[4]{x}$
    $= (8 - 7x + 3x) \sqrt[4]{x}$
    $= (8 - 4x) \sqrt[4]{x}$

37. $\dfrac{1}{\sqrt{5}} + 2$

$= \dfrac{1 \cdot \sqrt{5}}{\sqrt{5} \cdot \sqrt{5}} + 2$

$= \dfrac{\sqrt{5}}{5} + 2$

$= \dfrac{\sqrt{5}}{5} + \dfrac{10}{5}$

$= \dfrac{\sqrt{5} + 10}{5}$

39. $\dfrac{12}{\sqrt{6}} - 2\sqrt{6}$

$= \dfrac{12\sqrt{6}}{\sqrt{6}\sqrt{6}} - 2\sqrt{6}$

$= \dfrac{12\sqrt{6}}{6} - 2\sqrt{6}$

$= 2\sqrt{6} - 2\sqrt{6}$
$= 0$

41. $\sqrt{\dfrac{1}{2}} + \sqrt{2}$

$= \dfrac{1}{\sqrt{2}} + \sqrt{2}$

$= \dfrac{1 \cdot \sqrt{2}}{\sqrt{2}\sqrt{2}} + \sqrt{2}$

$= \dfrac{\sqrt{2}}{2} + \sqrt{2}$

$= \left(\dfrac{1}{2} + 1\right)\sqrt{2}$

$= \dfrac{3\sqrt{2}}{2}$

43. $\sqrt{\dfrac{5}{2}} + \sqrt{\dfrac{2}{5}}$

$= \dfrac{\sqrt{5}}{\sqrt{2}} + \dfrac{\sqrt{2}}{\sqrt{5}}$

$= \dfrac{\sqrt{5}\sqrt{2}}{\sqrt{2}\sqrt{2}} + \dfrac{\sqrt{2}\sqrt{5}}{\sqrt{5}\sqrt{5}}$

$= \dfrac{\sqrt{10}}{2} + \dfrac{\sqrt{10}}{5}$

$= \dfrac{5\sqrt{10}}{10} + \dfrac{2\sqrt{10}}{10}$

$= \dfrac{5\sqrt{10} + 2\sqrt{10}}{10}$

$= \dfrac{7\sqrt{10}}{10}$

45. $\sqrt{\dfrac{1}{7}} - 3\sqrt{\dfrac{1}{5}}$

$= \dfrac{1}{\sqrt{7}} - \dfrac{3}{\sqrt{5}}$

$= \dfrac{1 \cdot \sqrt{7}}{\sqrt{7}\sqrt{7}} - \dfrac{3\sqrt{5}}{\sqrt{5}\sqrt{5}}$

$= \dfrac{\sqrt{7}}{7} - \dfrac{3\sqrt{5}}{5}$

$= \dfrac{5\sqrt{7}}{35} - \dfrac{21\sqrt{5}}{35}$

$= \dfrac{5\sqrt{7} - 21\sqrt{5}}{35}$

47. $\dfrac{1}{\sqrt[3]{2}} - 6\sqrt[3]{4}$

$= \dfrac{1 \cdot \sqrt[3]{4}}{\sqrt[3]{2} \cdot \sqrt[3]{4}} - 6\sqrt[3]{4}$

$= \dfrac{\sqrt[3]{4}}{\sqrt[3]{8}} - 6\sqrt[3]{4}$

$= \dfrac{\sqrt[3]{4}}{2} - 6\sqrt[3]{4}$

$= \left(\dfrac{1}{2} - 6\right)\sqrt[3]{4}$

$= -\dfrac{11}{2}\sqrt[3]{4}$

$= \dfrac{-11\sqrt[3]{4}}{2}$

49. $\sqrt{\dfrac{1}{x}} + \sqrt{\dfrac{1}{y}}$

$= \dfrac{1}{\sqrt{x}} + \dfrac{1}{\sqrt{y}}$

$= \dfrac{1\sqrt{x}}{\sqrt{x}\sqrt{x}} + \dfrac{1\sqrt{y}}{\sqrt{y}\sqrt{y}}$

$= \dfrac{\sqrt{x}}{x} + \dfrac{\sqrt{y}}{y}$

$= \dfrac{y\sqrt{x}}{xy} + \dfrac{x\sqrt{y}}{xy}$

$= \dfrac{y\sqrt{x} + x\sqrt{y}}{xy}$

51. $\dfrac{1}{\sqrt[3]{9}} - \dfrac{3}{\sqrt[3]{3}}$

$= \dfrac{1 \cdot \sqrt[3]{3}}{\sqrt[3]{9}\sqrt[3]{3}} - \dfrac{3 \cdot \sqrt[3]{9}}{\sqrt[3]{3} \cdot \sqrt[3]{9}}$

$= \dfrac{\sqrt[3]{3}}{\sqrt[3]{27}} - \dfrac{3\sqrt[3]{9}}{\sqrt[3]{27}}$

$= \dfrac{\sqrt[3]{3}}{3} - \dfrac{3\sqrt[3]{9}}{3}$

$= \dfrac{\sqrt[3]{3} - 3\sqrt[3]{9}}{3}$

53. $3\sqrt{\dfrac{2}{49}} + 3\sqrt{7}$

$= \dfrac{3\sqrt{2}}{\sqrt{49}} + 3\sqrt{7}$

$= \dfrac{3\sqrt{2}}{7} + 3\sqrt{7}$

$= \dfrac{3\sqrt{2}}{7} + \dfrac{21\sqrt{7}}{7}$

$= \dfrac{3\sqrt{2} + 21\sqrt{7}}{7}$

55. $3\sqrt{10} - \dfrac{4}{\sqrt{10}} + \dfrac{2}{\sqrt{10}}$

$= 3\sqrt{10} + \dfrac{-4 + 2}{\sqrt{10}}$

$= 3\sqrt{10} - \dfrac{2}{\sqrt{10}}$

$= 3\sqrt{10} - \dfrac{2\sqrt{10}}{\sqrt{10}\sqrt{10}}$

$= 3\sqrt{10} - \dfrac{2\sqrt{10}}{10}$

$= \dfrac{30\sqrt{10}}{10} - \dfrac{2\sqrt{10}}{10}$

$= \dfrac{30\sqrt{10} - 2\sqrt{10}}{10}$

$= \dfrac{28\sqrt{10}}{10}$

$= \dfrac{14\sqrt{10}}{5}$

57. $6\sqrt[3]{25} - \dfrac{15}{\sqrt[3]{5}} + 5\sqrt[3]{\dfrac{1}{5}}$

$= 6\sqrt[3]{25} - \dfrac{15}{\sqrt[3]{5}} + \dfrac{5 \cdot 1}{\sqrt[3]{5}}$

$= 6\sqrt[3]{25} + \dfrac{-15 + 5}{\sqrt[3]{5}}$

$= 6\sqrt[3]{25} - \dfrac{10}{\sqrt[3]{5}}$

$= 6\sqrt[3]{25} - \dfrac{10\sqrt[3]{25}}{\sqrt[3]{5}\sqrt[3]{25}}$

$= 6\sqrt[3]{25} - \dfrac{10\sqrt[3]{25}}{\sqrt[3]{125}}$

$= 6\sqrt[3]{25} - \dfrac{10\sqrt[3]{25}}{5}$

$= 6\sqrt[3]{25} - 2\sqrt[3]{25}$

$= 4\sqrt[3]{25}$

59. $y = 0.1x^2 - 0.6x + 3.7$

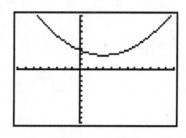

minimum value: 2.8

61. $3x^3 - 75x$
$= 3x(x^2 - 25)$
$= 3x(x^2 - 5^2)$
$= 3x(x - 5)(x + 5)$

7.6 Exercises

1. $5(\sqrt{5} - 3)$
$= 5\sqrt{5} - 15$

3. $2(\sqrt{3} - \sqrt{5}) - 4(\sqrt{3} + \sqrt{5})$
$= 2\sqrt{3} - 2\sqrt{5} - 4\sqrt{3} - 4\sqrt{5}$
$= -2\sqrt{3} - 6\sqrt{5}$

5. $\sqrt{a}(\sqrt{a} + \sqrt{b})$
$= \sqrt{a}\sqrt{a} + \sqrt{a}\sqrt{b}$
$= a + \sqrt{ab}$

7. $\sqrt{2}(\sqrt{5} + \sqrt{2})$
$= \sqrt{2}\sqrt{5} + \sqrt{2}\sqrt{2}$
$= \sqrt{10} + 2$

9. $3\sqrt{5}(2\sqrt{3} - 4\sqrt{5})$
$= 3\sqrt{5} \cdot 2\sqrt{3} - 3\sqrt{5} \cdot 4\sqrt{5}$
$= 6\sqrt{15} - 12 \cdot 5$
$= 6\sqrt{15} - 60$

11. $\sqrt{2}(\sqrt{3} + \sqrt{2}) - 3(2\sqrt{6} - 4)$
$= \sqrt{2}\sqrt{3} + \sqrt{2}\sqrt{2} - 6\sqrt{6} + 12$
$= \sqrt{6} + 2 - 6\sqrt{6} + 12$
$= 14 - 5\sqrt{6}$

13. $(\sqrt{5} - 2)(\sqrt{3} + 1)$
$= \sqrt{5}\sqrt{3} + \sqrt{5} - 2\sqrt{3} - 2$
$= \sqrt{15} + \sqrt{5} - 2\sqrt{3} - 2$

15. $(\sqrt{5} - \sqrt{3})(\sqrt{5} + \sqrt{3})$
$= \sqrt{5}\sqrt{5} - \sqrt{3}\sqrt{3}$
$= 5 - 3$
$= 2$

17. $(\sqrt{5} - \sqrt{3})^2$
$= (\sqrt{5})^2 - 2\sqrt{5}\sqrt{3} + (\sqrt{3})^2$
$= 5 - 2\sqrt{15} + 3$
$= 8 - 2\sqrt{15}$

19. $(2\sqrt{7} - 5)(2\sqrt{7} + 5)$
$= (2\sqrt{7})^2 - 5^2$
$= 4 \cdot 7 - 25$
$= 3$

21. $(5\sqrt{2} - 3\sqrt{5})(5\sqrt{2} + 3\sqrt{5})$
$= (5\sqrt{2})^2 - (3\sqrt{5})^2$
$= 5^2(\sqrt{2})^2 - 3^2(\sqrt{5})^2$
$= 25 \cdot 2 - 9 \cdot 5$
$= 5$

23. $(2\sqrt{a} - \sqrt{b})^2$
$= (2\sqrt{a})^2 - 2(2\sqrt{a})(\sqrt{b}) + (\sqrt{b})^2$
$= 2^2(\sqrt{a})^2 - 4\sqrt{ab} + b$
$= 4a - 4\sqrt{ab} + b$

25. $(3\sqrt{x} - 2\sqrt{y})(3\sqrt{x} + 2\sqrt{y})$
$= (3\sqrt{x})^2 - (2\sqrt{y})^2$
$= 3^2(\sqrt{x})^2 - 2^2(\sqrt{y})^2$
$= 9x - 4y$

27. $(\sqrt{x} - 3)^2$
$= (\sqrt{x})^2 - 2\sqrt{x} \cdot 3 + 3^2$
$= x - 6\sqrt{x} + 9$

29. $(\sqrt{x - 3})^2 = x - 3$

31. $(\sqrt{x + 1})^2 - (\sqrt{x} + 1)^2$
$= x + 1 - [(\sqrt{x})^2 + 2\sqrt{x} \cdot 1 + 1^2]$
$= x + 1 - (x + 2\sqrt{x} + 1)$
$= x + 1 - x - 2\sqrt{x} - 1$
$= -2\sqrt{x}$

33. $(\sqrt[3]{2} - \sqrt[3]{3})(\sqrt[3]{4} + \sqrt[3]{6} + \sqrt[3]{9})$
$= \sqrt[3]{8} + \sqrt[3]{12} + \sqrt[3]{18} - \sqrt[3]{12} - \sqrt[3]{18} - \sqrt[3]{27}$
$= \sqrt[3]{8} - \sqrt[3]{27}$
$= 2 - 3$
$= -1$

35. $\dfrac{4\sqrt{2} - 6\sqrt{3}}{2}$

$= \dfrac{2(2\sqrt{2} - 3\sqrt{3})}{2}$

$= 2\sqrt{2} - 3\sqrt{3}$

37. $\dfrac{5\sqrt{8} - 2\sqrt{7}}{8}$

$= \dfrac{10\sqrt{2} - 2\sqrt{7}}{8}$

$= \dfrac{2(5\sqrt{2} - \sqrt{7})}{8}$

$= \dfrac{5\sqrt{2} - \sqrt{7}}{4}$

39. $\dfrac{3\sqrt{50} + 5\sqrt{5}}{5}$

$= \dfrac{15\sqrt{2} + 5\sqrt{5}}{5}$

$= \dfrac{5(3\sqrt{2} + \sqrt{5})}{5}$

$= 3\sqrt{2} + \sqrt{5}$

41. $\dfrac{4 + \sqrt{28}}{4}$

$= \dfrac{4 + 2\sqrt{7}}{4}$

$= \dfrac{2(2 + \sqrt{7})}{4}$

$= \dfrac{2 + \sqrt{7}}{2}$

43. $\dfrac{1}{\sqrt{2} - 3}$

$= \dfrac{1(\sqrt{2} - 3)}{(\sqrt{2} - 3)(\sqrt{2} + 3)}$

$= \dfrac{\sqrt{2} + 3}{(\sqrt{2})^2 - 3^2}$

$= \dfrac{\sqrt{2} + 3}{2 - 9}$

$= \dfrac{\sqrt{2} + 3}{-7}$

$= -\dfrac{\sqrt{2} + 3}{7}$

45. $\dfrac{10}{\sqrt{5} + 1}$

$= \dfrac{10(\sqrt{5} - 1)}{(\sqrt{5} + 1)(\sqrt{5} - 1)}$

$= \dfrac{10(\sqrt{5} - 1)}{(\sqrt{5})^2 - 1^2}$

$= \dfrac{10(\sqrt{5} - 1)}{5 - 1}$

$= \dfrac{10(\sqrt{5} - 1)}{4}$

$= \dfrac{5(\sqrt{5} - 1)}{2}$

$= \dfrac{5\sqrt{5} - 5}{2}$

47. $\dfrac{2}{\sqrt{3} - \sqrt{a}}$

$= \dfrac{2(\sqrt{3} + \sqrt{a})}{(\sqrt{3} - \sqrt{a})(\sqrt{3} + \sqrt{a})}$

$= \dfrac{2(\sqrt{3} + \sqrt{a})}{(\sqrt{3})^2 - (\sqrt{a})^2}$

$= \dfrac{2(\sqrt{3} + \sqrt{a})}{3 - a}$

$= \dfrac{2\sqrt{3} + 2\sqrt{a}}{3 - a}$

49. $\dfrac{\sqrt{x}}{\sqrt{x} - \sqrt{y}}$

$= \dfrac{\sqrt{x}(\sqrt{x} + \sqrt{y})}{(\sqrt{x} - \sqrt{y})(\sqrt{x} + \sqrt{y})}$

$= \dfrac{\sqrt{x}(\sqrt{x} + \sqrt{y})}{(\sqrt{x})^2 - (\sqrt{y})^2}$

$= \dfrac{\sqrt{x}(\sqrt{x} + \sqrt{y})}{x - y}$

$= \dfrac{x + \sqrt{xy}}{x - y}$

51. $\dfrac{\sqrt{2}}{\sqrt{5} - \sqrt{2}}$

$= \dfrac{\sqrt{2}(\sqrt{5} + \sqrt{2})}{(\sqrt{5} - \sqrt{2})(\sqrt{5} + \sqrt{2})}$

$= \dfrac{\sqrt{2}(\sqrt{5} + \sqrt{2})}{(\sqrt{5})^2 - (\sqrt{2})^2}$

$= \dfrac{\sqrt{2}(\sqrt{5} + \sqrt{2})}{5 - 2}$

$= \dfrac{\sqrt{10} + 2}{3}$

53. $\dfrac{2\sqrt{2}}{2\sqrt{5} - \sqrt{2}}$

$= \dfrac{2\sqrt{2}(2\sqrt{5} + \sqrt{2})}{(2\sqrt{5} - \sqrt{2})(2\sqrt{5} + \sqrt{2})}$

$= \dfrac{2\sqrt{2}(2\sqrt{5} + \sqrt{2})}{(2\sqrt{5})^2 - (\sqrt{2})^2}$

$= \dfrac{2\sqrt{2}(2\sqrt{5} + \sqrt{2})}{20 - 2}$

$= \dfrac{2\sqrt{2}(2\sqrt{5} + \sqrt{2})}{18}$

$= \dfrac{\sqrt{2}(2\sqrt{5} + \sqrt{2})}{9}$

$= \dfrac{2\sqrt{10} + 2}{9}$

55. $\dfrac{2\sqrt{5} - \sqrt{2}}{2\sqrt{2}}$

$= \dfrac{(2\sqrt{5} - \sqrt{2})\sqrt{2}}{2\sqrt{2}\sqrt{2}}$

$= \dfrac{2\sqrt{10} - 2}{4}$

$= \dfrac{2(\sqrt{10} - 1)}{4}$

$= \dfrac{\sqrt{10} - 1}{2}$

57. $\dfrac{\sqrt{3} + \sqrt{2}}{\sqrt{3} - \sqrt{2}}$

$= \dfrac{(\sqrt{3} + \sqrt{2})(\sqrt{3} - \sqrt{2})}{(\sqrt{3} - \sqrt{2})(\sqrt{3} + \sqrt{2})}$

$= \dfrac{(\sqrt{3})^2 + 2(\sqrt{3})(\sqrt{2}) + (\sqrt{2})^2}{(\sqrt{3})^2 - (\sqrt{2})^2}$

$= \dfrac{3 + 2\sqrt{6} + 2}{3 - 2}$

$= 5 + 2\sqrt{6}$

59. $\dfrac{3\sqrt{5} - 2\sqrt{2}}{2\sqrt{5} - 3\sqrt{2}}$

$= \dfrac{(3\sqrt{5} - 2\sqrt{2})(2\sqrt{5} + 3\sqrt{2})}{(2\sqrt{5} - 3\sqrt{2})(2\sqrt{5} + 3\sqrt{2})}$

$= \dfrac{6 \cdot 5 + 9\sqrt{10} - 4\sqrt{10} - 6 \cdot 2}{(2\sqrt{5})^2 - (3\sqrt{2})^2}$

$= \dfrac{30 + 5\sqrt{10} - 12}{20 - 18}$

$= \dfrac{18 + 5\sqrt{10}}{2}$

61. $\dfrac{x - y}{\sqrt{x} - \sqrt{y}}$

$= \dfrac{(x - y)(\sqrt{x} + \sqrt{y})}{(\sqrt{x} - \sqrt{y})(\sqrt{x} + \sqrt{y})}$

$= \dfrac{(x - y)(\sqrt{x} + \sqrt{y})}{(\sqrt{x})^2 - (\sqrt{y})^2}$

$= \dfrac{(x - y)(\sqrt{x} + \sqrt{y})}{x - y}$

$= \sqrt{x} + \sqrt{y}$

63. $\dfrac{x^2 - x - 2}{\sqrt{x} - \sqrt{2}}$

$= \dfrac{(x^2 - x - 2)(\sqrt{x} + \sqrt{2})}{(\sqrt{x} - \sqrt{2})(\sqrt{x} + \sqrt{2})}$

$= \dfrac{x^2 - x - 2)(\sqrt{x} + \sqrt{2})}{(\sqrt{x})^2 - (\sqrt{2})^2}$

$= \dfrac{(x - 2)(x + 1)(\sqrt{x} + \sqrt{2})}{x - 2}$

$= (x + 1)(\sqrt{x} + \sqrt{2})$

65. $\dfrac{12}{\sqrt{6} - 2} - \dfrac{36}{\sqrt{6}}$

$= \dfrac{12(\sqrt{6} + 2)}{(\sqrt{6} - 2)(\sqrt{6} + 2)} - \dfrac{36\sqrt{6}}{\sqrt{6}\sqrt{6}}$

$= \dfrac{12(\sqrt{6} + 2)}{(\sqrt{6})^2 - 2^2} - \dfrac{36\sqrt{6}}{6}$

$= \dfrac{12(\sqrt{6} + 2)}{6 - 4} - 6\sqrt{6}$

$= \dfrac{12(\sqrt{6} + 2)}{2} - 6\sqrt{6}$

$= 6(\sqrt{6} + 2) - 6\sqrt{6}$

$= 6\sqrt{6} + 12 - 6\sqrt{6}$

$= 12$

67. $\dfrac{20}{\sqrt{7} + \sqrt{3}} + \dfrac{28}{\sqrt{7}}$

$= \dfrac{20(\sqrt{7} - \sqrt{3})}{(\sqrt{7} + \sqrt{3})(\sqrt{7} - \sqrt{3})} + \dfrac{28\sqrt{7}}{\sqrt{7}\sqrt{7}}$

$= \dfrac{20(\sqrt{7} - \sqrt{3})}{(\sqrt{7})^2 - (\sqrt{3})^2} + \dfrac{28\sqrt{7}}{7}$

$= \dfrac{20(\sqrt{7} - \sqrt{3})}{7 - 3} + 4\sqrt{7}$

$= \dfrac{20(\sqrt{7} - \sqrt{3})}{4} + 4\sqrt{7}$

$= 5(\sqrt{7} - \sqrt{3}) + 4\sqrt{7}$

$= 5\sqrt{7} - 5\sqrt{3} + 4\sqrt{7}$

$= 9\sqrt{7} - 5\sqrt{3}$

71. $(2x - 1)(2x + 3) = 12$
$4x^2 + 4x - 3 = 12$
$4x^2 + 4x - 15 = 0$
$(2x - 3)(2x + 5) = 0$
$2x - 3 = 0 \quad \text{or} \quad 2x + 5 = 0$
$\quad 2x = 3 \qquad\qquad 2x = -5$
$\qquad x = \dfrac{3}{2} \quad \text{or} \qquad x = \dfrac{-5}{2}$

73. $12x^2 - 46x + 40$
$= 2(6x^2 - 23x + 20)$
$= 2(6x^2 - 8x - 15x + 20)$
$= 2[2x(3x - 4) - 5(3x - 4)]$
$= 2(3x - 4)(2x - 5)$

7.7 Exercises

1. $\sqrt{a} = 7$
$(\sqrt{a})^2 = 7^2$
$\quad a = 49$

3. $6 = \sqrt{a + 5}$
$6^2 = (\sqrt{a + 5})^2$
$36 = a + 5$
$31 = a$

5. $\sqrt{3x - 1} = 5$
$(\sqrt{3x - 1})^2 = 5^2$
$3x - 1 = 25$
$\quad 3x = 26$
$\quad\quad x = \dfrac{26}{3}$

7. $\sqrt{y + 5} - 4 = 7$
$\sqrt{y + 5} = 11$
$(\sqrt{y + 5})^2 = 11^2$
$y + 5 = 121$
$\quad y = 116$

9. $\quad 2 = 5 - \sqrt{3x - 1}$
$\quad -3 = -\sqrt{3x - 1}$
$(-3)^2 = (-\sqrt{3x - 1})^2$
$\quad 9 = 3x - 1$
$\quad 10 = 3x$
$\dfrac{10}{3} = x$

11. $3\sqrt{x} = 12$
$\sqrt{x} = 4$
$(\sqrt{x})^2 = 4^2$
$\quad x = 16$

13. $6\sqrt{a} + 3.4 = 21.3$
$6\sqrt{a} = 17.9$
$(6\sqrt{a})^2 = (17.9)^2$
$36a = 320.41$
$\quad a = 8.90$

15. $7 + \sqrt{x - 1} = 3$
$\sqrt{x - 1} = -4$

Since square root is nonnegative, there is no solution.

17. $\sqrt{2x - 3} - \sqrt{x + 5} = 0$
$\sqrt{2x - 3} = \sqrt{x + 5}$
$(\sqrt{2x - 3})^2 = (\sqrt{x + 5})^2$
$2x - 3 = x + 5$
$x - 3 = 5$
$\quad x = 8$

19. $\sqrt{5 - 4x} - \sqrt{6 - 3x} = 0$

$\qquad\sqrt{5 - 4x} = \sqrt{6 - 3x}$

$\qquad (\sqrt{5 - 4x})^2 = (\sqrt{6 - 3x})^2$

$\qquad\qquad 5 - 4x = 6 - 3x$

$\qquad\qquad\qquad 5 = 6 + x$

$\qquad\qquad\quad -1 = x$

21. $\quad 4 = \sqrt[3]{x}$

$\quad 4^3 = \left(\sqrt[3]{x}\right)^3$

$\quad 64 = x$

23. $\quad \sqrt[5]{s} = -3$

$\quad \left(\sqrt[5]{s}\right)^5 = (-3)^5$

$\qquad\quad s = -243$

25. $\quad -3 = \sqrt[3]{y + 5}$

$\quad (-3)^3 = \left(\sqrt[3]{y + 5}\right)^3$

$\quad -27 = y + 5$

$\quad -32 = y$

27. $\quad x^{1/3} = 5$

$\quad (x^{1/3})^3 = 5^3$

$\qquad\quad x = 125$

29. $\quad x^{1/3} + 7 = 5$

$\qquad\quad x^{1/3} = -2$

$\quad (x^{1/3})^3 = (-2)^3$

$\qquad\qquad x = -8$

31. $\quad (x + 7)^{1/3} = 5$

$\quad [(x + 7)^{1/3}]^3 = 5^3$

$\qquad\qquad x + 7 = 125$

$\qquad\qquad\quad x = 118$

33. $\quad x^{-1/4} = 4$

$\quad (x^{-1/4})^{-4} = 4^{-4}$

$\qquad\qquad x = \dfrac{1}{4^4}$

$\qquad\qquad x = \dfrac{1}{256}$

35. $\quad (x + 3)^{1/3} + 4 = 2$

$\qquad\quad (x + 3)^{1/3} = -2$

$\quad [(x + 3)^{1/3}]^3 = (-2)^3$

$\qquad\qquad x + 3 = -8$

$\qquad\qquad\quad x = -11$

37. $\sqrt{x} = 7.8$

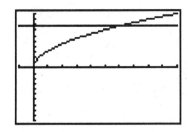

$x = 60.84$

39. $9.38 = \sqrt{x - 4.3}$

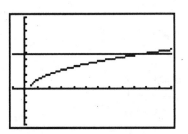

$x = 92.28$

41. $\sqrt{3x - 1} = 5.24$

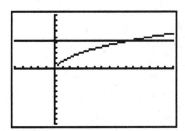

$x = 9.49$

43. $6\sqrt{a} + 3.4 = 21$
$\qquad 6\sqrt{a} = 17.6$

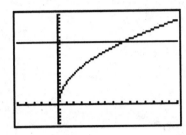

$a = 8.60$

45. $7 - \sqrt{x - 2} = 4$
$\qquad -\sqrt{x - 2} = -3$
$\qquad \sqrt{x - 2} = 3$

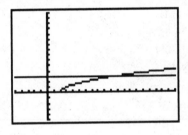

$x = 11$

47. $x^{1/3} - 3 = 5$
$\qquad x^{1/3} = 8$

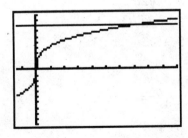

$x = 512$

49. $\sqrt{3a} = 5$
$\quad (\sqrt{3a})^2 = 5^2$
$\qquad 3a = 25$
$\qquad a = \dfrac{25}{3}$

51. $\quad 8 = \sqrt{3x - 2}$
$\quad 8^2 = (\sqrt{3x - 2})^2$
$\quad 64 = 3x - 2$
$\quad 66 = 3x$
$\quad 22 = x$

53. $\sqrt{y + 4} + 8 = 15.6$
$\qquad \sqrt{y + 4} = 7.6$
$\quad (\sqrt{y + 4})^2 = (7.6)^2$
$\qquad y + 4 = 57.76$
$\qquad y = 53.76$

55. $3.5\sqrt{2x} = 8.4$
$\qquad \sqrt{2x} = 2.4$
$\quad (\sqrt{2x})^2 = (2.4)^2$
$\qquad 2x = 5.76$
$\qquad x = 2.88$

57. $\sqrt{7 - 9x} - \sqrt{13 - 6x} = 0$
$\qquad \sqrt{7 - 9x} = \sqrt{13 - 6x}$
$\quad (\sqrt{7 - 9x})^2 = (\sqrt{13 - 6x})^2$
$\qquad 7 - 9x = 13 - 6x$
$\qquad 7 = 13 + 3x$
$\qquad -6 = 3x$
$\qquad -2 = x$

59. $v = 2\sqrt{5s}$
$\quad 55 = 2\sqrt{5s}$

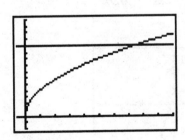

$s = 151.25$

$$55 = 2\sqrt{5s}$$
$$27.5 = \sqrt{5s}$$
$$(27.5)^2 = \left(\sqrt{5s}\right)^2$$
$$756.25 = 5s$$
$$151.25 = s$$

The skid marks are 151.25 ft.

61.
$$T = 2\pi\sqrt{\dfrac{L}{980}}$$

$$8 = 2\pi\sqrt{\dfrac{L}{980}}$$

$$\dfrac{4}{\pi} = \sqrt{\dfrac{L}{980}}$$

$$\left(\dfrac{4}{\pi}\right)^2 = \left(\sqrt{\dfrac{L}{980}}\right)^2$$

$$1.6211 = \dfrac{L}{980}$$
$$1588.68 = L$$

It is 1588.68 cm.

65.
$$\dfrac{x-3}{4} - \dfrac{x+2}{5} = \dfrac{x-5}{2}$$

$$20\left(\dfrac{x-3}{4} - \dfrac{x+2}{5}\right) = 20\left(\dfrac{x-5}{2}\right)$$

$$5(x-3) - 4(x+2) = 10(x-5)$$
$$5x - 15 - 4x - 8 = 10x - 50$$
$$x - 23 = 10x - 50$$
$$-23 = 9x - 50$$
$$27 = 9x$$
$$3 = x$$

67. $f(x) = 1.4 - 2.3x - 0.9x^2$

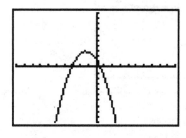

$(-1.3, 2.9)$

1. $i^3 = i^2 \cdot i$
    $= -1 \cdot i$
    $= -i$

3. $-i^{21} = -(i^4)^5 \cdot i$
    $= -(1)^5 \cdot i$
    $= -i$

5. $(-i)^{29} = (-1)^{29}i^{29}$
    $= -(i^4)^7 \cdot i$
    $= -(1)^7 \cdot i$
    $= -i$

7. $i^{72} = (i^4)^{18}$
    $= 1^{18}$
    $= 1$

9. $i^{67} = (i^4)^{16} \cdot i^3$
    $= 1^{16} \cdot (-i)$
    $= -i$

11. $-i^{16} = -(i^4)^4$
    $= -(1)^4$
    $= -1$

13. $\sqrt{-5} = \sqrt{5(-1)}$
    $= \sqrt{5i^2}$
    $= \sqrt{5}\sqrt{i^2}$
    $= \sqrt{5}i \ \text{or} \ i\sqrt{5}$

15. $3 - \sqrt{-2} = 3 - \sqrt{2(-1)}$
    $= 3 - \sqrt{2i^2}$
    $= 3 - \sqrt{2}\sqrt{i^2}$
    $= 3 - \sqrt{2}i \ \text{or} \ 3 - i\sqrt{2}$

17. $2\sqrt{-4} - \sqrt{28}$
    $= 2\sqrt{4(-1)} - \sqrt{4 \cdot 7}$
    $= 2\sqrt{4i^2} - 2\sqrt{7}$
    $= 2\sqrt{4}\sqrt{i^2} - 2\sqrt{7}$
    $= 2(2)i - 2\sqrt{7}$
    $= 4i - 2\sqrt{7}$
    $= -2\sqrt{7} + 4i$

19. $\dfrac{3 - \sqrt{-2}}{5} = \dfrac{3 - i\sqrt{2}}{5}$

$\qquad = \dfrac{3}{5} - \dfrac{\sqrt{2}}{5}i$

21. $\dfrac{6 - \sqrt{-3}}{3} = \dfrac{6 - i\sqrt{3}}{3}$

$\qquad = \dfrac{6}{3} - \dfrac{\sqrt{3}}{3}i$

$\qquad = 2 - \dfrac{\sqrt{3}}{3}i$

23. $(3 + 2i) - (2 + 3i)$
$= (3 - 2) + (2 - 3)i$
$= 1 - i$

25. $(2 - i) + (2 + i)$
$= (2 + 2) + (-1 + 1)i$
$= 4$

27. $5(3 - 7i) = 5(3) - 5(7i)$
$\qquad\qquad = 15 - 35i$

29. $(2 - i)(-2) = 2(-2) - i(-2)$
$\qquad\qquad = -4 + 2i$

31. $(5i)(3i) = 15i^2$
$\qquad\qquad = -15$

33. $(-3i)^2 = 9i^2$
$\qquad\quad = -9$

35. $2i(3 + 2i) = 6i + 4i^2$
$\qquad\qquad = 6i - 4$
$\qquad\qquad = -4 + 6i$

37. $-2i(-3 + 5i) = 6i - 10i^2$
$\qquad\qquad\quad = 6i + 10$
$\qquad\qquad\quad = 10 + 6i$

39. $(3 + 9i)(0) = 0$

41. $(7 - 4i)(3 + i)$
$= 21 + 7i - 12i - 4i^2$
$= 21 + 7i - 12i + 4$
$= 25 - 5i$

43. $(3 + i)(3 - i)$
$= 3^2 - i^2$
$= 9 + 1$
$= 10$

45. $(2 - 7i)(2 + 7i)$
$= 2^2 - (7i)^2$
$= 4 - 49i^2$
$= 4 + 49$
$= 53$

47. $(i + 1)^2$
$= i^2 + 2(i)(1) + 1^2$
$= -1 + 2i + 1$
$= 2i$

49. $(2 - i)^2 - 4(2 - i)$
$= 2^2 - 2(2)(i) + i^2 - 8 + 4i$
$= 4 - 4i - 1 - 8 + 4i$
$= -5$

51. $(1 + i)^2 - 2(1 + i) + 1$
$= 1^2 + 2(1)(i) + i^2 - 2 - 2i + 1$
$= 1 + 2i - 1 - 2 - 2i + 1$
$= -1$

53. $\dfrac{6 - 5i}{3} = \dfrac{6}{3} - \dfrac{5i}{3}$

$\qquad\qquad = 2 - \dfrac{5}{3}i$

55. $\dfrac{3 - i}{i}$

$= \dfrac{(3 - i) \cdot i}{i \cdot i}$

$= \dfrac{3i - i^2}{i^2}$

$= \dfrac{3i + 1}{-1}$

$= \dfrac{1 + 3i}{-1}$

$= -1 - 3i$

57. $\dfrac{5 - 2i}{2i}$

$= \dfrac{(5 - 2i) \cdot i}{2i \cdot i}$

$= \dfrac{5i - 2i^2}{2i^2}$

$= \dfrac{5i + 2}{-2}$

$= \dfrac{2 + 5i}{-2}$

$= -1 - \dfrac{5}{2}i$

59. $\dfrac{2i}{5 - 2i}$

$= \dfrac{2i(5 + 2i)}{(5 - 2i)(5 + 2i)}$

$= \dfrac{10i + 4i^2}{5^2 - (2i)^2}$

$= \dfrac{10i + 4i^2}{5^2 - (2i)^2}$

$= \dfrac{10i - 4}{25 + 4}$

$= \dfrac{-4 + 10i}{29}$

$= -\dfrac{4}{29} + \dfrac{10}{29}i$

61. $\dfrac{2}{5 - 2i}$

$= \dfrac{2(5 + 2i)}{(5 - 2i)(5 + 2i)}$

$= \dfrac{10 + 4i}{5^2 - (2i)^2}$

$= \dfrac{10 + 4i}{25 + 4}$

$= \dfrac{10 + 4i}{29}$

$= \dfrac{10}{29} + \dfrac{4}{29}i$

63. $\dfrac{2 + i}{2 - i}$

$= \dfrac{(2 + i)(2 + i)}{(2 - i)(2 + i)}$

$= \dfrac{2^2 + 2(2)i + i^2}{2^2 - i^2}$

$= \dfrac{4 + 4i - 1}{4 + 1}$

$= \dfrac{3 + 4i}{5}$

$= \dfrac{3}{5} + \dfrac{4}{5}i$

65. $\dfrac{2 - 5i}{2 + 5i}$

$= \dfrac{(2 - 5i)(2 - 5i)}{(2 + 5i)(2 - 5i)}$

$= \dfrac{2^2 - 2(2)(5i) + (5i)^2}{2^2 - (5i)^2}$

$= \dfrac{4 - 20i + 25i^2}{4 - 25i^2}$

$= \dfrac{4 - 20i - 25}{4 + 25}$

$= \dfrac{-21 - 20i}{29}$

$= -\dfrac{21}{29} - \dfrac{20}{29}i$

67. $\dfrac{3 - 7i}{5 + 2i}$

$= \dfrac{(3 - 7i)(5 - 2i)}{(5 + 2i)(5 - 2i)}$

$= \dfrac{15 - 6i - 35i + 14i^2}{5^2 - (2i)^2}$

$= \dfrac{15 - 6i - 35i - 14}{25 - 4i^2}$

$$= \frac{1 - 41i}{25 + 4}$$

$$= \frac{1 - 41i}{29}$$

$$= \frac{1}{29} - \frac{41}{29}i$$

69. $\dfrac{5 - \sqrt{-4}}{3 + \sqrt{-25}}$

$$= \frac{5 - 2i}{3 + 5i}$$

$$= \frac{(5 - 2i)(3 - 5i)}{(3 + 5i)(3 - 5i)}$$

$$= \frac{16 - 25i - 6i + 10i^2}{3^2 - (5i)^2}$$

$$= \frac{15 - 25i - 6i - 10}{9 - 25i^2}$$

$$= \frac{5 - 31i}{9 + 25}$$

$$= \frac{5 - 31i}{34}$$

$$= \frac{5}{34} - \frac{31}{34}i$$

71. $x^2 + 4 = 0$
$(2i)^2 + 4 = 0$
$4i^2 + 4 = 0$
$-4 + 4 = 0$
$\qquad 0 = 0$
$\qquad$ True
$x = 2i$ is a solution.

73. $\qquad\qquad\quad x^2 - 4x + 5 = 0$
$\qquad\qquad (2 - i)^2 - 4(2 - i) + 5 = 0$
$2^2 - 2(2)(i) + i^2 - 8 + 4i + 5 = 0$
$\qquad 4 - 4i - 1 - 8 + 4i + 5 = 0$
$\qquad\qquad\qquad\qquad\qquad\qquad 0 = 0$
$\qquad\qquad\qquad\qquad\qquad\qquad$ True
$x = 2 - i$ is a solution.

79. $\dfrac{x + 3}{x - 6} \div (x^2 - 3x - 18)$

$$= \frac{x + 3}{x - 6} \cdot \frac{1}{x^2 - 3x - 18}$$

$$= \frac{x + 3}{x - 6} \cdot \frac{1}{(x - 6)(x + 3)}$$

$$= \frac{1}{(x - 6)^2}$$

81. $\qquad\quad x^2 - 25 \neq 0$
$\quad (x - 5)(x + 5) \neq 0$
$x - 5 \neq 0 \qquad x + 5 \neq 0$
$\qquad x \neq 5 \qquad\qquad x \neq -5$

Domain: $\{x \mid x \neq \pm 5\}$

CHAPTER 7 REVIEW EXERCISES

1. $(x^2 x^5)(x^4 x)$
$= (x^{2+5})(x^{4+1})$
$= (x^7)(x^5)$
$= x^{7+5}$
$= x^{12}$

3. $(-3x^2 y)(-2xy^4)(-x)$
$= -6x^{2+1+1}y^{1+4}$
$= -6x^4 y^5$

5. $(a^3)^4 = a^{(3)(4)}$
$\qquad\quad = a^{12}$

7. $(a^2 b^3)^7 = a^{(2)(7)}b^{(3)(7)}$
$\qquad\qquad = a^{14}b^{21}$

9. $(a^2 b^3)^2(a^2 b)^3$
$= a^{(2)(2)}b^{(3)(2)}a^{(2)(3)}b^3$
$= a^4 b^6 a^6 b^3$
$= a^{4+6}b^{6+3}$
$= a^{10}b^9$

11. $(a^2 bc^2)^2(ab^2 c)^3$
$= a^{(2)(2)}b^2 c^{(2)(2)}a^3 b^{(2)(3)}c^3$
$= a^4 b^2 c^4 a^3 b^6 c^3$
$= a^{4+3}b^{2+6}c^{4+3}$
$= a^7 b^8 c^7$

13. $\dfrac{a^5}{a^6} = a^{5-6}$

   $= a^{-1}$

   $= \dfrac{1}{a}$

15. $\dfrac{x^2 x^5}{x^4 x^3} = \dfrac{x^{2+5}}{x^{4+3}}$

   $= \dfrac{x^7}{x^7}$

   $= x^{7-7}$
   $= x^0$
   $= 1$

17. $\dfrac{(x^3 y^2)^3}{(x^5 y^4)^5} = \dfrac{x^9 y^6}{x^{25} y^{20}}$

   $= x^{9-25} y^{6-20}$
   $= x^{-16} y^{-14}$

   $= \dfrac{1}{x^{16} y^{14}}$

19. $\left(\dfrac{a^2 b}{ab}\right)^4 = (a^{2-1} b^{1-1})^4$

   $= (ab^0)^4$
   $= a^4$

21. $\dfrac{(2ax^2)^2(3ax)^2}{(-2x)^2}$

   $= \dfrac{2^2 a^2 x^4 3^2 a^2 x^2}{(-1)^2 (2)^2 x^2}$

   $= \dfrac{a^2 x^4 3^2 a^2}{1}$
   $= 9a^4 x^4$

23. $\left(\dfrac{-3xy}{x^2}\right)^2 \left(\dfrac{-2xy^2}{x}\right)^3$

   $= \left(\dfrac{-3y}{x}\right)^2 (-2y^2)^3$

   $= \dfrac{(-3)^2 y^2}{x^2} \cdot (-2)^3 y^6$

   $= \dfrac{9y^2}{x^2} \cdot \dfrac{-8y^6}{1}$

   $= \dfrac{-72y^8}{x^2}$

25. $a^{-3} a^{-4} a^5 = a^{-3-4+5}$
   $= a^{-2}$
   $= \dfrac{1}{a^2}$

27. $(x^{-2} y^5)^{-4} = x^{(-2)(-4)} y^{5(-4)}$

   $= x^8 y^{-20}$

   $= \dfrac{x^8}{y^{20}}$

29. $(-3)^{-4}(-2)^{-1}$

   $= \dfrac{1}{(-3)^4} \cdot \dfrac{1}{(-2)}$

   $= \dfrac{1}{81} \cdot \dfrac{1}{(-2)}$

   $= -\dfrac{1}{162}$

31. $\left(\dfrac{3x^{-5} y^2 z^{-4}}{2x^{-7} y^{-4}}\right)^0 = 1$

33. $\dfrac{x^{-3} x^{-6}}{x^{-5} x^0} = \dfrac{x^{-3-6}}{x^{-5+0}}$

   $= \dfrac{x^{-9}}{x^{-5}}$

   $= x^{-9-(-5)}$
   $= x^{-4}$
   $= \dfrac{1}{x^4}$

35. $\left(\dfrac{x^{-2}y^{-3}}{y^{-3}x^2}\right)^{-2}$

$= (x^{-2-2}y^{-3-(-3)})^{-2}$

$= (x^{-4}y^0)^{-2}$

$= (x^{-4})^{-2}$

$= x^{(-4)(-2)}$

$= x^8$

37. $\left(\dfrac{r^{-2}s^{-3}r^{-2}}{s^{-4}}\right)^{-2}\left(\dfrac{r^{-1}}{s^{-1}}\right)^{-3}$

$= (r^{-2-2}s^{-3-(-4)})^{-2}\left(\dfrac{r^{(-1)(-3)}}{s^{(-1)(-3)}}\right)$

$= (r^{-4}s)^{-2}\left(\dfrac{r^3}{s^3}\right)$

$= r^{(-4)(-2)}s^{-2}\left(\dfrac{r^3}{s^3}\right)$

$= \dfrac{r^8s^{-2}r^3}{s^3}$

$= r^{8+3}s^{-2-3}$
$= r^{11}s^{-5}$
$= \dfrac{r^{11}}{s^5}$

39. $\left(\dfrac{2}{5}\right)^{-2} = \left(\dfrac{5}{2}\right)^2$

$= \dfrac{25}{4}$

41. $\dfrac{(2x^2y^{-1}z)^{-2}}{(3xy^2)^{-3}}$

$= \dfrac{2^{-2}x^{-4}y^2z^{-2}}{3^{-3}x^{-3}y^{-6}}$

$= \dfrac{3^3}{2^2}x^{-4-(-3)}y^{2-(-6)}z^{-2}$

$= \dfrac{27}{4}x^{-1}y^8z^{-2}$

$= \dfrac{27y^8}{4xz^2}$

43. $(x^{-1} + y^{-1})(x - y)$
$= x^{-1+1} - x^{-1}y + xy^{-1} - y^{-1+1}$

$= x^0 - \dfrac{y}{x} + \dfrac{x}{y} - y^0$

$= 1 - \dfrac{y}{x} + \dfrac{x}{y} - 1$

$= \dfrac{x}{y} - \dfrac{y}{x}$

$= \dfrac{x \cdot x}{xy} - \dfrac{y \cdot y}{xy}$

$= \dfrac{x^2 - y^2}{xy}$

45. $\dfrac{x^{-1}y^{-3}}{x^{-1}y^2}$

$= \dfrac{\dfrac{1}{x} + \dfrac{1}{y^3}}{\dfrac{y^2}{x}}$

$= \dfrac{\left(\dfrac{1}{x} + \dfrac{1}{y^3}\right)xy^3}{\dfrac{y^2}{x} \cdot xy^3}$

$= \dfrac{y^3 + x}{y^5}$

47. move 4 places right
$2.83 \times 10^4 = 28,300$

49. move 5 places left
$7.96 \times 10^{-5} = 0.0000796$

51. $92.59 = 9.259 \times 10^?$
$= 9.259 \times 10^1$

53. $625,897 = 6.25897 \times 10^?$
$= 6.25897 \times 10^5$

55. $\dfrac{(0.0014)(9,000)}{(20,000)(63,000)}$

$= \dfrac{(1.4 \times 10^{-3})(9 \times 10^3)}{(2 \times 10^4)(6.3 \times 10^4)}$

$= \dfrac{(1.4)(9)}{(2)(6.3)} \times \dfrac{(10^{-3})(10^3)}{(10^4)(10^4)}$

$= 1 \times \dfrac{10^0}{10^8}$

$= 1 \times 10^{-8}$
$= 0.00000001$

57. $x^{1/2}x^{1/3} = x^{1/2 + 1/3}$
$= x^{5/6}$

59. $(x^{-1/2}x^{1/3})^{-6}$
$= (x^{-1/2 + 1/3})^{-6}$
$= (x^{-1/6})^{-6}$
$= x^{(-1/6)(-6)}$
$= x^1 = x$

61. $\dfrac{x^{1/2}x^{1/3}}{x^{2/5}} = \dfrac{x^{1/2 + 1/3}}{x^{2/5}}$

$= \dfrac{x^{5/6}}{x^{2/5}}$

$= x^{5/6 - 2/5}$

$= x^{13/30}$

63. $\left(\dfrac{a^{1/2}a^{-1/3}}{a^{1/2}b^{1/5}}\right)^{-15}$

$= \left(\dfrac{a^{1/2 - 1/3}}{a^{1/2}b^{1/5}}\right)^{-15}$

$= \left(\dfrac{a^{1/6}}{a^{1/2}b^{1/5}}\right)^{-15}$

$= \left(\dfrac{a^{1/6 - 1/2}}{b^{1/5}}\right)^{-15}$

$= \left(\dfrac{a^{-1/3}}{b^{1/5}}\right)^{-15}$

$= \dfrac{a^{(-1/3)(-15)}}{b^{(1/5)(-15)}}$

$= \dfrac{a^5}{b^{-3}}$

$= a^5 b^3$

65. $\dfrac{(x^{-1/2}y^{1/2})^{-2}}{(x^{-1/3}y^{-1/3})^{-1/2}}$

$= \dfrac{x^{(-1/2)(-2)}y^{(1/2)(-2)}}{x^{(-1/3)(-1/2)}y^{(-1/3)(-1/2)}}$

$= \dfrac{xy^{-1}}{x^{1/6}y^{1/6}}$

$= x^{1 - 1/6}y^{-1 - 1/6}$
$= x^{5/6}y^{-7/6}$
$= \dfrac{x^{5/6}}{y^{7/6}}$

67. $\left(\dfrac{4^{-1/2} \cdot 16^{-3/4}}{8^{1/3}}\right)^{-2}$

$= \left[\dfrac{(2^2)^{-1/2}(2^4)^{-3/4}}{(2^3)^{1/3}}\right]^{-2}$

$= \left(\dfrac{2^{-1}2^{-3}}{2}\right)^{-2}$

$= \left(\dfrac{2^{-5}}{2}\right)^{-2}$

$= (2^{-5})^{-2}$
$= 2^{10}$

69. $(a^{1/3} + b^{1/3})(a^{1/3} - b^{1/3})$
$= (a^{1/3})^2 - (b^{1/3})^2$
$= a^{2/3} - b^{2/3}$

71. $x^{1/2} = \sqrt{x}$

73. $xy^{1/2} = x\sqrt{y}$

75. $m^{2/3} = \sqrt[3]{m^2}$

77. $(5x)^{3/4} = \sqrt[4]{(5x)^3}$

79. $\sqrt[3]{a} = a^{1/3}$

81. $-\sqrt[5]{n^4} = -n^{4/5}$

83. $\dfrac{1}{\sqrt[5]{t^7}} = \dfrac{1}{t^{7/5}}$

$\qquad = t^{-7/5}$

85. $\sqrt{54} = \sqrt{9 \cdot 6}$

$\qquad = \sqrt{9}\sqrt{6}$

$\qquad = 3\sqrt{6}$

87. $\sqrt{x^{60}} = x^{30}$

89. $\sqrt[3]{48x^4y^8} = \sqrt[3]{8 \cdot 6x^3xy^6y^2}$

$\qquad = 2xy^2 \sqrt[3]{6xy^2}$

91. $\sqrt{75xy}\sqrt{3x} = \sqrt{(75xy)(3x)}$

$\qquad = \sqrt{225x^2y}$

$\qquad = 15x\sqrt{y}$

93. $(x\sqrt{xy})(2x^2y\sqrt{xy^2})$

$\qquad = (x\sqrt{xy})(2x^2y^2\sqrt{x})$

$\qquad = 2x^3y^2\sqrt{x^2y}$

$\qquad = 2x^4y^2\sqrt{y}$

95. $\dfrac{\sqrt{28}}{\sqrt{63}} = \sqrt{\dfrac{28}{63}}$

$\qquad = \sqrt{\dfrac{4}{9}}$

$\qquad = \dfrac{\sqrt{4}}{\sqrt{9}}$

$\qquad = \dfrac{2}{3}$

97. $\dfrac{y}{x\sqrt{y}}$

$\qquad = \dfrac{y\sqrt{y}}{x\sqrt{y}\sqrt{y}}$

$\qquad = \dfrac{y\sqrt{y}}{x\sqrt{y^2}}$

$\qquad = \dfrac{y\sqrt{y}}{xy} = \dfrac{\sqrt{y}}{x}$

99. $\sqrt{\dfrac{48a^2b}{3a^5b^2}}$

$\qquad = \sqrt{\dfrac{16}{a^3b}}$

$\qquad = \dfrac{\sqrt{16}}{\sqrt{a^3b}}$

$\qquad = \dfrac{4}{a\sqrt{ab}}$

$\qquad = \dfrac{4\sqrt{ab}}{a\sqrt{ab}\sqrt{ab}}$

$\qquad = \dfrac{4\sqrt{ab}}{a\sqrt{(ab)^2}}$

$\qquad = \dfrac{4\sqrt{ab}}{a(ab)}$

$\qquad = \dfrac{4\sqrt{ab}}{a^2b}$

101. $\sqrt{\dfrac{5}{a}}$

$\qquad = \dfrac{\sqrt{5}}{\sqrt{a}}$

$\qquad = \dfrac{\sqrt{5}\sqrt{a}}{\sqrt{a}\sqrt{a}}$

$\qquad = \dfrac{\sqrt{5a}}{\sqrt{a^2}}$

$\qquad = \dfrac{\sqrt{5a}}{a}$

103. $\dfrac{4}{\sqrt[3]{2a}}$

$\qquad = \dfrac{4 \cdot \sqrt[3]{4a^2}}{\sqrt[3]{2a} \cdot \sqrt[3]{4a^2}}$

$\qquad = \dfrac{4\sqrt[3]{4a^2}}{\sqrt[3]{8a^3}}$

$\qquad = \dfrac{4\sqrt[3]{4a^2}}{2a}$

$\qquad = \dfrac{2\sqrt[3]{4a^2}}{a}$

105. $\sqrt[6]{x^4} = \sqrt[3]{x^2}$

107. $\sqrt{2}\,\sqrt[3]{2}$

$\qquad = 2^{1/2} \cdot 2^{1/3}$

$\qquad = 2^{1/2 + 1/3}$

$\qquad = 2^{5/6}$

$\qquad = \sqrt[6]{2^5}$

109. $4\sqrt{2} + \sqrt{2} - 5\sqrt{2}$

$\qquad = (4 + 1 - 5)\sqrt{2}$

$\qquad = 0$

111. $6\sqrt{12} - 4\sqrt{27}$

$\qquad = 12\sqrt{3} - 12\sqrt{3}$

$\qquad = 0$

113. $\sqrt{\dfrac{3}{2}} + \sqrt{\dfrac{5}{3}}$

$\qquad = \dfrac{\sqrt{3}}{\sqrt{2}} + \dfrac{\sqrt{5}}{\sqrt{3}}$

$\qquad = \dfrac{\sqrt{3}\sqrt{2}}{\sqrt{2}\sqrt{2}} + \dfrac{\sqrt{5}\sqrt{3}}{\sqrt{3}\sqrt{3}}$

$\qquad = \dfrac{\sqrt{6}}{2} + \dfrac{\sqrt{15}}{3}$

$\qquad = \dfrac{3\sqrt{6}}{6} + \dfrac{2\sqrt{15}}{6}$

$\qquad = \dfrac{3\sqrt{6} + 2\sqrt{15}}{6}$

115. $\sqrt{3}(\sqrt{6} - \sqrt{2}) + \sqrt{2}(\sqrt{3} - 3)$
 $= \sqrt{18} - \sqrt{6} + \sqrt{6} - 3\sqrt{2}$
 $= 3\sqrt{2} - \sqrt{6} + \sqrt{6} - 3\sqrt{2}$
 $= (3 - 3)\sqrt{2} + (-1 + 1)\sqrt{6}$
 $= 0$

117. $(3\sqrt{7} - \sqrt{3})(\sqrt{7} - 2\sqrt{3})$
 $= 3(7) - 6\sqrt{21} - \sqrt{21} + 2(3)$
 $= 21 - 6\sqrt{21} - \sqrt{21} + 6$
 $= 27 - 7\sqrt{21}$

119. $(\sqrt{x} - 5)^2$
 $= (\sqrt{x})^2 - 2\sqrt{x}(5) + 5^2$
 $= x - 10\sqrt{x} + 25$

121. $(\sqrt{a + 7})^2 - (\sqrt{a} + 7)^2$
 $= a + 7 - [(\sqrt{a})^2 + 2\sqrt{a}(7) + 7^2]$
 $= a + 7 - (a + 14\sqrt{a} + 49)$
 $= a + 7 - a - 14\sqrt{a} - 49$
 $= -42 - 14\sqrt{a}$

123. $\dfrac{12}{\sqrt{5} + \sqrt{3}}$

 $= \dfrac{12(\sqrt{5} - \sqrt{3})}{(\sqrt{5} + \sqrt{3})(\sqrt{5} - \sqrt{3})}$

 $= \dfrac{12(\sqrt{5} - \sqrt{3})}{5 - 3}$

 $= \dfrac{12(\sqrt{5} - \sqrt{3})}{2}$

 $= 6(\sqrt{5} - \sqrt{3})$
 $= 6\sqrt{5} - 6\sqrt{3}$

125. $\dfrac{8x - 20y}{\sqrt{2x} - \sqrt{5y}}$

 $= \dfrac{(8x - 20y)(\sqrt{2x} + \sqrt{5y})}{(\sqrt{2x} - \sqrt{5y})(\sqrt{2x} + \sqrt{5y})}$

 $= \dfrac{4(2x - 5y)(\sqrt{2x} + \sqrt{5y})}{2x - 5y}$

 $= 4(\sqrt{2x} + \sqrt{5y})$
 $= 4\sqrt{2x} + 4\sqrt{5y}$

127. $\sqrt{2x} - 5 = 7$
 $\sqrt{2x} = 12$
 $(\sqrt{2x})^2 = 12^2$
 $2x = 144$
 $x = 72$

129. $\sqrt{2x - 5} = 7$
 $(\sqrt{2x - 5})^2 = 7^2$
 $2x - 5 = 49$
 $2x = 54$
 $x = 27$

131. $\sqrt[4]{x - 4} = 3$
 $(\sqrt[4]{x - 4})^4 = 3^4$
 $x - 4 = 81$
 $x = 85$

133. $x^{1/4} + 1 = 3$
 $x^{1/4} = 2$
 $(x^{1/4})^4 = 2^4$
 $x = 16$

135. $\sqrt{3x} = 7.25$

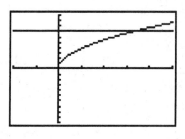

 $x = 17.5$

137. $2\sqrt{x - 4} = 6.2$
 $\sqrt{x - 4} = 3.1$

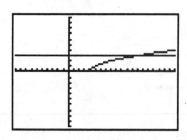

 $x = 13.6$

139.
$$G = 34.2d^2\sqrt{p}$$
$$450 = 34.2(3)^2\sqrt{p}$$
$$450 = 307.8\sqrt{p}$$
$$1.4620 = \sqrt{p}$$
$$(1.45620)^2 = \left(\sqrt{p}\right)^2$$
$$2.1 = p$$

The pressure is 2.1 *psi*.

141. $i^{11} = (i^4)^2 \cdot i^3$
$$= 1^2 \cdot (-i)$$
$$= -i$$

143. $-i^{14} = -(i^4)^3 i^2$
$$= -(1)^3(-1)$$
$$= 1$$

145. $(5 + i) + (4 - 2i)$
$$= (5 + 4) + (1 - 2)i$$
$$= 9 - i$$

147. $(7 - 2i)(2 - 3i)$
$$= 14 - 21i - 4i + 6i^2$$
$$= 14 - 21i - 4i - 6$$
$$= 8 - 25i$$

149. $2i(3i - 4)$
$$= 6i^2 - 8i$$
$$= -6 - 8i$$

151. $(3 - 2i)^2$
$$= 3^2 - 2(3)(2i) + (2i)^2$$
$$= 9 - 12i + 4i^2$$
$$= 9 - 12i - 4$$
$$= 5 - 12i$$

153. $(6 - i)^2 - 12(6 - i)$
$$= 6^2 - 2(6)i + i^2 - 72 + 12i$$
$$= 36 - 12i - 1 - 72 + 12i$$
$$= -37$$

155. $\dfrac{4 - 3i}{3 + i}$

$$= \frac{(4 - 3i)(3 - i)}{(3 + i)(3 - i)}$$

$$= \frac{12 - 4i - 9i + 3i^2}{9 - i^2}$$

$$= \frac{12 - 4i - 9i - 3}{9 + 1}$$

$$= \frac{9 - 13i}{10}$$

$$= \frac{9}{10} - \frac{13}{10}i$$

157.
$$x^2 - 2x + 5 = 0$$
$$(1 - 2i)^2 - 2(1 - 2i) + 5 = 0$$
$$1^2 - 2(1)(2i) + (2i)^2 - 2 + 4i + 5 = 0$$
$$1 - 4i + 4i^2 - 2 + 4i + 5 = 0$$
$$1 - 4i - 4 - 2 + 4i + 5 = 0$$
$$0 = 0$$
True

$x = 1 - 2i$ is a solution.

CHAPTER 7 PRACTICE TEST

1. $(a^2b^3)(ab^4)^2$
$$= (a^2b^3)(a^2b^8)$$
$$= a^4b^{11}$$

2. $(-2x^2y)^3(-5x)^2$
$$= (-8x^6y^3)(25x^2)$$
$$= -200x^8y^3$$

3. $(-3ab^{-2})^{-1}(-2x^{-1}y)^2$
$$= (-3)^{-1}b^2(4x^{-2}y^2)$$

$$= \frac{b^2}{3a} \cdot \frac{4y^2}{x^2}$$

$$= -\frac{4b^2y^2}{3ax^2}$$

4. $\dfrac{(-2x^2y)^3}{(-6xy^4)^2}$

$$= \frac{-8x^6y^3}{36x^2y^8}$$

$$= -\frac{2}{9}x^{6-2}y^{3-8}$$

$$= -\frac{2}{9}x^4y^{-5}$$

$$= -\frac{2x^4}{9y^5}$$

5. $\left(\dfrac{5r^{-1}s^{-3}}{3rs^2}\right)^{-2}$

$= \left(\dfrac{5}{3}r^{-1-1}s^{-3-2}\right)^{-2}$

$= \left(\dfrac{5}{3}r^{-2}s^{-5}\right)^{-2}$

$= \left(\dfrac{5}{3}\right)^{-2}r^4s^{10}$

$= \left(\dfrac{3}{5}\right)^{-2}r^4s^{10}$

$= \dfrac{9}{25}r^4s^{10}$

6. $\dfrac{27^{2/3} \cdot 3^{-4}}{9^{-1/2}}$

$= \dfrac{(3^3)^{2/3} \cdot 3^{-4}}{(3^2)^{-1/2}}$

$= \dfrac{3^2 \cdot 3^{-4}}{3^{-1}}$

$= \dfrac{3^{-2}}{3^{-1}}$

$= 3^{-2-(-1)}$
$= 3^{-1}$
$= \dfrac{1}{3}$

7. $\left(\dfrac{x^{1/4}x^{-2/3}}{x^{-1}}\right)^4$

$= \left(\dfrac{x^{1/4-2/3}}{x^{-1}}\right)^4$

$= \left(\dfrac{x^{-5/12}}{x^{-1}}\right)^4$

$= (x^{-5/12-(1)})^4$
$= (x^{7/12})^4$
$= x^{(7/12)(4)}$
$= x^{7/3}$

8. $\dfrac{x^{-3} + x^{-1}}{yx^{-2}}$

$= \dfrac{\dfrac{1}{x^3} + \dfrac{1}{x}}{\dfrac{y}{x^2}}$

$= \dfrac{\left(\dfrac{1}{x^3} + \dfrac{1}{x}\right)(x^3)}{\left(\dfrac{y}{x^2}\right)(x^3)}$

$= \dfrac{1 + x^2}{yx}$

9. $x^{1/3}(x^{2/3} - x)$
$= x^{1/3+2/3} - x^{1/3+1}$
$= x - x^{4/3}$

10. (a) $(-125)^{1/3} = -5$

(b) $(-128)^{-3/7} = \left[(-128)^{1/7}\right]^{-3}$
$= (-2)^{-3}$

$= \dfrac{1}{(-2)^3}$

$= -\dfrac{1}{8}$

11. $3a^{2/5} = 3\sqrt[3]{a^2}$

12. $\dfrac{(64)(28)}{(8000)(70000)}$

$= \dfrac{(6.4 \times 10^1)(2.8 \times 10^1)}{(8 \times 10^3)(7 \times 10^4)}$

$= \dfrac{(6.4)(2.8)}{(8)(7)} \times \dfrac{(10^1)(10^1)}{(10^3)(10^4)}$

$= 0.32 \times 10^{-5}$
$= 0.0000032$

13. 1 hr = 60 min = (60)(60) sec
        = 3600 sec

$D = r \cdot t$
$\quad = (186000)(3600)$
$\quad = (1.86 \times 10^5)(3.6 \times 10^3)$
$\quad = (1.86)(3.6) \times (10^5 \cdot 10^3)$
$\quad = 6.696 \times 10^8$

It travels 6.696×10^8 miles.

14. $\sqrt[3]{27x^6y^9} = 3x^2y^3$

15. $\sqrt[3]{4x^2y^2}\,\sqrt[3]{2x}$
$= \sqrt[3]{8x^3y^2}$
$= 2x\,\sqrt[3]{y^2}$

16. $(2x^2\sqrt{x})(3x\sqrt{xy^2})$
$= 6x^3\sqrt{x^2y^2}$
$= 6x^4y$

17. $\sqrt{\dfrac{5}{7}} = \dfrac{\sqrt{5}}{\sqrt{7}}$

$\quad\quad = \dfrac{\sqrt{5}\sqrt{7}}{\sqrt{7}\sqrt{7}}$

$\quad\quad = \dfrac{\sqrt{35}}{7}$

18. $\dfrac{\sqrt[3]{8}}{9}$

$= \dfrac{\sqrt[3]{8}}{\sqrt[3]{9}}$

$= \dfrac{2}{\sqrt[3]{9}}$

$= \dfrac{2\,\sqrt[3]{3}}{\sqrt[3]{9}\,\sqrt[3]{3}}$

$= \dfrac{2\,\sqrt[3]{3}}{\sqrt[3]{27}}$

$= \dfrac{2\,\sqrt[3]{3}}{3}$

19. $\dfrac{(xy\sqrt{2xy})(3x\sqrt{y})}{\sqrt{4x^3}}$

$= \dfrac{3x^2y\sqrt{2xy^2}}{2x\sqrt{x}}$

$= \dfrac{3x^2y^2\sqrt{2x}}{2x\sqrt{x}}$

$= \dfrac{3xy^2}{2}\sqrt{\dfrac{2x}{x}}$

$= \dfrac{3xy^2\sqrt{2}}{2}$

20. $\sqrt[4]{5}\sqrt{5}$
$= 5^{1/4} \cdot 5^{1/2}$
$= 5^{1/4 + 1/2}$
$= 5^{1/4 + 1/2}$
$= 5^{3/4}$
$= \sqrt[4]{5^3}$
$= \sqrt[4]{125}$

21. $\sqrt{50} - 3\sqrt{8} + 2\sqrt{18}$
$= 5\sqrt{2} - 6\sqrt{2} + 6\sqrt{2}$
$= (5 - 6 + 6)\sqrt{2}$
$= 5\sqrt{2}$

22. $(\sqrt{x} - 3)^2$
$= (\sqrt{x})^2 - 2\sqrt{x}(3) + 3^2$
$= x - 6\sqrt{x} + 9$

23. $\dfrac{\sqrt{6}}{\sqrt{6} - 2}$

$= \dfrac{\sqrt{6}(\sqrt{6} + 2)}{(\sqrt{6} - 2)(\sqrt{6} + 2)}$

$= \dfrac{\sqrt{6}(\sqrt{6} + 2)}{6 - 4}$

$= \dfrac{6 + 2\sqrt{6}}{2}$

$= \dfrac{2(3 + \sqrt{6})}{2}$

$= 3 + \sqrt{6}$

24. $\sqrt{x - 3} + 4 = 8$

$\qquad \sqrt{x - 3} = 4$

$\qquad \left(\sqrt{x - 3}\right)^2 = 4^2$

$\qquad x - 3 = 16$

$\qquad x = 19$

25. $\sqrt{x} - 3 = 8$

$\qquad \sqrt{x} = 11$

$\qquad \left(\sqrt{x}\right)^2 = 11^2$

$\qquad x = 121$

26. $3.2\sqrt{x - 3} = 5.8$

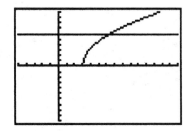

$\qquad x = 6.3$

27. $i^{51} = (i^4)^{12} \cdot i^3$

$\qquad = 1^{12} \cdot (-i)$

$\qquad = -i$

28. $(3i + 1)(i - 2)$

$= 3i^2 - 6i + i - 2$

$= -3 - 6i + i - 2$

$= -5 - 5i$

29. $\dfrac{2i + 3}{i - 2}$

$= \dfrac{(2i + 3)(i + 2)}{(i - 2)(i + 2)}$

$= \dfrac{2i^2 + 4i + 3i + 6}{i^2 - 4}$

$= \dfrac{-2 + 4i + 3i + 6}{-1 - 4}$

$= \dfrac{4 + 7i}{-5}$

$= -\dfrac{4}{5} - \dfrac{7}{5}i$

8.1 Exercises

1. $P(x) = -3.5x^2 + 600.4x$

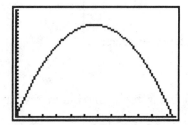

(a) When $x = 124$, $P = \$20,633.60$

(b) When $P = 18,500$, $x = 40$ and 131 items

(c) Maximum: $P = \$25,748.58$ when $x = 86$ items

3. $s(t) = -16t^2 + 725t$

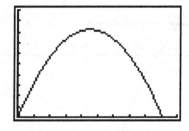

(a) When $t = 20$, $S = 8100$ feet

(b) When $S = 3500$, $t = 5.5$ and 39.8 sec

(c) Maximum: $S = 8212.9$ feet when $t = 22.7$ sec

5. (a) $2x + 2y = 200$
$$2y = -2x + 200$$
$$y = -x + 100$$

Area $= xy$
$$A(x) = x(-x + 100)$$
$$A(x) = -x^2 + 100x$$

(b)

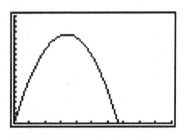

Maximum: $x = 50$
$$y = -50 + 100 = 50$$
The dimensions are 50 ft by 50 ft.

7. Let $x = 1^{st}$ number
then $225 - x = 2^{nd}$ number

$$P(x) = x(225 - x)$$
$$P(x) = 225x - x^2$$

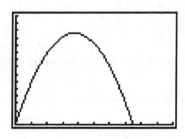

$x = 112.5$
$y = 225 - 112.5 = 112.5$

The numbers are 112.5 and 112.5.

9. (a) $4x + 2y = 1000$
$$2y = 1000 - 4x$$
$$y = 500 - 2x$$

$$A(x) = (2x)(y)$$
$$A(x) = 2x(500 - 2x)$$
$$A(x) = 1000x - 4x^2$$

(b)

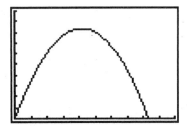

$x = 125$ ft
$y = 500 - 2(125) = 250$ ft

11. $p = 10 - \dfrac{x}{6}$

(a) $\quad 6 = 10 - \dfrac{x}{6}$

$6(6) = 6\left(10 - \dfrac{x}{6}\right)$

$36 = 60 - x$
$-24 = -x$
$24 = x$

24 items would be sold.

(b) $p = 10 - \dfrac{30}{6}$

$p = 5$

The price should be $5 per item.

(c)

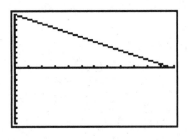

(d) $R = xp$

$R = x\left(10 - \dfrac{x}{6}\right)$

$R = 10x - \dfrac{x^2}{6}$

(e)

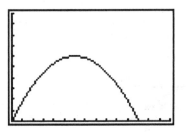

(f) $x = 30$ items for a maximum revenue

$p = 10 - \dfrac{30}{6} = 5$

The corresponding unit price is $5.

13. (a) Revenue $= \begin{pmatrix} \text{number} \\ \text{of cubes} \\ \text{sold} \end{pmatrix} \cdot \begin{pmatrix} \text{price} \\ \text{per} \\ \text{cube} \end{pmatrix}$

$R(x) = (200000 - 1000x)(3 + 0.25x)$

(b)

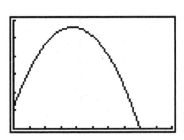

The maximum revenue is $2,809,000 when $x = 94$. Hence the price should be $3 + 0.25(94) = \$26.50$.

15. $\quad 3x^3y^2 - x^2 + y^3$
$= 3(-2)^3(-3)^2 - (-2)^2 + (-3)^3$
$= 3(-8)(9) - 4 - 27$
$= -247$

17. $\quad |5 - 2x| \geq 6$
$\quad 5 - 2x \geq 6 \quad$ or $\quad 5 - 2x \leq -6$
$\quad\quad -2x \geq 1 \quad\quad\quad\quad\quad -2x \leq -11$
$\quad\quad\quad x \leq -\dfrac{1}{2} \quad$ or $\quad\quad x \geq \dfrac{11}{2}$

1. $(x + 2)(x - 3) = 0$
   $x + 2 = 0$ or $x - 3 = 0$
   $\quad x = -2$ or $\quad\quad x = 3$

3. $x^2 = 25$
   $\quad x = \pm\sqrt{25}$
   $\quad x = \pm 5$

5. $x(x - 4) = 0$
   $x = 0$ or $x - 4 = 0$
   $x = 0$ or $\quad\quad x = 4$

7. $12 = x(x - 4)$
   $12 = x^2 - 4x$
   $\quad 0 = x^2 - 4x - 12$
   $\quad 0 = (x - 6)(x + 2)$
   $x - 6 = 0$ or $x + 2 = 0$
   $\quad\quad x = 6$ or $\quad\quad x = -2$

9. $5y(y - 7) = 0$
   $5y = 0$ or $y - 7 = 0$
   $y = 0$ or $\quad\quad y = 7$

11. $x^2 - 16 = 0$
    $\quad x^2 = 16$
    $\quad\quad x = \pm\sqrt{16}$
    $\quad\quad x = \pm 4$

13. $\quad\quad 0 = 9c^2 - 16$
    $\quad\quad 16 = c^2$

    $\quad\dfrac{16}{9} = c^2$

    $\pm\sqrt{\dfrac{16}{9}} = c$

    $\pm\dfrac{4}{3} = c$

15. $8a^2 - 18 = 0$
    $\quad 8a^2 = 18$

    $\quad\quad a^2 = \dfrac{9}{4}$

    $\quad\quad a = \pm\sqrt{\dfrac{9}{4}}$

    $\quad\quad a = \pm\dfrac{3}{2}$

17. $0 = x^2 - x - 6$
    $0 = (x - 3)(x + 2)$
    $x - 3 = 0$ or $x + 2 = 0$
    $\quad\quad x = 3$ or $\quad\quad x = -2$

19. $\quad 2y^2 - 3y + 1 = 0$
    $(2y - 1)(y - 1) = 0$
    $2y - 1 = 0$ or $y - 1 = 0$
    $\quad 2y = 1$ $\quad\quad\quad y = 1$

    $\quad\quad y = \dfrac{1}{2}$ or $\quad\quad y = 1$

21. $4x^2 = 24$
    $\quad x^2 = 6$
    $\quad\quad x = \pm\sqrt{6}$

23. $7a^2 - 19 = 0$
    $\quad 7a^2 = 19$

    $\quad\quad a^2 = \dfrac{19}{7}$

    $\quad\quad a = \pm\sqrt{\dfrac{19}{7}}$

    $\quad\quad a = \pm\dfrac{\sqrt{19} \cdot \sqrt{7}}{\sqrt{7} \cdot \sqrt{7}}$

    $\quad\quad a = \pm\dfrac{\sqrt{133}}{7}$

25. $10 = x^2 - 3x$
    $\quad 0 = x^2 - 3x - 10$
    $\quad 0 = (x - 5)(x + 2)$
    $x - 5 = 0$ or $x + 2 = 0$
    $\quad\quad x = 5$ or $\quad\quad x = -2$

27. $\quad 6a^2 + 3a - 1 = 4a$
    $\quad\quad 6a^2 - a - 1 = 0$
    $(3a + 1)(2a - 1) = 0$
    $3a + 1 = 0$ or $2a - 1 = 0$
    $\quad 3a = -1$ $\quad\quad\quad 2a = 1$

    $\quad\quad a = -\dfrac{1}{3}$ or $\quad\quad a = \dfrac{1}{2}$

29. $3y^2 - 5y + 8 = 9y^2 - 10y + 2$

$0 = 6y^2 - 5y - 6$

$0 = (2y - 3)(3y + 2)$

$2y - 3 = 0 \quad \text{or} \quad 3y + 2 = 0$

$2y = 3 \qquad\qquad 3y = -2$

$y = \dfrac{3}{2} \quad \text{or} \qquad y = -\dfrac{2}{3}$

31. $0 = 3x^2 + 5$

$-5 = 3x^2$

$-\dfrac{5}{3} = x^2$

$\pm\sqrt{-\dfrac{5}{3}} = x$

$\pm i\dfrac{\sqrt{5}}{\sqrt{3}} = x$

$\pm\dfrac{i\sqrt{5}\sqrt{3}}{\sqrt{3}\sqrt{3}} = x$

$\pm\dfrac{\sqrt{15}}{3}i = x$

33. $(2s - 3)(3s + 1) = 7$

$6s^2 - 7s - 3 = 7$

$6s^2 - 7s - 10 = 0$

$(6s + 5)(s - 2) = 0$

$6s + 5 = 0 \quad \text{or} \quad s - 2 = 0$

$6s = -5 \qquad\qquad s = 2$

$s = -\dfrac{5}{6} \quad \text{or} \qquad s = 2$

35. $(x + 2)^2 = 25$

$x + 2 = \pm\sqrt{25}$

$x + 2 = \pm 5$

$x + 2 = 5 \quad \text{or} \quad x + 2 = -5$

$x = 3 \quad \text{or} \qquad x = -7$

37. $(x - 2)(3x - 1) = (2x - 3)(x + 1)$

$3x^2 - 7x + 2 = 2x^2 - x - 3$

$x^2 - 6x + 5 = 0$

$(x - 5)(x - 1) = 0$

$x - 5 = 0 \quad \text{or} \quad x - 1 = 0$

$x = 5 \quad \text{or} \qquad x = 1$

39. $(x + 3)^2 = x(x + 5)$

$x^2 + 6x + 9 = x^2 + 5x$

$6x + 9 = 5x$

$9 = -x$

$-9 = x$

41. $x^2 + 3 = 1$

$x^2 = -2$

$x = \pm\sqrt{-2}$

$x = \pm i\sqrt{2}$

43. $8 = (x - 8)^2$

$\pm\sqrt{8} = x - 8$

$\pm 2\sqrt{2} = x - 8$

$8 \pm 2\sqrt{2} = x$

45. $(y - 5)^2 = -16$

$y - 5 = \pm\sqrt{-16}$

$y - 5 = \pm 4i$

$y = 5 \pm 4i$

47. $\dfrac{2}{x - 1} + x = 4$

$(x - 1)\left(\dfrac{2}{x - 1} + x\right) = (x - 1)4$

$2 + x(x - 1) = 4x - 4$

$2 + x^2 - x = 4x - 4$

$x^2 - 5x + 6 = 0$

$(x - 2)(x - 3) = 0$

$x - 2 = 0 \quad \text{or} \quad x - 3 = 0$

$x = 2 \quad \text{or} \qquad x = 3$

49. $2a - 5 = \dfrac{3(a + 2)}{0 + 4}$

$(a + 4)(2a - 5) = (a + 4)\left[\dfrac{3(a + 2)}{a + 4}\right]$

$2a^2 + 3a - 20 = 3a + 6$

$2a^2 = 26$

$a^2 = 13$

$a = \pm\sqrt{13}$

51. $\dfrac{x}{x + 2} - \dfrac{3}{x} = \dfrac{x + 1}{x}$

$x(x + 2)\left(\dfrac{x}{x + 2} - \dfrac{3}{x}\right) = x(x + 2)\left(\dfrac{x + 1}{x}\right)$

$x^2 - 3(x + 2) = (x + 2)(x + 1)$

$x^2 - 3x - 6 = x^2 + 3x + 2$

$-3x - 6 = 3x + 2$

$-6 = 6x + 2$

$-8 = 6x$

$-\dfrac{4}{3} = x$

53.
$$\frac{1}{x} + x = 2$$

$$x\left(\frac{1}{x} + x\right) = x(2)$$

$$1 + x^2 = 2x$$
$$x^2 - 2x + 1 = 0$$
$$(x - 1)^2 = 0$$
$$x - 1 = \pm\sqrt{0}$$
$$x - 1 = 0$$
$$x = 1$$

55.
$$\frac{3}{x - 2} + \frac{7}{x + 2} = \frac{x + 1}{x - 2}$$

$$(x - 2)(x + 2)\left(\frac{3}{x - 2} + \frac{7}{x + 2}\right) = (x - 2)(x + 2)\left(\frac{x + 1}{x - 2}\right)$$

$$3(x + 2) + 7(x - 2) = (x + 2)(x + 1)$$
$$3x + 6 + 7x - 14 = x^2 + 3x + 2$$
$$10x - 8 = x^2 + 3x + 2$$
$$0 = x^2 - 7x + 10$$
$$0 = (x - 5)(x - 2)$$
$$x - 5 = 0 \quad \text{or} \quad x - 2 = 0$$
$$x = 5 \quad \text{or} \quad x = 2$$

The only solution is $x = 5$, since $x = 2$ causes a denominator to equal 0 in the original equation.

57.
$$\frac{2}{x - 1} + \frac{3x}{x + 2} = \frac{2(5x + 9)}{x^2 + x - 2}$$

$$\frac{2}{x - 1} + \frac{3x}{x + 2} = \frac{2(5x + 9)}{(x + 2)(x - 1)}$$

$$(x - 1)(x + 2)\left(\frac{2}{x - 1} + \frac{3x}{x + 2}\right) = (x - 1)(x + 2)\left[\frac{2(5x + 9)}{(x + 2)(x - 1)}\right]$$

$$2(x + 2) + 3x(x - 1) = 2(5x + 9)$$
$$2x + 4 + 3x^2 - 3x = 10x + 18$$
$$3x^2 - x + 4 = 10x + 18$$
$$3x^2 - 11x - 14 = 0$$
$$(3x - 14)(x + 1) = 0$$
$$3x - 14 = 0 \quad \text{or} \quad x + 1 = 0$$
$$3x = 14 \qquad\qquad x = -1$$
$$x = \frac{14}{3} \quad \text{or} \quad x = -1$$

59.
$$8a^2 + 3b = 5b$$
$$8a^2 = 2b$$
$$a^2 = \frac{b}{4}$$

$$a = \pm\sqrt{\frac{b}{4}}$$

$$a = \pm\frac{\sqrt{b}}{2}$$

61.
$$5x^2 + 7y^2 = 9$$
$$7y^2 = 9 - 5x^2$$

$$y^2 = \frac{9 - 5x^2}{7}$$

$$y = \pm\sqrt{\frac{9 - 5x^2}{7}}$$

$$y = \pm\frac{\sqrt{9 - 5x^2}}{\sqrt{7}}$$

$$y = \pm\frac{\sqrt{9 - 5x^2} \cdot \sqrt{7}}{\sqrt{7} \cdot \sqrt{7}}$$

$$y = \pm\frac{\sqrt{7(9 - 5x^2)}}{7}$$

63.
$$V = \frac{2}{3}\pi r^2$$

$$3V = 2\pi r^2$$

$$\frac{3V}{2\pi} = r^2$$

$$\sqrt{\frac{3v}{2\pi}} = r$$

$$\frac{\sqrt{3V}}{\sqrt{2\pi}} = r$$

$$\frac{\sqrt{3V}\sqrt{2\pi}}{\sqrt{2\pi}\sqrt{2\pi}} = r$$

$$\frac{\sqrt{6V\pi}}{2\pi} = r$$

65.
$$a^2 - 4b^2 = 0$$
$$a^2 = 4b^2$$
$$a = \pm\sqrt{4b^2}$$
$$a = \pm 2b$$

67.
$$x^2 - xy - 6y^2 = 0$$
$$(x - 3y)(x + 2y) = 0$$
$$x - 3y = 0 \quad \text{or} \quad x + 2y = 0$$
$$x = 3y \quad \text{or} \quad x = -2y$$

69. Let x = the number

$$x^2 - 5 = 1 + 5x$$
$$x^2 - 5x - 6 = 0$$
$$(x - 6)(x + 1) = 0$$
$$x - 6 = 0 \quad \text{or} \quad x + 1 = 0$$
$$x = 6 \quad \text{or} \quad x = -1$$

$x = 6$, since the number is positive.

71. Let x = other number

$$x^2 + 8^2 = 68$$
$$x^2 + 64 = 68$$
$$x^2 = 4$$
$$x = \pm\sqrt{4}$$
$$x = \pm 2$$

$x = 2$, since the number is positive.

73. Let x = the number

$$(x + 6)^2 = 169$$
$$x + 6 = \pm\sqrt{169}$$
$$x + 6 = \pm 13$$
$$x + 6 = 13 \quad \text{or} \quad x + 6 = -13$$
$$x = 7 \quad \text{or} \quad x = -19$$

The numbers are 7 and −19.

75. Let x = the number

$$x + \frac{1}{x} = \frac{13}{6}$$

$$6x\left(x + \frac{1}{x}\right) = 6x\left(\frac{13}{6}\right)$$

$$6x^2 + 6 = 13x$$
$$6x^2 - 13x + 6 = 0$$
$$(3x - 2)(2x - 3) = 0$$

$$3x - 2 = 0 \quad \text{or} \quad 2x - 3 = 0$$
$$3x = 2 \qquad\qquad 2x = 3$$
$$x = \frac{2}{3} \quad \text{or} \qquad x = \frac{3}{2}$$

The numbers are $\dfrac{2}{3}$ and $\dfrac{3}{2}$.

77.
$$P(d) = 10000(-d^2 + 12d - 35)$$

(a) $\quad P(5) = 10000\left[-5^2 + 12(5) - 35\right]$
$$= 0$$

The profit is \$0.

(b) $\quad 10000 = 10000(-d^2 + 12d - 35)$
$$1 = -d^2 + 12d - 35$$
$$d^2 - 12d + 36 = 0$$
$$(d - 6)^2 = 0$$
$$d - 6 = \pm\sqrt{0}$$
$$d - 6 = 0$$
$$d = 6$$

The price must be \$6.

79.
$$h(t) = -16t^2 + 40$$

(a) $\quad h(1) = -16(1)^{22} + 40$
$$= 24$$

He is 24 feet above the pool.

(b) $\quad 0 = -16t^2 + 40$
$$16t^2 = 40$$
$$t^2 = \frac{5}{2}$$

$$t = \sqrt{\frac{5}{2}}$$

$$t = \frac{\sqrt{10}}{2} \approx 1.58$$

It will take 1.58 sec.

(c) $\quad h(0) = -16(0)^2 + 40$
$$= 40$$

It is 40 feet high.

81. Let x = width
then $x + 2$ = length

Area = wl
$$80 = x(x + 2)$$
$$80 = x^2 + 2x$$
$$0 = x^2 + 2x - 80$$
$$0 = (x + 10)(x - 8)$$
$$x + 10 = 0 \quad \text{or} \quad x - 8 = 0$$
$$x = -10 \quad \text{or} \quad x = 8$$

Since x is positive, $x = 8$ and
$x + 2 = 8 + 2 = 10$.
The dimensions are 8 ft by 10 ft.

83. Let x = side length

Area = s^2
$$60 = x^2$$
$$\pm\sqrt{60} = x$$
$$\pm 2\sqrt{15} = x$$

Since x is positive, $x = 2\sqrt{15}$.
The square is $2\sqrt{15}$ in. by $2\sqrt{15}$ in.

85. Let x = 1st number,
then $x - 2$ = 2nd number

$$x(x - 2) = 120$$
$$x^2 - 2x = 120$$
$$x^2 - 2x - 120 = 0$$
$$(x - 12)(x + 10) = 0$$
$$x - 12 = 0 \quad \text{or} \quad x + 10 = 0$$
$$x = 12 \quad \text{or} \quad x = -10$$
$$x - 2 = 10 \qquad x - 2 = -12$$

The numbers are 12 and 10 or −10 and −12.

87. Let x = 1st number,
then $3x + 2$ = 2nd number

$$x(3x + 2) = 85$$
$$3x^2 + 2x = 85$$
$$3x^2 + 2x - 85 = 0$$
$$(3x + 17)(x - 5) = 0$$
$$3x + 17 = 0 \quad \text{or} \quad x - 5 = 0$$
$$3x = -17 \qquad x = 5$$
$$x = -\frac{17}{3} \quad \text{or} \quad x = 5$$
$$3x + 2 = -15 \qquad 3x + 2 = 17$$

The numbers are $-\dfrac{17}{3}$ and −15 or 5 and 17.

89.
$$a^2 + b^2 = c^2$$
$$4^2 + b^2 = 7^2$$
$$16 + b^2 = 49$$
$$b^2 = 33$$
$$b = \sqrt{33}$$

The other leg is $\sqrt{33}$ in.

91.
$$a^2 + b^2 = c^2$$
$$8^2 + (x - 2)^2 = (x + 4)^2$$
$$64 + x^2 - 4x + 4 = x^2 + 8x + 16$$
$$x^2 - 4x + 68 = x^2 + 8x + 16$$
$$-4x + 68 = 8x + 16$$
$$68 = 12x + 16$$
$$52 = 12x$$

$$\frac{13}{3} = x$$

$$x - 2 = \frac{13}{3} - 2 = \frac{7}{3}$$

$$x + 4 = \frac{13}{3} + 4 = \frac{25}{3}$$

They are 8, $\dfrac{7}{3}$ and $\dfrac{25}{3}$ in.

93.

$$7^2 + 12^2 = x^2$$
$$49 + 144 = x^2$$
$$193 = x^2$$
$$\sqrt{193} = x$$

The diagonal is $\sqrt{193}$ in.

95. Let x = height

$$8^2 + x^2 = 30^2$$
$$64 + x^2 = 900$$
$$x^2 = 836$$
$$x = \sqrt{836}$$
$$x = 2\sqrt{209}$$

It reaches $2\sqrt{209}$ ft.

97.

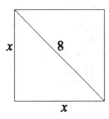

$$x^2 + x^2 = 8^2$$
$$2x^2 = 64$$
$$x^2 = 32$$
$$x = 4\sqrt{2}$$

$$\text{Area} = x^2$$
$$= \left(4\sqrt{2}\right)^2$$
$$= 32$$

The area is 32 sq in.

99.

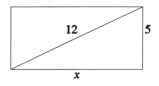

$$x^2 + 5^2 = 12^2$$
$$x^2 + 25 = 144$$
$$x^2 = 119$$
$$x = \sqrt{119}$$

$$\text{Area} = wl$$
$$= 5\sqrt{119}$$

The area is $5\sqrt{119}$ sq. in.

101. $(1, n)$ and $(3, n^2)$

$$m_1 = \frac{n^2 - n}{3 - 1} = \frac{n^2 - n}{2}$$

$(-6, 0)$ and $(-5, 6)$

$$m_2 = \frac{6 - 0}{-5 - (-6)} = 6$$

$$m_1 = m_2$$

$$\frac{n^2 - n}{2} = 6$$

$$n^2 - n = 12$$
$$n^2 - n - 12 = 0$$
$$(n - 4)(n + 3) = 0$$
$$n - 4 = 0 \quad \text{or} \quad n + 3 = 0$$
$$n = 4 \quad \text{or} \quad n = -3$$

103.

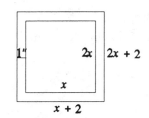

$$(x + 2)(2x + 2) = 60$$
$$2x^2 + 6x + 4 = 60$$
$$2x^2 + 6x - 56 = 0$$
$$x^2 + 3x - 28 = 0$$
$$(x + 7)(x - 4) = 0$$
$$x + 7 = 0 \quad \text{or} \quad x - 4 = 0$$
$$x = -7 \quad \text{or} \quad x = 4$$

Since x is positive, $x = 4$ and $2x = 8$.
The dimensions are $4''$ by $8''$.

105.

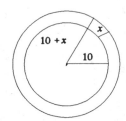

Area of = area of – area of
path   outer circle inner circle

$44\pi = \pi(10 + x)^2 - \pi(10)^2$
$44\pi = \pi(100 + 20x + x^2) - 100\pi$
$44\pi = 100\pi + 20\pi x + \pi x^2 - 100\pi$
$44\pi = 20\pi x + \pi x^2$
$44 = 20x + x^2$
$0 = x^2 + 20x - 44$
$0 = (x + 22)(x - 2)$
$x + 22 = 0 \quad$ or $\quad x - 2 = 0$
$x = -22 \quad$ or $\quad x = 2$

The path is 2 ft wide.

107. Let x = boat's rate in still water,

then rate upstream = $x - 5$
and rate downstream = $x + 5$

(time upstream) + (time downstream) = $\dfrac{4}{3}$

$$\frac{20}{x - 5} + \frac{10}{x + 5} = \frac{4}{3}$$

$$3(x - 5)(x + 5)\left(\frac{20}{x - 5} + \frac{10}{x + 5}\right) = 3(x - 5)(x + 5)\left(\frac{4}{3}\right)$$

$60(x + 5) + 30(x - 5) = 4(x - 5)(x + 5)$
$60x + 300 + 30x - 150 = 4(x^2 - 25)$
$90x + 150 = 4x^2 - 100$
$0 = 4x^2 - 90x - 250$
$0 = 2x^2 - 45x - 125$
$0 = (2x + 5)(x - 25)$
$2x + 5 = 0$ or $x - 25 = 0$
$2x = -5 \qquad x = 25$
$x = -\dfrac{5}{2}$

Since x is positive, $x = 25$.
The boat's rate is 25 kph.

1. $x^2 - 6x - 1 = 0$
$x^2 - 6x = 1$
$x^2 - 6x + 9 = 1 + 9$
$(x - 3)^2 = 10$
$x - 3 = \pm\sqrt{10}$
$x = 3 \pm \sqrt{10}$

3. $0 = c^2 - 2c - 5$
$5 = c^2 - 2c$
$5 + 1 = c^2 - 2c + 1$
$6 = (c - 1)^2$
$\pm\sqrt{6} = c - 1$
$1 \pm \sqrt{6} = c$

5. $y^2 + 5y - 2 = 0$
$y^2 + 5y = 2$

$$y^2 + 5y + \frac{25}{4} = 2 + \frac{25}{4}$$

$$\left(y + \frac{5}{2}\right)^2 = \frac{33}{4}$$

$$y + \frac{5}{2} = \pm\sqrt{\frac{33}{4}}$$

$$y + \frac{5}{2} = \pm\frac{\sqrt{33}}{2}$$

$$y = -\frac{5}{2} \pm \frac{\sqrt{33}}{2}$$

$$\text{or } \ y = \frac{-5 \pm \sqrt{33}}{2}$$

7. $2x^2 + 3x - 1 = x^2 - 2$
$x^2 + 3x = -1$

$$x^2 + 3x + \frac{9}{4} = -1 + \frac{9}{4}$$

$$\left(x + \frac{3}{2}\right)^2 = \frac{5}{4}$$

$$x + \frac{3}{2} = \pm\sqrt{\frac{5}{4}}$$

$$x + \frac{3}{2} = \pm\frac{\sqrt{5}}{2}$$

$$x = -\frac{3}{2} \pm \frac{\sqrt{5}}{2}$$

$$\text{or } x = \frac{-3 \pm \sqrt{5}}{2}$$

9. $(a - 2)(a + 1) = 2$
$$a^2 - a - 2 = 2$$
$$a^2 - a = 4$$

$$a^2 - a + \frac{1}{4} = 4 + \frac{1}{4}$$

$$\left(a - \frac{1}{2}\right)^2 = \frac{17}{4}$$

$$a - \frac{1}{2} = \pm\sqrt{\frac{17}{4}}$$

$$a - \frac{1}{2} = \pm\frac{\sqrt{17}}{2}$$

$$a = \frac{1}{2} \pm \frac{\sqrt{17}}{2}$$

$$\text{or } a = \frac{1 \pm \sqrt{17}}{2}$$

11. $$10 = 5a^2 + 10a + 20$$
$$-10 = 5a^2 + 10a$$
$$-2 = a^2 + 2a$$
$$-2 + 1 = a^2 + 2a + 1$$
$$-1 = (a + 1)^2$$
$$\pm\sqrt{-1} = a + 1$$
$$\pm i = a + 1$$
$$-1 \pm i = a$$

13. $$3x^2 + 3x = x^2 - 5x + 4$$
$$2x^2 + 8x = 4$$
$$x^2 + 4x = 2$$
$$x^2 + 4x + 4 = 2 + 4$$
$$(x + 2)^2 = 6$$
$$x + 2 = \pm\sqrt{6}$$
$$x = -2 \pm \sqrt{6}$$

15. $(a - 2)(a + 1) = 6$
$$a^2 - a - 2 = 6$$
$$a^2 - a = 8$$

$$a^2 - a + \frac{1}{4} = 8 + \frac{1}{4}$$

$$\left(a - \frac{1}{2}\right)^2 = \frac{33}{4}$$

$$a - \frac{1}{2} = \pm\sqrt{\frac{33}{4}}$$

$$a - \frac{1}{2} = \pm\frac{\sqrt{33}}{2}$$

$$a = \frac{1}{2} \pm \frac{\sqrt{33}}{2}$$

$$\text{or } a = \frac{1 \pm \sqrt{33}}{2}$$

17. $$2x - 7 = x^2 - 3x + 4$$
$$-11 = x^2 - 5x$$

$$-11 + \frac{25}{4} = x^2 - 5x + \frac{25}{4}$$

$$-\frac{19}{4} = \left(x - \frac{5}{2}\right)^2$$

$$\pm\sqrt{-\frac{19}{4}} = x - \frac{5}{2}$$

$$\pm\frac{\sqrt{19}}{2}i = x - \frac{5}{2}$$

$$\frac{5}{2} \pm \frac{\sqrt{19}}{2}i = x$$

$$\text{or } \frac{5 \pm i\sqrt{19}}{2} = x$$

19. $5x^2 + 10x - 14 = 20$
$$5x^2 + 10x = 34$$

$$x^2 + 2x = \frac{34}{5}$$

$$x^2 + 2x + 1 = \frac{34}{5} + 1$$

$$(x + 1)^2 = \frac{39}{5}$$

$$x + 1 = \pm\sqrt{\frac{39}{5}}$$

$$x + 1 = \pm\frac{\sqrt{195}}{5}$$

$$x = -1 \pm \frac{\sqrt{195}}{5}$$

$$\text{or } x = \frac{-5 \pm \sqrt{195}}{5}$$

21. $\quad 2t^2 + 3t - 4 = 2t - 1$
$$2t^2 + t = 3$$

$$t^2 + \frac{1}{2}t = \frac{3}{2}$$

$$t^2 + \frac{1}{2}t + \frac{1}{16} = \frac{3}{2} + \frac{1}{16}$$

$$\left(t + \frac{1}{4}\right)^2 = \frac{25}{16}$$

$$t + \frac{1}{4} = \pm\sqrt{\frac{25}{16}}$$

$$t + \frac{1}{4} = \pm\frac{5}{4}$$

$$t + \frac{1}{4} = \frac{5}{4} \quad \text{or} \quad t + \frac{1}{4} = -\frac{5}{4}$$

$$t = 1 \quad \text{or} \quad t = -\frac{3}{2}$$

23. $\quad (2x + 5)(x - 3) = (x + 4)(x - 1)$
$$2x^2 - x - 15 = x^2 + 3x - 4$$
$$x^2 - 4x = 11$$
$$x^2 - 4x + 4 = 11 + 4$$
$$(x - 2)^2 = 15$$
$$x - 2 = \pm\sqrt{15}$$
$$x = 2 \pm \sqrt{15}$$

25. $\qquad \dfrac{2x}{2x - 3} = \dfrac{3x - 1}{x + 1}$

$$(2x - 3)(x + 1)\left(\frac{2x}{2x - 3}\right) = (2x - 3)(x + 1)\left(\frac{3x - 1}{x + 1}\right)$$

$$2x(x + 1) = (2x - 3)(3x - 1)$$
$$2x^2 + 2x = 6x^2 - 11x + 3$$
$$-4x^2 + 13x = 3$$

$$x^2 - \frac{13}{4}x = -\frac{3}{4}$$

$$x^2 - \frac{13}{4}x + \frac{169}{64} = -\frac{3}{4} + \frac{169}{64}$$

$$\left(x - \frac{13}{8}\right)^2 = \frac{121}{64}$$

$$x - \frac{13}{8} = \pm\sqrt{\frac{121}{64}}$$

$$x - \frac{13}{8} = \pm\frac{11}{8}$$

$$x - \frac{13}{8} = \frac{11}{8} \quad \text{or} \quad x - \frac{13}{8} = \frac{11}{8}$$

$$x = 3 \quad \text{or} \quad x = \frac{1}{4}$$

27. $\quad a^2 - a + 1 = 0$
$$a^2 - a = -1$$

$$a^2 - a + \frac{1}{4} = -1 + \frac{1}{4}$$

$$\left(a - \frac{1}{2}\right)^2 = -\frac{3}{4}$$

$$a - \frac{1}{2} = \pm\sqrt{-\frac{3}{4}}$$

$$a - \frac{1}{2} = \pm\frac{\sqrt{3}}{2}i$$

$$a = \frac{1}{2} \pm \frac{\sqrt{3}}{2}i$$

$$\text{or } a = \frac{1 \pm i\sqrt{3}}{2}$$

29.

$$0 = 2y^2 + 2y + 5$$
$$-5 = 2y^2 + 2y$$
$$-\frac{5}{2} = y^2 + y$$
$$-\frac{5}{2} + \frac{1}{4} = y^2 + y + \frac{1}{4}$$
$$-\frac{9}{4} = \left(y + \frac{1}{2}\right)^2$$
$$\pm\sqrt{-\frac{9}{4}} = y + \frac{1}{2}$$
$$\pm\frac{3}{2}i = y + \frac{1}{2}$$
$$-\frac{1}{2} \pm \frac{3}{2}i = y$$
$$\text{or } y = \frac{-1 \pm 3i}{2}$$

31.

$$5n^2 - 3n = 2n^2 - 6$$
$$3n^2 - 3n = -6$$
$$n^2 - n = -2$$
$$n^2 - n + \frac{1}{4} = -2 + \frac{1}{4}$$
$$\left(n - \frac{1}{2}\right)^2 = -\frac{7}{4}$$
$$n - \frac{1}{2} = \pm\sqrt{-\frac{7}{4}}$$
$$n - \frac{1}{2} = \pm\frac{\sqrt{7}}{2}i$$
$$n = \frac{1}{2} \pm \frac{\sqrt{7}}{2}i$$
$$\text{or } n = \frac{1 \pm i\sqrt{7}}{2}$$

33.

$$(3t + 5)(t + 1) = (t + 4)(t + 2)$$
$$3t^2 + 8t + 5 = t^2 + 6t + 8$$
$$2t^2 + 2t = 3$$
$$t^2 + t = \frac{3}{2}$$
$$t^2 + t + \frac{1}{4} = \frac{3}{2} + \frac{1}{4}$$

$$\left(t + \frac{1}{2}\right)^2 = \frac{7}{4}$$
$$t + \frac{1}{2} = \pm\sqrt{\frac{7}{4}}$$
$$t + \frac{1}{2} = \pm\frac{\sqrt{7}}{2}$$
$$t = -\frac{1}{2} \pm \frac{\sqrt{7}}{2}$$
$$\text{or } t = \frac{-1 \pm \sqrt{7}}{2}$$

35.

$$\frac{3}{x + 2} - \frac{2}{x - 1} = 5$$
$$(x + 2)(x - 1)\left(\frac{3}{x + 2} - \frac{2}{x - 1}\right) = (x + 2)(x - 1)(5)$$
$$3(x - 1) - 2(x + 2) = 5(x^2 + x - 2)$$
$$3x - 3 - 2x - 4 = 5x^2 + 5x - 10$$
$$x - 7 = 5x^2 + 5x - 10$$
$$3 = 5x^2 + 5x - 10$$
$$3 = 5x^2 + 4x$$
$$\frac{3}{5} + \frac{4}{25} = x^2 + \frac{4}{5}x + \frac{4}{25}$$
$$\frac{19}{25} = \left(x + \frac{2}{5}\right)^2$$
$$\pm\sqrt{\frac{19}{25}} = x + \frac{2}{5}$$
$$\pm\frac{\sqrt{19}}{5} = x + \frac{2}{5}$$
$$-\frac{2}{5} \pm \frac{\sqrt{19}}{5} = x$$
$$\frac{-2 \pm \sqrt{19}}{5} = x$$

41.
$$\begin{cases} 0.3x + 0.4y = 15 \\ 0.5x + 0.6y = 24 \end{cases}$$

Multiply both equations by 10:

$$\begin{cases} 3x + 4y = 150 \\ 5x + 6y = 240 \end{cases}$$

Multiply 1ˢᵗ equation by -3 and 2ⁿᵈ equation by 2, then add:

$$-9x - 12y = -450$$
$$\underline{10x + 12y = 480}$$
$$x \qquad\quad = 30$$

$$3(30) + 4y = 150$$
$$90 + 4y = 150$$
$$4y = 60$$
$$y = 15$$

$$x = 30, \quad y = 15$$

43. $\quad 5n^0 + n^{-1} + 6n^{-2}$

$$= 5(1) + \frac{1}{n} + \frac{6}{n^2}$$

$$= 5 + \frac{1}{n} + \frac{6}{n^2}$$

When $n = 3$:

$$5 + \frac{1}{3} + \frac{6}{3^2}$$

$$= 5 + \frac{1}{3} + \frac{2}{3}$$

$$= 6$$

8.4 Exercises

1. $x^2 - 4x - 5 = 0$
$A = 1, \quad B = -4, \quad C = -5$

$$x = \frac{-B \pm \sqrt{B^2 - 4AC}}{2A}$$

$$= \frac{-(-4) \pm \sqrt{(-4)^2 - 4(1)(-5)}}{2(1)}$$

$$= \frac{4 \pm \sqrt{16 + 20}}{2}$$

$$= \frac{4 \pm \sqrt{36}}{2}$$

$$= \frac{4 \pm 6}{2}$$

$$x = \frac{4 + 6}{2} = 5 \quad \text{or} \quad x = \frac{4 - 6}{2} = -1$$

3. $2a^2 - 3a - 4 = 0$
$A = 2, \quad B = -3, \quad C = -4$

$$a = \frac{-B \pm \sqrt{B^2 - 4AC}}{2A}$$

$$= \frac{-(-3) \pm \sqrt{(-3)^2 - 4(2)(-4)}}{2(2)}$$

$$= \frac{3 \pm \sqrt{9 + 32}}{4}$$

$$= \frac{3 \pm \sqrt{41}}{4}$$

5. $(3y - 1)(2y - 3) = y$
$6y^2 - 11y + 3 = y$
$6y^2 - 12y + 3 = 0$
$2y^2 - 4y + 1 = 0$

$A = 2, \quad B = -4, \quad C = 1$

$$y = \frac{-B \pm \sqrt{B^2 - 4AC}}{2A}$$

$$= \frac{-(-4) \pm \sqrt{(-4)^2 - 4(2)(1)}}{2(2)}$$

$$= \frac{4 \pm \sqrt{16 - 8}}{4}$$

$$= \frac{4 \pm \sqrt{8}}{4}$$

$$= \frac{4 \pm 2\sqrt{2}}{4}$$

$$= \frac{2(2 \pm \sqrt{2})}{4}$$

$$= \frac{2 \pm \sqrt{2}}{2}$$

7. $y^2 - 3y + 4 = 2y^2 + 4y - 3$
$0 = y^2 + 7y - 7$

$A = 1, \quad B = 7, \quad C = -7$

$$y = \frac{-B \pm \sqrt{B^2 - 4AC}}{2A}$$

$$= \frac{-7 \pm \sqrt{7^2 - 4(1)(-7)}}{2(1)}$$

$$= \frac{-7 \pm \sqrt{49 + 28}}{2}$$

$$= \frac{-7 \pm \sqrt{77}}{2}$$

9. $3a^2 + a + 2 = 0$
$A = 3, \quad B = 1, \quad C = 2$

$$a = \frac{-B \pm \sqrt{B^2 - 4AC}}{2A}$$

$$= \frac{-1 \pm \sqrt{1^2 - 4(3)(2)}}{2(3)}$$

$$= \frac{-1 \pm \sqrt{1 - 24}}{6}$$

$$= \frac{-1 \pm \sqrt{-23}}{6}$$

$$= \frac{-1 \pm i\sqrt{23}}{6}$$

11. $(s - 3)(s + 4) = (2s - 1)(s + 2)$
$s^2 + s - 12 = 2s^2 + 3s - 2$
$0 = s^2 + 2s + 10$
$A = 1, \quad B = 2, \quad C = 10$

$$s = \frac{-B \pm \sqrt{B^2 - 4AC}}{2A}$$

$$= \frac{-2 \pm \sqrt{2^2 - 4(1)(10)}}{2(1)}$$

$$= \frac{-2 \pm \sqrt{4 - 40}}{2}$$

$$= \frac{-2 \pm \sqrt{-36}}{2}$$

$$= \frac{-2 \pm 6i}{2}$$

$$= \frac{2(-1 \pm 3i)}{2}$$

$$= -1 \pm 3i$$

13. $3a^2 - 2a + 5 = 0$
$A = 3, \quad B = -2, \quad C = 5$

$B^2 - 4AC$
$= (-2)^2 - 4(3)(5)$
$= 4 - 60$
$= -56 < 0$
The roots are not real.

15. $(3y + 5)(2y - 8) = (y - 4)(y + 1)$
$6y^2 - 14y - 40 = y^2 - 3y - 4$
$5y^2 - 11y - 36 = 0$
$A = 5, \quad B = -11, \quad C = -36$

$B^2 - 4AC$
$= (-11)^2 - 4(5)(-36)$
$= 121 + 720$
$= 841 > 0$
The roots are real and distinct.

17. $2a^2 + 4a = 0$
$A = 2, \quad B = 4, \quad C = 0$

$B^2 - 4AC$
$= 4^2 - 4(2)(0)$
$= 16 > 0$
The roots are real and distinct.

19. $(2y + 3)(y - 1) = y + 5$
$2y^2 + y - 3 = y + 5$
$2y^2 - 8 = 0$
$y^2 - 4 = 0$
$A = 1, \quad B = 0, \quad C = -4$

$B^2 - 4AC$
$= 0^2 - 4(1)(-4)$
$= 16 > 0$
The roots are real and distinct.

21. $5x^2 - 2x - 4 = 0$

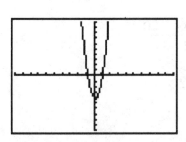

$x = -0.72$ and $x = 1.12$

23. $3.1x^2 - 2x + 8.5 = 9.9$

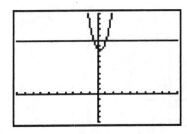

$x = -0.42$ and $x = 1.07$

25. $2x^2 - 5x + 8 = 0$

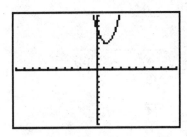

No real solution

27. $2.2x^2 - 16.3x = -24.9$

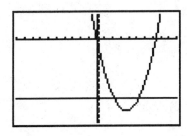

$x = 2.15$ and $x = 5.26$

29. $a^2 - 3a - 4 = 0$
$(a - 4)(a + 1) = 0$

$a - 4 = 0$ or $a + 1 = 0$
$\quad\quad a = 4$ or $\quad\quad a = -1$

31. $8y^2 = 3$

$y^2 = \dfrac{3}{8}$

$y = \pm\sqrt{\dfrac{3}{8}}$

$y = \pm\dfrac{\sqrt{3}}{2\sqrt{2}}$

$y = \pm\dfrac{\sqrt{6}}{4}$

33. $(5x - 4)(2x - 3) = 0$
$5x - 4 = 0$ or $2x - 3 = 0$
$\quad 5x = 4 \quad\quad\quad 2x = 3$
$\quad\quad x = \dfrac{4}{5}$ or $\quad\quad x = \dfrac{3}{2}$

35. $(5x - 4) + (2x - 3) = 0$
$\quad\quad\quad 7x - 7 = 0$
$\quad\quad\quad\quad 7x = 7$
$\quad\quad\quad\quad\; x = 1$

37. $(5x - 4)(2x - 3) = 17$
$10x^2 - 23x + 12 = 17$
$\;10x^2 - 23x - 5 = 0$
$(5x + 1)(2x - 5) = 0$
$5x + 1 = 0$ or $2x - 5 = 0$
$\quad 5x = -1 \quad\quad\quad 2x = 5$
$\quad\quad x = -\dfrac{1}{5}$ or $\quad\quad x = \dfrac{5}{2}$

39. $(a - 1)(a + 2) = -2$
$\quad a^2 + a - 2 = -2$
$\quad\quad a^2 + a = 0$
$\quad\quad a(a + 1) = 0$
$a = 0$ or $a + 1 = 0$
$a = 0$ or $\quad\quad a = -1$

41.

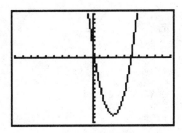

$x = 0.3$ and $x = 5$

43. $x^2 - 3x + 5 = 0$
$A = 1, \quad B = -3, \quad C = 5$

$$x = \frac{-B \pm \sqrt{B^2 - 4AC}}{2A}$$

$$= \frac{-(-3) \pm \sqrt{(-3)^2 - 4(1)(5)}}{2(1)}$$

$$= \frac{3 \pm \sqrt{9 - 20}}{2}$$

$$= \frac{3 \pm \sqrt{-11}}{2}$$

$$= \frac{3 \pm i\sqrt{11}}{2}$$

45. $2s^2 - 5s - 12 = -5s$
$\qquad 2s^2 - 12 = 0$
$\qquad \quad 2s^2 = 12$
$\qquad \quad\; s^2 = 6$
$\qquad \quad\;\; s = \pm\sqrt{6}$

47. $3x^2 - 2x + 9 = 2x^2 - 3x - 1$
$\qquad x^2 + x + 10 = 0$
$A = 1, \quad B = 1, \quad C = 10$

$$x = \frac{-B \pm \sqrt{B^2 - 4AC}}{2A}$$

$$= \frac{-1 \pm \sqrt{1^2 - 4(1)(10)}}{2(1)}$$

$$= \frac{-1 \pm \sqrt{1 - 40}}{2}$$

$$= \frac{-1 \pm \sqrt{-39}}{2}$$

$$= \frac{-1 \pm i\sqrt{39}}{2}$$

49. $x^2 + 3x - 8 = x^2 - x + 11$
$\qquad 3x - 8 = -x + 11$
$\qquad\qquad 4x = 19$
$\qquad\qquad\; x = \dfrac{19}{4}$

51. $3x^2 + 2.7x = 14.58$

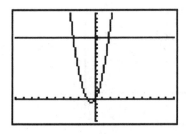

$x = -2.7$ and $x = 1.8$

53. $3a^2 - 4a + 2 = 0$
$0A = 3, \quad B = -4, \quad C = 2$

$$a = \frac{-B \pm \sqrt{B^2 - 4AC}}{2A}$$

$$= \frac{-(-4) \pm \sqrt{(-4)^2 - 4(3)(2)}}{2(3)}$$

$$= \frac{4 \pm \sqrt{16 - 24}}{6}$$

$$= \frac{4 \pm \sqrt{-8}}{6}$$

$$= \frac{4 \pm 2i\sqrt{2}}{6}$$

$$= \frac{2(2 \pm i\sqrt{2})}{6}$$

$$= \frac{2 \pm i\sqrt{2}}{3}$$

55. $(x - 4)(2x + 3) = x^2 - 4$
$\qquad 2x^2 - 5x - 12 = x^2 - 4$
$\qquad\quad x^2 - 5x - 8 = 0$
$A = 1, \quad B = -5, \quad C = -8$

$$x = \frac{-B \pm \sqrt{B^2 - 4AC}}{2A}$$

$$= \frac{-(-5) \pm \sqrt{(-5)^2 - 4(1)(-8)}}{2(1)}$$

$$= \frac{5 \pm \sqrt{25 + 32}}{2}$$

$$= \frac{5 \pm \sqrt{57}}{2}$$

57. $(a + 1)(a - 3) = (3a + 1)(a - 2)$
$a^2 - 2a - 3 = 3a^2 - 5a - 2$
$0 = 2a^2 - 3a + 1$
$0 = (2a - 1)(a - 1)$
$2a - 1 = 0$ or $a - 1 = 0$
$2a = 1$ $\qquad a = 1$
$a = \dfrac{1}{2}$ or $\qquad a = 1$

59. $(z + 3)(z - 1) = (z - 2)^2$
$z^2 + 2z - 3 = z^2 - 4z + 4$
$2z - 3 = -4z + 4$
$6z - 3 = 4$
$6z = 7$
$z = \dfrac{7}{6}$

61. $\dfrac{y}{y - 2} = \dfrac{y - 3}{y}$

$y(y - 2)\left(\dfrac{y}{y - 2}\right) = y(y - 2)\left(\dfrac{y - 3}{y}\right)$

$y^2 = (y - 2)(y - 3)$
$y^2 = y^2 - 5y + 6$
$0 = -5y + 6$
$5y = 6$
$y = \dfrac{6}{5}$

63. $0.001x^2 - 2x = 0.1$
$0.001x^2 - 2x - 0.1 = 0$

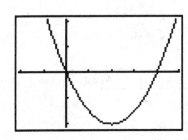

$x = -0.05$ and $x = 2000.05$

65. $\dfrac{3a}{a + 1} + \dfrac{2}{a - 2} = 5$

$(a + 1)(a - 2)\left(\dfrac{3a}{a + 1} + \dfrac{2}{a - 2}\right) = (a + 1)(a - 2)(5)$

$3a(a - 2) + 2(a + 1) = 5(a^2 - a - 2)$
$3a^2 - 6a + 2a + 2 = 5a^2 - 5a - 10$
$3a^2 - 4a + 2 = 5a^2 - 5a - 10$
$0 = 2a^2 - a - 12$
$A = 2, \quad B = -1, \quad C = -12$

$a = \dfrac{-B \pm \sqrt{B^2 - 4AC}}{2A}$

$= \dfrac{-(-1) \pm \sqrt{(-1)^2 - 4(2)(-12)}}{2(2)}$

$= \dfrac{1 \pm \sqrt{1 + 96}}{4}$

$= \dfrac{1 \pm \sqrt{97}}{4}$

67. $\dfrac{3}{a+2} - \dfrac{5}{a-2} = 2$

$(a+2)(a-2)\left(\dfrac{3}{a+2} - \dfrac{5}{a-2}\right) = (a+2)(a-2)(2)$

$3(a - 2) - 5(a + 2) = 2(a^2 - 4)$
$3a - 6 - 5a - 10 = 2a^2 - 8$
$-2a - 16 = 2a^2 - 8$
$0 = 2a^2 + 2a + 8$
$0 = a^2 + a + 4$
$A = 1, \quad B = 1, \quad C = 4$

$a = \dfrac{-B \pm \sqrt{B^2 - 4AC}}{2A}$

$= \dfrac{-1 \pm \sqrt{1^2 - 4(1)(4)}}{2(1)}$

$= \dfrac{-1 \pm \sqrt{1 - 16}}{2}$

$= \dfrac{-1 \pm \sqrt{-15}}{2}$

$= \dfrac{-1 \pm i\sqrt{15}}{2}$

69. Let x = 1st number,

then $\dfrac{40}{x}$ = 2nd number

$$x - \dfrac{40}{x} = 3$$

$$x\left(x - \dfrac{40}{x}\right) = x(3)$$

$$x^2 - 40 = 3x$$
$$x^2 - 3x - 40 = 0$$
$$(x - 8)(x + 5) = 0$$
$$x - 8 = 0 \quad \text{or} \quad x + 5 = 0$$
$$x = 8 \quad \text{or} \qquad x = -5$$
$$\dfrac{40}{x} = 5 \qquad\qquad \dfrac{40}{x} = -8$$

The numbers are 8 and 5 or -5 and -8.

71. $a^2 + b^2 = c^2$
$3^2 + 8^2 = c^2$
$9 + 64 = c^2$
$73 = c^2$
$\sqrt{73} = c$

The hypotenuse is $\sqrt{73} \approx 8.54''$.

73.

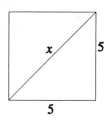

$5^2 + 5^2 = x^2$
$25 + 25 = x^2$
$50 = x^2$
$\sqrt{50} = x$
$5\sqrt{2} = x$

The diagonals are $5\sqrt{2} \approx 7.07''$.

75.

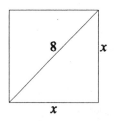

$x^2 + x^2 = 8^2$
$2x^2 = 64$
$x^2 = 32$
$x = \sqrt{32}$
$x = 4\sqrt{2}$

The sides are $4\sqrt{2} \approx 5.66''$.

77. $(0, 1)$ and (c, c)

$$m_1 = \dfrac{c - 1}{c - 0} = \dfrac{c - 1}{c}$$

$(0, 2)$ and (c, c)

$$m_2 = \dfrac{c - 2}{c - 0} = \dfrac{c - 2}{c}$$

$$m_1 = -\dfrac{1}{m_2}$$

$$\dfrac{c - 1}{c} = -\dfrac{c}{c - 2}$$

$$c(c - 2)\left(\dfrac{c - 1}{c}\right) = c(c - 2)\left(-\dfrac{c}{c - 2}\right)$$

$$(c - 2)(c - 1) = -c^2$$
$$c^2 - 3c + 2 = -c^2$$
$$2c^2 - 3c + 2 = 0$$

$$c = \dfrac{-(-3) \pm \sqrt{(-3)^2 - 4(2)(2)}}{2(2)}$$

$$= \dfrac{3 \pm \sqrt{9 - 16}}{4}$$

$$= \dfrac{3 \pm \sqrt{-7}}{4} = \dfrac{3 \pm i\sqrt{7}}{4}$$

Since c is not real, no solution.

79.
$$P(w) = 20w - w^2$$
$$-2400 = 20w - w^2$$
$$w^2 - 20w - 2400 = 0$$
$$(w - 60)(w + 40) = 0$$
$$w - 60 = 0 \quad \text{or} \quad w + 40 = 0$$
$$w = 60 \quad \text{or} \quad w = -40$$

Since w is positive, $w = 60$.

They sold 60 items.

81. Let x = price per roll during 1st week,
then $x - 2$ = price per roll during 2nd week

$$(x - 2)\left(2250 - \frac{10000}{x}\right) = 10000$$

$$x\left[(x - 2)\left(2250 - \frac{10000}{x}\right)\right] = x(10000)$$
$$2250x(x - 2) - 10000(x - 2) = 10000x$$
$$2250x^2 - 4500x - 10000x + 20000 = 10000x$$
$$2250x^2 - 24500x + 20000 = 0$$
$$9x^2 - 98x + 80 = 0$$
$$(9x - 8)(x - 10) = 0$$
$$9x - 8 = 0 \quad \text{or} \quad x - 10 = 0$$
$$9x = 8 \qquad\qquad x = 10$$
$$x = \frac{8}{9} \quad \text{or} \qquad x = 10$$

$$x - 2 = -\frac{10}{9} \qquad x - 2 = 8$$

Since both prices are positive, $x = 10$ and $x - 2 = 8$. The price during the 1st week was \$10 and during the 2nd week, \$8.

87. $(-27)^{-2/3} = \left[(-27)^{1/3}\right]^{-2}$
$$= (-3)^{-2}$$

$$= \frac{1}{(-3)^2}$$

$$= \frac{1}{9}$$

89. Let x = amount invested at 8%
and y = amount invested at 10%

$$\begin{cases} 0.08x + 0.10y = 730 \\ 0.08y + 0.10x = 710 \end{cases}$$

$$\begin{cases} 8x + 10y = 73000 \\ 10x + 8y = 71000 \end{cases}$$

$$\begin{array}{r} -40x - 50y = -365000 \\ \underline{40x + 32y = 284000} \\ -18y = -81000 \\ y = 4500 \end{array}$$
$$8x + 10(4500) = 73000$$
$$8x + 45000 = 73000$$
$$8x = 28000$$
$$x = 3500$$
$$x + y = 3500 + 4500$$
$$= 8000$$

\$8000 was invested.

8.5 Exercises

Note: Be sure to check all solutions. Only the checks of those numbers leading to extraneous solutions have been shown in this manual.

1. $\sqrt{x} + 3 = 2x$
$$\sqrt{x} = 2x - 3$$
$$\left(\sqrt{x}\right)^2 = (2x - 3)^2$$
$$x = 4x^2 - 12x + 9$$
$$0 = 4x^2 - 13x + 9$$
$$0 = (4x - 9)(x - 1)$$
$$4x - 9 = 0 \quad \text{or} \quad x - 1 = 0$$
$$4x = 9 \qquad\qquad x = 1$$
$$x = \frac{9}{4}$$

When you check $x = 1$, you find it to be extraneous.

$$\sqrt{1} + 3 = 2(1)$$
$$1 + 3 = 2$$
$$4 = 2$$
$$\text{False}$$

The only solution is $\frac{9}{4}$.

3. $\sqrt{x + 5} = 7 - x$
$$\left(\sqrt{x + 5}\right)^2 = (7 - x)^2$$
$$x + 5 = 49 - 14x + x^2$$
$$0 = x^2 - 15x + 44$$
$$0 = (x - 4)(x - 11)$$
$$x - 4 = 0 \quad \text{or} \quad x - 11 = 0$$
$$x = 4 \quad \text{or} \qquad x = 11$$

When you check $x = 11$, you find it to be extraneous.

$$\sqrt{11 + 5} = 7 - 11$$
$$\sqrt{16} = -4$$
$$4 = -4$$
$$\text{False}$$

The only solution is $x = 4$.

5. $\sqrt{5a - 1} + 5 = a$
$$\sqrt{5a - 1} = a - 5$$
$$\left(\sqrt{5a - 1}\right)^2 = (a - 5)^2$$
$$5a - 1 = a^2 - 10a + 25$$
$$0 = a^2 - 15a + 26$$
$$0 = (a - 2)(a - 13)$$
$$a - 2 = 0 \quad \text{or } a - 13 = 0$$
$$a = 2 \quad \text{or} \quad a = 13$$

When you check $a = 2$, you find it to be extraneous.

$$\sqrt{5(2) - 1} + 5 = 2$$
$$\sqrt{9} + 5 = 2$$
$$3 + 5 = 2$$
$$8 = 2$$
$$\text{False}$$

The only solution is $a = 13$.

7. $\sqrt{a + 1} + a = 11$
$$\sqrt{a + 1} = 11 - a$$
$$\left(\sqrt{a + 1}\right)^2 = (11 - a)^2$$
$$a + 1 = 121 - 22a + a^2$$
$$0 = a^2 - 23a + 120$$
$$0 = (a - 8)(a - 15)$$
$$a - 8 = 0 \quad \text{or } a - 15 = 0$$
$$a = 8 \quad \text{or} \quad a = 15$$

When you check $a = 15$, you find it to be extraneous.

$$\sqrt{15 + 1} + 15 = 11$$
$$\sqrt{16} + 15 = 11$$
$$4 + 15 = 11$$
$$19 = 11$$
$$\text{False}$$

The only solution is $a = 8$.

9. $\sqrt{3x + 1} + 3 = x$
$$\sqrt{3x + 1} = x - 3$$
$$\left(\sqrt{3x + 1}\right)^2 = (x - 3)^2$$
$$3x + 1 = x^2 - 6x + 9$$
$$0 = x^2 - 9x + 8$$
$$0 = (x - 8)(x - 1)$$
$$x - 8 = 0 \quad \text{or } x - 1 = 0$$
$$x = 8 \quad \text{or} \quad x = 1$$

When you check $x = 1$, you find it to be extraneous.

$$\sqrt{3(1) + 1} + 3 = 1$$
$$\sqrt{4} + 3 = 1$$
$$2 + 3 = 1$$
$$5 = 1$$
$$\text{False}$$

The only solution is $x = 8$.

11. $5a - 2\sqrt{a + 3} = 2a - 1$
$$-2\sqrt{a + 3} = -3a - 1$$
$$2\sqrt{a + 3} = 3a + 1$$
$$\left(2\sqrt{a + 3}\right)^2 = (3a + 1)^2$$
$$4(a + 3) = 9a^2 + 6a + 1$$
$$4a + 12 = 9a^2 + 6a + 1$$
$$4a + 12 = 9a^2 + 6a + 1$$
$$0 = 9a^2 + 2a - 11$$
$$0 = (9a + 11)(a - 1)$$
$$9a + 11 = 0 \quad \text{or } a - 1 = 0$$
$$a = -\frac{11}{9} \quad \text{or} \quad a = 1$$

When you check $a = -\dfrac{11}{9}$, you find it to be extraneous.

$$5\left(-\frac{11}{9}\right) - 2\sqrt{-\frac{11}{9} + 3} = 2\left(-\frac{11}{9}\right) - 1$$

$$-\frac{55}{4} - 2\sqrt{-\frac{2}{9}} = -\frac{22}{9} - 1$$

$\sqrt{-\dfrac{2}{9}}$ is not a real number.

The only solution is $a = 1$.

13.
$$\sqrt{y+3} = 1 + \sqrt{y}$$
$$\left(\sqrt{y+3}\right)^2 = \left(1 + \sqrt{y}\right)^2$$
$$y + 3 = 1 + 23\sqrt{y} + y$$
$$2 = 2\sqrt{y}$$
$$1 = \sqrt{y}$$
$$1^2 = \left(\sqrt{y}\right)^2$$
$$1 = y$$

15.
$$\sqrt{a+7} = 1 + \sqrt{2a}$$
$$\left(\sqrt{a+7}\right)^2 = \left(1 + \sqrt{2a}\right)^2$$
$$a + 7 = 1 + 2\sqrt{2a} + 2a$$
$$-a + 6 = 2\sqrt{2a}$$
$$(-a+6)^2 = \left(2\sqrt{2a}\right)^2$$
$$a^2 - 12a + 36 = 4(2a)$$
$$a^2 - 12a + 36 = 8a$$
$$a^2 - 20a + 36 = 0$$
$$(a-2)(a-18) = 0$$
$$a - 2 = 0 \quad \text{or} \quad a - 18 = 0$$
$$a = 2 \quad \text{or} \qquad a = 18$$

When you check $a = 18$, you find it to be extraneous.

$$\sqrt{18+7} = 1 + \sqrt{2(18)}$$
$$\sqrt{25} = 1 + \sqrt{36}$$
$$5 = 1 + 6$$
$$5 = 7$$
$$\text{False}$$

The only solution is $a = 2$.

17.
$$\sqrt{7s+1} - 2\sqrt{s} = 2$$
$$\sqrt{7s+1} = 2 + 2\sqrt{s}$$
$$\left(\sqrt{7s+1}\right)^2 = \left(2 + 2\sqrt{s}\right)^2$$
$$7s + 1 = 4 + 8\sqrt{s} + 4s$$
$$3s - 3 = 8\sqrt{s}$$
$$(3s-3)^2 = \left(8\sqrt{s}\right)^2$$
$$9s^2 - 18s + 9 = 64s$$
$$9s^2 - 82s + 9 = 0$$
$$(9s-1)(s-9) = 0$$
$$9s - 1 = 0 \quad \text{or} \quad s - 9 = 0$$
$$s = \frac{1}{9} \quad \text{or} \qquad s = 9$$

When you check $s = \dfrac{1}{9}$, you find it to be extraneous.

$$\sqrt{7\left(\frac{1}{9}\right) + 1} - 2\sqrt{\frac{1}{9}} = 2$$
$$\frac{4}{3} - 2\left(\frac{1}{3}\right) = 2$$
$$\frac{2}{3} = 2$$
$$\text{False}$$

The only solution is $s = 9$.

19.
$$\sqrt{7-a} - \sqrt{3+a} = 2$$
$$\sqrt{7-a} = 2 + \sqrt{3+a}$$
$$\left(\sqrt{7-a}\right)^2 = \left(2 + \sqrt{3+a}\right)^2$$
$$7 - a = 4 + 4\sqrt{3+a} + 3 + a$$
$$7 - a = 7 + a + 4\sqrt{3+a}$$
$$-2a = 4\sqrt{3+a}$$
$$(-2a)^2 = \left(4\sqrt{3+a}\right)^2$$
$$4a^2 = 16(3+a)$$
$$4a^2 = 48 + 16a$$
$$4a^2 - 16a - 48 = 0$$
$$a^2 - 4a - 12 = 0$$
$$(a-6)(a+2) = 0$$
$$a - 6 = 0 \quad \text{or} \quad a + 2 = 0$$
$$a = 6 \quad \text{or} \qquad a = -2$$

When you check $a = 6$, you find it to be extraneous.

$$\sqrt{7-6} - \sqrt{3+6} = 2$$
$$\sqrt{1} - \sqrt{9} = 2$$
$$1 - 3 = 2$$
$$-2 = 2$$
$$\text{False}$$

The only solution is $a = -2$.

21.
$$\sqrt{x} + a = b$$
$$\sqrt{x} = b - a$$
$$\left(\sqrt{x}\right)^2 = (b-a)^2$$
$$x = (b-a)^2$$

23.
$$\frac{\sqrt{\pi L}}{g} = T$$
$$\sqrt{\pi L} = gT$$

$$\left(\sqrt{\pi L}\right)^2 = (gT)^2$$

$$\pi L = g^2 T^2$$

$$L = \frac{g^2 T^2}{\pi}$$

25.
$$\sqrt{5x + b} = 6 + b$$
$$\left(\sqrt{5x + b}\right)^2 = (6 + b)^2$$
$$5x + b = 36 + 12b + b^2$$
$$5x = b^2 + 11b + 36$$
$$x = \frac{b^2 + 11b + 36}{5}$$

27.
$$t = \frac{\overline{X} - a}{\dfrac{s}{\sqrt{n}}}$$

$$\frac{s}{\sqrt{n}}(t) = \frac{s}{\sqrt{n}}\left(\frac{\overline{X} - a}{\dfrac{s}{\sqrt{n}}}\right)$$

$$\frac{st}{\sqrt{n}} = \overline{X} - a$$

$$st = \sqrt{n}(\overline{X} - a)$$

$$\frac{st}{\overline{X} - a} = \sqrt{n}$$

$$\left(\frac{st}{\overline{X} - a}\right)^2 = \left(\sqrt{n}\right)^2$$

$$\left(\frac{st}{\overline{X} - a}\right)^2 = n$$

29. $x^3 - 2x^2 - 15x = 0$
$$x(x - 5)(x + 3) = 0$$
$$x = 0 \quad \text{or} \quad x - 5 = 0 \quad \text{or} \quad x + 3 = 0$$
$$x = 0 \quad \text{or} \qquad x = 5 \quad \text{or} \qquad x = -3$$

31. $6a^3 - a^2 - 2a = 0$
$$a(3a - 2)(2a + 1) = 0$$
$$a = 0 \quad \text{or} \quad 3a - 2 = 0 \quad \text{or} \quad 2a + 1 = 0$$
$$a = 0 \quad \text{or} \qquad a = \frac{2}{3} \quad \text{or} \qquad a = -\frac{1}{2}$$

33. $y^4 - 17y^2 + 16 = 0$
$$(y^2 - 16)(y^2 - 1) = 0$$
$$y^2 - 16 = 0 \quad \text{or} \quad y^2 - 1 = 0$$
$$y^2 = 16 \qquad\qquad y^2 = 1$$
$$y = \pm 4 \quad \text{or} \qquad y = \pm 1$$

35.
$$3a^4 + 24 = 18a^2$$
$$3a^4 - 18a^2 + 24 = 0$$
$$a^4 - 6a^2 + 8 = 0$$
$$(a^2 - 4)(a^2 - 2) = 0$$
$$a^2 - 4 = 0 \quad \text{or} \quad a^2 - 2 = 0$$
$$a^2 = 4 \qquad\qquad a^2 = 2$$
$$a = \pm 2 \quad \text{or} \qquad a = \pm\sqrt{2}$$

37.
$$b^4 + 112 = 23b^2$$
$$b^4 - 23b^2 + 112 = 0$$
$$(b^2 - 7)(b^2 - 16) = 0$$
$$b^2 - 7 = 0 \quad \text{or} \quad b^2 - 16 = 0$$
$$b^2 = 7 \qquad\qquad b^2 = 16$$
$$b = \pm\sqrt{7} \quad \text{or} \qquad b = \pm 4$$

39.
$$9 - \frac{8}{x^2} = x^2$$

$$x^2\left(9 - \frac{8}{x^2}\right) = x^2(x^2)$$

$$9x^2 - 8 = x^4$$
$$0 = x^4 - 9x^2 + 8$$
$$0 = (x^2 - 8)(x^2 - 1)$$
$$x^2 - 8 = 0 \quad \text{or} \quad x^2 - 1 = 0$$
$$x^2 = 8 \qquad\qquad x^2 = 1$$
$$x = \pm 2\sqrt{2} \quad \text{or} \qquad x = \pm 1$$

41. $x^3 + x^2 - x - 1 = 0$
$$(x^2 - 1)(x + 1) = 0$$
$$x^2 - 1 = 0 \quad \text{or} \quad x + 1 = 0$$
$$x^2 = 1 \qquad\qquad x = -1$$
$$x = \pm 1$$

43. $x^{2/3} - 4 = 0$
$$x^{2/3} = 4$$
$$(x^{2/3})^3 = 4^3$$
$$x^2 = 64$$
$$x = \pm 8$$

45. $x + x^{1/2} - 6 = 0$
$(x^{1/2})^2 + x^{1/2} - 6 = 0$
Let $u = x^{1/2}$.

$u^2 + u - 6 = 0$
$(u + 3)(u - 2) = 0$
$u + 3 = 0$ or $u - 2 = 0$
$u = -3$ or $u = 2$
$x^{1/2} = -3$ $x^{1/2} = 2$
No solution $(x^{1/2})^2 = 2^2$
$x = 4$

The solution is $x = 4$.

47. $y^{2/3} - 4y^{1/3} - 5 = 0$
$(y^{1/3})^2 - 4y^{1/3} - 5 = 0$
Let $u = y^{1/3}$.

$u^2 - 4u - 5 = 0$
$(u - 5)(u + 1) = 0$
$u - 5 = 0$ or $u + 1 = 0$
$u = 5$ $u = -1$
$y^{1/3} = 5$ $y^{1/3} = -1$
$y = 5^3$ $y = (-1)^3$
$y = 125$ or $y = -1$

49. $x^{1/2} + 8x^{1/4} + 7 = 0$
$(x^{1/4})^2 + 8x^{1/4} + 7 = 0$
Let $u = x^{1/4}$.

$u^2 + 8u + 7 = 0$
$(u + 7)(u + 1) = 0$
$u + 7 = 0$ or $u + 1 = 0$
$u = -7$ $u = -1$
$x^{1/4} = -7$ $x^{1/4} = -1$
No solution No solution

No solution

51. $x^{-2} - 5x^{-1} + 6 = 0$
$(x^{-1})^2 - 5x^{-1} + 6 = 0$
Let $u = x^{-1}$.

$u^2 - 5u + 6 = 0$
$(u - 2)(u - 3) = 0$
$u - 2 = 0$ or $u - 3 = 0$
$u = 2$ $u = 3$
$x^{-1} = 2$ $x^{-1} = 3$
$(x^{-1})^{-1} = 2^{-1}$ $(x^{-1})^{-1} = 3^{-1}$
$x = \dfrac{1}{2}$ or $x = \dfrac{1}{3}$

53. $6x^{-2} + x^{-1} - 1 = 0$
$6(x^{-1})^2 + x^{-1} - 1 = 0$
Let $u = x^{-1}$.

$6u^2 + u - 1 = 0$
$(3u - 1)(2u + 1) = 0$
$3u - 1 = 0$ or $2u + 1 = 0$

$u = \dfrac{1}{3}$ $u = -\dfrac{1}{2}$

$x^{-1} = \dfrac{1}{3}$ $x^{-1} = -\dfrac{1}{2}$

$(x^{-1})^{-1} = \left(\dfrac{1}{3}\right)^{-1}$ $(x^{-1})^{-1} = \left(-\dfrac{1}{2}\right)^{-1}$

$x = 3$ $x = -2$

55. $x^{-4} - 13x^{-2} + 36 = 0$
$(x^{-2})^2 - 13x^{-2} + 36 = 0$
Let $u = x^{-2}$.

$u^2 - 13u + 36 = 0$
$(u - 9)(u - 4) = 0$
$u - 9 = 0$ or $u - 4 = 0$
$u = 9$ $u = 4$
$x^{-2} = 9$ $x^{-2} = 4$
$(x^{-2})^{-1/2} = \pm 9^{-1/2}$ $(x^{-2})^{-1/2} = \pm 4^{-1/2}$

$x = \pm \dfrac{1}{3}$ $x = \pm \dfrac{1}{2}$

57. $\sqrt{a} - \sqrt[4]{a} - 6 = 0$
$a^{1/2} - a^{1/4} - 6 = 0$
$(a^{1/4})^2 - a^{1/4} - 6 = 0$
Let $u = a^{1/4}$.

$u^2 - u - 6 = 0$
$(u - 3)(u + 2) = 0$
$u - 3 = 0$ or $u + 2 = 0$
$u = 3$ $u = -2$
$a^{1/4} = 3$ $a^{1/4} = -2$
$(a^{1/4})^4 = 3^4$ No solution
$a = 81$

The only solution is $a = 81$.

59.
$$\sqrt{x} - 4\sqrt[4]{x} = 5$$
$$x^{1/2} - 4x^{1/4} = 5$$
$$(x^{1/4})^2 - 4x^{1/4} - 5 = 0$$
Let $u = x^{1/4}$

$$u^2 - 4u - 5 = 0$$
$$(u - 5)(u + 1) = 0$$
$$u - 5 = 0 \quad \text{or} \quad u + 1 = 0$$
$$u = 5 \qquad\qquad u = -1$$
$$(x^{1/4})^4 = 5^4 \qquad\qquad \text{No solution}$$
$$x = 625$$

The only solution is 625.

61. $(a + 4)^2 + 6(a + 4) + 9 = 0$
Let $u = a + 4$.

$$u^2 + 6u + 9 = 0$$
$$(u + 3)^2 = 0$$
$$u + 3 = \pm\sqrt{0}$$
$$u + 3 = 0$$
$$u = -3$$
$$a + 4 = -3$$
$$a = -7$$

63. $2(3x + 1)^2 - 5(3x + 1) - 3 = 0$
Let $u = 3x + 1$.

$$2u^2 - 5u - 3 = 0$$
$$(2u + 1)(u - 3) = 0$$
$$2u + 1 = 0 \quad \text{or} \quad u - 3 = 0$$
$$u = -\frac{1}{2} \qquad\qquad u = 3$$

$$3x + 1 = -\frac{1}{2} \qquad 3x + 1 = 3$$

$$3x = -\frac{3}{2} \qquad\qquad 3x = 2$$

$$x = -\frac{1}{2} \qquad\qquad x = \frac{2}{3}$$

65. $(x^2 + x)^2 - 4 = 0$
$$(x^2 + x)^2 = 4$$
$$x^2 + x = \pm 2$$

$$x^2 + x = 2$$
$$x^2 + x - 2 = 0$$
$$(x + 2)(x - 1) = 0$$
$$x + 2 = 0 \quad \text{or} \quad x - 1 = 0$$
$$x = -2 \quad \text{or} \quad x = 1$$

or $\quad x^2 + x = -2$
$$x^2 + x + 2 = 0$$

$$x = \frac{-1 \pm \sqrt{1^2 - 4(1)(2)}}{2(1)}$$

$$= \frac{-1 \pm \sqrt{1 - 8}}{2}$$

$$= \frac{-1 \pm \sqrt{-7}}{2}$$

$$= \frac{-1 \pm i\sqrt{7}}{2}$$

The solutions are -2, 1, $\dfrac{-1 \pm i\sqrt{7}}{2}$.

67. $\left(a - \dfrac{10}{a}\right)^2 - 12\left(a - \dfrac{10}{a}\right) + 27 = 0$

Let $u = a - \dfrac{10}{a}$.

$$u^2 - 12u + 27 = 0$$
$$(u - 9)(u - 3) = 0$$
$$u - 9 = 0$$
$$u = 9$$
$$a - \frac{10}{a} = 9$$
$$a^2 - 10 = 9a$$
$$a^2 - 9a - 10 = 0$$
$$(a - 10)(a + 1) = 0$$
$$a - 10 = 0 \quad \text{or} \quad a + 1 = 0$$
$$a = 10 \quad \text{or} \quad a = -1$$

or $\qquad u - 3 = 0$
$$u = 3$$

$$5 \; a - \frac{10}{a} = 3$$

$$a^2 - 10 = 3a$$
$$a^2 - 3a - 10 = 0$$
$$(a - 5)(a + 2) = 0$$
$$a - 5 = 0 \quad \text{or} \quad a + 2 = 0$$
$$a = 5 \quad \text{or} \quad a = -2$$

$a = 10$ or $a = -1$ or $a = 5$ or $a = -2$

69. $s_e = s_y\sqrt{1 - r^2}$

$1.4 = 2.2\sqrt{1 - r^2}$

$(1.4)^2 = \left(2.2\sqrt{1 - r^2}\right)^2$

$1.96 = 4.84(1 - r^2)$

$0.4050 = 1 - r^2$

$-0.595 = -r^2$

$0.595 = r^2$

$\pm 0.77 = r$

71. $x^2 - 1 = \sqrt{x + 5}$

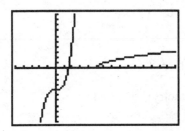

$x = -1.680$ and $x = 1.905$

73. $x^3 - 4 = \sqrt{x - 5}$

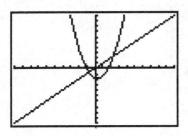

No solution

75. $x^2 - 3\sqrt{x + 5} + 5 = x$

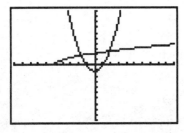

$x = -0.710$ and $x = 2.337$

79. $(x - 2)^2 = (x + 1)^2 - 15$

$x^2 - 4x + 4 = x^2 + 2x + 1 - 15$

$x^2 - 4x + 4 = x^2 + 2x - 14$

$-4x + 4 = 2x - 14$

$4 = 6x - 14$

$18 = 6x$

$3 = x$

81. $\dfrac{3}{a} = \dfrac{r}{1 - a}$

$a(1 - a)\left(\dfrac{3}{a}\right) = a(1 - a)\left(\dfrac{r}{1 - a}\right)$

$3(1 - a) = ar$

$3 - 3a = ar$

$3 = 3a + ar$

$3 = a(3 + r)$

$\dfrac{3}{3 + r} = a$

8.6 Exercises

1. $y = 3x^2$

x	y
-2	12
-1	3
0	0
1	3
2	12

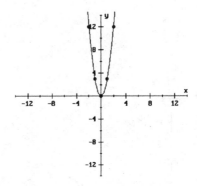

3. $y = -4x^2$

x	y
-2	-16
-1	-4
0	0
1	-4
2	-16

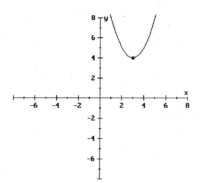

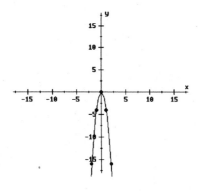

9. $y = -3(x - 1)^2 - 8$
   Vertex: $(1, -8)$
   Axis of symmetry: $x = 1$

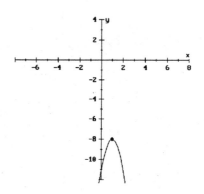

5. $6x^2 - y = 0$
   $6x^2 = y$

x	y
-2	24
-1	6
0	0
1	6
2	24

11. $y = 2(x + 4)^2 - 2$
    Vertex: $(-4, -2)$
    Axis of symmetry: $x = -4$

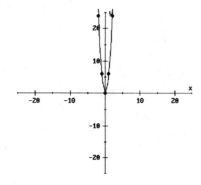

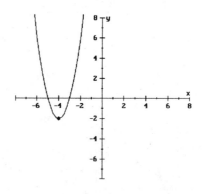

7. $y = (x - 3)^2 + 4$
   Vertex: $(3, 4)$
   Axis of symmetry: $x = 3$

13. $y = 2x^2 - 12x + 19$
    $y = 2(x^2 - 6x) + 19$
    $y = 2(x^2 - 6x + 9) + 19 - 18$
    $y = 2(x - 3)^2 + 1$
    Vertex: $(3, 1)$
    Axis of symmetry: $x = 3$

15. $y = -x^2 + 4x - 1$
    $y = -(x^2 - 4x) - 1$
    $y = -(x^2 - 4x + 4) - 1 + 4$
    $y = -(x - 2)^2 + 3$
    Vertex: $(2, 3)$
    Axis of symmetry: $x = 2$

17. $y = -3x^2 + 30x - 70$
    $y = -3(x^2 - 10x) - 70$
    $y = -3(x^2 - 10x + 25) - 70 + 75$
    $y = -3(x - 5)^2 + 5$
    Vertex: $(5, 5)$
    Axis of symmetry: $x = 5$

19. $y = x^2 - 4$
    $y = (x - 0)^2 - 4$

    Vertex: $(0, 4)$
    Axis of symmetry: $x = 0$
    x-intercepts: Let $y = 0$
     $0 = x^2 - 4$
     $4 = x^2$
    $\pm 2 = x$

    y-intercept: Let $x = 0$
    $y = 0^2 - 4$
    $y = -4$

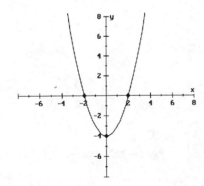

21. $y = 2x^2 + 8$
    $y = 2(x - 0)^2 + 8$

    Vertex: $(0, 8)$
    Axis of symmetry: $x = 0$
    x-intercepts: Let $y = 0$
     $0 = 2x^2 + 8$
     $-8 = 2x^2$
     $-4 = x^2$
     none

    y-intercept: Let $x = 0$
    $y = 2(0)^2 + 8$
    $y = 8$

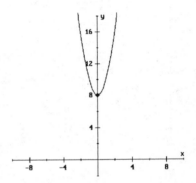

23. $y = -x^2 - 10x - 25$
    $y = -(x^2 + 10x) - 25$
    $y = -(x^2 + 10x + 25) - 25 + 25$
    $y = -(x + 5)^2$

    Vertex: $(-5, 0)$
    Axis of symmetry: $x = -5$
    x-intercepts: Let $y = 0$
     $0 = -(x + 5)^2$
     $0 = (x + 5)^2$
     $0 = x + 5$
    $-5 = x$

    y-intercept: Let $x = 0$
    $y = -0^2 - 10(0) - 25$
    $y = -25$

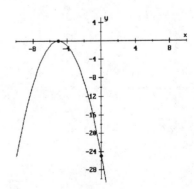

25. $y = x^2 - 2x - 35$
   $y = (x^2 - 2x + 1) - 35 - 1$
   $y = (x - 1)^2 - 36$

   Vertex: (1, –36)
   Axis of symmetry: $x = 1$
   x-intercepts: Let $y = 0$
    $0 = (x - 1)^2 - 36$
   $36 = (x - 1)^2$
   $\pm 6 = x - 1$
   $x - 1 = 6$ or $x - 1 = -6$
        $x = 7$ or $x = -5$

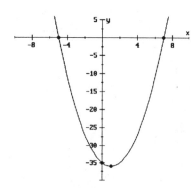

27. $y = 2x^2 - 4x + 4$
   $y = 2(x^2 - 2x) + 4$
   $y = 2(x^2 - 2x + 1) + 4 - 2$
   $y = 2(x - 1)^2 + 2$

   Vertex: (1, 2)
   Axis of symmetry: $x = 1$
   x-intercepts: Let $y = 0$
    $0 = 2(x - 1)^2 + 2$
   $-2 = 2(x - 1)^2$
   $-1 = (x - 1)^2$
        none

   y-intercept: Let $x = 0$
   $y = 2(0)^2 - 4(0) + 4$
   $y = 4$

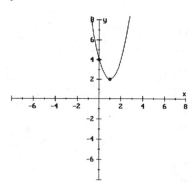

29. $y = -x^2 - 3x - 4$
   $y = -(x^2 + 3x) - 4$

   $y = -\left(x^2 + 3x + \dfrac{9}{4}\right) - 4 + \dfrac{9}{4}$

   $y = -\left(x + \dfrac{3}{2}\right)^2 - \dfrac{7}{4}$

   Vertex: $\left(-\dfrac{3}{2},\ -\dfrac{7}{4}\right)$

   Axis of symmetry: Let $y = 0$
   x-intercepts: Let $y = 0$
    $0 = -\left(x + \dfrac{3}{2}\right)^2 - \dfrac{7}{4}$

   $\left(x + \dfrac{3}{2}\right)^2 = -\dfrac{7}{4}$

   y-intercept: Let $x = 0$
   $y = -0^2 - 3(0) - 4$
   $y = -4$

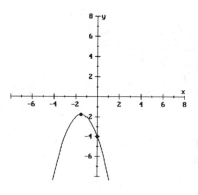

31. $y = 2x^2 - 3x + 2$

   $y = 2\left(x^2 - \dfrac{3}{2}x\right) + 2$

   $y = 2\left(x^2 - \dfrac{3}{2}x + \dfrac{9}{16}\right) + 2 - \dfrac{9}{8}$

   $y = 2\left(x - \dfrac{3}{4}\right)^2 + \dfrac{7}{8}$

   Vertex: $\left(\dfrac{3}{4},\ \dfrac{7}{8}\right)$

   Axis of symmetry: $x = \dfrac{3}{4}$

x-intercepts: Let $y = 0$

$$0 = 2\left(x - \frac{3}{4}\right)^2 + \frac{7}{8}$$

$$-\frac{7}{8} = 2\left(x - \frac{3}{4}\right)^2$$

$$-\frac{7}{16} = \left(x - \frac{3}{4}\right)^2$$

y-intercept: Let $x = 0$

$y = 2(0)^2 - 3(0) + 2$

$y = 2$

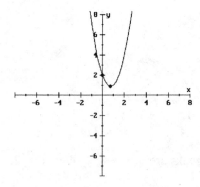

y-intercept: Let $x = 0$

$y = -2(0)^2 + 4(0) - 1$

$y = -1$

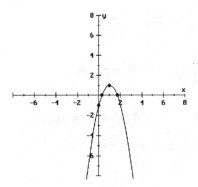

35. $P(x) = -x^2 + 70x$

$P(x) = -(x^2 - 70x)$

$P(x) = -(x^2 - 70x + 1225) + 1225$

$P(x) = -(x - 35)^2 + 1225$

Vertex: (35, 1225), maximum

3500 bagels must be produced and the maximum profit is \$1225.

33. $y = -2x^2 + 4x - 1$

$y = -2(x^2 - 2x) - 1$

$y = -2(x^2 - 2x + 1) - 1 + 2$

$y = -2(x - 1)^2 + 1$

Vertex: (1, 1)

Axis of symmetry: $x = 1$

x-intercepts: Let $y = 0$

$$0 = -2(x - 1)^2 + 1$$

$$2(x - 1)^2 = 1$$

$$(x - 1)^2 = \frac{1}{2}$$

$$x - 1 = \pm\sqrt{\frac{1}{2}}$$

$$x = 1 \pm \frac{\sqrt{2}}{2}$$

37. $P(x) = -x^2 + 112x - 535$

$P(x) = -(x^2 - 112x) - 535$

$P(x) = -(x^2 - 112x + 3136) - 535 + 3136$

$P(x) = -(x - 56)^2 + 2601$

Vertex: (56, 2601), maximum

56 cases of candy canes made daily will yield a maximum profit of \$2601.

39. $y = -16t^2 + 400t$

$y = -16(t^2 - 25t)$

$y = -16\left(t^2 - 25t + \dfrac{625}{4}\right) + 2500$

$y = -16\left(t - \dfrac{25}{2}\right)^2 + 2500$

Vertex: $\left(\dfrac{25}{2}, 2500\right)$, maximum

It will take $\dfrac{25}{2}$ = 12.5 sec to reach a maximum height of 2500 feet.

41.

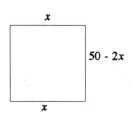

$A = x(50 - 2x)$

$A = 50x - 2x^2$

$A = -2(x^2 - 25x)$

$A = -2\left(x^2 - 25x + \dfrac{625}{4}\right) + \dfrac{625}{2}$

$A = -2\left(x - \dfrac{25}{2}\right)^2 + \dfrac{625}{2}$

Vertex: $\left(\dfrac{25}{2}, \dfrac{625}{2}\right)$, maximum

$x = \dfrac{25}{2}$ and $50 - 2x = 50 - 2\left(\dfrac{25}{2}\right) = 25$

The dimensions are $\dfrac{25}{2}$ = 12.5 ft and 25 ft.

49. $\dfrac{x - 3}{x + 3} \div (x^2 - 9)$

$= \dfrac{x - 3}{x + 3} \cdot \dfrac{1}{x^2 - 9}$

$= \dfrac{x - 3}{x + 3} \cdot \dfrac{1}{(x + 3)(x - 3)}$

$= \dfrac{1}{(x + 3)^2}$

51. $m = -\dfrac{2}{3} = \dfrac{-2}{3}$

Start at $(0, -2)$ and move 2 units down, then 3 units to the right to locate the point $(3, -4)$ on the line.

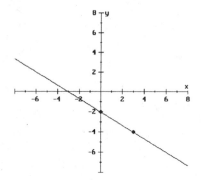

8.7 Exercises

1. $(x + 4)(x - 2) < 0$

$x + 4 = 0 \qquad x - 2 = 0$

$\qquad x = -4 \qquad\quad x = 2$

The intervals are $x < -4$, $-4 < x < 2$, and $x > 2$.
Test points:

	$(x + 4)(x - 2)$	
-5	$(-5 + 4)(-5 - 2)$	pos
0	$(0 + 4)(0 - 2)$	neg
3	$(3 + 4)(3 - 2)$	pos

We want $(x + 4)(x - 2) < 0$: neg

Solution: $-4 < x < 2$

3. $(x + 4)(x - 2) > 0$: positive
See exercise #1.
$x < -4$ or $x > 2$

5. $(x + 2)(x - 5) \leq 0$
   $x + 2 = 0 \qquad x - 5 = 0$
   $\qquad x = -2 \qquad\qquad x = 5$

   The intervals are $x < -2$, $-2 < x < 5$, and $x > 5$

   Test points:

	$(x + 2)(x - 5)$	
-3	$(-3 + 2)(-3 - 5)$	pos
0	$(0 + 2)(0 - 5)$	neg
6	$(6 + 2)(6 - 5)$	pos

   We want $(x + 2)(x - 5) \leq 0$: neg or 0

   Solution: $-2 \leq x \leq 5$

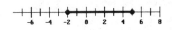

7. $(x - 3)(2x - 1) \geq 0$
   $x - 3 = 0 \qquad 2x - 1 = 0$
   $\qquad x = 3 \qquad\qquad x = \dfrac{1}{2}$

   The intervals are $x < \dfrac{1}{2}$, $\dfrac{1}{2} < x < 3$, $x > 3$

   Test points:

	$(x - 3)(2x - 1)$	
0	$(0 - 3)(2 \cdot 0 - 1)$	pos
1	$(1 - 3)(2 \cdot 1 - 1)$	neg
4	$(4 - 3)(2 \cdot 4 - 1)$	pos

   We want $(x - 3)(2x - 1) \geq 0$; pos or 0

   Solution: $x \leq \dfrac{1}{2}$ or $x \geq 3$

9. $a^2 - a - 20 < 0$
   $(a - 5)(a + 4) = 0$
   $a - 5 = 0 \qquad a + 4 = 0$
   $\qquad a = 5 \qquad\qquad a = -4$

   The intervals are $a < -4$, $-4 < a < 5$, $a > 5$

   Test points:

	$(a - 5)(a + 4)$	
-5	$(-5 - 5)(-5 + 4)$	pos
0	$(0 - 5)(0 + 4)$	neg
6	$(6 - 5)(6 + 4)$	pos

   We want $(a - 5)(a + 4) < 0$: neg

   Solution $-4 < a < 5$

11. $x^2 + x - 12 \geq 0$
    $(x + 4)(x - 3) \geq 0$
    $x + 4 = 0 \qquad x - 3 = 0$
    $\qquad x = -4 \qquad\qquad x = 3$

    The intervals are $x < -4$, $-4 < x < 3$, $x > 3$

    Test points:

	$(x + 4)(x - 3)$	
-5	$(-5 + 4)(-5 - 3)$	pos
0	$(0 + 4)(0 - 3)$	neg
4	$(4 + 4)(4 - 3)$	pos

    We want $(x + 4)(x - 3) \geq 0$: pos or 0

    Solution: $x \leq -4$ or $x \geq 3$

13. $2a^2 - 9a \leq 5$
    $2a^2 - 9a - 5 \leq 0$
    $(2a + 1)(a - 5) \leq 0$
    $(2a + 1) = 0 \qquad a - 5 = 0$
    $\qquad a = -\dfrac{1}{2} \qquad\qquad a = 5$

The intervals are $a < -\dfrac{1}{2}$, $-\dfrac{1}{2} < a < 5$, $a > 5$

Test points:

	$(2a + 1)(a - 5)$	
-1	$(2(-1) + 1) + (-1 - 5)$	pos
0	$(2 \cdot 0 + 1)(0 - 5)$	neg
6	$(2 \cdot 6 + 1)(6 - 5)$	pos

We want $(2a + 1)(a - 5) \le 0$; neg or 0
Solution: $-\dfrac{1}{2} \le a \le 5$

15. $6y^2 - y > 1$
$6y^2 - y - 1 > 0$
$(3y + 1)(2y - 1) > 0$
$3y + 1 = 0 \qquad 2y - 1 = 0$
$\qquad y = -\dfrac{1}{3} \qquad\qquad y = \dfrac{1}{2}$

The intervals are $y < -\dfrac{1}{3}$, $-\dfrac{1}{3} < y < \dfrac{1}{2}$, $y > \dfrac{1}{2}$

Test points:

	$(3y + 1)(2y - 1)$	
-1	$[3(-1) + 1][2(-1) - 1]$	pos
0	$(3 \cdot 0 + 1)(2 \cdot 0 - 1)$	neg
1	$(3 \cdot 1 + 1)(2 \cdot 1 - 1)$	pos

We want $(3y + 1)(2y - 1) > 0$; pos
Solution: $y < -\dfrac{1}{3}$ or $y > \dfrac{1}{2}$

17. $3x^2 \le 10 - 13x$
$3x^2 + 13x - 10 \le 0$
$(3x - 2)(x + 5) \le 0$
$3x - 2 = 0 \qquad x + 5 = 0$
$\qquad x = \dfrac{2}{3} \qquad\qquad x = -5$

The intervals are $x < -5$, $-5 < x < \dfrac{2}{3}$, $x > \dfrac{2}{3}$

Test points:

	$(3x - 2)(x + 5)$	
-6	$[3(-6) - 2](-6 + 5)$	pos
0	$(3 \cdot 0 - 2)(0 + 5)$	neg
1	$(3 \cdot 1 - 2)(1 + 5)$	pos

We want $(3x - 2)(x + 5) \le 0$; neg or 0
Solution: $-5 \le x \le \dfrac{2}{3}$

19. $x^2 + 2x + 1 \ge 0$
$(x + 1)^2 \ge 0$

A quantity squared is always ≥ 0.
Solution: All real numbers

21. $x^2 - 6x + 9 < 0$
$(x - 3)^2 < 0$

A quantity squared is never < 0.
No solution

23. $2x^2 - 13x > -15$
$2x^2 - 13x + 15 > 0$
$(2x - 3)(x - 5) > 0$
$2x - 3 = 0 \qquad x - 5 = 0$
$\qquad x = \dfrac{3}{2} \qquad\qquad x = 5$

The intervals are $x < \dfrac{3}{2}$, $\dfrac{3}{2} < x < 5$, $x > 5$

Test points:

	$(2x - 3)(x - 5)$	
0	$(2 \cdot 0 - 3)(0 - 5)$	pos
4	$(2 \cdot 4 - 3)(4 - 5)$	neg
6	$(2 \cdot 6 - 3)(6 - 5)$	pos

We want $(2x - 3)(x - 5) > 0$; pos
Solution: $x < \dfrac{3}{2}$ or $x > 5$

25.
$$3y^2 \geq 5y + 2$$
$$3y^2 - 5y - 2 \geq 0$$
$$(3y + 1)(y - 2) \geq 0$$
$$3y + 1 = 0 \qquad y - 2 = 0$$
$$y = -\frac{1}{3} \qquad y = 2$$

The intervals are $y < -\frac{1}{3}$, $-\frac{1}{3} < y < 2$, $y > 2$

Test points:

	$(3y + 1)(y - 2)$	
-1	$[3(-1) + 1](-1 - 2)$	pos
0	$(3 \cdot 0 + 1)(0 - 2)$	neg
3	$(3 \cdot 3 + 1)(3 - 2)$	pos

We want $(3y + 1)(y - 2) \geq 0$; pos or 0
Solution: $y \leq -\frac{1}{3}$ or $y \geq 2$

27.
$$x^2 + 2x \leq -1$$
$$x^2 + 2x + 1 \leq 0$$
$$(x + 1)^2 \leq 0$$

A quantity squared cannot be < 0.
Hence $x + 1 = 0$
$$x = -1$$

29. $\dfrac{x - 2}{x + 1} < 0$
$$x - 2 = 0 \qquad x + 1 = 0$$
$$x = 2 \qquad x = -1$$

The intervals are $x < -1$, $-1 < x < 2$, $x > 2$

Test points:

	$(x - 2)/(x + 1)$	
-2	$(-2 - 2)/(-2 + 1)$	pos
0	$(0 - 2)/(0 + 1)$	neg
3	$(3 - 2)/(3 + 1)$	pos

We want $\dfrac{x - 2}{x + 1} < 0$; neg
Solution: $-1 < x < 2$

31. $\dfrac{y + 3}{y - 5} \geq 0$

$$y + 3 = 0 \qquad y - 5 = 0$$
$$y = -3 \qquad y = 5$$

The intervals are $y < -3$, $-3 < y < 5$, $y > 5$

Test points:

	$(y + 3)/(y - 5)$	
-4	$(-4 + 3)/(-4 - 5)$	pos
0	$(0 + 3)/(0 - 5)$	neg
6	$(6 + 3)/(6 - 5)$	pos

We want $\dfrac{y + 3}{y - 5} \geq 0$; pos or 0

Solution: $y \leq -3$ or $y > 5$

33. $\dfrac{a - 6}{a + 4} < 0$

$$a - 6 = 0 \qquad a + 4 = 0$$
$$a = 6 \qquad a = -4$$

The intervals are $a < -4$, $-4 < a < 6$, $a > 6$

Test points:

	$(a - 6)/(a + 4)$	
-5	$(-5 - 6)/(-5 + 4)$	pos
0	$(0 - 6)/(0 + 4)$	neg
7	$(7 - 6)/(7 + 4)$	pos

We want $\dfrac{a - 6}{a + 4} < 0$; neg
Solution: $-4 < a < 6$

35. $\dfrac{5}{y - 4} > 0$

$$y - 4 = 0$$
$$y = 4$$

The intervals are $y < 4$, $y > 4$

Test points:

	(5)/(y - 4)	
0	(5)/(0 - 4)	neg
5	(5)/(5 - 4)	pos

We want $\dfrac{5}{y - 4} < 0$; neg

Solution: $y < 4$

37. $\dfrac{3}{y - 1} < 1$

$\dfrac{3}{y - 1} - 1 < 0$

$\dfrac{3 - (y - 1)}{y - 1} < 0$

$\dfrac{-y + 4}{y - 1} < 0$

$-y + 4 = 0 \qquad y - 1 = 0$
$\qquad 4 = y \qquad \qquad y = 1$

The intervals are $y < 1$, $1 < y < 4$, $y > 4$

Test points:

	(-y + 4)/(y - 1)	
0	(-0 + 4)/(0 - 1)	neg
2	(-2 + 4)/(2 - 1)	pos
5	(-5 + 4)/(5 - 1)	neg

We want $\dfrac{-y + 4}{y - 1} < 0$; neg

Solution: $y < 1$ or $y > 4$

39. $\dfrac{y + 1}{y + 2} > 3$

$\dfrac{y + 1}{y + 2} - 3 > 0$

$\dfrac{y + 1 - 3(y + 2)}{y + 2} > 0$

$\dfrac{-2y - 5}{y + 2} > 0$

$-2y - 5 = 0 \qquad y + 2 = 0$
$\qquad y = -\dfrac{5}{2} \qquad \qquad y = -2$

The intervals are $y < -\dfrac{5}{2}$, $-\dfrac{5}{2} < -2$, $y > -2$

Test points:

	(-2y - 5)/(y + 2)	
-3	(-2(-3) - 5)/(-3 + 2)	neg
-9/4	(-2(-9/4) - 5)/(-9/4 + 2)	pos
0	(-2(0) - 5)/(0 + 2)	neg

We want $\dfrac{-2y - 5}{y + 2} > 0$; pos

Solution: $-\dfrac{5}{2} < y < -2$

41. $\dfrac{2y + 3}{y - 1} \le 2$

$\dfrac{2y + 3}{y - 1} - 2 \le 0$

$\dfrac{2y + 3 - 2(y - 1)}{y - 1} \le 0$

$\dfrac{5}{y - 1} \le 0$

$y - 1 = 0$
$\qquad y = 1$

The intervals are $y < 1$, $y > 1$

Test points:

	$(5)/(y - 1)$	
0	$(5)/(0 - 1)$	neg
2	$(5)/(2 - 1)$	pos

We want $\dfrac{5}{y - 1} \leq 0$; neg or 0

Solution: $y < 1$

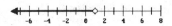

43.
$$\frac{y - 3}{y + 1} > -2$$

$$\frac{y - 3}{y + 1} + 2 > 0$$

$$\frac{y - 3 + 2(y + 1)}{y + 1} > 0$$

$$\frac{3y - 1}{y + 1} > 0$$

$$3y - 1 = 0 \qquad y + 1 = 0$$
$$y = \frac{1}{3} \qquad\quad y = -1$$

The intervals are $y < -1$, $-1 < y < \dfrac{1}{3}$, $y > \dfrac{1}{3}$

Test points:

	$(3y - 1)/(y + 1)$	
-2	$(3(-2) - 1)/(-2 + 1)$	pos
0	$(3(0) - 1)/(0 + 1)$	neg
1	$(3(1) - 1)/(1 + 1)$	pos

We want $\dfrac{3y - 1}{y + 1} > 0$; pos

Solution: $y < -1$ or $y > \dfrac{1}{3}$

45.
$$\frac{x}{x - 1} \leq \frac{3}{x - 1}$$

$$\frac{x}{x - 1} - \frac{3}{x - 1} \leq 0$$

$$\frac{x - 3}{x - 1} \leq 0$$

$$x - 3 = 0 \qquad x - 1 = 0$$
$$x = 3 \qquad\quad x = 1$$

The intervals are $x < 1$, $1 < x < 3$, $x > 3$

Test points:

	$(x - 3)/(x - 1)$	
0	$(0 - 3)/(0 - 1)$	pos
2	$(2 - 3)/(2 - 1)$	neg
4	$(4 - 3)/(4 - 1)$	pos

We want $\dfrac{x - 3}{x - 1} \leq 0$; neg or 0

Solution: $1 < x \leq 3$

47.
$$\frac{x - 2}{x + 3} > \frac{x + 4}{x + 3}$$

$$\frac{x - 2}{x + 3} - \frac{x + 4}{x + 3} > 0$$

$$\frac{x - 2 - (x + 4)}{x + 3} > 0$$

$$\frac{-6}{x + 3} > 0$$

$$x + 3 = 0$$
$$x = -3$$

The intervals are $x < -3$, $x > -3$

Test points:

	$(-6)/(x + 3)$	
-4	$(-6)/(-4 + 3)$	pos
0	$(-6)/(0 + 3)$	neg

We want $\dfrac{-6}{x + 3} > 0$; pos

Solution: $x < -3$

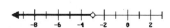

49.
$$\frac{1}{x - 2} + \frac{2}{x + 3} \leq \frac{3}{x + 3}$$

$$\frac{1}{x - 2} + \frac{2}{x + 3} - \frac{3}{x + 3} \leq 0$$

$$\frac{1}{x - 2} - \frac{1}{x + 3} \leq 0$$

$$\frac{x + 3 - (x - 2)}{(x - 2)(x + 3)} \leq 0$$

$$\frac{5}{(x - 2)(x + 3)} \leq 0$$

$$x - 2 = 0 \qquad x + 3 = 0$$
$$x = 2 \qquad\quad x = -3$$

The intervals are $x < -3$, $-3 < x < 2$, $x > 2$

Test points:

	$(5)/\{(x - 2)(x + 3)\}$	
-4	$(5)/\{(-4 - 2)(-4 + 3)\}$	pos
0	$(5)/\{(0 - 2)(0 + 3)\}$	neg
3	$(5)/\{(3 - 2)(3 + 3)\}$	pos

We want $\dfrac{5}{(x - 2)(x + 3)} \leq 0$; neg or 0

Solution: $-3 < x < 2$

53. $(x, P)(50, 600)$ and $(65, 750)$

(a) $m = \dfrac{750 - 600}{65 - 50} = 10$

$$P - 600 = 10(x - 50)$$
$$P - 600 = 10x - 500$$
$$P = 10x + 100$$

(b) When $x = 90$:
$$P = 10(90) + 100 = 1000$$

The profit would be $1000.

55. $(x^{-2} + y)^{-1}$

$$= \frac{1}{x^{-2} + y}$$

$$= \frac{1}{\dfrac{1}{x^2} + y}$$

$$= \frac{1 \cdot x^2}{\left(\dfrac{1}{x^2} + y\right) \cdot x^2}$$

$$= \frac{x^2}{1 + x^2 y}$$

8.8 Exercises

1. $d = \sqrt{(6 - 3)^2 + (9 - 5)^2}$
$= \sqrt{3^2 + 4^2}$
$= \sqrt{9 + 16}$
$= \sqrt{25}$
$= 5$

3. $d = \sqrt{(6 - 3)^2 + (3 - 6)^2}$
$= \sqrt{3^2 + (-3)^2}$
$= \sqrt{9 + 9}$
$= \sqrt{18}$
$= 3\sqrt{2}$

5. $d = \sqrt{[6 - (-6)]^2 + (-9 - 9)^2}$
   $= \sqrt{12^2 + (-18)^2}$
   $= \sqrt{144 + 324}$
   $= \sqrt{468}$
   $= 6\sqrt{13}$

7. $d = \sqrt{[-8 - (-7)]^2 + [-3 - (-3)]^2}$
   $= \sqrt{(-1)^2 + 0^2}$
   $= \sqrt{1}$
   $= 1$

9. $d = \sqrt{\left(\dfrac{1}{2} - \dfrac{1}{3}\right)^2 + (0 - 2)^2}$

   $= \sqrt{\left(\dfrac{1}{6}\right)^2 + (-2)^2}$

   $= \sqrt{\dfrac{1}{36} + 4}$

   $= \sqrt{\dfrac{145}{36}}$

   $= \dfrac{\sqrt{145}}{6}$

11. $d = \sqrt{(1.7 - 1.4)^2 + (1.2 - 0.8)^2}$
    $= \sqrt{(0.3)^2 + (0.4)^2}$
    $= \sqrt{0.09 + 0.16}$
    $= \sqrt{0.25}$
    $= 0.5$

13. midpoint $= \left(\dfrac{0 + 0}{2}, \dfrac{5 + 7}{2}\right)$

    $= (0, 6)$

15. midpoint $= \left(\dfrac{-3 + 3}{2}, \dfrac{1 + 1}{2}\right)$

    $= (0, 1)$

17. midpoint $= \left(\dfrac{-3 + 3}{2}, \dfrac{4 + (-4)}{2}\right)$

    $= (0, 0)$

19. midpoint $= \left(\dfrac{\dfrac{2}{5} + \dfrac{1}{3}}{2}, \dfrac{\dfrac{3}{4} + 2}{2}\right)$

    $= \left(\dfrac{\dfrac{11}{15}}{2}, \dfrac{\dfrac{11}{4}}{2}\right)$

    $= \left(\dfrac{11}{30}, \dfrac{11}{8}\right)$

21. $d(P, Q) = |5 - 8| = 3$
    $d(Q, R) = |2 - 6| = 4$
    $d(P, R) = \sqrt{(5 - 8)^2 + (2 - 6)^2}$
    $ = \sqrt{9 + 16}$
    $ = \sqrt{25}$
    $ = 5$

    $[d(P, Q)]^2 + [d(Q, R)]^2 = [d(P, R)]^2$
    $3^2 + 4^2 = 5^2$
    $9 + 16 = 25$
    $25 = 25$
    True

    They are vertices of a right triangle.

23. $d(P, Q) = \sqrt{[4 - (-1)]^2 + (2 - 5)^2}$
    $ = \sqrt{25 + 9}$
    $ = \sqrt{34}$

    $d(Q, R) = \sqrt{(-1 - 3)^2 + (5 - 9)^2}$
    $ = \sqrt{16 + 16}$
    $ = \sqrt{32}$
    $ = 4\sqrt{2}$

    $d(P, R) = \sqrt{(4 - 3)^2 + (2 - 9)^2}$
    $ = \sqrt{1 + 49}$
    $ = \sqrt{50}$
    $ = 5\sqrt{2}$

    $[d(P, Q)]^2 + [d(Q, R)]^2 = [d(P, R)]^2$
    $\left(\sqrt{34}\right)^2 + \left(4\sqrt{2}\right)^2 = \left(5\sqrt{2}\right)^2$
    $34 + 32 = 50$
    $66 = 50$
    False

    They are not vertices of a right triangle.

25. $P(5, 3)$, $Q(7, 4)$, $R(9, 7)$, $S(7, 6)$

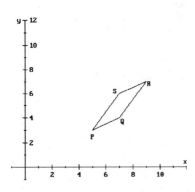

$$\text{midpoint}_{PR} = \left(\frac{5 + 9}{2}, \frac{3 + 7}{2}\right) = (7, 5)$$

$$\text{midpoint}_{SQ} = \left(\frac{7 + 7}{2}, \frac{4 + 6}{2}\right) = (7, 5)$$

Yes, since both diagonals have the same midpoint $(7, 5)$.

27. $P(-2, 3)$, $Q(5, -4)$, $R(-6, 5)$, $S(3, -4)$

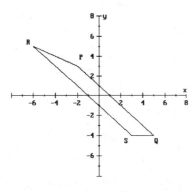

$$\text{midpoint}_{RQ} = \left(\frac{-6 + 5}{2}, \frac{5 - 4}{2}\right) = \left(-\frac{1}{2}, \frac{1}{2}\right)$$

$$\text{midpoint}_{PS} = \left(\frac{-2 + 3}{2}, \frac{3 - 4}{2}\right) = \left(\frac{1}{2}, -\frac{1}{2}\right)$$

No, since the diagonals have different midpoints.

29. $(x - h)^2 + (y - k)^2 = r^2$
$(x - 0)^2 + (y - 0)^2 = 1^2$
$x^2 + y^2 = 1$

31. $(x - h)^2 + (y - k)^2 = r^2$
$(x - 1)^2 + (y - 0)^2 = 7^2$
$(x - 1)^2 + y^2 = 49$

33. $(x - h)^2 + (y - k)^2 = r^2$
$(x - 2)^2 + (y - 5)^2 = 6^2$
$(x - 2)^2 + (y - 5)^2 = 36$

35. $(x - h)^2 + (y - k)^2 = r^2$
$(x - 6)^2 + [y - (-2)]^2 = 5^2$
$(x - 6)^2 + (y + 2)^2 = 25$

37. $(x - h)^2 + (y - k)^2 = r^2$
$[x - (-3)]^2 + [y - (-2)]^2 = 1^2$
$(x + 3)^2 + (y + 2)^2 = 1$

39. $x^2 + y^2 = 16$
$(x - 0)^2 + (y - 0)^2 = 4^2$
Center: $(0, 0)$
$r = 4$

41. $x^2 + y^2 = 24$
$(x - 0)^2 + (y - 0)^2 = (\sqrt{24})^2$
Center: $(0, 0)$
$r = \sqrt{24} = 2\sqrt{6}$

43. $(x - 3)^2 + y^2 = 16$
$(x - 3)^2 + (y - 0)^2 = 4^2$
Center: $(3, 0)$
$r = 4$

45. $(x - 2)^2 + (y - 1)^2 = 1$
Center: $(2, 1)$
$r = \sqrt{1} = 1$

47. $(x + 1)^2 + (y - 3)^2 = 25$
 Center: $(-1, 3)$
 $r = \sqrt{25} = 5$

55. $x^2 + y^2 - 4x - 2y = 20$
 $(x^2 - 4x + 4) + (y^2 - 2y + 1) = 20 + 4 + 1$
 $(x - 2)^2 + (y - 1)^2 = 25$
 Center: $(2, 1)$
 $r = 5$

49. $(x + 2)^2 + (y + 3)^2 = 32$
 Center: $(-2, -3)$
 $r = \sqrt{32} = 4\sqrt{2}$

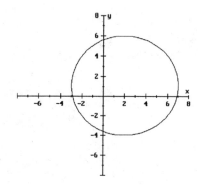

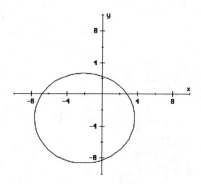

57. $x^2 + y^2 - 2x = 20 + 4y$
 $x^2 - 2x + y^2 - 4y = 20$
 $(x^2 - 2x + 1) + (y^2 - 4y + 4) = 20 + 1 + 4$
 $(x - 1)^2 + (y - 2)^2 = 25$
 Center: $(1, 2)$
 $r = 5$

51. $(x + 7)^2 + (y + 1)^2 = 2$
 Center: $(-7, -1)$
 $r = \sqrt{2}$

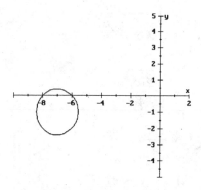

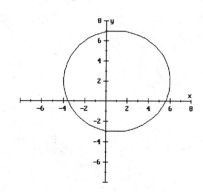

53. $x^2 + y^2 - 2x = 15$
 $(x^2 - 2x + 1) + y^2 = 15 + 1$
 $(x - 1)^2 + (y - 0)^2 = 16$
 Center: $(1, 0)$
 $r = 4$

59. $x^2 + 10y = 71 - y^2 + 4x$
 $x^2 - 4x + y^2 + 10y = 71$
 $(x^2 - 4x + 4) + (y^2 + 10y + 25) = 71 + 4 + 25$
 $(x - 2)^2 + (y + 5)^2 = 100$
 Center: $(2, -5)$
 $r = 10$

61. $x^2 + y^2 = 2y - 6x - 2$
 $x^2 + 6x + y^2 - 2y = -2$
 $(x^2 + 6x + 9) + (y^2 - 2y + 1) = -2 + 9 + 1$
 $(x + 3)^2 + (y - 1)^2 = 8$
 Center: $(-3, 1)$
 $r = \sqrt{8} = 2\sqrt{2}$

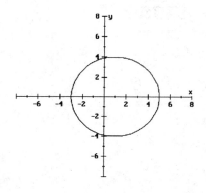

63. $2x^2 + 2y^2 - 4x + 4y = 22$
       $x^2 + y^2 - 2x + 2y = 11$
$(x^2 - 2x + 1) + (y^2 + 2y + 1) = 11 + 1 + 1$
          $(x - 1)^2 + (y + 1)^2 = 13$
Center: $(1, -1)$
          $r = \sqrt{13}$

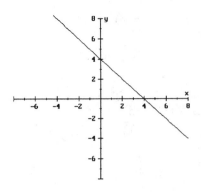

65. $x^2 + y^2 - x + 2y = \dfrac{59}{4}$

$\left(x^2 - x + \dfrac{1}{4}\right) + (y^2 + 2y + 1) = \dfrac{59}{4} + \dfrac{1}{4} + 1$

$\left(x - \dfrac{1}{2}\right)^2 + (y + 1)^2 = 16$

    Center: $\left(\dfrac{1}{2}, -1\right)$

          $r = 4$

67. $x^2 + y^2 = 16$
     Center: $(0, 0)$
     $r = 4$

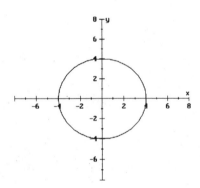

69. $x + y = 4$

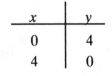

x	y
0	4
4	0

71. Center = midpoint

          $= \left(\dfrac{-2 + 4}{2}, \dfrac{8 - 5}{2}\right)$

          $= \left(1, \dfrac{3}{2}\right)$

    $r = \sqrt{(-2 - 1)^2 + \left(8 - \dfrac{3}{2}\right)^2}$

          $= \sqrt{9 + \dfrac{169}{4}}$

          $= \sqrt{\dfrac{205}{4}}$

       $r^2 = \dfrac{205}{4}$

    $(x - h)^2 + (y - k)^2 = r^2$

    $(x - 1)^2 + \left(y - \dfrac{3}{2}\right)^2 = \dfrac{205}{4}$

73. $r = \sqrt{(3 - 2)^2 + (-5 - 6)^2}$
         $= \sqrt{1 + 121}$
         $= \sqrt{122}$
     $r^2 = 122$

       $(x - h)^2 + (y - k)^2 = r^2$
     $(x - 3)^2 + [y - (-5)]^2 = 122$
       $(x - 3)^2 + (y + 5)^2 = 122$

75. $r = \sqrt{(-3 - 5)^2 + (4 - 2)^2}$

    $= \sqrt{64 + 4}$

    $= \sqrt{68}$

    $= 2\sqrt{17}$

    $C = 2\pi r$

      $= 2\pi(2\sqrt{17})$

      $= 4\pi\sqrt{17}$

77. It will touch the x-axis at (3, 0).

    $r = |-2 - 0| = 2$

        $(x - h)^2 + (y - k)^2 = r^2$

  $(x - 3)^2 + [y - (-2)]^2 = 2^2$

     $(x - 3)^2 + (y + 2)^2 = 4$

79. The center will be the point of intersection of the horizontal line through (0, −3) with the vertical line through (3, 0).

    $\begin{cases} y = -3 \\ x = 3 \end{cases}$

    $C(3, -3)$

    $r = |3 - 0| = 3$

        $(x - h)^2 + (y - k)^2 = r^2$

  $(x - 3)^2 + [y - (-3)]^2 = 3^2$

     $(x - 3)^2 + (y + 3)^2 = 9$

83. $\dfrac{83700}{0.0042} = \dfrac{8.37 \times 10^4}{4.2 \times 10^{-3}}$

         $= \dfrac{8.37}{4.2} \times \dfrac{10^4}{10^{-3}}$

         $= 1.99 \times 10^7$

It is closest to 10^7.

85. $2x^{1/2} - (5x)^{2/3}$

    $= 2\sqrt{x} - \sqrt[3]{(5x)^2}$

    $= 2\sqrt{x} - \sqrt[3]{25x^2}$

1. $d = 18v - \left(\dfrac{v}{2.3}\right)^2$, $10 \le v \le 90$

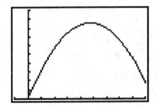

(a) When $v = 40$, $d = 417.5$
It can travel 417.5 miles

(b) When $d = 400$, $v = 35$ mph
and $v = 60$ mph

(c) 48 mph

3. $\quad 2y^2 - y - 1 = 0$
$(2y + 1)(y - 1) = 0$
$2y + 1 = 0 \quad$ or $\quad y - 1 = 0$
$\qquad 2y = -1 \qquad\qquad y = 1$
$\qquad\quad y = -\dfrac{1}{2} \quad$ or $\qquad y = 1$

5. $\qquad 3x^2 - 17x = 28$
$\quad 3x^2 - 17x - 28 = 0$
$\quad (3x + 4)(x - 7) = 0$
$3x + 4 = 0 \quad$ or $\quad x - 7 = 0$
$\qquad 3x = -4 \qquad\qquad x = 7$
$\qquad\quad x = -\dfrac{4}{3} \quad$ or $\qquad x = 7$

7. $\qquad 81 = a^2$
$\quad \pm\sqrt{81} = \sqrt{a^2}$
$\qquad \pm 9 = a$

9. $z^2 + 7 = 2$
$\qquad z^2 = -5$
$\quad \sqrt{z^2} = \pm\sqrt{-5}$
$\qquad z = \pm i\sqrt{5}$

11.
$$4x^2 + 36 = 24x$$
$$4x^2 - 24x + 36 = 0$$
$$x^2 - 6x + 9 = 0$$
$$(x - 3)^2 = 0$$
$$x - 3 = \pm\sqrt{0}$$
$$x - 3 = 0$$
$$x = 3$$

13.
$$(a + 7)(a + 3) = (3a + 1)(a + 1)$$
$$a^2 + 10a + 21 = 3a^2 + 4a + 1$$
$$0 = 2a^2 - 6a - 20$$
$$0 = a^2 - 3a - 10$$
$$0 = (a - 5)(a + 2)$$
$$a - 5 = 0 \quad \text{or} \quad a + 2 = 0$$
$$a = 5 \quad \text{or} \qquad a = -2$$

15.
$$x - 2 = \frac{1}{x + 2}$$

$$(x + 2)(x - 1) = 1$$
$$x^2 - 4 = 1$$
$$x^2 = 5$$
$$x = \pm\sqrt{5}$$

17.
$$\frac{2}{x - 2} - \frac{5}{x + 2} = 1$$

$$(x - 2)(x + 2)\left(\frac{2}{x - 2} - \frac{5}{x + 2}\right) = (x - 2)(x + 2)(1)$$

$$2(x + 2) - 5(x - 2) = x^2 - 4$$
$$2x + 4 - 5x + 10 = x^2 - 4$$
$$-3x + 14 = x^2 - 4$$
$$0 = x^2 + 3x - 18$$
$$0 = (x + 6)(x - 3)$$
$$x + 6 = 0 \quad \text{or} \quad x - 3 = 0$$
$$x = -6 \quad \text{or} \qquad x = 3$$

19.
$$x^2 + 2x - 4 = 0$$
$$x^2 + 2x = 4$$
$$x^2 + 2x + 1 = 4 + 1$$
$$(x + 1)^2 = 5$$
$$x + 1 = \pm\sqrt{5}$$
$$x = -1 \pm \sqrt{5}$$

21.
$$2y^2 + 4y - 3 = 0$$
$$y^2 + 2y - \frac{3}{2} = 0$$
$$y^2 + 2y = \frac{3}{2}$$
$$y^2 + 2y + 1 = \frac{3}{2} + 1$$
$$(y + 1)^2 = \frac{5}{2}$$
$$y + 1 = \pm\sqrt{\frac{5}{2}}$$
$$y = -1 \pm \frac{\sqrt{10}}{2}$$
$$\text{or} \quad y = \frac{-2 \pm \sqrt{10}}{2}$$

23.
$$3a^2 + 6a - 5 = 0$$
$$a^2 + 2a - \frac{5}{3} = 0$$
$$a^2 + 2a = \frac{5}{3}$$
$$a^2 + 2a = \frac{5}{3}$$
$$a^2 + 2a + 1 = \frac{5}{3} + 1$$
$$(a + 1)^2 = \frac{8}{3}$$
$$a + 1 = \pm\sqrt{\frac{8}{3}}$$
$$a = -1 \pm \frac{2\sqrt{2}}{3}$$
$$\text{or} \quad a = \frac{-3 \pm 2\sqrt{2}}{3}$$

25.
$$\frac{1}{a-5} + \frac{3}{a+2} = 4$$

$$(a-5)(a+2)\left(\frac{1}{a-5} + \frac{3}{a+2}\right) = (a-5)(a+2)(4)$$

$$a + 2 + 3(a-5) = 4(a^2 - 3a - 10)$$
$$a + 2 + 3a - 15 = 4a^2 - 12a - 40$$
$$4a - 13 = 4a^2 - 12a - 40$$
$$27 = 4a^2 - 16a$$

$$\frac{27}{4} = a^2 - 4a$$

$$\frac{27}{4} + 4 = a^2 - 4a + 4$$

$$\frac{43}{4} = (a-2)^2$$

$$\pm\sqrt{\frac{43}{2}} = a - 2$$

$$\frac{\pm\sqrt{43}}{2} = a - 2$$

$$2 \pm \frac{\sqrt{43}}{2} = a$$

$$\text{or} \quad \frac{4 \pm \sqrt{43}}{2} = a$$

27.
$$6a^2 - 13a = 5$$
$$6a^2 - 13a - 5 = 0$$
$$(3a + 1)(2a - 5) = 0$$
$$3a + 1 = 0 \quad \text{or} \quad 2a - 5 = 0$$
$$3a = -1 \qquad 2a = 5$$
$$a = -\frac{1}{3} \quad \text{or} \quad a = \frac{5}{2}$$

29. $3.2a^2 - 5.8a + 4 = 9.6$

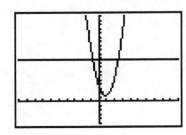

$$a = -0.70 \text{ and } a = 2.51$$

31.
$$5a^2 - 3a = 3 - 3a + 2a^2$$
$$3a^2 = 3$$
$$a^2 = 1$$
$$a = \pm 1$$

33.
$$(x-4)(x+1) = x - 2$$
$$x^2 - 3x - 4 = x - 2$$
$$x^2 - 4x - 2 = 0$$

$$x = \frac{-(-4) \pm \sqrt{(-4)^2 - 4(1)(-2)}}{2(1)}$$

$$= \frac{4 \pm \sqrt{16 + 8}}{2}$$

$$= \frac{4 \pm \sqrt{24}}{2}$$

$$= \frac{4 \pm 2\sqrt{6}}{2}$$

$$= \frac{2(2 \pm \sqrt{6})}{2}$$

$$= 2 \pm \sqrt{6}$$

35.
$$(t+3)(t-4) = t(t+2)$$
$$t^2 - t - 12 = t^2 + 2t$$
$$-12 = 3t$$
$$-4 = t$$

37.
$$8x^2 = 12$$
$$x^2 = \frac{3}{2}$$
$$x = \pm\sqrt{\frac{3}{2}}$$
$$x = \pm\frac{\sqrt{6}}{2}$$

39.
$$3x^2 - 2x + 5 = 7x^2 - 2x + 5$$
$$0 = 4x^2$$
$$0 = x^2$$
$$\pm\sqrt{0} = x$$
$$0 = x$$

41. $(x + 2)(x - 4) = 2x - 10$

$\qquad x^2 - 2x - 8 = 2x - 10$

$\qquad x^2 - 4x + 2 = 0$

$$x = \frac{-(-4) \pm \sqrt{(-4)^2 - 4(1)(2)}}{2(1)}$$

$$= \frac{4 \pm \sqrt{16 - 8}}{2}$$

$$= \frac{4 \pm \sqrt{8}}{2}$$

$$= \frac{4 \pm 2\sqrt{2}}{2}$$

$$= \frac{2(2 \pm \sqrt{2})}{2}$$

$$= 2 \pm \sqrt{2}$$

43. $\dfrac{1}{z + 2} = z - 4$

$\qquad 1 = (z + 2)(z - 4)$

$\qquad 1 = z^2 - 2z - 8$

$\qquad 0 = z^2 - 2z - 9$

$$z = \frac{-(-2) \pm \sqrt{(-2)^2 - 4(1)(-9)}}{2(1)}$$

$$= \frac{2 \pm \sqrt{4 + 36}}{2}$$

$$= \frac{2 \pm \sqrt{40}}{2}$$

$$= \frac{2 \pm 2\sqrt{10}}{2}$$

$$= \frac{2(1 \pm \sqrt{10})}{2}$$

$$= 1 \pm \sqrt{10}$$

45. $\qquad \dfrac{1}{x + 4} - \dfrac{3}{x + 2} = 5$

$$(x + 4)(x + 2)\left(\frac{1}{x + 4} - \frac{3}{x + 2}\right) = (x + 4)(x + 2)(5)$$

$\qquad x + 2 - 3(x + 4) = 5(x^2 + 6x + 8)$

$\qquad x + 2 - 3x - 12 = 5x^2 + 30x + 40$

$\qquad -2x - 10 = 5x^2 + 30x + 40$

$\qquad 0 = 5x^2 + 32x + 50$

$$x = \frac{-32 \pm \sqrt{32^2 - 4(5)(50)}}{2(5)}$$

$$= \frac{-32 \pm \sqrt{1024 - 1000}}{10}$$

$$= \frac{-32 \pm \sqrt{24}}{10}$$

$$= \frac{-32 \pm 2\sqrt{6}}{10}$$

$$= \frac{2(-16 \pm \sqrt{6})}{10}$$

$$= \frac{-16 \pm \sqrt{6}}{5}$$

47. $\qquad \dfrac{3}{x - 4} + \dfrac{2x}{x - 5} = \dfrac{3}{x - 5}$

$$(x - 4)(x - 5)\left(\frac{3}{x - 4} + \frac{2x}{x - 5}\right) = (x - 4)(x - 5)\left(\frac{3}{x - 5}\right)$$

$\qquad 3(x - 5) + 2x(x - 4) = 3(x - 4)$

$\qquad 3x - 15 + 2x^2 - 8x = 3x - 12$

$\qquad 2x^2 - 5x - 15 = 3x - 12$

$\qquad 2x^2 - 8x - 3 = 0$

$$x = \frac{-(-8) \pm \sqrt{(-8)^2 - 4(2)(-3)}}{2(2)}$$

$$= \frac{8 \pm \sqrt{64 + 24}}{4}$$

$$= \frac{8 \pm \sqrt{88}}{4}$$

$$= \frac{8 \pm 2\sqrt{22}}{4}$$

$$= \frac{2(4 \pm \sqrt{22})}{4}$$

$$= \frac{4 \pm \sqrt{22}}{2}$$

49. $A = \pi r^2 h$

$\frac{A}{\pi h} = r^2$

$\sqrt{\frac{A}{\pi h}} = r$

$\frac{\sqrt{A\pi h}}{\pi h} = r$

51. $2x^2 + xy - 3y^2 = 0$
$(2x + 3y)(x - y) = 0$
$2x + 3y = 0$ or $x - y = 0$
$2x = -3y$ $\quad x = y$
$x = -\frac{3}{2}y$ or $x = y$

53. $\sqrt{2a + 3} = a$
$(\sqrt{2a + 3})^2 = a^2$
$2a + 3 = a^2$
$0 = a^2 - 2a - 3$
$0 = (a - 3)(a + 1)$
$a - 3 = 0$ or $a + 1 = 0$
$a = 3$ or $a = -1$

When you check $a = -1$, you find it to be extraneous. The only solution is $a = 3$.

55. $\sqrt{3a + 1} + 1 = a$
$\sqrt{3a + 1} = a - 1$
$(\sqrt{3a + 1})^2 = (a - 1)^2$
$3a + 1 = a^2 - 2a + 1$
$0 = a^2 - 5a$
$0 = a(a - 5)$
$a = 0$ or $a - 5 = 0$
$a = 5$

When you check a = 0, you find it to be extraneous. The only solution is $a = 5$.

57. $\sqrt{2x} + 1 - \sqrt{x - 3} = 4$
$\sqrt{2x} = 3 + \sqrt{x - 3}$
$(\sqrt{2x})^2 = (3 + \sqrt{x - 3})^2$
$2x = 9 + 6\sqrt{x - 3} + x - 3$
$2x = 6 + 6\sqrt{x - 3} + x$
$x - 6 = 6\sqrt{x - 3}$
$(x - 6)^2 = (6\sqrt{x - 3})^2$
$x^2 - 12x + 36 = 36(x - 3)$
$x^2 - 12x + 36 = 36x - 108)$
$x^2 - 48x + 144 = 0$
$x = \frac{-(-48) \pm \sqrt{(-48)^2 - 4(1)(144)}}{2(1)}$
$= \frac{48 \pm \sqrt{1728}}{2}$
$x = 3.2$ or $x = 44.8$

When you check $x = 3.2$, you find it to be extraneous. The only solution is $x = 44.8$.

59. $\sqrt{3x + 4} - \sqrt{x - 3} = 3$
$\sqrt{3x + 4} = 3 + \sqrt{x - 3}$
$(\sqrt{3x + 4})^2 = (3 + \sqrt{x - 3})^2$
$3x + 4 = 9 + 6\sqrt{x - 3} + x - 3$
$3x + 4 = 6 + x + 6\sqrt{x - 3}$
$2x - 2 = 6\sqrt{x - 3}$
$x - 1 = 3\sqrt{x - 3}$
$(x - 1)^2 = (3\sqrt{x - 3})^2$
$x^2 - 2x + 1 = 9(x - 3)$
$x^2 - 2x + 1 = 9x - 27$
$x^2 - 11x + 28 = 0$
$(x - 4)(x - 7) = 0$
$x - 4 =$ or $x - 7 = 0$
$x = 4$ or $x = 7$

61. $\sqrt{3y + z} = x$
$(\sqrt{3y + z})^2 = x^2$
$3y + z = x^2$
$3y = x^2 - z$
$y = \frac{x^2 - z}{3}$

63. $\sqrt{3y} + z = x$
$\sqrt{3y} = x - z$
$(\sqrt{3y})^2 = (x - z)^2$
$3y = (x - z)^2$
$y = \frac{(x - z)^2}{3}$

65. $x^3 - 2x^2 - 15x = 0$
$x(x - 5)(x + 3) = 0$
$x = 0$ or $x - 5 = 0$ or $x + 3 = 0$
$x = 0$ or $x = 5$ or $x = -3$

67. $4x^3 - 10x^2 - 6x = 0$
$2x(2x + 1)(x - 3) = 0$
$2x = 0$ or $2x + 1 = 0$ or $x - 3 = 0$
$x = 0$ or $x = -\frac{1}{2}$ or $x = 3$

69.

$$a^4 - 17a^2 = -16$$
$$a^4 - 17a^2 + 16 = 0$$
$$(a^2 - 1)(a^2 - 16) = 0$$
$$a^2 - 1 = 0 \quad \text{or} \quad a^2 - 16 = 0$$
$$a^2 = 1 \qquad\qquad a^2 = 16$$
$$a = \pm 1 \quad \text{or} \qquad a = \pm 4$$

71.

$$y^4 - 3y^2 = 4$$
$$y^4 - 3y^2 - 4 = 0$$
$$(y^2 - 4)(y^2 + 1) = 0$$
$$y^2 - 4 = 0 \quad \text{or} \quad y^2 + 1 = 0$$
$$y^2 = 4 \qquad\qquad y^2 = -1$$
$$y = \pm 2 \quad \text{or} \qquad y = \pm i$$

73.

$$z^4 = 6z^2 - 5$$
$$z^4 - 6z^2 + 5 = 0$$
$$(z^2 - 5)(z^2 - 1) = 0$$
$$z^2 - 5 = 0 \quad \text{or} \quad z^2 - 1 = 0$$
$$z^2 = 5 \qquad\qquad z^2 = 1$$
$$z = \pm\sqrt{5} \quad \text{or} \qquad z = \pm 1$$

75.

$$a^{1/2} - a^{1/4} - 6 = 0$$
$$(a^{1/4})^2 - a^{1/4} - 6 = 0$$
Let $u = a^{1/4}$.
$$u^2 - u - 6 = 0$$
$$(u - 3)(u + 2) = 0$$
$$u - 3 = 0 \quad \text{or} \quad u + 2 = 0$$
$$u = 3 \qquad\qquad u = -2$$
$$a^{1/4} = 3 \qquad\qquad a^{1/4} = -2$$
$$(a^{1/4})^4 = 3^4 \qquad \text{No solution}$$
$$a = 81$$

The only solution is $a = 81$.

77.

$$2x^{2/3} = 5x^{1/3} + 3$$
$$2(x^{1/3})^2 - 5x^{1/3} - 3 = 0$$
Let $u = x^{1/3}$.
$$2u^2 - 5u - 3 = 0$$
$$(2u + 1)(u - 3) = 0$$
$$2u + 1 = 0 \qquad \text{or} \qquad u - 3 = 0$$
$$u = -\frac{1}{2} \qquad\qquad u = 3$$

$$x^{1/3} = -\frac{1}{2} \qquad\qquad x^{1/3} = 3$$

$$(x^{1/3})^3 = \left(-\frac{1}{2}\right)^3 \qquad (x^{1/3})^3 = 3^3$$

$$x = -\frac{1}{8} \qquad \text{or} \qquad x = 27$$

79.

$$\sqrt{x} + 2\sqrt[4]{x} - 35 = 0$$
$$x^{1/2} + 2x^{1/4} - 35 = 0$$
$$(x^{1/4})^2 + 2x^{1/4} - 35 = 0$$
Let $u = x^{1/4}$.
$$u^2 + 2u - 35 = 0$$
$$(u + 7)(u - 5) = 0$$
$$u + 7 = 0 \quad \text{or} \quad u - 5 = 0$$
$$u = -7 \qquad\qquad u = 5$$
$$x^{1/4} = -7 \qquad\qquad x^{1/4} = 5$$
No solution $\qquad (x^{1/4})^4 = 5^4$
$$x = 625$$

The only solution is $x = 625$.

81.

$$3x^{-2} + x^{-1} - 2 = 0$$
$$3(x^{-1})^2 + x^{-1} - 2 = 0$$
Let $u = x^{-1}$.
$$3u^2 + u - 2 = 0$$
$$(3u - 2)(u + 1) = 0$$
$$3u - 2 = 0 \qquad \text{or} \qquad u + 1 = 0$$
$$u = \frac{2}{3} \qquad\qquad u = -1$$

$$x^{-1} = \frac{2}{3} \qquad\qquad x^{-1} = -1$$

$$(x^{-1})^{-1} = \left(\frac{2}{3}\right)^{-1} \qquad (x^{-1})^{-1} = (-1)^{-1}$$

$$x = \frac{3}{2} \qquad \text{or} \qquad x = -1$$

83. $x^2 - 8 = \sqrt{x + 1}$

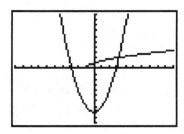

$$x = 3.169$$

85. $y = (x - 2)^2 + 1$
Vertex: $(2, 1)$
Axis of symmetry: $x = 2$

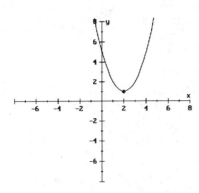

87. $y = -3(x - 2)^2 - 4$
Vertex: $(2, -4)$
Axis of symmetry: $x = 2$

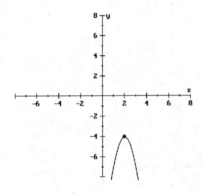

89. $y = 2x^2 - 12x + 4$
$y = 2(x^2 - 6x) + 4$
$y = 2(x^2 - 6x + 9) + 4 - 18$
$y = 2(x - 3)^2 - 14$
Vertex: $(3, -14)$
Axis of symmetry: $x = 3$

91. $y = -x^2 + 4x - 12$
$y = -(x^2 - 4x) - 12$
$y = -(x^2 - 4x + 4) - 12 + 4$
$y = -(x - 2)^2 - 8$
Vertex: $(2, -8)$
Axis of symmetry: $x = 2$

93. $y = 7x^2$
Vertex: $(0, 0)$
Axis of symmetry: $x = 0$
x-intercept: Let $y = 0$
$0 = 7x^2$
$0 = x^2$
$0 = x$

y-intercept: Let $x = 0$
$y = 7(0)^2$
$y = 0$

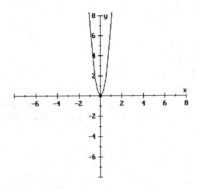

95. $y = -7x^2 + 3$
Vertex: $(0, 3)$
Axis of symmetry: $x = 0$
x-intercepts: Let $y = 0$
$0 = -7x^2 + 3$
$7x^2 = 3$
$x^2 = \dfrac{3}{7}$

$x = \pm\sqrt{\dfrac{3}{7}}$

$x = \pm\dfrac{\sqrt{21}}{7}$

y-intercept: Let $x = 0$
$y = -7(0)^2 + 3$
$ = 3$

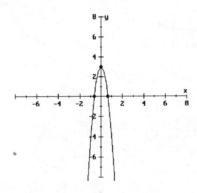

97. $y = x^2 - 6x$
$y = (x^2 - 6x + 9) - 9$
$y = (x - 3)^2 - 9$

Vertex: $(3, -9)$
Axis of symmetry: $x = 3$
x-intercepts: Let $y = 0$
$$0 = (x - 3)^2 - 9$$
$$9 = (x - 3)^2$$
$$\pm 3 = x - 3$$
$x - 3 = 3$ or $x - 3 = -3$
$\quad x = 6$ or $\quad\quad x = 0$
y-intercept: Let $x = 0$
$$y = 0^2 - 6(0)$$
$$= 0$$

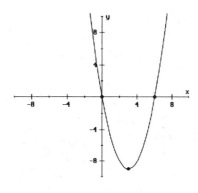

99. $y = x^2 - 2x - 8$
$\quad y = (x^2 - 2x + 1) - 8 - 1$
$\quad y = (x - 1)^2 - 9$
Vertex: $(1, -9)$
Axis of symmetry: $x = 1$
x-intercepts: Let $y = 0$
$$0 = (x - 1)^2 - 9$$
$$9 = (x - 1)^2$$
$$\pm 3 = x - 1$$
$x - 1 = 3$ or $x - 1 = -3$
$\quad x = 4$ or $\quad\quad x = -2$
y-intercept: Let $x = 0$
$$y = 0^2 - 2(0) - 8$$
$$= -8$$

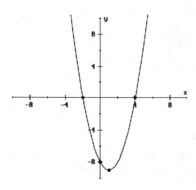

101. $y = -x^2 + 2x - 5$
$\quad y = -(x^2 - 2x + 1) - 5 + 1$
$\quad y = -(x - 1)^2 - 4$
Vertex: $(1, -4)$
Axis of symmetry: $x = 1$
x-intercepts: Let $y = 0$
$$0 = -(x - 1)^2 - 4$$
$$(x - 1)^2 = -4$$
$$\text{none}$$
y-intercept: Let $x = 0$
$$y = -0^2 + 2(0) - 5$$
$$= -5$$

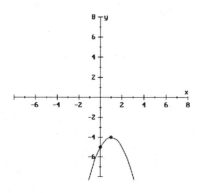

103. $(x - 2)(x + 1) > 0$
$\quad x - 2 = 0 \quad\quad x + 1 = 0$
$\quad\quad x = 2 \quad\quad\quad x = -1$
The intervals are $x < -1$, $-1 < x < 2$, $x > 2$

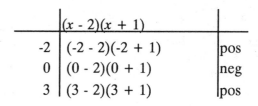

	$(x - 2)(x + 1)$	
-2	$(-2 - 2)(-2 + 1)$	pos
0	$(0 - 2)(0 + 1)$	neg
3	$(3 - 2)(3 + 1)$	pos

We want $(x - 2)(x + 1) > 0$: pos
Solution: $x < -1$ or $x > 2$

105. $(3x + 1)(x - 2) \le 0$
$\quad 3x + 1 = 0 \quad\quad x - 2 = 0$
$\quad\quad x = -\dfrac{1}{3} \quad\quad x = 2$

The intervals are $x < -\dfrac{1}{3}$, $-\dfrac{1}{3} < x < 2$, $x > 2$

Test points:

	$(3x + 1)(x - 2)$	
-1	$[3(-1) + 1](-1 - 2)$	pos
0	$(3 \cdot 0 + 1)(0 - 2)$	neg
3	$(3 \cdot 3 + 1)(3 - 2)$	pos

We want $(3x + 1)(x - 2) \le 0$ neg or 0

Solution: $-\dfrac{1}{3} \le x \le 2$

107. $y^2 - 5y + 4 > 0$
$(y - 4)(y - 1) > 0$
$y - 4 = 0 \qquad y - 1 = 0$
$\qquad y = 4 \qquad\qquad y = 1$

The intervals are $y < 1,\ 1 < y < 4,\ y > 4$

Test points:

	$(y - 4)(y - 1)$	
0	$(0 - 4)(0 - 1)$	pos
2	$(2 - 4)(2 - 1)$	neg
5	$(5 - 4)(5 - 1)$	pos

We want $(y - 4)(y - 1) > 0$; pos

Solution: $y < 1$ or $y > 4$

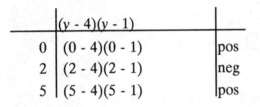

109. $a^2 < 81$
$a^2 - 81 < 0$
$(a - 9)(a + 9) < 0$
$a - 9 = 0 \qquad a + 9 = 0$
$\qquad a = 9 \qquad\qquad a = -9$

The intervals are $a < -9,\ -9 < a < 9,\ a > 9$

Test points:

	$(a - 9)(a + 9)$	
-10	$(-10 - 9)(-10 + 9)$	pos
0	$(0 - 9)(0 + 9)$	neg
10	$(10 - 9)(10 + 9)$	pos

We want $(a - 9)(a + 9) < 0$; neg

$-9 < a < 9$

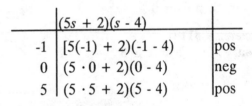

111. $5s^2 - 18s \ge 8$
$5s^2 - 18s - 8 \ge 0$
$(5s + 2)(s - 4) \ge 0$
$5s + 2 = 0 \qquad s - 4 = 0$
$\qquad s = -\dfrac{2}{5} \qquad\qquad s = 4$

The intervals are $s < -\dfrac{2}{5},\ -\dfrac{2}{5} < s < 4,\ s > 4$

Test points:

	$(5s + 2)(s - 4)$	
-1	$[5(-1) + 2](-1 - 4)$	pos
0	$(5 \cdot 0 + 2)(0 - 4)$	neg
5	$(5 \cdot 5 + 2)(5 - 4)$	pos

We want $(5s + 2)(s - y) \ge 0$; pos or 0

Solution: $s \le -\dfrac{2}{5}$ or $s \ge 4$

113. $\dfrac{x-3}{x+2} < 0$

$x - 3 = 0 \qquad x + 2 = 0$
$\qquad x = 3 \qquad\quad x = -2$

The intervals are $x < -2,\ -2 < x < 3,\ x > 3$
Test points:

	$(x-3)/(x+2)$	
-3	$(-3 - 3)/(-3 + 2)$	pos
0	$(0 - 3)/(0 + 2)$	neg
4	$(4 - 3)/(4 + 2)$	pos

We want $\dfrac{x-3}{x+2} < 0$; neg

Solution: $-2 < x < 3$

115. $\dfrac{x-3}{x+2} \geq 0$; pos or 0
(See exercise #113)
Solution: $x < -2$ or $x \geq 3$

117. $\dfrac{2x+1}{x-3} < 2$

$\dfrac{2x+1}{x-3} - 2 < 0$

$\dfrac{2x + 1 - 2(x - 3)}{x - 3} < 0$

$\dfrac{4}{x - 3} < 0$

$x - 3 = 0$
$\qquad x = 3$

The intervals are $x < 3,\ x > 3$
Test points:

	$4/(x - 3)$	
0	$4/(0 - 3)$	neg
4	$4/(4 - 3)$	pos

We want $\dfrac{4}{x-3} < 0$; neg

Solution: $x < 3$

119. $\dfrac{5}{x+4} \geq 4$

$\dfrac{5}{x+4} - 4 \geq 0$

$\dfrac{5 - 4(x + 4)}{x + 4} \geq 0$

$\dfrac{-4x - 11}{x + 4} \geq 0$

$-4x - 11 = 0 \qquad x + 4 = 0$
$\qquad x = -\dfrac{11}{4} \qquad\quad x = -4$

The intervals are $x < -4,\ -4 < x < -\dfrac{11}{4},\ x > -\dfrac{11}{4}$

Test points:

	$(-4x - 11)/(x + 4)$	
-5	$(-4(-5) - 11)/(5 + 4)$	neg
-3	$(-4(-3) - 11)/(-3 + 4)$	pos
0	$(-4(0) - 11)/(0 + 4)$	neg

We want $\dfrac{-4x - 11}{x+4} \geq 0$; pos or 0

Solution: $-4 < x \leq -\dfrac{11}{4}$

121. $d = \sqrt{(2 - 0)^2 + (6 - 0)^2}$
$\qquad = \sqrt{2^2 + 6^2}$
$\qquad = \sqrt{4 + 36}$
$\qquad = \sqrt{40}$
$\qquad = 2\sqrt{10}$

midpoint $= \left(\dfrac{0 + 2}{2}, \dfrac{0 + 6}{2}\right) = (1, 3)$

123. $d = |2 - (-2)| = 4$

$$\text{midpoint} = \left(\frac{2 - 2}{2}, \frac{5 + 5}{2}\right)$$

$$= (0, 5)$$

125. $d = \sqrt{(6 - 4)^2 + (-4 - (-6)]^2}$

$\quad = \sqrt{2^2 + 2^2}$

$\quad = \sqrt{4 + 4}$

$\quad = \sqrt{8}$

$\quad = 2\sqrt{2}$

$$\text{midpoint} = \left(\frac{6 + 4}{2}, \frac{-4 - 6}{2}\right)$$

$$= (5, -5)$$

127. $d = \sqrt{(-2 - 2)^2 + (-5 - 5)^2}$

$\quad = \sqrt{(-4)^2 + (-10)^2}$

$\quad = \sqrt{16 + 100}$

$\quad = \sqrt{116}$

$\quad = 2\sqrt{29}$

$$\text{midpoint} = \left(\frac{-2 + 2}{2}, \frac{-5 + 5}{2}\right)$$

$$= (0, 0)$$

129. $d(P, Q) = \sqrt{(5 - 3)^2 + (9 - 6)^2}$

$\qquad\quad = \sqrt{4 + 9}$

$\qquad\quad = \sqrt{13}$

$d(Q, R) = \sqrt{(8 - 5)^2 + (7 - 9)^2}$

$\qquad\quad = \sqrt{9 + 4}$

$\qquad\quad = \sqrt{13}$

$d(P, R) = \sqrt{(8 - 3)^2 + (7 - 6)^2}$

$\qquad\quad = \sqrt{25 + 1}$

$\qquad\quad = \sqrt{26}$

$[d(P, Q)]^2 + [d(Q, R)]^2 = [d(P, R)]^2$

$\left(\sqrt{13}\right)^2 + \left(\sqrt{13}\right)^2 = \left(\sqrt{26}\right)^2$

$\qquad\qquad 13 + 13 = 26$

$\qquad\qquad\qquad 26 = 26$

$\qquad\qquad\qquad$ True

Yes, they form a right triangle.

131. $x^2 + y^2 = 100$

Center: $(0, 0)$

$r = \sqrt{100} = 10$

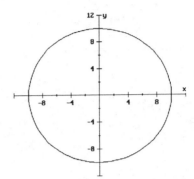

133. $\quad x^2 + y^2 - 4x - 14y = -52$

$\quad x^2 - 4x + y^2 - 14y = -52$

$(x^2 - 4x + 4) + (y^2 - 14y + 49) = -52 + 4 + 49$

$\quad (x - 2)^2 + (y - 7)^2 = 1$

Center: $(2, 7)$

$r = \sqrt{1} = 1$

135. $\quad x^2 + y^2 - 6x + 4y = 68$

$\quad x^2 - 6x + y^2 + 4y = 68$

$(x^2 - 6x + 9) + (y^2 + 4y + 4) = 68 + 9 + 4$

$\quad (x - 3)^2 + (y + 2)^2 = 81$

Center: $(3, -2)$

$r = \sqrt{81} = 9$

137. Center = midpoint

$$= \left(\frac{-2 + 6}{2}, \frac{4 + 8}{2}\right)$$

$$= (2, 6)$$

$r = \sqrt{(6 - 2)^2 + (8 - 6)^2}$

$\quad = \sqrt{16 + 4}$

$\quad = \sqrt{20}$

$r^2 = \left(\sqrt{20}\right)^2 = 20$

$(x - h)^2 + (y - k)^2 = r^2$

$(x - 2)^2 + (y - 6)^2 = 20$

139. $P(x) = 10000(-x^2 + 12x - 35)$

$$-\frac{b}{2a} = -\frac{12}{2(-1)}$$

$$= 6$$

$6 per ticket would produce the maximum profit.

141. Let x = the number
$$x^2 + 4 = 36$$
$$x^2 = 32$$
$$x = \pm\sqrt{32}$$
$$x = \pm 4\sqrt{2}$$
The numbers are $\pm 4\sqrt{2}$.

143.   Let x = the number

$$x + \frac{1}{x} = \frac{53}{14}$$

$$14x\left(x + \frac{1}{x}\right) = 14x\left(\frac{53}{14}\right)$$

$$14x^2 + 14 = 53x$$
$$14x^2 - 53x + 14 = 0$$
$$(2x - 7)(7x - 2) = 0$$
$$2x - 7 = 0 \quad \text{or} \quad 7x - 2 = 0$$
$$2x = 7 \qquad\qquad 7x = 2$$
$$x = \frac{7}{2} \qquad\qquad x = \frac{2}{7}$$

The numbers are $\frac{7}{2}$ and $\frac{2}{7}$.

145. width: x
      length: $2x$

$$A = wl$$
$$50 = x(2x)$$
$$50 = 2x^2$$
$$25 = x^2$$
$$5 = x$$
$$2x = 2(5) = 10$$

The dimensions are 5 ft by 10 ft.

147.

$$(8 + 2x)(5 + 2x) - (8)(5) = 114$$
$$40 + 26x + 4x^2 - 40 = 114$$
$$26x + 4x^2 = 114$$
$$4x^2 + 26x - 114 = 0$$
$$2x^2 + 13x - 57 = 0$$
$$(2x + 19)(x - 3) = 0$$
$$2x + 19 = 0 \quad \text{or} \quad x - 3 = 0$$
$$x = -\frac{19}{2} \quad \text{or} \quad x = 3$$

Since $x > 0$, $x = 3$. The width of the frame is $3''$.

149. $a^2 + b^2 = c^2$
$$5^2 + 15^2 = c^2$$
$$25 + 225 = c^2$$
$$250 = c^2$$
$$5\sqrt{10} = c$$
The hypotenuse is $5\sqrt{10}''$.

151. $a^2 + b^2 = c^2$
$$5^2 + 4^2 = c^2$$
$$25 + 16 = c^2$$
$$41 = c^2$$
$$\sqrt{41} = c$$
The length of the diagonal is $\sqrt{41} \approx 6.4''$.

153. Let x = rate of the wind

then $200 - x$ = rate into the wind
and $200 + x$ = rate with the wind

(time into the wind) + (time with the wind) = $\dfrac{25}{8}$

$$\frac{300}{200-x} + \frac{300}{200+x} = \frac{25}{8}$$

$$8(200-x)(200+x)\left(\frac{300}{200-x} + \frac{300}{200+x}\right) = 8(200-x)(200+x)\left(\frac{25}{8}\right)$$

$$2400(200 + x) + 2400(200 - x) = 25(40000 - x^2)$$
$$480000 + 2400x + 480000 - 2400x = 1000000 - 25x^2$$

$$960000 = 1000000 - 25x^2$$
$$-40000 = -25x^2$$
$$1600 = x^2$$
$$40 = x$$
The wind's rate is 40 mph.

CHAPTER 8 PRACTICE TEST

1.

(a) $(3z - 1)(z - 4) = z^2 - 8z + 7$
$$3z^2 - 13z + 4 = z^2 - 8z + 7$$
$$2z^2 - 5z - 3 = 0$$
$$(2z + 1)(z - 3) = 0$$
$$2z + 1 = 0 \quad \text{or} \quad z - 3 = 0$$
$$2z = -1 \qquad\qquad z = 3$$
$$z = -\frac{1}{2} \quad \text{or} \quad z = 3$$

(b) $3 + \dfrac{5}{x^2} = 4$

$$\frac{5}{x^2} = 1$$

$$5 = x^2$$

$$\pm\sqrt{5} = x$$

2.

(a) $y^2 + 4y - 1 = 0$

$$y = \frac{-4 \pm \sqrt{4^2 - 4(1)(-1)}}{2(1)}$$

$$= \frac{-4 \pm \sqrt{16 + 4}}{2}$$

$$= \frac{-4 \pm \sqrt{20}}{2}$$

$$= \frac{-4 \pm 2\sqrt{5}}{2}$$

$$= \frac{2(-2 \pm \sqrt{5})}{2}$$

$$= 2\frac{(-2 \pm \sqrt{5})}{2}$$

$$= -2 \pm \sqrt{5}$$

(b) $(a - 2)(a + 1) = 3a^2 - 4$
$$a^2 - a - 2 = 3a^2 - 4$$
$$0 = 2a^2 + a - 2$$

$$a = \frac{-1 \pm \sqrt{1^2 - 4(2)(-2)}}{2(2)}$$

$$= \frac{-1 \pm \sqrt{1 + 16}}{4}$$

$$= \frac{-1 \pm \sqrt{17}}{4}$$

(c) $\dfrac{3}{x - 2} + \dfrac{3x}{x + 2} = \dfrac{66}{x^2 - 4}$

$$\frac{3}{x - 2} + \frac{3x}{x + 2} = \frac{66}{(x-2)(x+2)}$$

$$(x-2)(x+2)\left(\frac{3}{x-2} + \frac{3x}{x+2}\right) = (x-2)(x+2)\left[\frac{66}{(x-2)(x+2)}\right]$$

$$3(x + 2) + 3x(x - 2) = 66$$
$$3x + 6 + 3x^2 - 6x = 66$$
$$3x^2 - 3x + 6 = 66$$
$$3x^2 - 3x - 60 = 0$$
$$x^2 - x - 20 = 0$$
$$(x - 5)(x + 4) = 0$$
$$x - 5 = 0 \quad \text{or} \quad x + 4 = 0$$
$$x = 5 \quad \text{or} \qquad x = -4$$

(d) $-4.8x^2 = 3.1x - 5.1$
$$0 = 4.8x^2 + 3.1x - 5.1$$

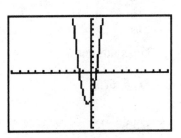

$$x = -1.40 \quad \text{or} \quad x = 0.76$$

3. $2x^2 - 3x + 5 = 0$

$$b^2 - 4ac = (-3)^2 - 4(2)(5)$$
$$= -3 < 0$$
The roots are not real.

4. $\quad V = 2\pi r^2 h$

$$\frac{V}{2\pi h} = r^2$$

$$\sqrt{\frac{V}{2\pi h}} = r$$

$$\frac{\sqrt{2V\pi h}}{2\pi h} = r$$

	$(x - 9)(x + 4)$	
-5	$(-5 - 9)(-5 + 4)$	pos
0	$(0 - 9)(0 + 4)$	neg
10	$(10 - 9)(10 + 4)$	pos

We want $(x - 9)(x + 4) \geq 0$; pos or 0
Solution: $x \leq -4$ or $x \geq 9$

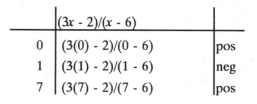

5. $\quad$
$$\sqrt{3x} = 2 + \sqrt{x + 4}$$
$$(\sqrt{3x})^2 = (2 + \sqrt{x + 4})^2$$
$$3x = 4 + 4\sqrt{x + 4} + x + 4$$
$$3x = 8 + x + 4\sqrt{x + 4}$$
$$2x - 8 = 4\sqrt{x + 4}$$
$$x - 4 = 2\sqrt{x + 4}$$
$$(x - 4)^2 = (2\sqrt{x + 4})^2$$
$$x^2 - 8x + 16 = 4(x + 4)$$
$$x^2 - 8x + 16 = 4x + 16$$
$$x^2 - 12x = 0$$
$$x(x - 12) = 0$$
$$x = 0 \quad \text{or} \quad x - 12 = 0$$
$$x = 12$$

When you check $x = 0$, you find it to be extraneous. The only solution is $x = 12$.

6. $\quad x^{1/2} - x^{1/4} - 20 = 0$
$(x^{1/4})^2 - x^{1/4} - 20 = 0$
Let $u = x^{1/4}$.
$u^2 - u - 20 = 0$
$(u - 5)(u + 4) = 0$
$u - 5 = 0 \quad$ or $\quad u + 4 = 0$
$\quad u = 5 \qquad\qquad u = -4$
$\quad x^{1/4} = 5 \qquad\quad x^{1/4} = -4$
$(x^{1/4})^4 = 5^4 \qquad$ No solution
$\quad x = 625$
The only solution is $x = 625$.

(b) $\quad \dfrac{3x - 2}{x - 6} < 0$

$$3x - 2 = 0 \qquad x - 6 = 0$$
$$x = \frac{2}{3} \qquad\qquad x = 6$$

The intervals are $x < \dfrac{2}{3}$, $\dfrac{2}{3} < x < 6$, $x > 6$
Test points:

	$(3x - 2)/(x - 6)$	
0	$(3(0) - 2)/(0 - 6)$	pos
1	$(3(1) - 2)/(1 - 6)$	neg
7	$(3(7) - 2)/(7 - 6)$	pos

We want $\dfrac{3x - 2}{x - 6} < 0$; neg

Solution: $\dfrac{2}{3} < x < 6$

7. (a) $\quad x^2 - 5x \geq 36$
$x^2 - 5x - 36 \geq 0$
$(x - 9)(x + 4) \geq 0$
$x - 9 = 0 \qquad x + 4 = 0$
$\quad x = 9 \qquad\qquad x = -4$
The intervals are $x < -4$, $-4 < x < 9$, $x > 9$
Test points:

8. Let $x =$ length of diagonal

$$9^2 + 9^2 = x^2$$
$$81 + 81 = x^2$$
$$162 = x^2$$
$$\sqrt{162} = x$$
$$9\sqrt{2} = x$$

The diagonal is $9\sqrt{2} \approx 12.7$ ft.

9. Let x = rate of the boat in still water,

then $x - 3$ = rate upstream
and $x + 3$ = rate downstream

$$\frac{15}{x-3} + \frac{12}{x+3} = \frac{9}{4}$$

$$4(x-3)(x+3)\left(\frac{15}{x-3} + \frac{12}{x+3}\right) = 4(x-3)(x+3)\left(\frac{9}{4}\right)$$

$$60(x + 3) + 48(x - 3) = 9(x^2 - 9)$$
$$60x + 180 + 48x - 144 = 9x^2 - 81$$
$$108x + 36 = 9x^2 - 81$$
$$0 = 9x^2 - 108x - 117$$
$$0 = x^2 - 12x - 13$$
$$0 = (x - 13)(x + 1)$$
$$x - 13 = 0 \quad \text{or} \quad x + 1 = 0$$
$$x = 13 \qquad\qquad x = -1$$

Since $x > 0$, $x = 13$
The boat's rate is 13 mph.

10. (a) $y = 2x^2 + 3x + 1$

$$\frac{-b}{2a} = \frac{-3}{2(2)} = -\frac{3}{4}$$

$$y = 2\left(-\frac{3}{4}\right)^2 + 3\left(-\frac{3}{4}\right) + 1$$

$$= -\frac{1}{8}$$

Vertex: $\left(-\frac{3}{4}, -\frac{1}{8}\right)$

Axis of symmetry: $x = -\frac{3}{4}$

x-intercepts: Let $y = 0$
$$0 = 2x^2 + 3x + 1$$
$$0 = (2x + 1)(x + 1)$$
$$2x + 1 = 0 \quad \text{or} \quad x + 1 = 0$$
$$x = -\frac{1}{2} \quad \text{or} \qquad x = -1$$

y-intercept: Let $x = 0$
$$y = 2(0)^2 + 3(0) + 1$$
$$= 1$$

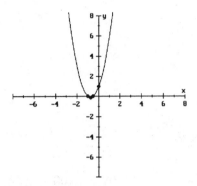

(b) $y = -2x^2 + x - 3$

$$y = -2\left(x^2 - \frac{1}{2}x\right) - 3$$

$$y = -2\left(x^2 - \frac{1}{2}x + \frac{1}{16}\right) - 3 + \frac{1}{8}$$

$$y = -2\left(x - \frac{1}{4}\right)^2 - \frac{23}{8}$$

Vertex: $\left(\frac{1}{4}, -\frac{23}{8}\right)$

Axis of symmetry: $x = \frac{1}{4}$

x-intercepts: Let $y = 0$

$$0 = -2\left(x - \frac{1}{4}\right)^2 - \frac{23}{8}$$

$$2\left(x - \frac{1}{4}\right)^2 = -\frac{23}{8}$$

$$\left(x - \frac{1}{4}\right)^2 = -\frac{23}{16}$$

y-intercept: Let $x = 0$
$$y = -2(0)^2 + 0 - 3 = -3$$

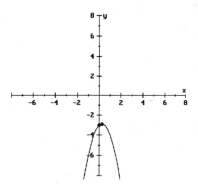

11. $C(x) = \dfrac{-x^2}{10} + 100x - 24000$

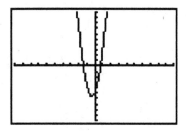

maximum: (500, 1000)

500 widgets would produce the maximum cost of $1000.

12. $d = \sqrt{(-4 - 2)^2 + (7 - 5)^2}$
$ = \sqrt{(-6)^2 + 2^2}$
$ = \sqrt{36 + 4}$
$ = \sqrt{40}$
$ = 2\sqrt{10}$

$\text{midpoint} = \left(\dfrac{-4 + 2}{2}, \dfrac{7 + 5}{2} \right)$

$\phantom{\text{midpoint}} = (-1, 6)$

13.
$$x^2 + y^2 - 8x + 12y = -2$$
$$x^2 - 8x + y^2 + 12y = -2$$
$$(x^2 - 8x + 16) + (y^2 + 12y + 36) = -2 + 16 + 36$$
$$(x - 4)^2 + (y + 6)^2 = 50$$
Center: (4, -6)
$r = \sqrt{50} = 5\sqrt{2}$

9.1 Exercises

1. $f(x) = 7x - 2$
   $f(-2) = 7(-2) - 2 = -16$
   $f(0) = 7(0) - 2 = -2$
   $f(2) = 7(2) - 2 = 12$

3. $f(x) = \sqrt{3x - 2}$
   -2 and 0 are not in the domain of $f(x)$.
   $f(2) = \sqrt{3(2) - 2} = \sqrt{4} = 2$

5. $f(x) = \sqrt{2 - x}$
   $f(-2) = \sqrt{2 - (-2)}$
   $\quad = \sqrt{4}$
   $\quad = 2$
   $f(0) = \sqrt{2 - 0}$
   $\quad = \sqrt{2}$
   $f(2) = \sqrt{2 - 2}$
   $\quad = \sqrt{0}$
   $\quad = 0$

7. $f(x) = \dfrac{2}{x}$
   0 is not in the domain of $f(x)$.
   $f(-2) = \dfrac{2}{-2} = -1$
   $f(2) = \dfrac{2}{2} = 1$

9. $f(x) = \dfrac{x + 2}{x - 2}$
   2 is not in the domain of $f(x)$.
   $f(-2) = \dfrac{-2 + 2}{-2 - 2}$

   $\quad = \dfrac{0}{-4}$

   $\quad = 0$

   $f(0) = \dfrac{0 + 2}{0 - 2}$

   $\quad = -1$

11. all real numbers

13. all real numbers

15. all real numbers

17. $2x - 1 \geq 0$
    $2x \geq 1$
    $x \geq \dfrac{1}{2}$
    $\left\{ x \mid x \geq \dfrac{1}{2} \right\}$

19. $2 - 3x \geq 0$
    $-3x \geq -2$
    $x \leq \dfrac{2}{3}$
    $\left\{ x \mid x \leq \dfrac{2}{3} \right\}$

21. $\{x \mid x \neq 0\}$

23. $x - 4 \neq 0$
    $x \neq 4$
    $\{x \mid x \neq 4\}$

25. $x - 1 \neq 0 \qquad x + 2 \neq 0$
    $\qquad x \neq 1 \qquad\qquad x \neq -2$
    $\{x \mid x \neq 1, -2\}$

27. $x^2 - x - 12 \neq 0$
    $(x - 4)(x + 3) \neq 0$
    $x - 4 \neq 0 \qquad x + 3 \neq 0$
    $\qquad x \neq 4 \qquad\qquad x \neq -3$
    $\{x \mid x \neq -3, 4\}$

29. $\{x \mid x > 0\}$

31. $2x - 1 > 0$
    $2x > 1$
    $x > \dfrac{1}{2}$
    $\left\{ x \mid x > \dfrac{1}{2} \right\}$

33. $f(x) = 2x - 3$

    (a) $\qquad f(5) = 2(5) - 3 = 7$
    $f(x) + f(5) = 2x - 3 + 7$
    $\qquad\qquad = 2x + 4$

    (b) $f(x + 5) = 2(x + 5) - 3$
    $\qquad\qquad = 2x + 10 - 3$
    $\qquad\qquad = 2x + 7$

(c) $f(2x) = 2(2x) - 3$
$= 4x - 3$

(d) $2f(x) = 2(2x - 3)$
$= 4x - 6$

(e) $\quad f(x + 5) - f(x) \quad \text{(see part b)}$
$= (2x + 7) - (2x - 3)$
$= 2x + 7 - 2x + 3$
$= 10$

(f) $\dfrac{f(x + 5) - f(x)}{5} = \dfrac{10}{5} = 2$

(see part e)

(g) $f(x + h) = 2(x + h) - 3$
$= 2x + 2h - 3$

(h) $\qquad f(h) = 2h - 3$
$f(x) + f(h) = (2x - 3) + (2h - 3)$
$= 2x + 2h - 6$

(i) $\dfrac{f(x + h) - f(x)}{h}$

$= \dfrac{2x + 2h - 3 - (2x - 3)}{h} \quad \text{(see part g)}$

$= \dfrac{2x + 2h - 3 - 2x + 3}{h}$

$= \dfrac{2h}{h}$

$= 2$

35. $f(x) = \dfrac{5}{x - 1}$

(a) $f(3) = \dfrac{5}{3 - 1} = \dfrac{5}{2}$

$f(x) + f(3) = \dfrac{5}{x - 1} + \dfrac{5}{2}$

$= \dfrac{5(2) + 5(x - 1)}{2(x - 1)}$

$= \dfrac{10 + 5x - 5}{2x - 2}$

$= \dfrac{5x + 5}{2x - 2}$

(b) $f(x + 3) = \dfrac{5}{x + 3 - 1}$

$= \dfrac{5}{x + 2}$

(c) $f(3x) = \dfrac{5}{3x - 1}$

(d) $3f(x) = 3\left(\dfrac{5}{x - 1}\right)$

$= \dfrac{15}{x - 1}$

(e) $f(x + 3) - f(x) \quad \text{(see part b)}$

$= \dfrac{5}{x + 2} - \dfrac{5}{x - 1}$

$= \dfrac{5(x - 1) - 5(x + 2)}{(x + 2)(x - 1)}$

$= \dfrac{5x - 5 - 5x - 10}{(x + 2)(x - 1)}$

$= -\dfrac{15}{(x + 2)(x - 1)}$

(f) $\dfrac{f(x + 3) - f(x)}{3} \quad \text{(see part e)}$

$= \dfrac{\dfrac{-15}{(x + 2)(x - 2)}}{3}$

$= \dfrac{-15}{3(x + 2)(x - 1)}$

$= \dfrac{-5}{(x + 2)(x - 1)}$

(g) $f(x + h)$

$= \dfrac{5}{x + h - 1}$

(h) $f(h) = \dfrac{5}{h-1}$

$f(x) + f(h)$

$= \dfrac{5}{x-1} + \dfrac{5}{h-1}$

$= \dfrac{5(h-1) + 5(x-1)}{(x-1)(h-1)}$

$= \dfrac{5h - 5 + 5x - 5}{(x-1)(h-1)}$

$= \dfrac{5x + 5h - 10}{(x-1)(h-1)}$

(i) $\dfrac{f(x+h) - f(x)}{h}$ (see part g)

$= \dfrac{\dfrac{5}{x+h-1} - \dfrac{5}{x-1}}{h}$

$= \dfrac{\dfrac{5(x-1) - 5(x+h-1)}{(x-1)(x+h-1)}}{h}$

$= \dfrac{5x - 5 - 5x - 5h + 5}{h(x-1)(x+h-1)}$

$= \dfrac{-5h}{h(x-1)(x+h-1)}$

$= \dfrac{-5}{(x-1)(x+h-1)}$

37. $f(x) = \begin{cases} 2x + 3 & \text{if } -4 \le x < 2 \\ 6 - x & \text{if } x \ge 2 \end{cases}$

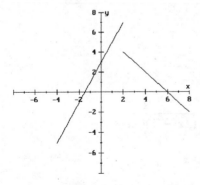

$f(-1) = 2(-1) + 3 = 1$
$f(2) = 6 - 2 = 4$
$f(5) = 6 - 5 = 1$

39. $h(x) = \begin{cases} x^2 - 1 & \text{if } -3 \le x \le 3 \\ 14 - 2x & \text{if } x > 3 \end{cases}$

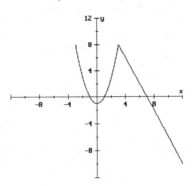

$h(-3) = (-3)^2 - 1 = 8$
$h(0) = 0^2 - 1 = -1$
$h(4) = 14 - 2(4) = 6$

43. $\dfrac{x}{x+1} - \dfrac{3}{x-1} + 1$

$= \dfrac{x(x-1)}{(x+1)(x-1)} - \dfrac{3(x+1)}{(x+1)(x-1)} + \dfrac{(x+1)(x-1)}{(x+1)(x-1)}$

$= \dfrac{x(x-1) - 3(x+1) + (x+1)(x-1)}{(x+1)(x-1)}$

$= \dfrac{x^2 - x - 3x - 3 + x^2 - 1}{(x+1)(x-1)}$

$= \dfrac{2x^2 - 4x - 4}{(x+1)(x-1)}$

45. Let x = rate going
then $x + 12$ = rate returning

$$\dfrac{60}{x+12} = \dfrac{60}{x} - \dfrac{1}{4}$$

$$4x(x+12)\left(\dfrac{60}{x+12}\right) = 4x(x+12)\left(\dfrac{60}{x} - \dfrac{1}{4}\right)$$

$240x = 240(x+12) - x(x+12)$
$240x = 240x + 2880 - x^2 - 12x$
$x^2 + 12x - 2880 = 0$
$(x+60)(x-48) = 0$
$x + 60 = 0 \quad$ or $\quad x - 48 = 0$
$x = -60 \quad$ or $\qquad x = 48$

Since $x > 0$, $\ x = 48$
$$x = 48 \qquad \text{going rate}$$
$$x + 12 = 48 + 12 = 60 \quad \text{returning rate}$$

$$\frac{60}{x} = \frac{60}{48} = 1\frac{1}{4} \qquad \text{going time}$$

$$\frac{60}{x + 12} = \frac{60}{60} = 1 \qquad \text{returning time}$$

9.2 Exercises

1. $(f + g)(0) = f(0) + g(0)$
    $$= [0^2 - 2(0) - 3] + (0 - 1)$$
    $$= -4$$

3. $(g - f)(2) = g(2) - f(2)$
    $$= (2 - 1) - [2^2 - 2(2) - 3]$$
    $$= 1 - (-3)$$
    $$= 4$$

5. $\left(\dfrac{f}{g}\right)(3) = \dfrac{f(3)}{g(3)}$
    $$= \frac{3^2 - 2(3) - 3}{3 - 1}$$
    $$= \frac{0}{2}$$
    $$= 0$$

7. $\left(\dfrac{f}{g}\right)(1) = \dfrac{f(1)}{g(1)}$
    $$= \frac{1^2 - 2(1) - 3}{1 - 1}$$
    $$= \frac{-4}{0}$$

    undefined

9. $(f + g)(x) = f(x) + g(x)$
    $$= (x^2 - 2x - 3) + (x - 1)$$
    $$= x^2 - x - 4$$

11. $\left(\dfrac{g}{f}\right) = \dfrac{g(x)}{f(x)}$
    $$= \frac{x - 1}{x^2 - 2x - 3}$$

13. $(h + g)(2) = h(2) + g(2)$
    $$= \frac{1}{2(2) - 1} + 2(2) + 1$$
    $$= \frac{1}{3} + 5$$
    $$= \frac{16}{3}$$

15. $(g - h)\left(\dfrac{1}{2}\right) = g\left(\dfrac{1}{2}\right) - h\left(\dfrac{1}{2}\right)$
    $$= 2\left(\frac{1}{2}\right) + 1 - \frac{1}{2\left(\dfrac{1}{2}\right) - 1}$$
    $$= 2 - \frac{1}{0}$$

    undefined

17. $\left(\dfrac{h}{g}\right)(3) = \dfrac{h(3)}{g(3)}$
    $$= \frac{\dfrac{1}{2(3) - 1}}{2(3) + 1}$$
    $$= \frac{\dfrac{1}{5}}{7}$$
    $$= \frac{1}{35}$$

19. $\left(\dfrac{h}{g}\right)(x) = \dfrac{h(x)}{g(x)}$
    $$= \frac{\dfrac{1}{2x - 1}}{2x + 1}$$
    $$= \frac{1}{(2x - 1)(2x + 1)}$$

21. $g(3) = \sqrt{3 + 1} = 2$
    $$f[g(3)] = [g(3)]^2 - 4$$
    $$= 2^2 - 4$$
    $$= 0$$

23. $f(3) = 3^2 - 4 = 5$
$$g[f(3)] = \sqrt{f(3) + 1}$$
$$= \sqrt{5 + 1}$$
$$= \sqrt{6}$$

25. $f[g(x)] = [g(x)]^2 - 4$
$$= \left(\sqrt{x + 1}\right)^2 - 4$$
$$= x + 1 - 4$$
$$= x - 3$$

27. $g[f(x)] = \sqrt{f(x) + 1}$
$$= \sqrt{x^2 - 4 + 1}$$
$$= \sqrt{x^2 - 3}$$

29. $h(3) = \dfrac{1}{3}$

$g[h(3)] = \sqrt{h(3) + 1}$

$$= \sqrt{\frac{1}{3} + 1}$$

$$= \sqrt{\frac{4}{3}}$$

$$= \frac{2}{\sqrt{3}}$$

$$= \frac{2\sqrt{3}}{3}$$

31. $g\left(\dfrac{1}{2}\right) = \sqrt{\dfrac{1}{2} + 1} = \sqrt{\dfrac{3}{2}} = \dfrac{\sqrt{6}}{2}$

$f\left[g\left(\dfrac{1}{2}\right)\right] = \left[g\left(\dfrac{1}{2}\right)\right]^2 - 4$

$$= \left(\frac{\sqrt{6}}{2}\right)^2 - 4$$

$$= \frac{3}{2} - 4$$

$$= -\frac{5}{2}$$

33. $g[h(x)] = \sqrt{h(x) + 1}$

$$= \sqrt{\frac{1}{x} + 1}$$

$$= \sqrt{\frac{1 + x}{x}}$$

$$= \frac{\sqrt{x + x^2}}{x}$$

35. $f(1) = 1^2 - 4 = -3$

This is not in the domain of $g(x)$, hence $x = 1$ is not in the domain of $g[f(x)]$.

37. $g(2) = 3(2) - 1 = 5$
$f[g(2)] = [g(2)]^2 - 3[g(2)] + 5$
$$= 5^2 - 3(5) + 5$$
$$= 15$$

39. $g(-1) = 3(-1) - 1 = -4$
$f[g(-1)] = [g(-1)]^2 - 3[g(-1)] + 5$
$$= (-4)^2 - 3(-4) + 5$$
$$= 33$$

41. $h\left(\dfrac{1}{3}\right) = \dfrac{\dfrac{1}{3} - 1}{4 - 2\left(\dfrac{1}{3}\right)}$

$$= \frac{-\dfrac{2}{3}}{\dfrac{10}{3}}$$

$$= -\frac{1}{5}$$

$f\left[h\left(\dfrac{1}{3}\right)\right] = \left[h\left(\dfrac{1}{3}\right)\right]^2 - 3\left[h\left(\dfrac{1}{3}\right)\right] + 5$

$$= \left(-\frac{1}{5}\right)^2 - 3\left(-\frac{1}{5}\right) + 5$$

$$= \frac{141}{25}$$

43. $h(2) = \dfrac{2 - 1}{4 - 2(2)} = \dfrac{1}{0}$

2 is not in the domain of $g[h(x)]$.

45. $f\left(\dfrac{1}{2}\right) = \left(\dfrac{1}{2}\right)^2 - 3\left(\dfrac{1}{2}\right) + 5 = \dfrac{15}{4}$

$g\left[f\left(\dfrac{1}{2}\right)\right] = 3\left[f\left(\dfrac{1}{2}\right)\right] - 1$

$\qquad = 3\left(\dfrac{15}{4}\right) - 1$

$\qquad = \dfrac{41}{4}$

47. $f[g(x)] = [g(x)]^2 - 3[g(x)] + 5$
$\qquad = (3x - 1)^2 - 3(3x - 1) + 5$
$\qquad = 9x^2 - 6x + 1 - 9x + 3 + 5$
$\qquad = 9x^2 - 15x + 9$

49. $g[h(x)] = 3[h(x)] - 1$

$\qquad = 3\left(\dfrac{x - 1}{4 - 2x}\right) - 1$

$\qquad = \dfrac{3(x - 1) - (4 - 2x)}{4 - 2x}$

$\qquad = \dfrac{3x - 3 - 4 + 2x}{4 - 2x}$

$\qquad = \dfrac{5x - 7}{4 - 2x}$

51. $g[g(x)] = 3[g(x)] - 1$
$\qquad = 3(3x - 1) - 1$
$\qquad = 9x - 3 - 1$
$\qquad = 9x - 4$

53. (a) $\quad r(t) = 3t$
$\qquad r(2) = 3(2) = 6$

$\qquad A(r) = \pi r^2$
$\qquad A[r(2)] = \pi[r(2)]^2$
$\qquad\qquad = \pi(6)^2$
$\qquad\qquad = 36\pi$
The area is $36\pi \approx 113.10$ sq. in.

(b) $A[r(t)] = \pi[r(t)]^2$
$\qquad\qquad = \pi(3t)^2$
$\qquad\qquad = 9\pi t^2$
The area is $9\pi t^2$ sq. in.

55. (a) $\quad r(t) = 3t$
$\qquad r(5) = 3(5) = 15$

$\qquad V[r(5)] = \dfrac{4}{3}\pi[r(5)]^3$

$\qquad\qquad = \dfrac{4}{3}\pi(15)^3$

$\qquad\qquad = 4500\pi$

The volume is $4500\pi \approx 14137.17$ cu. in.

(b) $\quad V[r(t)] = \dfrac{4}{3}\pi[r(t)]^3$

$\qquad\qquad = \dfrac{4}{3}\pi(3t)^3$

$\qquad\qquad = 36\pi t^3$

The volume is $36\pi t^3$ cu. in.

57. $C = 8P$

59. $A = 1000$
$\quad s^2 = 1000$
$\quad s = \sqrt{1000} = 10\sqrt{10}$

$\quad P = 4s$
$\qquad = 4\left(10\sqrt{10}\right)$
$\qquad = 40\sqrt{10}$

$\quad C = 40x\sqrt{10}$

61. $2w + 2l = 59$
$\qquad 2l = 59 - 2w$

$\qquad l = \dfrac{59}{2} - w$

$\qquad A = wl$

$\qquad 210 = w\left(\dfrac{59}{2} - w\right)$

$\qquad 210 = \dfrac{59}{2}w - w^2$

$\qquad 420 = 59w - 2w^2$

$\qquad 2w^2 - 59w + 420 = 0$
$\qquad (2w - 35)(w - 12) = 0$
$\qquad 2w - 35 = 0 \quad$ or $\quad w - 12 = 0$
$\qquad\quad 2w = 35 \qquad\qquad\quad w = 12$

$\qquad\quad w = \dfrac{35}{2} \quad$ or $\qquad\quad w = 12$

63. $\dfrac{2x^2 - 5x - 12}{x^2 - 7x + 12}$

$= \dfrac{(2x + 3)(x - 4)}{(x - 3)(x - 4)}$

$= \dfrac{2x + 3}{x - 3}$

9.3 Exercises

1. Linear function

3. Quadratic function

5. Polynomial function

7. Quadratic function

9. Square root function

11. Absolute value function

13. Linear function

15. Quadratic function

17. $f(x) = |4x - 1|$
$f(-2) = |4(-2) - 1| = 9$
$f(-1) = |4(-1) - 1| = 5$
$f(0) = |4(0) - 1| = 1$
$f(1) = |4(1) - 1| = 3$
$f(2) = |4(2) - 1| = 7$

19. $f(x) = |4x| - 1$
$f(-2) = |4(-2)| - 1 = 7$
$f(-1) = |4(-1)| - 1 = 3$
$f(0) = |4(0)| - 1 = -1$
$f(1) = |4(1)| - 1 = 3$
$f(2) = |4(2)| - 1 = 7$

21. $f(x) = |4x| + 1$
$f(-2) = |4(-2)| + 1 = 9$
$f(-1) = |4(-1)| + 1 = 5$
$f(0) = |4(0)| + 1 = 1$
$f(1) = |4(1)| + 1 = 5$
$f(2) = |4(2)| + 1 = 9$

23. $f(x) = |4x + 1|$
$f(-2) = |4(-2) + 1| = 7$
$f(-1) = |4(-1) + 1| = 3$
$f(0) = |4(0) + 1| = 1$
$f(1) = |4(1) + 1| = 5$
$f(2) = |4(2) + 1| = 9$

25. $f(x) = 4 - 3x$
Linear function
$(0, 4);\quad m = -3$

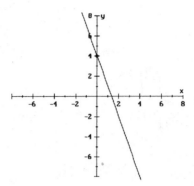

27. $f(x) = x^2 - 9$
Quadratic function

$\dfrac{-b}{2a} = -\dfrac{0}{2(1)} = 0$

$f(0) = 0^2 - 9 = -9$
Vertex: $(0, -9)$
x-intercepts: Let $y = 0$
$\quad 0 = x^2 - 9$
$\quad 9 = x^2$
$\pm 3 = x$
y-intercept: Let $x = 0$
$y = 0^2 - 9 = -9$

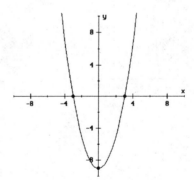

29. $f(x) = x^3$
Polynomial function

x	y
-2	-8
-1	-1
0	0
1	-1
2	-8

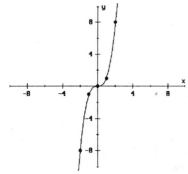

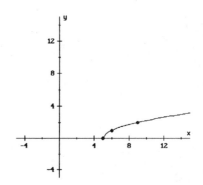

31. $f(x) = 2x^3$
Polynomial function

35. $f(x) = \sqrt{x + 5}$
Square root function

x	y
-2	-16
-1	-2
0	0
1	2
2	16

x	y
-5	0
-4	1
-1	2

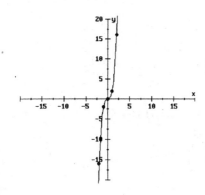

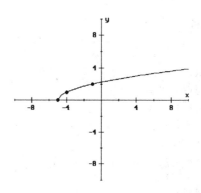

37. $f(x) = \sqrt{x} + 5$
Square root function

x	y
0	5
1	6
4	7

33. $f(x) = \sqrt{x - 5}$
Square root function

x	y
5	0
6	1
9	2

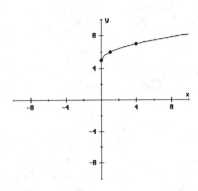

$$x = \frac{-(-4) \pm \sqrt{(-4)^2 - 4(1)(1)}}{2(1)}$$

$$= \frac{4 \pm \sqrt{12}}{2}$$

$$= \frac{4 \pm 2\sqrt{3}}{2}$$

$$= 2 \pm \sqrt{3}$$

y – intercept: Let $x = 0$
$y = 0^2 - 4(0) + 1 = 1$

39. $f(x) = \sqrt{x} - 5$
     Square root function

x	y
0	-5
1	-4
4	-3

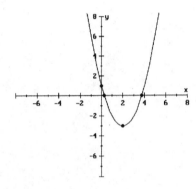

43. $f(x) = \sqrt{3x - 2}$
     Square root function

x	y
2/3	0
1	1
2	2

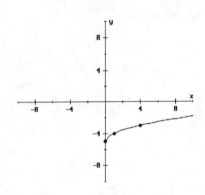

41. $f(x) = x^2 - 4x + 1$
     Quadratic function
     $$\frac{-b}{2a} = \frac{-(-4)}{2(1)} = 2$$
     $f(2) = 2^2 - 4(2) + 1 = -3$
     Vertex: (2, -3)
     x–intercepts: Let $y = 0$
     $0 = x^2 - 4x + 1$

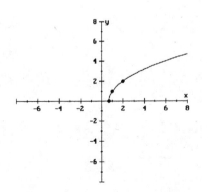

45. $f(x) = 8 - 2x - x^2$
Quadratic function

$$-\frac{b}{2a} = \frac{-(-2)}{2(-1)} = -1$$

$f(-1) = 8 - 2(-1) - (-1)^2 = 9$
Vertex: $(-1, 9)$

x-intercepts: Let $y = 0$
$0 = 8 - 2x - x^2$
$0 = x^2 + 2x - 8$
$0 = (x + 4)(x - 2)$
$x + 4 = 0, \quad x - 2 = 0$
$\qquad x = -4, \qquad x = 2$

y-intercept: Let $x = 0$
$y = 8 - 2(0) - 0^2 = 8$

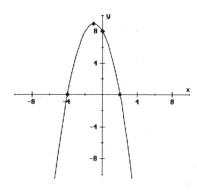

47. $f(x) = \sqrt{8 - 2x}$
Square root function

x	y
4	0
7/2	1
2	2

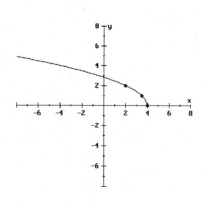

49. $f(x) = \sqrt{6 - 4x}$
Square root function

x	y
3/2	0
5/4	1
1/2	2

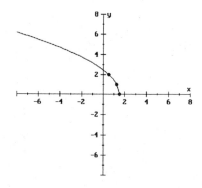

51. $f(x) = |x + 5|$
Absolute value function

x	y
-7	2
-5	0
-3	2

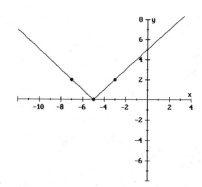

53. $f(x) = |x| + 5$
Absolute value function

x	y
-1	6
0	5
1	6

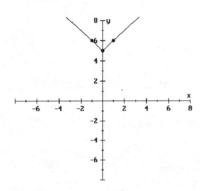

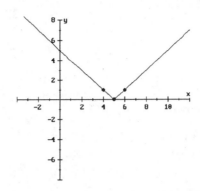

55. $f(x) = x + 5$
Linear function
$(0, 5)$; $m = 1$

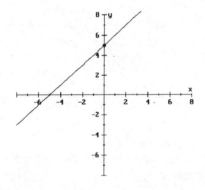

57. $f(x) = |5 - x|$
Absolute value function

x	y
4	1
5	0
6	1

65. Let x = rate of medium speed sorter,
then $+ 150$ = rate of high speed sorter

$$\frac{500}{x} = \frac{500}{x + 150} + \frac{1}{6}$$

$$6x(x + 150)\left(\frac{500}{x}\right) = 6x(x + 150)\left(\frac{500}{x + 150} + \frac{1}{6}\right)$$

$$3000(x + 150) = 3000x + x(x + 150)$$
$$3000x + 450000 = 3000x + x^2 + 150x$$
$$0 = x^2 + 150x - 450000$$
$$0 = (x + 750)(x - 600)$$
$$x + 750 = 0 \quad \text{or} \quad x - 600 = 0$$
$$x = -750 \quad \text{or} \quad x = 600$$

Since $x > 0$, $x = 600$
$$x + 150 = 750$$

$$\frac{500}{x} = \frac{500}{600} = \frac{5}{6}$$

$$\frac{500}{x + 150} = \frac{500}{750} = \frac{2}{3}$$

The medium speed sorter has a rate of 600
letters/min and takes $\frac{5}{6}$ hr = 50 min.
The high speed sorter has a rate of 750
letters/min and takes $\frac{2}{3}$ hr = 40 min.

67. $f(x) = 3x^2 - 2x + 5$

(a) $f(x - 2) = 3(x - 2)^2 - 2(x - 2) + 5$
$$= 3(x^2 - 4x + 4) - 2x + 4 + 5$$
$$= 3x^2 - 12x + 12 - 2x + 9$$
$$= 3x^2 - 14x + 21$$

(b) $f(x) - f(2) = (3x^2 - 2x + 5) - \left[3(2)^2 - 2(2) + 5\right]$
$$= 3x^2 - 2x + 5 - 13$$
$$= 3x^2 - 2x - 8$$

9.4 Exercises

1. $\{(3, 1), (5, 2)\}$

3. $\{(2, 3), (-1, -3), (3, -1)\}$

5. function has no inverse

7. function (not one-to-one since $y = 1$ is associated with 2 values of x).

9. one-to-one function

11. neither ($x = 3$ is assigned 2 values of y)

13. function (not one-to-one since $y = -2$ is associated with 2 values of x.)

15. one-to-one function

17. one-to-one function (passes horizontal line test)

19. function (not one-to-one since it fails horizontal line test)

21. neither (fails vertical line test)

23. one-to-one (passes horizontal line test)

25. domain: $\{3, 2\}$
    inverse: $\{(-2, 3), (-3, 2)\}$
    domain of inverse: $\{-2, -3\}$

27. domain: $\{6, 2\ -3\}$
    inverse: $\{(-3, 6), (-4, 2), (6, -3)\}$
    domain of inverse: $\{-3, -4, 6\}$

29. domain: $\{2, 3\ 4\}$
    inverse not a function

31. domain: all real numbers
$$f(x) = 3x + 4$$
$$y = 3x + 4$$
$$x = 3y + 4$$
$$x - 4 = 3y$$
$$\frac{x - 4}{3} = y$$
$$f^{-1}(x) = \frac{x - 4}{3}$$

domain of inverse: all real numbers

33. domain: all real numbers
$$g(x) = 2x - 3$$
$$y = 2x - 3$$
$$x = 2y - 3$$
$$x + 3 = 2y$$
$$\frac{x + 3}{2} = y$$
$$g^{-1}(x) = \frac{x + 3}{2}$$

domain of inverse: all real numbers

35. domain: all real numbers
$$h(x) = 4 - 5x$$
$$y = 4 - 5x$$
$$x = 4 - 5y$$
$$x - 4 = -5y$$
$$\frac{x - 4}{-5} = y$$
$$\frac{4 - x}{5} = y$$
$$h^{-1}(x) = \frac{4 - x}{5}$$

domain of inverse: all real numbers

37. domain: all real numbers
Since $f(x)$ is not one-to-one, the inverse is not a function.

39. domain: all real numbers
$$f(x) = x^3 + 4$$
$$y = x^3 + 4$$
$$x = y^3 + 4$$
$$x - 4 = y^3$$
$$\sqrt[3]{x - 4} = y$$
$$f^{-1}(x) = \sqrt[3]{x - 4}$$

domain: all real numbers

41. domain: $\{x \mid x \neq 0\}$
$$g(x) = \frac{1}{x}$$
$$y = \frac{1}{x}$$

$$x = \frac{1}{y}$$

$$xy = 1$$

$$y = \frac{1}{x}$$

$$g^{-1}(x) = \frac{1}{x}$$

domain of inverse: $\{x \mid x \neq 0\}$

43. domain: $x + 3 \neq 0$
$$x \neq -3$$
$$\{x \mid x \neq -3\}$$

$$g(x) = \frac{2}{x + 3}$$

$$y = \frac{2}{x + 3}$$

$$x = \frac{2}{y + 3}$$

$$x(y + 3) = 2$$

$$y + 3 = \frac{2}{x}$$

$$y = \frac{2}{x} - 3$$

$$g^{-1}(x) = \frac{2}{x} - 3 = \frac{2 - 3x}{x}$$

domain of inverse: $\{x \mid x \neq 0\}$

45. domain: $\{x \mid x \neq 0\}$

$$g(x) = \frac{x - 1}{x}$$

$$y = \frac{x - 1}{x}$$

$$x = \frac{y - 1}{y}$$

$$xy = y - 1$$
$$1 = y - xy$$
$$1 = y(1 - x)$$

$$\frac{1}{1 - x} = y$$

$$g^{-1}(x) = \frac{1}{1 - x}$$

domain of inverse: $1 - x \neq 0$
$$1 \neq x$$
$$\{x \mid x \neq 1\}$$

47. domain: $x - 1 \neq 0$
$$x \neq 1$$
$$\{x \mid x \neq 1\}$$

$$h(x) = \frac{x + 2}{x - 1}$$

$$y = \frac{x + 2}{x - 1}$$

$$x = \frac{y + 2}{y - 1}$$

$$x(y - 1) = y + 2$$

$$xy - x = y + 2$$

$$xy - y = x + 2$$

$$y(x - 1) = x + 2$$

$$y = \frac{x + 2}{x - 1}$$

$$h^{-1}(x) = \frac{x + 2}{x - 1}$$

domain of inverse: $\{x \mid x \neq 1\}$

49.

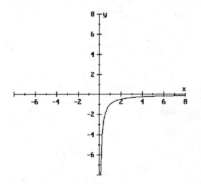

51.

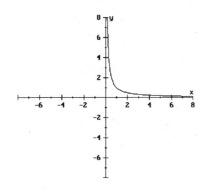

55. $\left(\sqrt{x} - 5\right)^2$

$= \left(\sqrt{x}\right)^2 - 2\left(\sqrt{x}\right)(5) + 5^2$

$= x - 10\sqrt{x} + 25$

57. Let x = other rate

$0.062(2800) + x(1800) = 300$

$173.6 + 1800x = 300$

$1800x = 126.4$

$x = 0.070$

The rate is 7%.

9.5 Exercises

1. $y = kx$

$8 = k(4)$

$2 = k$

$y = 2x$

$y = 2(3) = 6$

3. $y = kx$

$25 = k(15)$

$\dfrac{5}{3} = k$

$y = \dfrac{5}{3}x$

$y = \dfrac{5}{3}(8) = \dfrac{40}{3}$

5. $y = kx$

$22 = k(3)$

$\dfrac{22}{3} = k$

$y = \dfrac{22}{3}x$

$5 = \dfrac{22}{3}x$

$\dfrac{15}{22} = x$

7. $a = kb^2$

$4 = k(3)^2$

$4 = 9k$

$\dfrac{4}{9} = k$

$a = \dfrac{4}{9}b^2$

$a = \dfrac{4}{9}(9)^2 = 36$

9. $r = ks^4$

$12 = k(2)^4$

$12 = 16k$

$\dfrac{3}{4} = k$

$r = \dfrac{3}{4}s^4$

$r = \dfrac{3}{4}(3)^4 = \dfrac{243}{4}$

11. $y = \dfrac{k}{x}$

$20 = \dfrac{k}{4}$

$80 = k$

$y = \dfrac{80}{x}$

$y = \dfrac{80}{8} = 10$

13. $y = \dfrac{k}{x}$

$21 = \dfrac{k}{12}$

$252 = k$

$y = \dfrac{252}{x}$

$9x = 252$
$x = 28$

15. $y = \dfrac{k}{x}$

$25 = \dfrac{k}{10}$

$250 = k$

$y = \dfrac{250}{x}$

$12 = \dfrac{250}{x}$

$12x = 250$

$x = \dfrac{125}{6}$

17. $a = \dfrac{k}{b^3}$

$6 = \dfrac{k}{2^3}$

$6 = \dfrac{k}{8}$

$48 = k$

$a = \dfrac{48}{b^3}$

$a = \dfrac{48}{16^3} = \dfrac{3}{256}$

19. $a = \dfrac{k}{\sqrt{b}}$

$16 = \dfrac{k}{\sqrt{4}}$

$16 = \dfrac{k}{2}$

$32 = k$

$a = \dfrac{32}{\sqrt{b}}$

$a = \dfrac{32}{\sqrt{9}} = \dfrac{32}{3}$

21. $z = kxy$
$12 = k(2)(4)$
$12 = 8k$
$\dfrac{3}{2} = k$

$z = \dfrac{3}{2}xy$

$z = \dfrac{3}{2}(5)(2) = 15$

23. $z = kxy$
$20 = k(3)(4)$
$20 = 12k$
$\dfrac{5}{3} = k$

$z = \dfrac{5}{3}xy$

$z = \dfrac{5}{3}(2)(5)$

$z = \dfrac{50}{3}$

25. $a = kcd$
$20 = k(2)(4)$
$20 = 8k$
$\dfrac{5}{2} = k$

$a = \dfrac{5}{2}cd$

$25 = \dfrac{5}{2}(8)d$

$25 = 20d$

$\dfrac{5}{4} = d$

27. $z = kxy^2$
$20 = k(4)(2)^2$
$20 = 16k$
$\dfrac{5}{4} = k$

$z = \dfrac{5}{4}xy^2$

$z = \dfrac{5}{4}(2)(4)^2 = 40$

29. $z = \dfrac{kx}{y}$

$16 = \dfrac{k(3)}{2}$

$32 = 3k$

$\dfrac{32}{3} = k$

$z = \dfrac{\dfrac{32}{3}x}{y}$

$z = \dfrac{32x}{3y}$

$z = \dfrac{32(5)}{3(3)} = \dfrac{160}{9}$

31. $z = \dfrac{kx^2}{y}$

$20 = \dfrac{k(2)^2}{4}$

$80 = 4k$

$$20 = k$$

$$z = \frac{20x^2}{y}$$

$$z = \frac{20(4)^2}{2} = 160$$

33. $$z = \frac{kx}{y^2}$$

$$32 = \frac{k(4)}{2^2}$$

$$32 = k$$

$$z = \frac{32x}{y^2}$$

$$z = \frac{32(3)}{3^2} = \frac{32}{3}$$

35. $$V = kt$$
$$250 = k(30)$$

$$\frac{25}{3} = k$$

$$V = \frac{25}{3}t$$

$$V = \frac{25}{3}(40) = \frac{1000}{3}$$

The volume is $\dfrac{1000}{3}$ m^3.

37. $$V = kr^3$$
$$36\pi = k(3)^3$$

$$36\pi = 27k$$

$$\frac{4}{3}\pi = k$$

$$V = \frac{4}{3}\pi r^3$$

$$V = \frac{4}{3}\pi(4)^3 = \frac{256}{3}\pi$$

The volume is $\dfrac{256\pi}{3}$ cm^3.

39. $$d = kt^2$$
$$256 = k(4)^2$$
$$256 = 16k$$
$$16 = k$$

$$d = 16t^2$$
$$800 = 16t^2$$
$$50 = t^2$$
$$5\sqrt{2} = t$$

It takes $5\sqrt{2} \approx 7.07$ sec.

41. $$E = \frac{k}{d^2}$$

$$25 = \frac{k}{4^2}$$

$$400 = k$$

$$E = \frac{400}{d^2}$$

$$E = \frac{400}{8^2} = 6.25$$

The illumination is 6.25 footcandles.

43. $$V = khr^2$$
$$4\pi = k(3)(2)^2$$
$$4\pi = 12k$$
$$\frac{\pi}{3} = k$$

$$V = \frac{\pi}{3}hr^2$$

$$V = \frac{\pi}{3}(2)(3)^2 = 6\pi$$

The volume is 6π m^3.

45. $$R = \frac{kl}{d^2}$$

$$12 = \frac{k(80)}{(0.01)^2}$$

$$0.0012 = 80k$$
$$0.000015 = k$$

$$R = \frac{0.0000015l}{d^2}$$

$$R = \frac{0.000015(100)}{(0.02)^2}$$

$$= 3.75$$

The resistance is 3.75 ohms.

47. $C = 2\pi r$

radius doubled: $\begin{aligned} C_1 &= 2\pi(2r) \\ &= 2(2\pi r) \\ &= 2c \end{aligned}$

The circumference is doubled.

radiustripled: $\begin{aligned} C_2 &= 2\pi(3r) \\ &= 3(2\pi r) \\ &= 3C \end{aligned}$

The circumference is tripled.

radius halved: $\begin{aligned} C_3 &= 2\pi\left(\frac{1}{2}r\right) \\ &= \frac{1}{2}(2\pi r) \\ &= \frac{1}{2}C \end{aligned}$

The circumference is halved.

49. $y = \dfrac{k}{x}$

$\begin{aligned} y_1 &= \frac{c}{4x} \\ &= \frac{1}{4}\left(\frac{k}{x}\right) \\ &= \frac{1}{4}y \end{aligned}$

y is divided by 4.

51. $z = kxy$

$\begin{aligned} z_1 &= k(4x)(5y) \\ &= 20kxy \\ &= 20z \end{aligned}$

z is multiplied by 20.

53. $s = \dfrac{kr}{t}$

$\begin{aligned} s_1 &= \frac{k(2r)}{\frac{1}{2}t} \\ &= \frac{4kr}{t} \\ &= 4s \end{aligned}$

s is multiplied by 4.

55. $\sqrt{5x} + 5\sqrt[3]{x}$
$= (5x)^{1/2} + 5x^{1/3}$

57. Let x = amount invested at 6.5%,
then $x + 5000$ = amount invested at 7%

$\begin{aligned} 0.065x + 0.07(x + 5000) &= 1430 \\ 0.065x + 0.07x + 350 &= 1430 \\ 0.135x + 350 &= 1430 \\ 0.135x &= 1080 \\ x &= 8000 \\ x + 5000 &= 13000 \end{aligned}$

She invests \$8000 at 6.5% and \$13,000 at 7%.

CHAPTER 9 REVIEW EXERCISES

1. $f(x) = 5x + 4$
$f(-3) = 5(-3) + 4 = -11$
$f(1) = 5(1) + 4 = 9$
$f(3) = 5(3) + 4 = 19$

3. $f(x) = \sqrt{x - 1}$
-3 is not in the domain of $f(x)$.

$\begin{aligned} f(1) &= \sqrt{1 - 1} \\ &= \sqrt{0} \\ &= 0 \end{aligned}$

$f(3) = \sqrt{3 - 1} = \sqrt{2}$

5. $f(x) = \dfrac{3}{\sqrt{1 - x}}$

1 and 3 are not in the domain of $f(x)$.

$f(-3) = \dfrac{3}{\sqrt{1 - (-3)}}$

$= \dfrac{3}{\sqrt{4}}$

$= \dfrac{3}{2}$

7. all real numbers

9. all real numbers

11. $3 - 5x \geq 0$
$-5x \geq -3$
$x \leq \dfrac{3}{5}$
$\left\{ x \mid x \leq \dfrac{3}{5} \right\}$

13. $2x - 1 > 0$
$2x > 1$
$x > \dfrac{1}{2}$
$\left\{ x \mid x > \dfrac{1}{2} \right\}$

15. $2x - 3 > 0$
$2x > 3$
$x > \dfrac{3}{2}$
$\left\{ x \mid x > \dfrac{3}{2} \right\}$

17. $f(x + 3) = 5(x + 3) + 2$
$= 5x + 15 + 2$
$= 5x + 17$

19. $f(x) + 3 = (5x + 2) + 3$
$= 5x + 5$

21. $f(x) + f(3) = (5x + 2) + [5(3) + 2]$
$= 5x + 2 + 17$
$= 5x + 19$

23. $g(x + 3) = \sqrt{3 - 2(x + 3)}$
$= \sqrt{3 - 2x - 6}$
$= \sqrt{-2x - 3}$

25. $g(3x) = \sqrt{3 - 2(3x)}$
$= \sqrt{3 - 6x}$

27. $3g(x) = 3\sqrt{3 - 2x}$

29. $g(x + h) = \sqrt{3 - 2(x + h)}$
$= \sqrt{3 - 2x - 2h}$

$g(x + h) - g(x)$
$= \sqrt{3 - 2x - 2h} - \sqrt{3 - 2x}$

31. $g[f(x)] = \sqrt{3 - 2[f(x)]}$
$= \sqrt{3 - 2(5x + 2)}$
$= \sqrt{3 - 10x - 4}$
$= \sqrt{-10x - 1}$

33. $(f + g)(3) = f(3) + g(3)$
$= (3^2 - 3) + (3 - 8)$
$= 6 - 5$
$= 1$

35. $\left(\dfrac{f}{g} \right)(3) = \dfrac{f(3)}{g(3)}$

$= \dfrac{3^2 - 3}{3 - 8}$

$= -\dfrac{6}{5}$

37. $\left(\dfrac{f}{g} \right)(8) = \dfrac{f(8)}{g(8)}$

$= \dfrac{8^2 - 3}{8 - 8}$

$= \dfrac{61}{0}$

undefined

39. $g(3) = 3 + 1 = 4$

$f[g(3)] = [g(3)]^2 - [g(3)] - 5$
$= 4^2 - 4 - 5$
$= 7$

41. $g(-1) = -1 + 1 = 0$

$f[g(-1)] = [g(-1)]^2 - [g(-1)] - 5$
$= 0^2 - 0 - 5$
$= -5$

43. $h\left(\dfrac{1}{2}\right) = \dfrac{\frac{1}{2} + 1}{2\left(\frac{1}{2}\right)} = \dfrac{3}{2}$

$f\left[h\left(\dfrac{1}{2}\right)\right] = \left[h\left(\dfrac{1}{2}\right)\right]^2 - \left[h\left(\dfrac{1}{2}\right)\right] - 5$

$= \left(\dfrac{3}{2}\right)^2 - \dfrac{3}{2} - 5$

$= -\dfrac{17}{4}$

45. $h(-1) = \dfrac{-1 + 1}{2(-1)} = 0$

$g[h(-1)] = [h(-1)] + 1$
$= 0 + 1$
$= 1$

47. $g[f(x)] = [f(x)] + 1$
$= x^2 - x - 5 + 1$
$= x^2 - x - 4$

49. $g[h(x)] = [h(x)] + 1$

$= \dfrac{x + 1}{2x} + 1$

$= \dfrac{x + 1 + 2x}{2x}$

$= \dfrac{3x + 1}{2x}$

51. $A = 8t$

(a) $A = 8(4) = 32$
$\pi r^2 = 32$
$r^2 = \dfrac{32}{\pi}$

$r = \sqrt{\dfrac{32}{\pi}}$

$r = \dfrac{\sqrt{32\pi}}{\pi} = \dfrac{4\sqrt{2\pi}}{\pi}$

The radius is $\dfrac{4\sqrt{2\pi}}{\pi} \approx 3.2$ in.

(b) $A = 8t$
$\pi r^2 = 8t$

$r^2 = \dfrac{8t}{\pi}$

$r = \sqrt{\dfrac{8t}{\pi}}$

$r = \dfrac{2\sqrt{2t\pi}}{\pi}$

53. $A = \pi r^2$
$200\pi = \pi r^2$
$200 = r^2$
$\sqrt{200} = r^2$
$r = 10\sqrt{2}$

$C = 2\pi r$
$= 2\pi\left(10\sqrt{2}\right)$
$= 20\pi\sqrt{2}$

Cost $= x\left(20\pi\sqrt{2}\right)$ dollars

55. $f(x) = 2x - 3$
Linear function
$(0, -3)$; $m = 2$

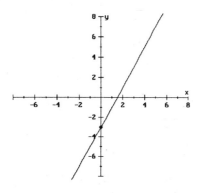

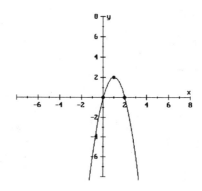

57. $f(x) = \sqrt{2x - 3}$
    Square root function

x	y
3/2	0
2	1
7/2	2

61. $f(x) = |2x| - 3$
    Absolute value function

x	y
-1	-1
0	-3
1	-1

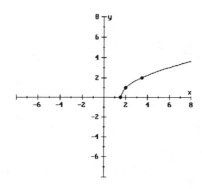

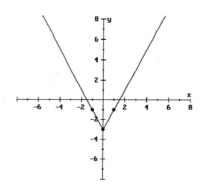

59. $f(x) = 4x - 2x^2$
    Quadratic function

$$-\frac{b}{2a} = -\frac{4}{2(-2)} = 1$$

$$f(1) = 4(1) - 2(1)^2 = 2$$

Vertex: (1, 2)
x-intercepts: Let $y = 0$
$0 = 4x - 2x^2$
$0 = 2x(2 - x)$
$2x = 0 \qquad 2 - x = 0$
$x = 0 \qquad\quad x = 2$

y-intercept: Let $x = 0$
$y = 4(0) - 2(0)^2 = 0$

63. function

65. one-to-one function

67. one-to-one function   (passes horizontal line test)

69. neither   (fails vertical line test)

71. {(3, 2), (4, 3)}

73. $y = 3x + 8$
        $x = 3y + 8$
    $x - 8 = 3y$
    $\dfrac{x - 8}{3} = y$

75.
$$y = x^3$$
$$x = y^3$$
$$x^{1/3} = y$$

77.
$$y = \frac{3}{x + 1}$$
$$x = \frac{3}{y + 1}$$
$$x(y + 1) = 3$$
$$xy + x = 3$$
$$xy = 3 - x$$
$$y = \frac{3 - x}{x}$$

79.

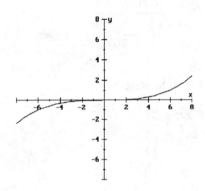

81.
$$x = ky$$
$$8 = k(3)$$
$$\frac{8}{3} = k$$
$$x = \frac{8}{3}y$$
$$x = \frac{8}{3}(2) = \frac{16}{3}$$

83.
$$r = \frac{k}{s^2}$$
$$8 = \frac{k}{2^2}$$
$$32 = k$$
$$r = \frac{32}{s^2}$$

$$4 = \frac{32}{s^2}$$
$$s^2 = 8$$
$$s = \pm\sqrt{8} = \pm 2\sqrt{2}$$

85.
$$z = kxy$$
$$16 = k(7)(2)$$
$$\frac{8}{7} = k$$
$$z = \frac{8}{7}xy$$
$$z = \frac{8}{7}(3)(4) = \frac{96}{7}$$

87.
$$V = kr^3$$
$$\frac{500\pi}{3} = k(5)^3$$
$$\frac{4\pi}{3} = k$$
$$V = \frac{4\pi}{3}r^3$$
$$V = \frac{4\pi}{3}(6)^3 = 288\pi$$

The volume is $288\pi \approx 904.78$ in.^{3}.

89.
$$V = \frac{kT}{P}$$
$$80 = \frac{k(20)}{30}$$
$$120 = k$$
$$V = \frac{120T}{P}$$
$$V = \frac{120(10)}{20} = 60$$

The volume is 60 m^3.

1. (a) all real numbers

   (b) $2x - 3 \neq 0$

$$2x \neq 3$$
$$x \neq \frac{3}{2}$$
$$\left\{ x \mid x \neq \frac{3}{2} \right\}$$

   (c) $8 - 3x \geq 0$

$$-3x \geq -8$$
$$x \leq \frac{8}{3}$$
$$\left\{ x \mid x \leq \frac{8}{3} \right\}$$

2. (a) $g(-1) = \dfrac{3}{-1 + 1}$

$$= \frac{3}{0}$$

undefined

   (b) $f(x + 2) = 3(x + 2) - 5$

$$= 3x + 6 - 5$$
$$= 3x + 1$$

   (c) $g(x + 3) = \dfrac{3}{x + 3 + 1} = \dfrac{3}{x + 4}$

$$g(x + 3) - g(x)$$
$$= \frac{3}{x + 4} - \frac{3}{x + 1}$$
$$= \frac{3(x + 1) - 3(x + 4)}{(x + 4)(x + 1)}$$
$$= \frac{3x + 3 - 3x - 12}{(x + 4)(x + 1)}$$
$$= \frac{-9}{(x + 4)(x + 1)}$$

3. (a) $\left(\dfrac{f}{h} \right)(3) = \dfrac{f(3)}{h(3)}$

$$= \frac{3^2 - 3}{\sqrt{3 - 2}}$$
$$= \frac{6}{1}$$
$$= 6$$

   (b) $h(6) = \sqrt{6 - 2} = 2$

$$f[h(6)] = [h(6)]^2 - 3$$
$$= 2^2 - 3$$
$$= 1$$

   (c) $f(6) = 6^2 - 3 = 33$

$$h[f(6)] = \sqrt{[f(6)] - 2}$$
$$= \sqrt{33 - 2}$$
$$= \sqrt{31}$$

   (d) $f[h(x)] = [h(x)]^2 - 3$

$$= \left(\sqrt{x - 2} \right)^2 - 3$$
$$= x - 2 - 3$$
$$= x - 5$$

   (e) $h[f(x)] = \sqrt{[f(x)] - 2}$

$$= \sqrt{x^2 - 3 - 2}$$
$$= \sqrt{x^2 - 5}$$

4. (a) Let x = side length

$$x^2 + x^2 = 8^2$$
$$2x^2 = 64$$
$$x^2 = 32$$
$$x = \sqrt{32} = 4\sqrt{2}$$

$$P = 4x$$
$$P = 4(4\sqrt{2}) = 16\sqrt{2}$$

The perimeter is $16\sqrt{2} \approx 22.6$ ft.

   (b) $x^2 + x^2 = d^2$

$$2x^2 = d^2$$
$$x^2 = \frac{d^2}{2}$$
$$x = \frac{d}{\sqrt{2}}$$

$$P = 4x$$
$$P = 4\left(\frac{d}{\sqrt{2}} \right)$$
$$P = 2d\sqrt{2}$$

5. $f(x) = |2x - 1|$

x	y
0	1
1/2	0
1	1

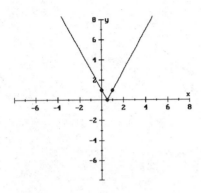

6. function has no inverse

7. No, it fails the horizontal line test.

8. (a) $f(x) = 2x - 4$
$$y = 2x - 4$$
$$x = 2y - 4$$
$$x + 4 = 2y$$

$$\frac{x + 4}{2} = y$$

$$f^{-1}(x) = \frac{x + 4}{2}$$

(b) $f(x) = \dfrac{x}{x + 3}$

$$y = \frac{x}{x + 3}$$

$$x = \frac{y}{y + 3}$$

$$x(y + 3) = y$$
$$xy + 3x = y$$
$$3x = y - xy$$
$$3x = y(1 - x)$$

$$\frac{3x}{1 - x} = y$$

$$f^{-1}(x) = \frac{3x}{1 - x}$$

9. $y = \dfrac{k}{x^2}$

$$24 = \frac{k}{2^2}$$

$$96 = k$$

$$y = \frac{96}{x^2}$$

$$y = \frac{96}{6^2} = \frac{8}{3}$$

10. $z = \dfrac{kx^2}{y}$

$$120 = \frac{k(2)^2}{3}$$

$$90 = k$$

$$z = \frac{90x^2}{y}$$

$$z = \frac{90(3)^2}{2} = 405$$

1. $(-2x^2y)(-3xy^2)^3$
   $= (-2x^2y)(-27x^3y^6)$
   $= 54x^5y^7$

3. $\left(\dfrac{2xy^2}{4x^2y^3}\right)^2$

   $= \left(\dfrac{1}{2xy}\right)^2$

   $= \dfrac{1}{4x^2y^2}$

5. $(x^{-5}y^{-2})(x^{-7}y)$

   $= x^{-12}y^{-1}$

   $= \dfrac{1}{x^{12}y}$

7. $\dfrac{r^{-3}s^{-2}}{r^{-2}s^0} = r^{-1}s^{-2}$

   $= \dfrac{1}{rs^2}$

9. $(-2r^{-3}s)^{-1}(-4r^{-2}s^{-3})^2$

   $= \dfrac{1}{-2r^{-3}s} \cdot 16r^{-4}s^{-6}$

   $= -8r^{-1}s^{-7}$

   $= -\dfrac{8}{rs^7}$

11. $(x^{-1} - y^{-1})(x + y)$

    $= x^0 - xy^{-1} + x^{-1}y - y^0$

    $= 1 - \dfrac{x}{y} + \dfrac{y}{x} - 1$

    $= \dfrac{y}{x} - \dfrac{x}{y}$

    $= \dfrac{y^2 - x^2}{xy}$

13. $56{,}429.32 = 5.642932 \times 10^4$

15. 1 day = 24 hr
    = 24(60 min)
    = 1440 min
    = 1440(60 sec)
    = 86400 sec

    d = (86400 sec)(186000 miles/sec)
    $= 1.60704 \times 10^{10}$ miles

17. $(-1000)^{1/3} = -10$

19. $a^{2/3}a^{-1/2} = a^{2/3 - 1/2}$
    $= a^{1/6}$

21. $\left[(81)^{-1/3}3^2\right]^{-3}$

    $= \left[(3^4)^{-1/3}3^2\right]^{-3}$

    $= (3^{-4/3} \cdot 3^2)^{-3}$

    $= (3^{2/3})^{-3}$

    $= 3^{-2}$

    $= \dfrac{1}{9}$

23. $x^{3/4} = \sqrt[4]{x^3}$

25. $\sqrt{16a^4b^8} = 4a^2b^4$

27. $\left(3x\sqrt{2x^2y}\right)\left(2x\sqrt{8xy^3}\right)$

    $= \left(3x^2\sqrt{2y}\right)\left(4xy\sqrt{2xy}\right)$

    $= 12x^3y\sqrt{4xy^2}$

    $= 24x^3y^2\sqrt{x}$

29. $\dfrac{\sqrt{48}}{\sqrt{3}} = \sqrt{\dfrac{48}{3}}$

    $= \sqrt{16}$
    $= 4$

31. $\sqrt[8]{x^4} = \sqrt{x}$

33. $2\sqrt{5} - 3\sqrt{5} + 8\sqrt{5}$
$= (2 - 3 + 8)\sqrt{5}$
$= 7\sqrt{5}$

35. $3a\sqrt[3]{a^4} - a^2\sqrt[3]{a}$
$= 3a^2\sqrt[3]{a} - a^2\sqrt[3]{a}$
$= (3a^2 - a^2)\sqrt[3]{a}$
$= 2a^2\sqrt[3]{a}$

37. $\sqrt{2}(\sqrt{2} - 1) + 2\sqrt{2}$
$= 2 - \sqrt{2} + 2\sqrt{2}$
$= 2 + \sqrt{2}$

39. $(\sqrt{5} - \sqrt{3})(\sqrt{5} + \sqrt{3})$
$= (\sqrt{5})^2 - (\sqrt{3})^2$
$= 5 - 3$
$= 2$

41. $\dfrac{5}{\sqrt{3} + \sqrt{2}} - \dfrac{3}{\sqrt{3}}$

$= \dfrac{5(\sqrt{3} - \sqrt{2})}{(\sqrt{3} + \sqrt{2})(\sqrt{3} - \sqrt{2})} - \dfrac{3 \cdot \sqrt{3}}{\sqrt{3} \cdot \sqrt{3}}$

$= \dfrac{5\sqrt{3} - 5\sqrt{2}}{3 - 2} - \dfrac{3\sqrt{3}}{3}$

$= 5\sqrt{3} - 5\sqrt{2} - \sqrt{3}$
$= 4\sqrt{3} - 5\sqrt{2}$

43. $\sqrt{3x + 2} = 7$
$(\sqrt{3x + 2})^2 = 7^2$
$3x + 2 = 49$
$3x = 47$
$x = \dfrac{47}{3}$

45. $\sqrt[3]{x - 1} = -2$
$(\sqrt[3]{x - 1})^3 = (-2)^3$
$x - 1 = -8$
$x = -7$

47. $i^{35} = (i^4)^8 \cdot i^3$
$= 1^8 \cdot (-i)$
$= -i$

49. $(3 - 2i)(5 + i)$
$= 15 + 3i - 10i - 2i^2$
$= 15 + 3i - 10i + 2$
$= 17 - 7i$

51. $a^2 - 2a - 15 = 0$
$(a - 5)(a + 3) = 0$
$a - 5 = 0$ or $a + 3 = 0$
$a = 5$ or $a = -3$

53. $2x^2 - 3x - 4 = 9 - 3(x - 2)$
$2x^2 - 3x - 4 = 9 - 3x + 6$
$2x^2 = 19$
$x^2 = \dfrac{19}{2}$

$x = \pm\sqrt{\dfrac{19}{2}}$

$x = \pm\dfrac{\sqrt{38}}{2}$

55. $y^2 + 6y - 1 = 0$
$y^2 + 6y = 1$
$y^2 + 6y + 9 = 1 + 9$
$(y + 3)^2 = 10$
$y + 3 = \pm\sqrt{10}$
$y = -3 \pm \sqrt{10}$

57. $3a^2 - 2a - 2 = 0$

$a = \dfrac{-(-2) \pm \sqrt{(-2)^2 - 4(3)(-2)}}{2(3)}$

$= \dfrac{2 \pm \sqrt{4 + 24}}{6}$

$= \dfrac{2 \pm \sqrt{28}}{6}$

$= \dfrac{2 \pm 2\sqrt{7}}{6}$

$= \dfrac{1 \pm \sqrt{7}}{3}$

59.

$$\frac{1}{x+2} = x+2$$

$$(x+2)\left(\frac{1}{x+2}\right) = (x+2)(x+2)$$

$$1 = (x+2)^2$$
$$\pm\sqrt{1} = x+2$$
$$\pm 1 = x+2$$
$$x+2 = 1 \quad \text{or} \quad x+2 = -1$$
$$x = -1 \quad \text{or} \quad x = -3$$

61.

$$\frac{2}{x+3} - \frac{3}{x} = -\frac{2}{3}$$

$$3x(x+3)\left(\frac{2}{x+3} - \frac{3}{x}\right) = 3x(x+3)\left(-\frac{2}{3}\right)$$

$$6x - 9(x+3) = -2x(x+3)$$
$$6x - 9x - 27 = -2x^2 - 6x$$
$$2x^2 + 3x - 27 = 0$$
$$(2x+9)(x-3) = 0$$
$$2x+9 = 0 \quad \text{or} \quad x-3 = 0$$
$$2x = -9 \qquad\qquad x = 3$$
$$x = -\frac{9}{2} \quad \text{or} \quad x = 3$$

63.

$$\frac{3}{x-3} + \frac{2x}{x+3} = \frac{5}{x-3}$$

$$(x-3)(x+3)\left(\frac{3}{x-3} + \frac{2x}{x+3}\right) = (x-3)(x+3)\left(\frac{5}{x-3}\right)$$

$$3(x+3) + 2x(x-3) = 5(x+3)$$
$$3x+9 + 2x^2 - 6x = 5x + 15$$
$$2x^2 - 8x - 6 = 0$$
$$x^2 - 4x - 3 = 0$$

$$x^2 - 4x - 3 = 0$$

$$x = \frac{-(-4) \pm \sqrt{(-4)^2 - 4(1)(-3)}}{2(1)}$$

$$= \frac{4 \pm \sqrt{16 + 12}}{2}$$

$$= \frac{4 \pm \sqrt{28}}{2}$$

$$= \frac{4 \pm 2\sqrt{7}}{2}$$

$$= 2 \pm \sqrt{7}$$

65. $P(x) = -200x^2 + 500x$

$$-\frac{b}{2a} = -\frac{500}{2(-200)} = \frac{5}{4} = 1.25$$

$$P\left(\frac{5}{4}\right) = -200\left(\frac{5}{4}\right)^2 + 500\left(\frac{5}{4}\right)$$

$$= 312.50$$

Vertex: (1.25, 312.50); maximum

The ticket price $1.25 would produce a maximum profit of $312.50.

67. $r = \sqrt{(1-0)^2 + (-3-2)^2}$
$$= \sqrt{1+25}$$
$$= \sqrt{26}$$

$$(x-h)^2 + (y-k)^2 = r^2$$
$$(x-0)^2 + (y-2)^2 = \left(\sqrt{26}\right)^2$$
$$x^2 + (y-2)^2 = 26$$

69. $x^2 + (y+2)^2 = 10$
Center: (0, -2)
$r = \sqrt{10}$

71. $f(x) \cdot g(x) = (2x^2 - 4x)(3x - 6)$
$$= 6x^3 - 24x^2 + 24x$$

73. $g[f(x)] = 3[f(x)] - 6$
$$= 3(2x^2 - 4x) - 6$$
$$= 6x^2 - 12x - 6$$

75. $h(x) - r(x) = \dfrac{2x}{x+1} - \dfrac{1}{x}$

$$= \frac{2x(x) - (x+1)}{x(x+1)}$$

$$= \frac{2x^2 - x - 1}{x(x+1)}$$

77. $\dfrac{g(x)}{r(x)} = \dfrac{3x-6}{\dfrac{1}{x}}$

$$= x(3x - 6)$$
$$= 3x^2 - 6x$$

79. $r[h(x)] = \dfrac{1}{h(x)}$

$\qquad = \dfrac{1}{\dfrac{2x}{x+1}}$

$\qquad = \dfrac{x+1}{2x}$

81. $\qquad y = \dfrac{2x-5}{3}$

$\qquad x = \dfrac{2y-5}{3}$

$\qquad 3x = 2y - 5$

$\qquad 3x + 5 = 2y$

$\qquad \dfrac{3x+5}{2} = y$

83. This function is not one-to-one, hence there is no inverse function.

85. $y = \sqrt{x} + 4$

x	y
0	4
1	5
4	6

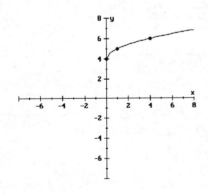

87. $y = |x - 3|$

x	y
0	3
1	2
3	0
6	3

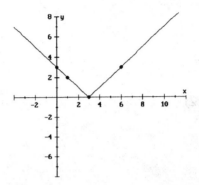

89. $\qquad x^2 + 6x + y^2 - 10y + 33 = 0$

$\qquad (x^2 + 6x + 9) + (y^2 - 10y + 25) = -33 + 9 + 25$

$\qquad (x + 3)^2 + (y - 5)^2 = 1$

Center: $(-3, 5)$

$r = \sqrt{1} = 1$

x-intercepts: Let $y = 0$

$\qquad (x + 3)^2 + (0 - 5)^2 = 1$

$\qquad (x + 3)^2 + 25 = 1$

$\qquad (x + 3)^2 = -24$

$\qquad$ none

y-intercepts: Let $x = 0$

$\qquad (0 + 3)^2 + (y - 5)^2 = 1$

$\qquad 9 + (y - 5)^2 = 1$

$\qquad (y - 5)^2 = -8$

$\qquad$ none

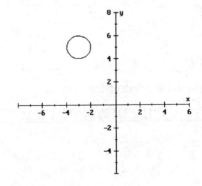

91. $y = \dfrac{k}{x}$

$8 = \dfrac{k}{6}$

$48 = k$

$y = \dfrac{48}{x}$

$y = \dfrac{48}{20} = \dfrac{12}{5}$

93. $x = \dfrac{ky}{z^2}$

$2 = \dfrac{k(12)}{6^2}$

$72 = 12k$

$6 = k$

$x = \dfrac{6y}{z^2}$

$x = \dfrac{6(20)}{5^2} = \dfrac{24}{5}$

95. $I = \dfrac{k}{d^2}$

$I_1 = \dfrac{k}{\left(\frac{1}{2}d\right)^2}$

$I_1 = \dfrac{k}{\frac{d^2}{4}}$

$I_1 = \dfrac{4k}{d^2}$

$I_1 = 4\left(\dfrac{k}{d^2}\right) = 4I$

The illumination quadruples.

CHAPTERS 7 - 9 CUMULATIVE PRACTICE TEST

1. (a) $\sqrt{24x^2y^5} = \sqrt{4 \cdot 6x^2y^4 \cdot y}$
   $\qquad\qquad = 2xy^2\sqrt{6y}$

(b) $\left(a^2\sqrt{2ab^2}\right)\left(b\sqrt{4a^2b^3}\right)$
   $= \left(a^2b\sqrt{2a}\right)\left(2ab^2\sqrt{b}\right)$
   $= 2a^3b^3\sqrt{2ab}$

(c) $\sqrt[3]{\dfrac{5}{a}} = \dfrac{\sqrt[3]{5}}{\sqrt[3]{a}}$

   $= \dfrac{\sqrt[3]{5} \cdot \sqrt[3]{a^2}}{\sqrt[3]{a} \cdot \sqrt[3]{a^2}}$

   $= \dfrac{\sqrt[3]{5a^2}}{\sqrt[3]{a^3}}$

   $= \dfrac{\sqrt[3]{5a^2}}{a}$

2. (a) $5\sqrt{8} - 5\sqrt{2} - \sqrt{50}$
   $= 10\sqrt{2} - 5\sqrt{2} - 5\sqrt{2}$
   $= (10 - 5 - 5)\sqrt{2}$
   $= 0$

(b) $\sqrt{\dfrac{x}{y}} - \sqrt{\dfrac{x}{2}}$

   $= \dfrac{\sqrt{x}}{\sqrt{y}} - \dfrac{\sqrt{x}}{\sqrt{2}}$

   $= \dfrac{\sqrt{x}\sqrt{y}}{\sqrt{y}\sqrt{y}} - \dfrac{\sqrt{x}\sqrt{2}}{\sqrt{2}\sqrt{2}}$

   $= \dfrac{\sqrt{xy}}{y} - \dfrac{\sqrt{2x}}{2}$

   $= \dfrac{2\sqrt{xy} - y\sqrt{2x}}{2y}$

(c) $(2 - \sqrt{3})(\sqrt{3} - 2)$
   $= 2\sqrt{3} - 4 - 3 + 2\sqrt{3}$
   $= 4\sqrt{3} - 7$

(d) $\dfrac{5\sqrt{2}}{\sqrt{5} - \sqrt{2}}$

$= \dfrac{5\sqrt{2}(\sqrt{5} + \sqrt{2})}{(\sqrt{5} - \sqrt{2})(\sqrt{5} + \sqrt{2})}$

$= \dfrac{5\sqrt{10} + 5 \cdot 2}{(\sqrt{5})^2 - (\sqrt{2})^2}$

$= \dfrac{5\sqrt{10} + 10}{5 - 2}$

$= \dfrac{5\sqrt{10} + 10}{3}$

3. $\sqrt{2x} + 3 = 4$

$\sqrt{2x} = 1$

$(\sqrt{2x})^2 = 1^2$

$2x = 1$

$x = \dfrac{1}{2}$

4. $\dfrac{2 - i}{3 - i}$

$= \dfrac{(2 - i)(3 + i)}{(3 - i)(3 + i)}$

$= \dfrac{6 + 2i - 3i - i^2}{3^2 - i^2}$

$= \dfrac{6 + 2i - 3i + 1}{9 + 1}$

$= \dfrac{7 - i}{10}$

$= \dfrac{7}{10} - \dfrac{1}{10}i$

5. (a) $2x^2 + x = 3$

$2x^2 + x - 3 = 0$

$(2x + 3)(x - 1) = 0$

$2x + 3 = 0 \quad\text{or}\quad x - 1 = 0$

$2x = -3 \qquad\qquad x = 1$

$x = -\dfrac{3}{2} \quad\text{or}\quad x = 1$

(b) $\dfrac{x}{3} = \dfrac{4}{x}$

$3x\left(\dfrac{x}{3}\right) = 3x\left(\dfrac{4}{x}\right)$

$x^2 = 12$

$x = \pm\sqrt{12}$

$x = \pm 2\sqrt{3}$

(c) $2x^2 + 4x - 3 = 0$

$x^2 + 2x - \dfrac{3}{2} = 0$

$x^2 + 2x = \dfrac{3}{2}$

$x^2 + 2x + 1 = \dfrac{3}{2} + 1$

$(x + 1)^2 = \dfrac{5}{2}$

$x + 1 = \pm\sqrt{\dfrac{5}{2}}$

$x = -1 \pm \dfrac{\sqrt{10}}{2}$

$\text{or}\quad x = \dfrac{-2 \pm \sqrt{10}}{2}$

(d) $\dfrac{x}{x + 1} + \dfrac{2}{x - 3} = 4$

$(x + 1)(x - 3)\left(\dfrac{x}{x + 1} + \dfrac{2}{x - 3}\right) = (x + 1)(x - 3)(4)$

$x(x - 3) + 2(x + 1) = 4(x^2 - 2x - 3)$

$x^2 - 3x + 2x + 2 = 4x^2 - 8x - 12$

$x^2 - x + 2 = 4x^2 - 8x - 12$

$0 = 3x^2 - 7x - 14$

$x = \dfrac{-(-7) \pm \sqrt{(-7)^2 - 4(3)(-14)}}{2(3)}$

$= \dfrac{7 \pm \sqrt{49 + 168}}{6}$

$= \dfrac{7 \pm \sqrt{217}}{6}$

6. (a) $(3x^2y)^2(-2xy^3)^3$
$$= (9x^4y^2)(-8x^3y^9)$$
$$= -72x^7y^{11}$$

(b) $(-2x^{-1}y^3)(-3x^2y^{-3})^2$
$$= (-2x^{-1}y^3)(9x^4y^{-6})$$
$$= -18x^3y^{-3}$$
$$= -\frac{18x^3}{y^3}$$

(c) $\left(\dfrac{3x^{-2}y^{-1}}{9xy^{-3}}\right)^{-2}$

$$= \left(\frac{1}{3}x^{-3}y^2\right)^{-2}$$

$$= \left(\frac{1}{3}\right)^{-2}x^6y^{-4}$$

$$= \frac{3^2x^6}{y^4}$$

$$= \frac{9x^6}{y^4}$$

(d) $\dfrac{a^{-1}}{a^{-1}+b^{-1}}$

$$= \frac{\dfrac{1}{a}}{\dfrac{1}{a}+\dfrac{1}{b}}$$

$$= \frac{\left(\dfrac{1}{a}\right)ab}{\left(\dfrac{1}{a}+\dfrac{1}{b}\right)ab}$$

$$= \frac{b}{b+a}$$

7. $0.000034 = 3.4 \times 10^?$
$$= 3.4 \times 10^{-5}$$

8. $\dfrac{(150000)(0.00028)}{(0.07)(0.0002)}$

$$= \frac{(1.5 \times 10^5)(2.8 \times 10^{-4})}{(7.0 \times 10^{-2})(2.0 \times 10^{-4})}$$

$$= \left(\frac{1.5 \cdot 2.8}{7 \cdot 2}\right) \times \left(\frac{10^5 \cdot 10^{-4}}{10^{-2} \cdot 10^{-4}}\right)$$

$$= 0.3 \times 10^7$$
$$= 3.0 \times 10^6$$

9. $(-128)^{-3/7} = [(-128)^{-1/7}]^{-3}$
$$= (-2)^{-3}$$
$$= \frac{1}{(-2)^3}$$
$$= -\frac{1}{8}$$

10. (a) $(x^{1/3}x^{-1/5})^5$
$$= (x^{2/15})^5$$
$$= x^{2/3}$$

(b) $\dfrac{x^{-2/3}y^{-3/4}}{x^{1/2}}$

$$= x^{-7/6}y^{-3/4}$$

$$= \frac{1}{x^{7/6}y^{3/4}}$$

11. (a) $f(x) \cdot g(x)$
$$= (3x^2 - 4x + 3)(3x - 1)$$
$$= 9x^3 - 12x^2 + 9x - 3x^2 + 4x - 3$$
$$= 9x^3 - 15x^2 + 13x - 3$$

(b) $f[g(x)] = 3[g(x)]^2 - 4[g(x)] + 3$
$$= 3(3x - 1)^2 - 4(3x - 1) + 3$$
$$= 3(9x^2 - 6x + 1) - 12x + 4 + 3$$
$$= 27x^2 - 18x + 3 - 12x + 7$$
$$= 27x^2 - 30x + 10$$

12. $r = \sqrt{(-2 - 5)^2 + [0 - (-1)]^2}$
$$= \sqrt{49 + 1}$$
$$= \sqrt{50}$$

$(x - h)^2 + (y - k)^2 = r^2$
$[x - (-2)]^2 + (y - 0)^2 = (\sqrt{50})^2$
$$(x + 2)^2 + y^2 = 50$$

13. (a) $y = \sqrt{x} - 2$

x	y
0	-2
1	-1
4	0

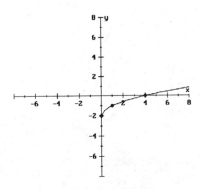

(b) $y = x^2 - 6x$

$$\frac{-b}{2a} = \frac{-(-6)}{2(1)} = 3$$

$$y = 3^2 - 6(3) = -9$$
Vertex: $(3, -9)$
x-intercepts: Let $y = 0$
$0 = x^2 - 6x$
$0 = x(x - 6)$
$x = 0$ or $x - 6 = 0$
$\qquad\qquad\qquad x = 6$

y-intercept: Let $x = 0$
$y = 0^2 - 6(0) = 0$

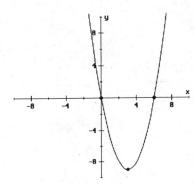

(c)
$$x^2 - 4x + y^2 = 0$$
$$x^2 - 4x + 4 + y^2 = 4$$
$$(x - 2)^2 + y^2 = 4$$
Center: $(2, 0)$
$r = \sqrt{4} = 2$

x-intercepts: Let $y = 0$
$x^2 - 4x + 0^2 = 0$
$\qquad x^2 - 4x = 0$
$\qquad x(x - 4) = 0$
$x = 0$ or $x - 4 = 0$
$\qquad\qquad\qquad x = 4$

y-intercept: Let $x = 0$
$0^2 - 4(0) + y^2 = 0$
$\qquad\qquad\qquad y^2 = 0$
$\qquad\qquad\qquad y = 0$

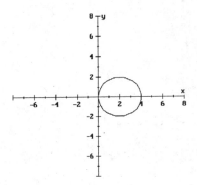

14. $y = \dfrac{kx^2}{\sqrt[3]{t}}$

$4 = \dfrac{k(3)^2}{\sqrt[3]{8}}$

$4 = \dfrac{9k}{2}$

$8 = 9k$

$\dfrac{8}{9} = k$

$y = \dfrac{\frac{8}{9}x^2}{\sqrt[3]{t}}$

$y = \dfrac{8x^2}{9\sqrt[3]{t}}$

$y = \dfrac{8(5)^2}{9\sqrt[3]{1}} = \dfrac{200}{9}$

1. $y = 4^x$

x	y
-2	1/16
-1	1/4
0	1
1	4
2	16

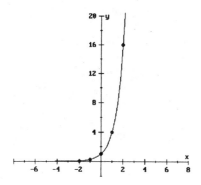

3. $y = \left(\dfrac{1}{5}\right)^x$

x	y
-2	25
-1	5
0	1
1	1/5
2	1/25

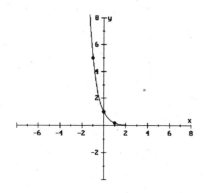

5. $y = 3^{-x} = \left(\dfrac{1}{3}\right)^x$

x	y
-2	9
-1	3
0	1
1	1/3
2	1/9

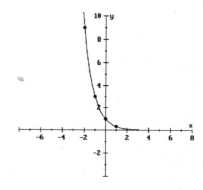

7. $y = 2^{x+1}$

x	y
-3	1/4
-2	1/2
-1	1
0	2
1	4

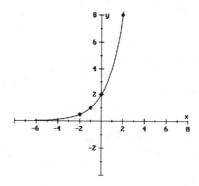

9. $y = 2^x + 1$

x	y
-2	5/4
-1	3/2
0	2
1	3
2	5

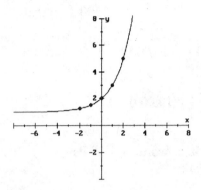

11. $2x = 2^{3x-2}$

$x = 3x - 2$

$-2x = -2$

$x = 1$

13. $5^x = 25^{x-1}$

$5x = (5^2)^{x-1}$

$5^x = 5^{2x-2}$

$x = 2x - 2$

$-x = -2$

$x = 2$

15. $4^{1-x} = 16$

$4^{1-x} = 4^2$

$1 - x = 2$

$-x = 1$

$x = -1$

17. $8^x = 4^{x+1}$

$(2^3)^x = (2^2)^{x+1}$

$2^{3x} = 2^{2x+2}$

$3x = 2x + 2$

$x = 2$

19. $\dfrac{9^{x^2}}{9^x} = 81$

$9^{x^2-x} = 9^2$

$x^2 - x = 2$

$x^2 - x - 2 = 0$

$(x - 2)(x + 1) = 0$

$x - 2 = 0$ or $x + 1 = 0$

$x = 2$ or $x = -1$

21. $4^{\sqrt{x}} = 2^{x-3}$

$(2^2)^{\sqrt{x}} = 2^{x-3}$

$2^{2\sqrt{x}} = 2^{x-3}$

$2\sqrt{x} = x - 3$

$(2\sqrt{x})^2 = (x - 3)^2$

$4x = x^2 - 6x + 9$

$0 = x^2 - 10x + 9$

$0 = (x - 9)(x - 1)$

$x - 9 = 0$ or $x - 1 = 0$

$x = 9$ or $x = 1$

When you check $x = 1$, you find it to be extraneous. The only solution is $x = 9$.

23. $16^{x^2-1} = 8^{x-1}$

$(2^4)^{x^2-1} = (2^3)^{x-1}$

$2^{4x^2-4} = 2^{3x-3}$

$4x^2 - 4 = 3x - 3$

$4x^2 - 3x - 1 = 0$

$(4x + 1)(x - 1) = 0$

$4x + 1 = 0$ or $x - 1 = 0$

$4x = -1$ $x = 1$

$x = -\dfrac{1}{4}$ or $x = 1$

25. $2^x = 5x^2$

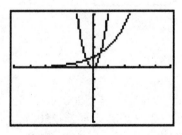

$x = -0.3906, \quad x = 0.5391$

27. $3^{2x-1} = 5x$

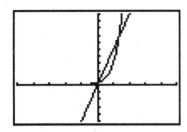

$x = 0.0794, \quad x = 1.3786$

29. $2^x = x^2 - 7x + 4$

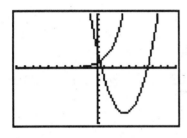

$x = 0.4057$

31. $2^x = 3x + 1$

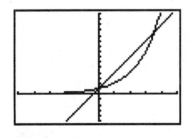

$x = 0, \quad x = 3.5377$

33. After 10 hours: $2(2500) = 5000$ bacteria
After 20 hours: $2[2(2500)] = 2^2(2500)$
$= 10,000$ bacteria
After 50 hours: $2^5(2500) = 80,000$ bacteria
After t hours: $2^{t/10}(2500)$

35. 2010 is 14 years after 1996:
$2(20,000) = 40,000$
2024 is 28 years after 1996:
$2[2(20,000)] = 2^2(20,000)$
$\qquad\qquad\quad = 80,000$

In year Y: $\quad 2^{(Y-1996)/14}(20,000)$

37. After 2 hours: $\dfrac{1}{2}\left(\dfrac{1}{2}\right)(10,000) = \left(\dfrac{1}{2}\right)^2(10,000)$

$\qquad\qquad\qquad\qquad = 2500$ bacteria

After 3 hours: $\left(\dfrac{1}{2}\right)^3(10,000) = 1250$ bacteria

After t hours: $\left(\dfrac{1}{2}\right)^t(10,000)$

39. After each year it is worth 5/6 of its previous year's value.

$V = \left(\dfrac{5}{6}\right)^t(16000)$

After 4 years: $\quad V = \left(\dfrac{5}{6}\right)^4(16000)$

$\qquad\qquad\qquad = \$7716.05$

41. Each year it is 4/5 of its previous year's value.

$V = \left(\dfrac{4}{5}\right)^t(24,000)$

In 3 years:

$V = \left(\dfrac{4}{5}\right)^3(24,000)$

$\quad = \$12,288$

43. Each year the population is 105% of the previous year's population.

$105\% = 1.05$
$P = (1.05)^t(100,000)$

4 years from now:

$P = (1.05)^4(100,000)$
$\quad = 121,551$

45. Each year the money is 108% of the previous year's value. This is interest compounded annually (not simple yearly interest.)

108% = 1.08

After 5 years:
$V = (1.08)^5(10,000) = 14,693.28$

47. Each year the money is 110% of the previous year's value. This is interest compounded annually (not simple yearly interest.)

110% = 1.10

After 5 years:
$V = (1.10)^5(5,000) = \$8,052.55$

51. $\dfrac{3}{2x} + \dfrac{6}{x+2}$

$= \dfrac{3(x+2) + 6(2x)}{2x(x+2)}$

$= \dfrac{3x + 6 + 12x}{2x(x+2)}$

$= \dfrac{15x + 6}{2x(x+2)}$

53. $\left(\sqrt{3}\right)^6 + \left(\sqrt[3]{5}\right)^6$
$= (3^{1/2})^6 + (5^{1/3})^6$
$= 3^3 + 5^2$
$= 27 + 25$
$= 52$

10.2 Exercises

1. $7^2 = 49$

3. $\log_3 81 = 4$

5. $10^4 = 10,000$

7. $\log_{10} 1000 = 3$

9. $9^2 = 81$

11. $81^{1/2} = 9$

13. $\log_6 \dfrac{1}{36} = -2$

15. $3^{-1} = \dfrac{1}{3}$

17. $\log_{25} 5 = \dfrac{1}{2}$

19. $8^1 = 8$

21. $\log_8 1 = 0$

23. $16^{3/4} = 8$

25. $\log_{27} \dfrac{1}{9} = -\dfrac{2}{3}$

27. $8^{-1/3} = \dfrac{1}{2}$

29. $\left(\dfrac{1}{2}\right)^{-2} = 4$

31. $\log_3 1 = 0$

33. $7^0 = 1$

35. $6^{1/2} = \sqrt{6}$

37. $\log_6 \sqrt{6} = \dfrac{1}{2}$

39. $\log_2 8 = \log_2 2^3$
$\qquad = 3$

41. $\log_9 81 = \log_9 9^2$
$\qquad = 2$

43. $\log_4 \dfrac{1}{4} = \log_4 4^{-1}$
$\qquad = -1$

45. $\log_5 \dfrac{1}{125} = \log_5 5^{-3}$
$\qquad = -3$

47. $\log_4 \dfrac{1}{2} = t$

$\qquad 4^t = \dfrac{1}{2}$

$\qquad (2^2)^t = 2^{-1}$
$\qquad 2^{2t} = 2^{-1}$
$\qquad 2t = -1$

$\qquad t = -\dfrac{1}{2}$

49. $\log_8 4 = t$

$\quad 8^t = 4$

$\quad (2^3)^t = 2^2$

$\quad 2^{3t} = 2^2$

$\quad 3t = 2$

$\quad t = \dfrac{2}{3}$

51. $\log_9 (-27) = t$

$\quad 9^t = -27$

Not defined

53. $\log_4 \dfrac{1}{8} = t$

$\quad 4^t = \dfrac{1}{8}$

$\quad (2^2)^t = 2^{-3}$

$\quad 2^{2t} = 2^{-3}$

$\quad 2t = -3$

$\quad t = -\dfrac{3}{2}$

55. $\log_6 \sqrt{6} = \log_6 6^{1/2}$

$\quad\quad\quad = \dfrac{1}{2}$

57. $\log_5 \sqrt[3]{25} = t$

$\quad 5^t = \sqrt[3]{25}$

$\quad 5^t = 5^{2/3}$

$\quad t = \dfrac{2}{3}$

59. $\log_5 (\log_3 243)$

$\quad = \log_5 (\log_3 3^5)$

$\quad = \log_5 5$

$\quad = 1$

61. $\log_8 (\log_7 7)$

$\quad = \log_8 1$

$\quad = 0$

63. $5^{\log_5 7} = 7$

65. $\log_5 x = 3$

$\quad 5^3 = x$

$\quad 125 = x$

67. $y = \log_{10} 1000$

$\quad 10^y = 1000$

$\quad 10^y = 10^3$

$\quad y = 3$

69. $\log_b 64 = 3$

$\quad b^3 = 64$

$\quad b^3 = 4^3$

$\quad b = 4$

71. $\log_6 x = -2$

$\quad 6^{-2} = x$

$\quad \dfrac{1}{36} = x$

73. $\log_4 x = \dfrac{3}{2}$

$\quad 4^{3/2} = x$

$\quad 8 = x$

75. $y = \log_8 32$

$\quad 8^y = 32$

$\quad (2^3)^y = 2^5$

$\quad 2^{3y} = 2^5$

$\quad 3y = 5$

$\quad y = \dfrac{5}{3}$

77. $\log_b \dfrac{1}{8} = -3$

$\quad b^{-3} = \dfrac{1}{8}$

$\quad b^{-3} = 2^{-3}$

$\quad b = 2$

79. $\log_5 x = 0$

$\quad 5^0 = x$

$\quad 1 = x$

81. $\log_{10} 23596 = 4.3728$

83. $\log_{10} 0.000925 = -3.0339$

85. $y = \log_2 x$ and $y = \log_5 x$

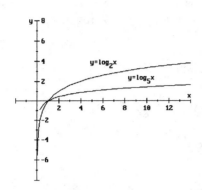

3. $\log_7 \dfrac{2}{3} = \log_7 2 - \log_7 3$

5. $\log_3 x^3 = 3 \log_3 x$

7. $\log_b a^{2/3} = \dfrac{2}{3} \log_b a$

9. $\log_b b^8 = 8$

11. $\log_s s^{-1/4} = -\dfrac{1}{4}$

13. $\log_b (x^2 y^3) = \log_b x^2 + \log_b y^3$
$$= 2 \log_b x + 3 \log_b y$$

15. $\log_b \dfrac{m^4}{n^2} = \log_b m^4 - \log_b n^2$
$$= 4 \log_b m - 2 \log_b n$$

17. $\log_b \sqrt{xy} = \log_b (xy)^{1/2}$
$$= \dfrac{1}{2} \log_b (xy)$$
$$= \dfrac{1}{2} (\log_b x + \log_b y)$$

91. $f(x) = \dfrac{\sqrt{x} + 2}{\sqrt{x} + 2}$

$f(7) = \dfrac{\sqrt{7} + 2}{\sqrt{7} + 2}$

$= \dfrac{\sqrt{9}}{\sqrt{7} + 2}$

$= \dfrac{3}{\sqrt{7} + 2}$

$= \dfrac{3(\sqrt{7} - 2)}{(\sqrt{7} + 2)(\sqrt{7} - 2)}$

$= \dfrac{3(\sqrt{7} - 2)}{7 - 4}$

$= \dfrac{3(\sqrt{7} - 2)}{3}$

$= \sqrt{7} - 2$

19. $\log_2 \sqrt[5]{\dfrac{x^2 y}{z^3}}$

$= \log_2 \left(\dfrac{x^2 y}{z^3}\right)^{1/5}$

$= \dfrac{1}{5} \log_2 \dfrac{x^2 y}{z^3}$

93.
$$
\begin{array}{r}
x^2 - 3x + 9 \\
x + 3 \overline{) x^3 + 0x^2 + 0x + 27} \\
\underline{-(x^3 + 3x^2)} \\
-3x^2 + 0x \\
\underline{-(-3x^2 - 9x)} \\
9x + 27 \\
\underline{-(9x + 27)} \\
0
\end{array}
$$

$= \dfrac{1}{5} \left(\log_2 x^2 y - \log_2 z^3 \right)$

$= \dfrac{1}{5} \left(\log_2 x^2 + \log_2 y - \log_2 z^3 \right)$

$= \dfrac{1}{5} \left(2 \log_2 x + \log_2 y - 3 \log_2 z \right)$

$= \dfrac{2}{5} \log_2 x + \dfrac{1}{5} \log_2 y - \dfrac{3}{5} \log_2 z$

10.3 Exercises

1. $\log_5 (xyz) = \log_5 x + \log_5 y + \log_5 z$

21. $\log_b (xy + z^2)$

23. $\log_b \dfrac{x^2}{yz} = \log_b x^2 - \log_b yz$

$= 2 \log_b x - (\log_b y + \log_b z)$

$= 2 \log_b x - \log_b y - \log_b z$

25. $\log_6 \sqrt{\dfrac{6m^2n}{p^5q}}$

$= \log_6 \left(\dfrac{6m^2n}{p^5q}\right)^{1/2}$

$= \dfrac{1}{2} \log_6 \dfrac{6m^2n}{p^5q}$

$= \dfrac{1}{2} (\log_6 6m^2n - \log_6 p^5q)$

$= \dfrac{1}{2} \left[\log_6 6 + \log_6 m^2 + \log_6 n - (\log_6 p^5 + \log_6 q)\right]$

$= \dfrac{1}{2} (1 + 2 \log_6 m + \log_6 n - \log_6 p^5 - \log_6 q)$

$= \dfrac{1}{2} + \log_6 m + \dfrac{1}{2} \log_6 n - \dfrac{5}{2} \log_6 p - \dfrac{1}{2} \log_6 q$

27. $\log_b x + \log_b y = \log_b xy$

29. $2 \log_b m - 3 \log_b n$

$= \log_b m^2 - \log_b n^3$

$= \log_b \dfrac{m^2}{n^3}$

31. $4 \log_b 2 + \log_b 5$

$= \log_b 2^4 + \log_b 5$

$= \log_b 2^4 \cdot 5$

$= \log_b 80$

33. $\dfrac{1}{3} \log_b x + \dfrac{1}{4} \log_b y - \dfrac{1}{5} \log_b z$

$= \log_b x^{1/3} + \log_b y^{1/4} - \log_b z^{1/5}$

$= \log_b (x^{1/3} y^{1/4}) - \log_b z^{1/5}$

$= \log_b \left(\dfrac{x^{1/3} y^{1/4}}{z^{1/5}}\right)$

35. $\dfrac{1}{2} (\log_b x + \log_b y) - 2 \log_b z$

$= \dfrac{1}{2} \log_b (xy) - 2 \log_b z$

$= \log_b (xy)^{1/2} - \log_b z^2$

$= \log_b \dfrac{(xy)^{1/2}}{z^2}$

$= \log_b \dfrac{\sqrt{xy}}{z^2}$

37. $2 \log_b x - (\log_b y + 3 \log_b z)$

$= \log_b x^2 - (\log_b y + \log_b z^3)$

$= \log_b x^2 - \log_b (yz^3)$

$= \log_b \left(\dfrac{x^2}{yz^3}\right)$

39. $\dfrac{2}{3} \log_p x + \dfrac{4}{3} \log_p y - \dfrac{3}{7} \log_p z$

$= \log_p x^{2/3} + \log_p y^{4/3} - \log_p z^{3/7}$

$= \log_p (x^{2/3} y^{4/3}) - \log_p z^{3/7}$

$= \log_p \left(\dfrac{x^{2/3} y^{4/3}}{z^{3/7}}\right)$

41. $\log_b 10$

$= \log_b (2 \cdot 5)$

$= \log_b 2 + \log_b 5$

$= 1.2 + 2.1$

$= 3.3$

43. $\log_b \dfrac{2}{5}$

$= \log_b 2 - \log_b 5$

$= 1.2 - 2.1$

$= -0.9$

45. $\log_b \dfrac{1}{3}$

$= \log_b 1 - \log_b 3$

$= 0 - 1.42$

$= -1.42$

47. $\log_b 32$

$= \log_b 2^5$

$= 5 \log_b 2$

$= 5(1.2)$

$= 6$

49. $\log_b 100$

$= \log_b 10^2$

$= 2 \log_b 10$

$= 2 \log_b (2 \cdot 5)$

$= 2(\log_b 2 + \log_b 5)$

$= 2(1.2 + 2.1)$

$= 6.6$

51. $\log_b \sqrt{20}$

$= \log_b 20^{1/2}$

$= \dfrac{1}{2} \log_b 20$

$= \dfrac{1}{2} \log_b (2^2 \cdot 5)$

$= \dfrac{1}{2} (\log_b 2^2 + \log_b 5)$

$= \dfrac{1}{2} (2 \log_b 2 + \log_b 5)$

$= \log_b 2 + \dfrac{\log_b 5}{2}$

$= 1.2 + \dfrac{2.1}{2}$

$= 2.25$

53. $\log_b \sqrt[3]{x^2}$

$= \log_b x^{2/3}$

$= \dfrac{2}{3} \log_b x$

$= \dfrac{2}{3} A$

55. $\log_b \dfrac{x^3 y^2}{z}$

$= \log_b x^3 + \log_b y^2 - \log_b z$

$= \log_b x^3 + \log_b y^2 - \log_b z$

$= 3 \log_b x + 2 \log_b y - \log_b z$

$= 3A + 2B - C$

57. $\log_b \sqrt{\dfrac{x^5 y}{z^3}}$

$= \log_b \left(\dfrac{x^5 y}{z^3} \right)^{1/2}$

$= \dfrac{1}{2} \log_b \left(\dfrac{x^5 y}{z^3} \right)$

$= \dfrac{1}{2} \left[\log_b (x^5 y) - \log_b z^3 \right]$

$= \dfrac{1}{2} (\log_b x^5 + \log_b y - \log_b z^3)$

$= \dfrac{1}{2} (5 \log_b x + \log_b y - 3 \log_b z)$

$= \dfrac{5}{2} \log_b x + \dfrac{1}{2} \log_b y - \dfrac{3}{2} \log_b z$

$= \dfrac{5}{2} A + \dfrac{1}{2} B - \dfrac{3}{2} C$

63. $(2\sqrt{x} - 5)(3\sqrt{x} + 4)$

$= 6x + 8\sqrt{x} - 15\sqrt{x} - 20$

$= 6x - 7\sqrt{x} - 20$

65. $(s, N);\ (15, 80)$ and $(18, 85)$

(a) $\quad m = \dfrac{85 - 80}{18 - 15} = \dfrac{5}{3}$

$N - 80 = \dfrac{5}{3}(s - 15)$

$N - 80 = \dfrac{5}{3}s - 25$

$N = \dfrac{5}{3}s + 55$

(b) When $s = 30$:

$N = \dfrac{5}{3}(30) + 55$

$= 105$

The heat rate should be 105 beats per minute.

1. $\log 584 = 2.7664$

3. $\log 0.00371 = -2.4306$

5. $\log 280{,}000 = 5.4472$

7. $\log 0.0000553 = -4.2573371$

9. $\log 8 = 0.9031$

11. $\text{antilog}(2.8420) = 695.0243$

13. $\text{antilog}(-2.2692) = 0.0054$

15. $\text{antilog}\ 4.1875 = 15399.2653$

17. $\ln 0.941 = -0.0608$

19. $\text{antilog}\ 0.941 = 8.7297$

21. $\ln 375 = 5.9269$

23. $\text{antilog}\ 4.85 = 70794.5784$

25. $\ln 0.0045 = -5.4037$

27. $e^{4.5} = 90.0171$

29. Undefined

31. $\log_5 x = \dfrac{\log x}{\log 5}$

33. $\log_7 8 = \dfrac{\log 8}{\log 7}$

35. $\log_5 8 = x$

$$5^x = 8$$
$$(5^x)^2 = 8^2$$
$$5^{2x} = 8^2$$
$$(5^2)^x = 8^2$$
$$25^x = 8^2$$

$$x = \log_{25} 8^2 = 2\log_{25} 8$$

37. $\log_5 87 = \dfrac{\log 87}{\log 5}$

≈ 2.7748

39. $\log_4 265 = \dfrac{\log 265}{\log 4}$

≈ 4.0249

41. $\log_3 821 = \dfrac{\log 821}{\log 3}$

≈ 6.1082

43. $\log_7 52 = \dfrac{\log 52}{\log 7}$

≈ 2.0305

45. (a) $y = \log_2 x$
(b) $a^y = x$
(c) $\log_b a^y = \log_b x$
(d) $y \log_b a = \log_b x$
(e) $y = \dfrac{\log_b x}{\log_b a}$

47. $y = \log_8 x = \dfrac{\log x}{\log 8}$

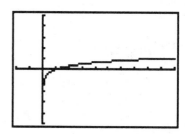

49. $y = -4 + \log_3 x$

$y = -4 + \dfrac{\log x}{\log 3}$

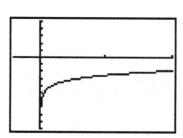

51. $y = \log_3 (x - 4)$

$y = \dfrac{\log (x - 4)}{\log 3}$

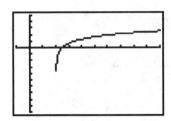

53. midpoint $= \left(\dfrac{2 + 5}{2}, \dfrac{4 - 6}{2} \right)$

$= \left(\dfrac{7}{2}, -1 \right)$

55. $\sqrt{7x + 4} - 2 = x$

$\sqrt{7x + 4} = x + 2$

$\left(\sqrt{7x + 4} \right)^2 = (x + 2)^2$

$7x + 4 = x^2 + 4x + 4$

$0 = x^2 + 4x + 4$

$0 = x^2 - 3x$

$0 = x(x - 3)$

$x = 0 \quad \text{or} \quad x - 3 = 0$

$x = 0 \quad \text{or} \qquad x = 3$

10.5 Exercises

1. (a) $N(m) = 10(2^m)$

(b) $N(2) = 10(2^{24})$

$= 167{,}772{,}160$ mice

(c)

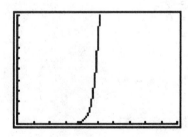

(d) When $y = 10{,}000$, $m \approx 10$.
It will take 10 months.

3. $\log_3 5 + \log_3 x = 2$

$\log_3 5x = 2$

$3^2 = 5x$

$9 = 5x$

$\dfrac{9}{5} = x$

5. $\qquad \log_2 x = 2 + \log_2 3$

$\log_2 x - \log_2 3 = 2$

$\log_2 \dfrac{x}{3} = 2$

$2^2 = \dfrac{x}{3}$

$12 = x$

7. $2 \log_5 x = \log_5 36$

$\log_5 x^2 = \log_5 36$

$x^2 = 36$

$x = 6$

(Remember: domain of log function is
$x > 0$.)

9. $\log_3 x + \log_3 (x - 8) = 2$

$\log_3 [x(x - 8)] = 2$

$3^2 = x(x - 8)$

$9 = x^2 - 8x$

$0 = x^2 - 8x - 9$

$0 = (x - 9)(x + 1)$

$x - 9 = 0 \quad \text{or} \quad x + 1 = 0$

$x = 9 \quad \text{or} \qquad x = -1$

$x = -1$ is not in the domain of $\log_3 x$, hence
the only solution is $x = 9$.

11. $\log_2 a + \log_2 (a + 2) = 3$

$\log_2 [a(a + 2)] = 3$

$2^3 = a(a + 2)$

$8 = a^2 + 2a$

$0 = a^2 + 2a - 8$

$0 = (a + 4)(a - 2)$

$a + 4 = 0 \quad \text{or} \quad a - 2 = 0$

$a = -4 \quad \text{or} \qquad a = 2$

$a = -4$ is not in the domain of $\log_2 a$, hence
the only solution is $a = 2$.

13. $\log_2 y - \log_2 (y - 2) = 3$

$\log_2 \dfrac{y}{y - 2} = 3$

$2^3 = \dfrac{y}{y - 2}$

$8 = \dfrac{y}{y - 2}$

$$8(y - 2) = y$$
$$8y - 16 = y$$
$$-16 = -7y$$
$$\frac{16}{7} = y$$

15. $\log_3 x - \log_3 (x + 3) = 5$

$$\log_3 \frac{x}{x + 3} = 5$$

$$3^5 = \frac{x}{x + 3}$$

$$243 = \frac{x}{x + 3}$$

$$243(x + 3) = x$$
$$243x + 729 = x$$
$$729 = -242x$$

$$-\frac{729}{242} = x$$

No solution, since $x = -\dfrac{729}{242}$ is not in the domain of $\log_3 x$.

17. $\log_b 5 + \log_b x = \log_b 10$
$$\log_b 5x = \log_b 10$$
$$5x = 10$$
$$x = 2$$

19. $\log_p x - \log_p 2 = \log_p 7$

$$\log_p \frac{x}{2} = \log_p 7$$

$$\frac{x}{2} = 7$$

$$x = 14$$

21. $\log_5 x + \log_5 (x + 1) = \log_5 2$
$$\log_5 [x(x + 2)] = \log_5 2$$
$$x(x + 1) = 2$$
$$x^2 + x = 2$$
$$x^2 + x - 2 = 0$$
$$(x + 2)(x - 1) = 0$$
$$x + 2 = 0 \quad \text{or} \quad x - 1 = 0$$
$$x = -2 \quad \text{or} \quad x = 1$$

$x = -2$ is not in the domain of $\log_5 x$, hence the only solution is $x = 1$.

23. $\log_3 x - \log_3 (x - 2) = \log_3 4$

$$\log_3 \frac{x}{x - 2} = \log_3 4$$

$$\frac{x}{x - 2} = 4$$

$$x = 4(x - 2)$$
$$x = 4x - 8$$
$$-3x = -8$$
$$x = \frac{8}{3}$$

25. $\log_4 x - \log_4 (x - 4) = \log_4 (x - 6)$

$$\log_4 \left(\frac{x}{x - 4} \right) = \log_4 (x - 6)$$

$$\frac{x}{x - 4} = x - 6$$
$$x = (x - 4)(x - 6)$$
$$x = x^2 - 10x + 24$$
$$0 = x^2 - 11x + 24$$
$$0 = (x - 8)(x - 3)$$
$$x - 8 = 0 \quad \text{or} \quad x - 3 = 0$$
$$x = 8 \quad \text{or} \quad x = 3$$
$x = 3$ is not in the domain of $\log_4 (x - 6)$, hence $x = 8$ is the only solution.

27. $2 \log_2 x = \log_2 (2x - 1)$
$$\log_2 x^2 = \log_2 (2x - 1)$$
$$x^2 = 2x - 1$$
$$x^2 - 2x + 1 = 0$$
$$(x - 1)^2 = 0$$
$$x - 1 = 0$$
$$x = 1$$

29. $\dfrac{1}{2} \log_3 x = \log_3 (x - 6)$

$$\log_3 x^{1/2} = \log_3 (x - 6)$$
$$x^{1/2} = x - 6$$
$$(x^{1/2})^2 = (x - 6)^2$$
$$x = x^2 - 12x + 36$$
$$0 = x^2 - 13x + 36$$
$$0 = (x - 9)(x - 4)$$
$$x - 9 = 0 \quad \text{or} \quad x - 4 = 0$$
$$x = 9 \quad \text{or} \quad x = 4$$
$x = 4$ is not in the domain of $\log_3 (x - 6)$, hence $x = 9$ is the only solution.

31.
$$2 \log_b x = \log_b (6x - 5)$$
$$\log_b x^2 = \log_b (6x - 5)$$
$$x^2 = 6x - 5$$
$$x^2 - 6x + 5 = 0$$
$$(x - 5)(x - 1) = 0$$
$$x - 5 = 0 \quad \text{or} \quad x - 1 = 0$$
$$x = 5 \quad \text{or} \qquad x = 1$$

33.
$$2^x = 5$$
$$\log 2^2 = \log 5$$
$$x \log 2 = \log 5$$
$$x = \frac{\log 5}{\log 2}$$
$$x \approx 2.3219$$

35.
$$2^{x+1} = 6$$
$$\log 2^{x+1} = \log 6$$
$$(x + 1) \log 2 = \log 6$$
$$x + 1 = \frac{\log 6}{\log 2}$$
$$x = \frac{\log 6}{\log 2} - 1$$
$$x \approx 1.5850$$

37.
$$4^{2x+3} = 5$$
$$\log 4^{2x+3} = \log 5$$
$$(2x + 3) \log 4 = \log 5$$
$$2x + 3 = \frac{\log 5}{\log 4}$$
$$2x = \frac{\log 5}{\log 4} - 3$$
$$x = \frac{\log 5}{2 \log 4} - \frac{3}{2}$$
$$x \approx -0.9195$$

39.
$$7^{y+1} = 3^y$$
$$\log 7^{y+1} = \log 3^y$$
$$(y + 1) \log 7 = y \log 3$$
$$y \log 7 + \log 7 = y \log 3$$
$$\log 7 = y \log 3 - y \log 7$$
$$\log 7 = y(\log 3 - \log 7)$$

$$\log 7 = y \left(\log \frac{3}{7} \right)$$
$$\frac{\log 7}{\log^{3/7}} = y$$
$$y \approx -2.2966$$

41.
$$6^{2x+1} = 5^{x+2}$$
$$\log 6^{2x+1} = \log 5^{x+2}$$
$$(2x + 1) \log 6 = (x + 2) \log 5$$
$$2x \log 6 + \log 6 = x \log 5 + 2 \log 5$$
$$2x \log 6 - x \log 5 = 2 \log 5 - \log 6$$
$$x(2 \log 6 - \log 5) = 2 \log 5 - \log 6$$
$$x = \frac{2 \log 5 - \log 6}{2 \log 6 - \log 5}$$
$$x \approx 0.7229$$

43.
$$8^{3x-2} = 9^{x+2}$$
$$\log 8^{3x-2} = \log 9^{x+2}$$
$$(3x - 2) \log 8 = (x + 2) \log 9$$
$$3x \log 8 - 2 \log 8 = x \log 9 + 2 \log 9$$
$$3x \log 8 - x \log 9 = 2 \log 9 + 2 \log 8$$
$$x(3 \log 8 - \log 9) = 2 \log 9 + 2 \log 8$$
$$x = \frac{2 \log 9 + 2 \log 8}{3 \log 8 - \log 9}$$
$$x \approx 2.1166$$

45.
$$3^x = 5 \cdot 2^x$$
$$\frac{3^x}{2^x} = 5$$
$$\left(\frac{3}{2} \right)^x = 5$$
$$\log \left(\frac{3}{2} \right)^x = \log 5$$
$$x \log \frac{3}{2} = \log 5$$
$$x = \frac{\log 5}{\log \frac{3}{2}}$$
$$x \approx 3.9694$$

47.
$$2^y 5^y = 3$$
$$(2 \cdot 5)^y = 3$$
$$10^y = 3$$
$$\log 10^y = \log 3$$
$$y = \log 3$$
$$y \approx 0.4771$$

49.
$$4^a 3^{a+1} = 2$$
$$4^a \cdot 3^a \cdot 3 = 2$$
$$4^a \cdot 3^a = \frac{2}{3}$$
$$(4 \cdot 3)^a = \frac{2}{3}$$
$$12^a = \frac{2}{3}$$
$$\log 12^a = \log \frac{2}{3}$$
$$a \log 12 = \log \frac{2}{3}$$
$$a = \frac{\log^{2/3}}{\log 12}$$
$$a \approx -0.1632$$

51.
$$pH = -\log\left[H_3O^+\right]$$
$$= -\log\left[3.98 \times 10^{-6}\right]$$
$$= 5.4$$

53.
$$pH = -\log\left[H_3O^+\right]$$
$$-7 = \log\left[H_3O^+\right]$$
$$-7 = \log\left[H_3O^+\right]$$
$$10^{-7} = H_3O^+$$

55.
$$N = 10 \log I + 160$$
$$200 = 10 \log I + 160$$
$$40 = 10 \log I$$
$$4 = \log I$$
$$10^4 = I$$

The intensity is 10^4 watts/cm^2.

57.
$$(x - h)^2 + (y - k)^2 = r^2$$
$$(x - 2)^2 + [y - (-3)]^2 = 6^2$$
$$(x - 2)^2 + (y + 3)^2 = 36$$

59.
$$2x^2 - 5x \le 3$$
$$2x^2 - 5x - 3 \le 0$$
$$2x^2 - 5x - 3 = 0$$
$$(2x + 1)(x - 3) = 0$$
$$2x + 1 = 0 \qquad x - 3 = 0$$
$$x = -\frac{1}{2} \qquad x = 3$$

The intervals are $x < -\dfrac{1}{2}$, $-\dfrac{1}{2} < x < 3$,
$x > 3$

Test points:

	$(2x + 1)(x - 3)$	
-1	$[2(-1) + 1](-1 - 3)$	pos
0	$(2 \cdot 0 + 1)(0 - 3)$	neg
4	$(2 \cdot 4 + 1)(4 - 3)$	pos

We want $(2x + 1)(x - 3) \le 0$: neg or 0
Solution: $-\dfrac{1}{2} \le x \le 3$

10.6 Exercises

1. $A = P\left(1 + \dfrac{r}{n}\right)^{nt}$
$$= 8000\left(1 + \frac{0.06}{2}\right)^{2(5)}$$
$$= \$10{,}751.33$$

3. $A = P\left(1 + \dfrac{r}{n}\right)^{nt}$
$$= 8000\left(1 + \frac{0.06}{12}\right)^{12(5)}$$
$$= \$10{,}790.80$$

5. $A = P\left(1 + \dfrac{r}{n}\right)^{nt}$
$$2(9000) = 9000\left(1 + \frac{0.062}{12}\right)^{12t}$$
$$2 = \left(1 + \frac{0.062}{12}\right)^{12t}$$

$$\ln 2 = \ln\left(1 + \frac{0.062}{12}\right)^{12t}$$

$$\ln 2 = 12t \ln\left(1 + \frac{0.062}{12}\right)$$

$$t = \frac{\ln 2}{12 \ln\left(1 + \frac{0.062}{12}\right)} \approx 11.21 \text{ years}$$

7. $2P = P\left(1 + \frac{0.062}{12}\right)^{12t}$

$2 = \left(1 + \frac{0.062}{12}\right)^{12t}$

See Exercise #5.
This is the same equation, hence it would take approximately 11.21 years.

9. $2P = P\left(1 + \frac{0.10}{12}\right)^{12t}$

$2 = \left(1 + \frac{0.10}{12}\right)^{12t}$

$\ln 2 = \ln\left(1 + \frac{0.10}{12}\right)^{12t}$

$\ln 2 = 12t \ln\left(1 + \frac{0.10}{12}\right)$

$t = \dfrac{\ln 2}{12 \ln\left(1 + \frac{0.10}{12}\right)} \approx 6.96 \text{ years}$

11. $A = Pe^{rt}$

$20000 = 5000e^{0.073t}$

$4 = e^{0.073t}$

$\ln 4 = \ln e^{0.073t}$

$\ln 4 = 0.073t$

$\dfrac{\ln 4}{0.073} = t$

$t \approx 18.99$

It would take approximately 19 years.

13. $P = P_0 e^{rt}$

$P = 2000e^{0.08(15)}$

$P \approx 6640$

There will be approximately 6640 inhabitants.

15. $A = 10000e^{0.0542t}$

$A(5) = 10000e^{0.0542(5)}$

≈ 13112.75

There are approximately 13113 bacteria present.

17. $A = 10000e^{0.0542t}$

$100000 = 10000e^{0.0542t}$

$10 = e^{0.0542t}$

$\ln 10 = \ln e^{0.0542t}$

$\ln 10 = 0.0542t$

$\dfrac{\ln 10}{0.0542} = t$

$t \approx 42.5 \text{ hr}$

$A = 10000e^{0.122t}$

$100000 = 10000e^{0.122t}$

$10 = e^{0.122t}$

$\ln 10 = \ln e^{0.122t}$

$\ln 10 = 0.122t$

$\dfrac{\ln 10}{0.122} = t$

$t \approx 18.9 \text{ hr}$

19. $P = P_0 e^{rt}$

$10000 = 1000e^{0.08t}$

$10 = e^{0.08t}$

$\ln 10 = \ln e^{0.08t}$

$\ln 10 = 0.08t$

$\dfrac{\ln 10}{0.08} = t$

$t \approx 28.8 \text{ hr}$

21. $P = P_0 e^{rt}$

$3000 = 1000e^{r(5)}$

$3 = e^{5r}$

$\ln 3 = \ln e^{5r}$

$\ln 3 = 5r$

$\dfrac{\ln 3}{5} = r$

$r = 0.2197$

$P = 1000e^{0.2197t}$

$50000 = 1000e^{0.2197t}$

$50 = e^{0.2197t}$

$\ln 50 = \ln e^{0.2197t}$

$\ln 50 = 0.2197t$

$\dfrac{\ln 50}{0.2197} = t$

$t \approx 17.8 \text{ hr}$

23. $$A = A_0 e^{-0.0004t}$$

$$\frac{1}{2}A_0 = A_0 e^{-0.0004t}$$

$$\frac{1}{2} = e^{-0.0004t}$$

$$\ln \frac{1}{2} = \ln e^{-0.0004t}$$

$$\ln \frac{1}{2} = -0.0004t$$

$$\frac{\ln \frac{1}{2}}{-0.0004} = t$$

$$t \approx 1732.87$$

25. $$A = A_0 e^{-rt}$$

$$\frac{1}{2}A_0 = A_0 e^{-r(4)}$$

$$\frac{1}{2} = e^{-4r}$$

$$\ln \frac{1}{2} = \ln e^{-4r}$$

$$\ln \frac{1}{2} = -4r$$

$$\frac{\ln \frac{1}{2}}{-4} = r$$

$$r = 0.173 = 17.3\% \text{ per day}$$

27. $$R = \log I$$
$$3.6 = \log I_1$$
$$10^{3.6} = I_1$$

$$R = \log I$$
$$7.2 = \log I_2$$
$$10^{7.2} = I_2$$

$$10^{7.2} = (10^{3.6})^2$$
Hence $I_2 = I_1^2$.

29. (a) $$A = A_0 e^{rt}$$

$$\frac{1}{2}A_0 = A_0 e^{r(5730)}$$

$$\frac{1}{2} = e^{5730r}$$

$$\ln \frac{1}{2} = \ln e^{5730r}$$

$$\ln \frac{1}{2} = 5730r$$

$$\frac{\ln \frac{1}{2}}{5730} = r$$

$$r = -0.000121$$

$$A = A_0 e^{-0.000121t}$$

(b) $$0.35A_0 = A_0 e^{-0.000121t}$$
$$0.35 = e^{0.000121t}$$
$$\ln 0.35 = \ln e^{-0.000121t}$$
$$\ln 0.35 = -0.000121t$$

$$\frac{\ln 0.35}{-0.000121} = t$$

$$t \approx 8676.2 \text{ years}$$

31. $y = 6 - x$

x	y
0	6
6	0

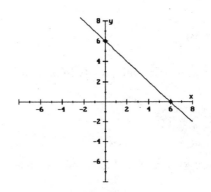

33. $(3x - 4)^2 = 5$

$\qquad 3x - 4 = \pm\sqrt{5}$

$\qquad\quad 3x = 4 \pm \sqrt{5}$

$\qquad\qquad x = \dfrac{4 \pm \sqrt{5}}{3}$

CHAPTER 10 REVIEW EXERCISES

1. $y = 6^x$

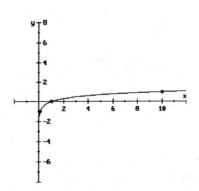

x	y
-2	1/36
-1	1/6
0	1
1	6
2	36

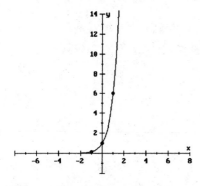

3. $y = \log_{10} x$

$\qquad 10_y = x$

x	y
1/100	-2
1/10	-1
1	0
10	1
100	2

5. $y = 2^{-x}$

x	y
-2	4
-1	2
0	1
1	1/2
2	1/4

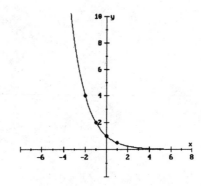

7. $\log_3 81 = 4$

$\qquad\quad 3^4 = 81$

9. $\qquad 4^{-3} = \dfrac{1}{64}$

$\qquad \log_4 \dfrac{1}{64} = -3$

11. $\log_8 4 = \dfrac{2}{3}$

$\qquad 8^{2/3} = 4$

13. $\qquad 25^{1/2} = 5$

$\qquad \log_{25} 5 = \dfrac{1}{2}$

15. $\log_7 \sqrt{7} = \dfrac{1}{2}$

$7^{1/2} = \sqrt{7}$

17. $\log_6 1 = 0$

$6^0 = 1$

19. $\log_{10} 1000 = x$

$10^x = 1000$

$10^x = 10^3$

$x = 3$

21. $\log_3 \dfrac{1}{9} = x$

$3^x = \dfrac{1}{9}$

$3^x = 3^{-2}$

$x = -2$

23. $\log_2 \dfrac{1}{4} = x$

$2^x = \dfrac{1}{4}$

$2^x = 2^{-2}$

$x = -2$

25. $\log_{1/3} 9 = x$

$\left(\dfrac{1}{3}\right)^x = 9$

$3^{-x} = 9$

$3^{-x} = 3^2$

$-x = 2$

$x = -2$

27. $\log_3 \dfrac{1}{9} = x$

$3^x = \dfrac{1}{9}$

$3^x = 3^{-2}$

$x = -2$

29. $\log_b \sqrt{b} = x$

$b^x = \sqrt{b}$

$b^x = b^{1/2}$

$x = \dfrac{1}{2}$

31. $\log_{16} 32 = x$

$16^x = 32$

$(2^4)^x = 2^5$

$2^{4x} = 2^5$

$4x = 5$

$x = \dfrac{5}{4}$

33. $\log_b (x^3 y^7) = \log_b x^3 + \log_b y^7$

$= 3 \log_b x + 7 \log_b y$

35. $\log_b \dfrac{u^2 v^5}{w^3} = \log_b (u^2 v^5) - \log_b w^3$

$= \log_b u^2 + \log_b v^5 - \log_b w^3$

$= 2 \log_b u + 5 \log_b v - 3 \log_b w$

37. $\log_b \sqrt[3]{xy} = \log_b (xy)^{1/3}$

$= \dfrac{1}{3} \log_b (xy)$

$= \dfrac{1}{3} (\log_b x + \log_b y)$

$= \dfrac{1}{3} \log_b x + \dfrac{1}{3} \log_b y$

39. $\log_b (x^3 + y^4)$

41. $\log_b \sqrt[4]{\dfrac{x^6 y^2}{z^2}}$

$= \log_b \left(\dfrac{x^6 y^2}{z^2}\right)^{1/4}$

$= \dfrac{1}{4} \log_b \left(\dfrac{x^6 y^2}{z^2}\right)$

$= \dfrac{1}{4} \left[\log_b (x^6 y^2) - \log_b z^2\right]$

$$= \frac{1}{4}(\log_b x^6 + \log_b y^2 - 2 \log_b z^2)$$

$$= \frac{1}{4}(6 \log_b x + 2 \log_b y - 2 \log_b z)$$

$$= \frac{3}{2} \log_b x + \frac{1}{2} \log_b y - \frac{1}{2} \log_b z$$

43. $\log_b 28 = \log_b 2^2 \cdot 7$
$$= \log_b 2^2 + \log_b 7$$
$$= 2 \log_b 2 + \log_b 7$$
$$= 2(1.1) + 1.32$$
$$= 3.52$$

45. $9^x = \dfrac{1}{81}$

$9^x = 9^{-2}$
$x = -2$

47. $16^x = 32$
$(2^4)^x = 2^5$
$2^{4x} = 2^5$
$4x = 5$
$x = \dfrac{5}{4}$

49. $\qquad 5^{x+1} = 3$
$\log 5^{x+1} = \log 3$
$(x + 1) \log 5 = \log 3$

$$x + 1 = \frac{\log 3}{\log 5}$$

$$x = \frac{\log 3}{\log 5} - 1$$

$$x \approx -0.3174$$

51. $\log (x + 10) - \log (x + 1) = 1$

$$\log \frac{x + 10}{x + 1} = 1$$

$$10^1 = \frac{x + 10}{x + 1}$$

$10(x + 1) = x + 10$
$10x + 10 = x + 10$
$9x = 0$
$x = 0$

53. $\log_2 (t + 1) + \log_2 (t - 1) = 3$
$\log_2 [(t + 1)(t - 1)] = 3$
$2^3 = (t + 1)(t - 1)$
$8 = t^2 - 1$
$9 = t^2$
$t = \pm 3$

$t = -3$ is not in the domains of the above log functions, hence $t = 3$ is the only solution.

55. $\log_b 3x + \log_b (x + 2) = \log_b 9$
$\log_b [3x(x + 2)] = \log_b 9$
$3x(x + 2) = 9$
$3x^2 + 6x = 9$
$3x^2 + 6x - 9 = 0$
$x^2 + 2x - 3 = 0$
$(x + 3)(x - 1) = 0$
$x + 3 = 0 \quad \text{or} \quad x - 1 = 0$
$x = -3 \quad \text{or} \qquad x = 1$

$x = -3$ is not in the domains of the above log functions, hence the only solution is $x = 1$.

57. $\log 783 \approx 2.8938$

59. antilog $(-3) = 0.001$

61. $\log 0.00499 \approx -2.3019$

63. $\ln 0.0063 \approx -5.0672$

65. $e^{7.8} \approx 2440.6020$

67. $\log_5 73 = \dfrac{\log 73}{\log 5}$

$$\approx 2.6658$$

69. $\log_{12} 764 = \dfrac{\log 764}{\log 12}$

$$\approx 2.6716$$

71. $\log_{0.2} 190 = \dfrac{\log 190}{\log 0.2}$

$$\approx -3.2602$$

73. $y = \log_7 x = \dfrac{\log x}{\log 7}$

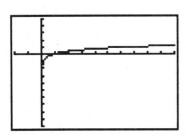

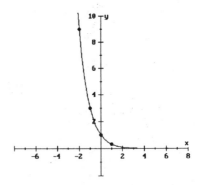

75. $A = P\left(1 + \dfrac{r}{n}\right)^{nt}$

$\quad = 6000\left(1 + \dfrac{0.082}{2}\right)^{2(8)}$

$\quad = \$11{,}412.03$

2. $\quad y = \log_7 x$

$\quad\quad 7^6 = x$

x	y
1/49	-2
1/7	-1
1	0
7	1
49	2

77. $\quad\quad A = A_0 e^{-0.045t}$

$\quad\quad 25 = 100 e^{-0.045t}$

$\quad\quad 0.25 = e^{-0.045t}$

$\quad\ln 0.25 = \ln e^{-0.045t}$

$\ln (0.25) = -0.045t$

$\quad\dfrac{\ln 0.25}{-0.045} = t$

$\quad\quad t \approx 30.8$

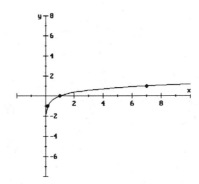

79. $pH = -\log\left[H_3O^+\right]$

$\quad\quad = -\log\left[6.21 \times 10^{-9}\right]$

$\quad\quad = 8.2069$

CHAPTER 10 PRACTICE TEST

1. $y = \left(\dfrac{1}{3}\right)^x$

x	y
-2	9
-1	3
0	1
1	1/3
2	1/9

3. (a) $\log_2 16 = 4$

$\quad\quad\quad 2^4 = 16$

(b) $\log_9 \dfrac{1}{3} = -\dfrac{1}{2}$

$\quad\quad 9^{-1/2} = \dfrac{1}{3}$

4. (a) $\quad 8^{2/3} = 4$

$\quad\quad \log_8 4 = \dfrac{2}{3}$

(b) $\quad\quad 10^{-2} = 0.01$

$\quad \log_{10} 0.01 = -2$

5. (a) $\log_3 \frac{1}{3} = x$

$$3^x = \frac{1}{3}$$
$$3^x = 3^{-1}$$
$$x = -1$$

(b) $\log_{81} 9 = x$

$$81^x = 9$$
$$(9^2)^x = 9$$
$$9^{2x} = 9^1$$
$$2x = 1$$
$$x = \frac{1}{2}$$

(c) $\log_8 32 = x$

$$8^x = 32$$
$$(2^3)^x = 2^5$$
$$2^{3x} = 2^5$$
$$3x = 5$$
$$x = \frac{5}{3}$$

6. (a) $\log_b (x^3 y^5 z)$

$$= \log_b x^3 + \log_b y^5 + \log_b z$$
$$= 3 \log_b x + 5 \log_b y + \log_b z$$

(b) $\log_4 \frac{64\sqrt{x}}{y^z}$

$$= \log_4 (64\sqrt{x}) - \log_4 (yz)$$
$$= \log_4 64 + \log_4 \sqrt{x}$$
$$\quad - [\log_4 y + \log_4 z]$$
$$= 3 + \log_4 x^{1/2} - \log_4 y - \log_4 z$$
$$= 3 + \frac{1}{2} \log_4 x - \log_4 y - \log_4 z$$

7. (a) $\log 27{,}900 \approx 4.4456$
   (b) antilog $(-2.4) \approx 0.0040$
   (c) $\ln 0.004 \approx -5.5215$
   (d) $e^{-0.02} \approx 0.9802$

8. (a) $\log_5 67 = \dfrac{\log 67}{\log 5} \approx 2.6125$

   (b) $\log_8 0.0034 = \dfrac{\log 0.0034}{\log 8} \approx -2.7334$

9. (a) $3^x = \dfrac{1}{81}$

$$3^x = 3^{-4}$$
$$x = -4$$

   (b) $\log_2 (x + 4) + \log_2 (x - 2) = 4$

$$\log_2 [(x + 4)(x - 2)] = 4$$
$$2^4 = (x + 4)(x - 2)$$
$$16 = x^2 + 2x - 8$$

$$0 = x^2 + 2x - 24$$
$$0 = (x + 6)(x - 4)$$
$$x + 6 = 0 \quad \text{or} \quad x - 4 = 0$$
$$x = -6 \quad \text{or} \quad x = 4$$

$x = -6$ is not in the domain of the above log functions, hence the only solution is $x = 4$.

(c) $4^{5x} = 32^{3x-4}$

$$(2^2)^{5x} = (2^5)^{3x-4}$$
$$2^{10x} = 2^{15x-20}$$
$$10x = 15x - 20$$
$$-5x = -20$$
$$x = 4$$

(d) $\log (5x) - \log (x - 5) = 1$

$$\log \frac{5x}{x - 5} = 1$$
$$10^1 = \frac{5x}{x - 5}$$
$$10(x - 5) = 5x$$
$$10x - 50 = 5x$$
$$-50 = -5x$$
$$10 = x$$

(e) $\qquad 9^{x+3} = 5$

$$\log 9^{x+3} = \log 5$$
$$(x + 3) \log 9 = \log 5$$
$$x + 3 = \frac{\log 5}{\log 9}$$
$$x = -3 + \frac{\log 5}{\log 9} \approx -2.2675$$

10. $A = P\left(1 + \dfrac{r}{n}\right)^{nt}$

$$= 5000\left(1 + \frac{0.084}{4}\right)^{4(7)}$$
$$= \$8947.27$$

11. $\qquad A = A_0 e^{-0.04t}$

$$25 = 100 e^{-0.04t}$$
$$0.25 = e^{-0.04t}$$
$$\ln 0.25 = \ln e^{-0.04t}$$
$$\ln 0.25 = -0.04t$$
$$\frac{\ln 0.25}{-0.04} = t$$
$$t \approx 34.7$$

11.1 Exercises

1. $\begin{cases} x + y + z = 9 \\ 2x - y + z = 9 \\ x - y + z = 3 \end{cases}$

$\begin{array}{r} x + y + z = 9 \\ \underline{2x - y + z = 9} \\ 3x + 2z = 18 \end{array}$

$\begin{array}{r} x + y + z = 9 \\ \underline{2x - y + z = 9} \\ 2x + 2z = 12 \end{array}$

$\begin{cases} 3x + 2z = 18 \\ 2x + 2z = 12 \end{cases}$

$\begin{array}{r} 3x + 2z = 18 \\ \underline{-2x - 2z = -12} \\ x = 6 \end{array}$

$3(6) + 2z = 18$
$18 + 2z = 18$
$2z = 0$
$z = 0$

$6 + y + 0 = 9$
$6 + y = 9$
$y = 3$

$x = 6, \; y = 3, \; z = 0$

3. $\begin{cases} -x + 2y + z = 0 \\ x - y + 2z = 1 \\ x + 3y + z = 5 \end{cases}$

$\begin{array}{r} -x + 2y + z = 0 \\ \underline{x - y + 2z = 1} \\ y + 3z = 1 \end{array}$

$\begin{array}{r} -x + 2y + z = 0 \\ \underline{x + 3y + z = 5} \\ 5y + 2z = 5 \end{array}$

$\begin{cases} y + 3z = 1 \\ 5y + 2z = 5 \end{cases}$

$\begin{array}{r} -5y - 15z = -5 \\ \underline{5y + 2z = 5} \\ -13z = 0 \\ z = 0 \end{array}$

$y + 3(0) = 1$
$y = 1$

$x - 1 + 2(0) = 1$
$x - 1 = 1$
$x = 2$
$x = 2, \; y = 1, \; z = 0$

5. $\begin{cases} x + y - z = 1 \\ 2x + 2y + 2z = 0 \\ x - y + z = 3 \end{cases}$

$\begin{cases} x + y - z = 1 \\ x + y + z = 0 \\ x - y + z = 3 \end{cases}$

$\begin{array}{r} x + y - z = 1 \\ \underline{x + y + z = 0} \\ 2x + 2y = 1 \end{array}$

$\begin{array}{r} x + y - z = 1 \\ \underline{x - y + z = 3} \\ 2x = 4 \\ x = 2 \end{array}$

$2(2) + 2y = 1$
$4 + 2y = 1$
$2y = -3$
$y = -\dfrac{3}{2}$

$2 - \left(-\dfrac{3}{2}\right) + z = 3$

$\dfrac{7}{2} + z = 3$

$z = -\dfrac{1}{2}$

$x = 2, \; y = -\dfrac{3}{2}, \; z = -\dfrac{1}{2}$

7. $\begin{cases} x + 2y + 3z = 1 \\ 3x + 6y + 9z = 3 \\ 4x + 8y + 12z = 4 \end{cases}$

$\begin{cases} x + 2y + 3z = 1 \\ x + 2y + 3z = 1 \\ x + 2y + 3z = 1 \end{cases}$

$\{(x, y, z) \mid x + 2y + 3z = 1\}$

9. $\begin{cases} 3x - 2y + 5z = 2 \\ 4x - 7y - z = 19 \\ 5x - 6y + 4z = 13 \end{cases}$

$$3x \ - \ 2y \ + \ 5z \ = \ 2$$
$$\underline{20x \ - \ 35y \ - \ 5z \ = \ 95}$$
$$23x \ - \ 37y \qquad = \ 97$$

$$16x \ - \ 28y \ - \ 4z \ = \ 76$$
$$\underline{5x \ - \ 6y \ + \ 4z \ = \ 13}$$
$$21x \ - \ 34y \qquad = \ 89$$

$$\begin{cases} 23x \ - \ 37y \ = \ 97 \\ 21x \ - \ 34y \ = \ 89 \end{cases}$$

$$-483x \ + \ 777y \ = \ -2037$$
$$\underline{483x \ - \ 782y \ = \ 2047}$$
$$-5y \ = \ 10$$
$$y \ = \ -2$$

$$21x \ - \ 34(-2) \ = \ 89$$
$$21x \ + \ 68 \ = \ 89$$
$$21x \ = \ 21$$
$$x \ = \ 1$$

$$3(1) \ - \ 2(-2) \ + \ 5z \ = \ 2$$
$$7 \ + \ 5z \ = \ 2$$
$$5z \ = \ -5$$
$$z \ = \ -1$$

$$x \ = \ 1, \ \ y \ = \ -2, \ \ z \ = \ -1$$

11. $$\begin{cases} 2a \ + \ b \ - \ 3c \ = \ -6 \\ 4a \ - \ 4b \ + \ 2c \ = \ 10 \\ 6a \ - \ 7b \ + \ c \ = \ 12 \end{cases}$$

$$8a \ + \ 4b \ - \ 12c \ = \ -24$$
$$\underline{4a \ - \ 4b \ + \ 2c \ = \ 10}$$
$$12a \qquad - \ 10c \ = \ -14$$
$$6a \qquad - \ 5c \ = \ -7$$

$$14a \ + \ 7b \ - \ 21c \ = \ -42$$
$$\underline{6a \ - \ 7b \ + \ c \ = \ 12}$$
$$20a \qquad - \ 20c \ = \ -30$$
$$2a \qquad - \ 2c \ = \ -3$$

$$\begin{cases} 6a \ - \ 5c \ = \ -7 \\ 2a \ - \ 2c \ = \ -3 \end{cases}$$

$$6a \ - \ 5c \ = \ -7$$
$$\underline{-6a \ + \ 6c \ = \ 9}$$
$$c \ = \ 2$$

$$2a \ - \ 2(2) \ = \ -3$$
$$2a \ - \ 4 \ = \ -3$$
$$2a \ = \ 1$$
$$a \ = \ \frac{1}{2}$$

$$2\left(\frac{1}{2}\right) \ + \ b \ - \ 3(2) \ = \ -6$$
$$1 \ + \ b \ - \ 6 \ = \ -6$$
$$b \ = \ -1$$

$$a \ = \ \frac{1}{2}, \ \ b \ = \ -1, \ \ c \ = \ 2$$

13. $$\begin{cases} x \ + \ 3y \ + \ 2z \ = \ 3 \\ x \qquad + \ 3z \ = \ 4 \\ x \ - \ 4y \ - \ z \ = \ 0 \end{cases}$$

$$4x \ + \ 12y \ + \ 8z \ = \ 12$$
$$\underline{3x \ - \ 12y \ - \ 3z \ = \ 0}$$
$$7x \qquad + \ 5z \ = \ 12$$

$$\begin{cases} x \ + \ 3z \ = \ 4 \\ 7x \ + \ 5z \ = \ 12 \end{cases}$$

$$-7x \ - \ 21z \ = \ -28$$
$$\underline{7x \ + \ 5z \ = \ 12}$$
$$-16z \ = \ -16$$
$$z \ = \ 1$$

$$x \ + \ 3(1) \ = \ 4$$
$$x \ = \ 1$$

$$1 \ + \ 3y \ + \ 2(1) \ = \ 3$$
$$3y \ = \ 0$$
$$y \ = \ 0$$

$$x \ = \ 1, \ \ y \ = \ 0, \ \ z \ = \ 1$$

15. $$\begin{cases} \dfrac{1}{2}s \ + \ \dfrac{1}{3}t \ + \ u \ = \ 3 \\[2mm] \dfrac{1}{3}s \ - \ \dfrac{1}{2}t \ - \ 2u \ = \ 1 \\[2mm] \dfrac{2}{3}s \ - \ \dfrac{1}{6}t \ + \ \dfrac{1}{2}u \ = \ 6 \end{cases}$$

$$\begin{cases} 3s \ + \ 2t \ + \ 6u \ = \ 18 \\ 2s \ - \ 3t \ - \ 12u \ = \ 6 \\ 4s \ - \ t \ + \ 3u \ = \ 36 \end{cases}$$

$$3s \ + \ 2t \ + \ 6u \ = \ 18$$
$$\underline{8s \ - \ 2t \ + \ 6u \ = \ 72}$$
$$11s \qquad + \ 12u \ = \ 90$$

$$2s \ - \ 3t \ - \ 12u \ = \ 6$$
$$\underline{-12s \ + \ 3t \ - \ 9u \ = \ -108}$$
$$-10s \qquad - \ 21u \ = \ -102$$

$$\begin{cases} 11s + 12u = 90 \\ -10s - 21u = -102 \end{cases}$$

$$\begin{array}{r} 77s + 84u = 630 \\ \underline{-40s - 84u = -408} \\ 37s \quad\quad = 222 \\ s = 6 \end{array}$$

$$\begin{array}{r} 11(6) + 12u = 90 \\ 12u = 24 \\ u = 2 \end{array}$$

$$\begin{array}{r} 4(6) - t + 3(2) = 36 \\ 30 - t = 36 \\ -t = 6 \\ t = -6 \end{array}$$

$$s = 6, \quad t = -6, \quad u = 2$$

17. $\begin{cases} p + q + r = 6 \\ 2q + r - p = 6 \\ r - p + q = 4 \end{cases}$

$$\begin{cases} p + q + r = 6 \\ -p + 2q + r = 6 \\ -p + q + r = 4 \end{cases}$$

$$\begin{array}{r} p + q + r = 6 \\ \underline{-p + 2q + r = 6} \\ 3q + 2r = 12 \end{array}$$

$$\begin{array}{r} p + q + r = 6 \\ \underline{-p + q + r = 4} \\ 2q + 2r = 10 \end{array}$$

$$\begin{cases} 3q + 2r = 12 \\ 2q + 2r = 10 \end{cases}$$

$$\begin{array}{r} 3q + 2r = 12 \\ \underline{-2q - 2r = -10} \\ q \quad\quad = 2 \end{array}$$

$$\begin{array}{r} 2(2) + 2r = 10 \\ 2r = 6 \\ r = 3 \end{array}$$

$$\begin{array}{r} p + 2 + 3 = 6 \\ p = 1 \end{array}$$

$$p = 1, \quad q = 2, \quad r = 3$$

19. $\begin{cases} x + y \quad\;\; = 0 \\ \quad\; y + z = 0 \\ x \quad\quad + z = 2 \end{cases}$

$$\begin{array}{r} -y - z = 0 \\ \underline{x \quad\quad + z = 2} \\ x - y \quad\quad = 2 \end{array}$$

$$\begin{cases} x + y = 0 \\ x - y = 2 \end{cases}$$

$$\begin{array}{r} x + y = 0 \\ \underline{x - y = 2} \\ 2x \quad\quad = 2 \\ x = 1 \end{array}$$

$$\begin{array}{r} 1 + y = 0 \\ y = -1 \end{array}$$

$$\begin{array}{r} -1 + z = 0 \\ z = 1 \end{array}$$

$$x = 1, \quad y = -1, \quad z = 1$$

21. $\begin{cases} a + b = 2b + c \\ a - 2b = c + 3 \\ 2a - b = 3c - 9 \end{cases}$

$$\begin{cases} a - b - c = 0 \\ a - 2b - c = 3 \\ 2a - b - 3c = -9 \end{cases}$$

$$\begin{array}{r} a - b - c = 0 \\ \underline{-a + 2b + c = -3} \\ b \quad\quad = -3 \end{array}$$

$$\begin{array}{r} -2a + 2b + 2c = 0 \\ \underline{2a - b - 3c = -9} \\ b - c = -9 \\ -3 - c = -9 \\ -c = -6 \\ c = 6 \end{array}$$

$$\begin{array}{r} a - (-3) - 6 = 0 \\ a = 3 \end{array}$$

$$a = 3, \quad b = -3, \quad c = 6$$

23. $\begin{cases} 12a + 5b + 3c = 24000 \\ 10a + 6b + 4c = 13300 \\ 8a + 7b + 5c = 8700 \end{cases}$

$$\begin{array}{r} 60a + 25b + 15c = 120000 \\ \underline{-60a - 36b - 24c = -79800} \\ -11b - 9c = 40200 \end{array}$$

$$\begin{array}{r} 24a + 10b + 6c = 48000 \\ \underline{-24a - 21b - 15c = -26100} \\ -11b - 9c = 21900 \end{array}$$

$$\begin{cases} -11b - 9c = 40200 \\ -11b - 9c = 21900 \end{cases}$$

$$\begin{aligned} -11b - 9c &= 40200 \\ \underline{11b + 9c} &= \underline{-21900} \\ 0 &= 18300 \end{aligned}$$

No solution

25. $$\begin{cases} 0.06x + 0.07y + 0.08z = 440 \\ 0.05x + 0.06y + 0.08z = 410 \\ 0.04x + 0.05y + 0.06z = 320 \end{cases}$$

$$\begin{cases} 6x + 7y + 8z = 44000 \\ 5x + 6y + 8z = 41000 \\ 4x + 5y + 6z = 32000 \end{cases}$$

$$\begin{aligned} 6x + 7y + 8z &= 44000 \\ \underline{-5x - 6y - 8z} &= \underline{-41000} \\ x + y \qquad &= 3000 \end{aligned}$$

$$\begin{aligned} 18x + 21y + 24z &= 132000 \\ \underline{-16x - 20y - 24z} &= \underline{-128000} \\ 2x + y \qquad &= 4000 \end{aligned}$$

$$\begin{cases} x + y = 3000 \\ 2x + y = 4000 \end{cases}$$

$$\begin{aligned} x + y &= 3000 \\ \underline{-2x - y} &= \underline{-4000} \\ -x \quad &= -1000 \\ x &= 1000 \end{aligned}$$

$$\begin{aligned} 1000 + y &= 3000 \\ y &= 2000 \end{aligned}$$

$$\begin{aligned} 4(1000) + 5(2000) + 6z &= 32000 \\ 14000 + 6z &= 32000 \\ 6z &= 18000 \\ z &= 3000 \end{aligned}$$

$x = 1000, \quad y = 2000, \quad z = 3000$

27. d = number of dimes
q = number of quarters
h = number of half-dollars

$$\begin{cases} d + q + h = 48 \\ 10d + 25q + 50h = 1055 \\ d = q + h - 2 \end{cases}$$

$$\begin{cases} d + q + h = 48 \\ 10d + 25q + 50h = 1055 \\ d - q - h = -2 \end{cases}$$

$$\begin{aligned} d + q + h &= 48 \\ \underline{d - q - h} &= \underline{-2} \\ 2d \qquad &= 46 \\ d &= 23 \end{aligned}$$

$$\begin{aligned} -50d - 50q - 50h &= -2400 \\ \underline{10d + 25q + 50h} &= \underline{1055} \\ -40d - 25q \qquad &= -1345 \\ -40(23) - 25q \qquad &= -1345 \\ -25q \qquad &= -425 \\ q &= 17 \\ 23 + 17 + h &= 48 \\ h &= 8 \end{aligned}$$

There are 23 dimes, 17 quarters and 8 half-dollars.

29. x = amount at 8.7%
y = amount at 9.3%
z = amount at 12.66%

$$\begin{cases} x + y + z = 12000 \\ 0.087x + 0.093y + 0.1266z = 1266 \\ 0.1266z = 0.087x + 0.093y \end{cases}$$

$$\begin{cases} x + y + z = 12000 \\ 870x + 930y + 1266z = 12660000 \\ -870x - 930y + 1266z = 0 \end{cases}$$

$$\begin{aligned} 870x + 930y + 1266z &= 12660000 \\ \underline{-870x - 930y + 1266z} &= \underline{0} \\ 2532z &= 12660000 \\ z &= 5000 \end{aligned}$$

$$\begin{aligned} -870x - 870y - 870z &= -10440000 \\ \underline{870x + 930y + 1266z} &= \underline{12660000} \\ 60y + 396z &= 2220000 \end{aligned}$$

$$\begin{aligned} 60y + 396(5000) &= 2220000 \\ 60y &= 240000 \\ y &= 4000 \\ x + 4000 + 5000 &= 12000 \\ x &= 3000 \end{aligned}$$

She has \$3000 at 8.7%, \$4000 at 9.3% and \$5000 at 12.66%.

31. x = number of orchestra seats
y = number of mezzanine seats
z = number of balcony seats

$$\begin{cases} x + y + z = 750 \\ 12x + 8y + 6z = 7290 \\ x = y + z + 100 \end{cases}$$

$$\begin{cases} x + y + z = 750 \\ 12x + 8y + 6z = 7290 \\ x - y - z = 100 \end{cases}$$

$$\begin{array}{r} x + y + z = 750 \\ \underline{x - y - z = 100} \\ 2x \qquad\qquad = 850 \\ x = 425 \end{array}$$

$$\begin{array}{r} 12x + 8y + 6z = 7290 \\ \underline{6x - 6y - 6z = 600} \\ 18x + 2y \qquad = 7890 \\ 18(425) + 2y = 7890 \\ 2y = 240 \\ y = 120 \end{array}$$

$$\begin{array}{r} 425 + 120 + z = 750 \\ z = 205 \end{array}$$

There are 425 orchestra seats, 120 mezzanine seats and 205 balcony seats.

33. $$\begin{cases} 2.1A + 2.8B + 3.2C = 721 \\ 3.2A + 3.6B + 4C = 974 \\ 0.5A + 0.6B + 0.8C = 168 \end{cases}$$

$$\begin{cases} 21A + 28B + 32C = 7210 \\ 32A + 36B + 40C = 9740 \\ 5A + 6B + 8C = 1680 \end{cases}$$

$$\begin{cases} 21A + 28B + 32C = 7210 \\ 8A + 9B + 10C = 2435 \\ 5A + 6B + 8C = 1680 \end{cases}$$

$$\begin{array}{r} 105A + 140B + 160C = 36050 \\ \underline{-128A - 144B - 160C = -38960} \\ -23A - 4B \qquad\qquad = -2910 \end{array}$$

$$\begin{array}{r} 32A + 36B + 40C = 9740 \\ \underline{-25A - 30B - 40C = -8400} \\ 7A + 6B \qquad\qquad = 1340 \end{array}$$

$$\begin{cases} -23A - 4B = -2910 \\ 7A + 6B = 1340 \end{cases}$$

$$\begin{array}{r} -69A - 12B = -8730 \\ \underline{14A + 12B = 2680} \\ -55A \qquad\quad = -6050 \\ A = 110 \end{array}$$

$$\begin{array}{r} 7(110) + 6B = 1340 \\ 6B = 570 \\ B = 95 \end{array}$$

$$\begin{array}{r} 5(110) + 6(95) + 8C = 1680 \\ 1120 + 8C = 1680 \\ 8C = 560 \\ C = 70 \end{array}$$

110 model A's, 95 model B's and 70 model C's

35. x = number of grams of Food A
y = number of grams of Food B
z = number of grams of Food C

$$\begin{cases} 0.05x + 0.06y + 0.04z = 9 \\ 0.20x + 0.15y + 0.10z = 28.5 \\ 0.40x + 0.60y + 0.70z = 97 \end{cases}$$

$$\begin{cases} 5x + 6y + 4z = 900 \\ 20x + 15y + 10z = 2850 \\ 40x + 60y + 70z = 9700 \end{cases}$$

$$\begin{cases} 5x + 6y + 4z = 900 \\ 4x + 3y + 2z = 570 \\ 4x + 6y + 7z = 970 \end{cases}$$

$$\begin{array}{r} 5x + 6y + 4z = 900 \\ \underline{-8x - 6y - 4z = -1140} \\ -3x \qquad\qquad = -240 \\ x = 80 \end{array}$$

$$\begin{array}{r} -5x - 6y - 4z = -900 \\ \underline{4x + 6y + 7z = 970} \\ -x \qquad + 3z = 70 \end{array}$$

$$\begin{array}{r} -80 + 3z = 70 \\ 3z = 150 \\ z = 50 \end{array}$$

$$\begin{array}{r} 5(80) + 6y + 4(50) = 900 \\ 6y + 600 = 900 \\ 6y = 300 \\ y = 50 \end{array}$$

80 gm of Food A, 50 gm of Food B, 50 gm of Food C

37. x = number of acres of A
y = number of acres of B
z = number of acres of C

$$\begin{cases} 90x + 110y + 75z = 94000 \\ 6x + 10y + 5z = 7200 \\ 400x + 500y + 600z = 555000 \end{cases}$$

$$\begin{cases} 18x + 22y + 15z = 18800 \\ 6x + 10y + 5z = 7200 \\ 4x + 5y + 6z = 5550 \end{cases}$$

$$18x + 22y + 15z = 18800$$
$$\underline{-18x - 30y - 15z = -21600}$$
$$-8y = -2800$$
$$y = 350$$

$$36x + 44y + 30z = 37600$$
$$\underline{-20x - 25y - 30z = -27750}$$
$$16x + 19y = 9850$$
$$16x + 19(350) = 9850$$
$$16x = 3200$$
$$x = 200$$

$$6(200) + 10(350) + 5z = 7200$$
$$4700 + 5z = 7200$$
$$5z = 2500$$
$$z = 500$$

200 acres of crop A, 350 acres of crop B, 500 acres of crop C

39. $$2x^2 - x - 3 \neq 0$$
$$(2x - 3)(x + 1) \neq 0$$
$$2x - 3 \neq 0 \quad x + 1 \neq 0$$
$$x \neq \frac{3}{2} \qquad x \neq -1$$

$$\left\{ x \,\middle|\, x \neq \frac{3}{2},\, -1 \right\}$$

41. $$y = 8 - 3x - x^2$$

$$-\frac{b}{2a} = \frac{-(-3)}{2(-1)} = -\frac{3}{2}$$

$$y = 8 - 3\left(-\frac{3}{2}\right) - \left(-\frac{3}{2}\right)^2 = \frac{41}{4}$$

Vertex: $\left(-\frac{3}{2}, \frac{41}{4}\right)$

x-intercepts: Let $y = 0$

$$0 = 8 - 3x - x^2$$

$$x = \frac{-(-3 \pm \sqrt{(-3)^2 - 4(-1)(8)}}{2(-1)}$$

$$= \frac{-3 \pm \sqrt{41}}{-2}$$

$$= \frac{-3 \pm \sqrt{41}}{2}$$

y-intercept: Let $x = 0$
$$y = 8 - 3(0) - 0^2 = 8$$

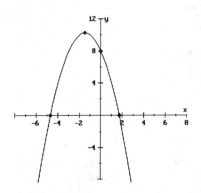

11.2 Exercises

1. $\begin{bmatrix} 3 & -2 & | & 5 \\ 1 & -1 & | & 8 \end{bmatrix}$

3. $\begin{bmatrix} 1 & -2 & 3 & | & 4 \\ 0 & 1 & -1 & | & -3 \\ 2 & 3 & 0 & | & 8 \end{bmatrix}$

5. $\begin{bmatrix} 1 & -2 & | & 7 \\ 2 & -3 & | & 12 \end{bmatrix}$

$$-2R_1 + R_2 \rightarrow R_2$$

$\begin{bmatrix} 1 & -2 & | & 7 \\ 0 & 1 & | & -2 \end{bmatrix}$

$$\begin{cases} x - 2y = 7 \\ y = -2 \end{cases}$$

$$x - 2(-2) = 7$$
$$x = 3$$

$$(3, -2)$$

7. $\begin{bmatrix} 1 & -3 & | & 6 \\ 3 & 5 & | & -10 \end{bmatrix}$

$$-3R_1 + R_2 \rightarrow R_2$$

$\begin{bmatrix} 1 & -3 & | & 6 \\ 0 & 14 & | & -28 \end{bmatrix}$

$$\begin{cases} x - 3y = 6 \\ 14y = -28 \end{cases}$$

$$14y = -28$$
$$y = -2$$

$$x - 3(-2) = 6$$
$$x = 0$$

$(0, -2)$

9. $\begin{bmatrix} 2 & -5 & | & -8 \\ 7 & 3 & | & -28 \end{bmatrix}$

$-7R_1 + 2R_2 \rightarrow R_2$

$\begin{bmatrix} 2 & -5 & | & -8 \\ 0 & 41 & | & 0 \end{bmatrix}$

$\begin{cases} 2x - 5y = -8 \\ 41y = 0 \end{cases}$

$$41y = 0$$
$$y = 0$$

$$2x - 5(0) = -8$$
$$2x = -8$$
$$x = -4$$

$(-4, 0)$

11. $\begin{bmatrix} 6 & 2 & | & 9 \\ 4 & -1 & | & -1 \end{bmatrix}$

$-2R_1 + 3R_2 \rightarrow R_2$

$\begin{bmatrix} 6 & 2 & | & 9 \\ 0 & -7 & | & -21 \end{bmatrix}$

$\begin{cases} 6x + 2y = 9 \\ -7y = -21 \end{cases}$

$$-7y = -21$$
$$y = 3$$

$$6x + 2(3) = 9$$
$$6x = 3$$
$$x = \frac{1}{2}$$

$\left(\frac{1}{2}, 3\right)$

13. $\begin{bmatrix} 6 & 2 & | & 5 \\ 3 & -4 & | & 0 \end{bmatrix}$

$-2R_2 + R_1 \rightarrow R_2$

$\begin{bmatrix} 6 & 2 & | & 5 \\ 0 & 10 & | & 5 \end{bmatrix}$

$\begin{cases} 6x + 2y = 5 \\ 10y = 5 \end{cases}$

$$10y = 5$$
$$y = \frac{1}{2}$$

$$6x + 2\left(\frac{1}{2}\right) = 5$$
$$6x = 4$$
$$x = \frac{2}{3}$$

$\left(\frac{2}{3}, \frac{1}{2}\right)$

15. $\begin{bmatrix} 1 & 3 & 1 & | & 8 \\ 1 & 2 & 1 & | & 7 \\ 1 & -2 & 2 & | & 6 \end{bmatrix}$

$R_1 - R_2 \rightarrow R_2$
$R_1 - R_3 \rightarrow R_3$

$\begin{bmatrix} 1 & 3 & 1 & | & 8 \\ 0 & 1 & 0 & | & 1 \\ 0 & 5 & -1 & | & 2 \end{bmatrix}$

$5R_2 + R_3 \rightarrow R_3$

$\begin{bmatrix} 1 & 3 & 1 & | & 8 \\ 0 & 1 & 0 & | & 1 \\ 0 & 0 & -1 & | & -3 \end{bmatrix}$

$\begin{cases} x + 3y + z = 8 \\ y = 1 \\ -z = -3 \end{cases}$

$$-z = -3$$
$$z = 3$$

$$x + 3(1) + 3 = 8$$
$$x = 2$$

$(2, 1, 3)$

17. $\begin{bmatrix} 1 & 1 & -1 & | & 2 \\ 3 & 1 & -1 & | & -2 \\ 4 & -2 & 1 & | & -13 \end{bmatrix}$

$-3R_1 + R_2 \rightarrow R_2$
$-4R_1 + R_3 \rightarrow R_3$

$\begin{bmatrix} 1 & 1 & -1 & | & 2 \\ 0 & -2 & 2 & | & -8 \\ 0 & -6 & 5 & | & -21 \end{bmatrix}$

$-3R_2 + R_3 \rightarrow R_3$

$\begin{bmatrix} 1 & 1 & -1 & | & 2 \\ 0 & -2 & 2 & | & -8 \\ 0 & 0 & -1 & | & 3 \end{bmatrix}$

$\begin{cases} x + y - z = 2 \\ \quad -2y + z = -8 \\ \qquad -z = 3 \end{cases}$

$-z = 3$
$z = -3$

$-2y + 2(-3) = -8$
$-2y = -2$
$y = 1$

$x + 1 - (-3) = 2$
$x = -2$

$(-2, 1, -3)$

19. $\begin{bmatrix} 1 & -2 & 3 & | & 7 \\ 2 & 3 & -1 & | & 0 \\ 1 & 1 & 1 & | & 1 \end{bmatrix}$

$-2R_1 + R_2 \rightarrow R_2$
$R_1 - R_3 \rightarrow R_3$

$\begin{bmatrix} 1 & -2 & 3 & | & 7 \\ 0 & 7 & -7 & | & -14 \\ 0 & -3 & 2 & | & 6 \end{bmatrix}$

$\frac{1}{7} R_2 \rightarrow R_2$

$\begin{bmatrix} 1 & -2 & 3 & | & 7 \\ 0 & 1 & -1 & | & -2 \\ 0 & -3 & 2 & | & 6 \end{bmatrix}$

$3R_2 + R_3 \rightarrow R_3$

$\begin{bmatrix} 1 & -2 & 3 & | & 7 \\ 0 & 1 & -1 & | & -2 \\ 0 & 0 & -1 & | & 0 \end{bmatrix}$

$\begin{cases} x - 2y + 3z = 7 \\ \quad y - z = -2 \\ \qquad -z = 0 \end{cases}$

$-z = 0$
$z = 0$

$y - 0 = -2$
$y = -2$

$x - 2(-2) + 3(0) = 7$
$x + 4 = 7$
$x = 3$

$(3, -2, 0)$

21. $\begin{bmatrix} 4 & -1 & 2 & | & 6 \\ 2 & 3 & -1 & | & 4 \\ 2 & -2 & 1 & | & 0 \end{bmatrix}$

$-2R_2 + R_1 \rightarrow R_2$
$R_2 - R_3 \rightarrow R_3$

$\begin{bmatrix} 4 & -1 & 2 & | & 6 \\ 0 & -7 & 4 & | & -2 \\ 0 & 5 & -2 & | & 4 \end{bmatrix}$

$5R_2 + 7R_3 \rightarrow R_3$

$\begin{bmatrix} 4 & -1 & 2 & | & 6 \\ 0 & -7 & 4 & | & -2 \\ 0 & 0 & 6 & | & 18 \end{bmatrix}$

$\begin{cases} 4x - y + 2z = 6 \\ \quad -7y + 4z = -2 \\ \qquad 6z = 18 \end{cases}$

$6z = 18$
$z = 3$

$$-7y + 4(3) = -2$$
$$-7y = -14$$
$$y = 2$$

$$4x - 2 + 2(3) = 6$$
$$4x = 2$$
$$x = \frac{1}{2}$$

$$\left(\frac{1}{2}, 2, 3\right)$$

23.
$$\begin{bmatrix} 2 & 3 & 2 & | & 4 \\ 4 & 6 & 4 & | & 8 \\ 2 & 3 & 5 & | & 13 \end{bmatrix}$$

$$-2R_1 + R_2 \rightarrow R_2$$
$$R_1 - R_3 \rightarrow R_3$$

$$\begin{bmatrix} 2 & 3 & 2 & | & 4 \\ 0 & 0 & 0 & | & 0 \\ 0 & 0 & -3 & | & -9 \end{bmatrix}$$

$$\begin{cases} -3z = -9 \\ 2x + 3y + 2z = 4 \end{cases}$$

$$-3z = -9$$
$$z = 3$$

$$2x + 3y + 2(3) = 4$$
$$2x + 3y = -2$$

$$\{(x, y, 3) \,|\, 2x + 3y = -2\}$$

25.
$$\begin{bmatrix} 1 & -2 & 1 & -1 & | & 2 \\ 1 & 2 & 2 & 1 & | & 0 \\ 2 & -2 & 1 & -1 & | & 3 \\ 2 & 0 & -2 & 1 & | & 5 \end{bmatrix}$$

$$R_1 - R_2 \rightarrow R_2$$
$$-2R_1 + R_3 \rightarrow R_3$$
$$R_3 - R_4 \rightarrow R_4$$

$$\begin{bmatrix} 1 & -2 & 1 & -1 & | & 2 \\ 0 & -4 & -1 & -2 & | & 2 \\ 0 & 2 & -1 & 1 & | & -1 \\ 0 & -2 & 3 & -2 & | & -2 \end{bmatrix}$$

$$R_2 + 2R_3 \rightarrow R_3$$
$$R_3 + R_4 \rightarrow R_4$$

$$\begin{bmatrix} 1 & -2 & 1 & -1 & | & 2 \\ 0 & -4 & -1 & -2 & | & 2 \\ 0 & 0 & -3 & 0 & | & 0 \\ 0 & 0 & 2 & -1 & | & -3 \end{bmatrix}$$

$$2R_3 + 3R_4 \rightarrow R_4$$

$$\begin{bmatrix} 1 & -2 & 1 & -1 & | & 2 \\ 0 & -4 & -1 & -2 & | & 2 \\ 0 & 0 & -3 & 0 & | & 0 \\ 0 & 0 & 0 & -3 & | & -9 \end{bmatrix}$$

$$\begin{cases} w - 2x + y - z = 2 \\ -4x - y - 2z = 2 \\ -3y = 0 \\ -3z = -9 \end{cases}$$

$$-3z = -9$$
$$z = 3$$

$$-3y = 0$$
$$y = 0$$

$$-4x - 0 - 2(3) = 2$$
$$-4x = 8$$
$$x = -2$$

$$w - 2(-2) + 0 - 3 = 2$$
$$w = 1$$
$$(1, -2, 0, 3)$$

27.
$$\begin{bmatrix} 2 & 3 & | & 5 \\ 1 & 4 & | & 0 \end{bmatrix}$$

$$-2R_2 + R_1 \rightarrow R_2$$
$$1/2R_1 \rightarrow R_1$$

$$\begin{bmatrix} 1 & 3/2 & | & 5/2 \\ 0 & -5 & | & 5 \end{bmatrix}$$

$$(-1/5)R_2 \rightarrow R_2$$

$$\begin{bmatrix} 1 & 3/2 & | & 5/2 \\ 0 & 1 & | & -1 \end{bmatrix}$$

$$(-3/2)R_2 + R_1 \rightarrow R_1$$

$$\begin{bmatrix} 1 & 0 & | & 4 \\ 0 & 1 & | & -1 \end{bmatrix}$$

29. $\begin{bmatrix} 1 & 1 & 2 & | & 1 \\ 2 & 4 & 2 & | & 6 \\ 3 & 1 & 2 & | & 5 \end{bmatrix}$

$-2R_1 + R_2 \rightarrow R_2$
$-3R_1 + R_3 \rightarrow R_3$

$$\begin{bmatrix} 1 & 1 & 2 & | & 1 \\ 0 & 2 & -2 & | & 4 \\ 0 & -2 & -4 & | & 2 \end{bmatrix}$$

$(1/2)R_2 \rightarrow R_2$
$R_2 + R_3 \rightarrow R_3$

$$\begin{bmatrix} 1 & 1 & 2 & | & 1 \\ 0 & 1 & -1 & | & 2 \\ 0 & 0 & -6 & | & 6 \end{bmatrix}$$

$(-1/6)R_3 \rightarrow R_3$

$$\begin{bmatrix} 1 & 1 & 2 & | & 1 \\ 0 & 1 & -1 & | & 2 \\ 0 & 0 & 1 & | & -1 \end{bmatrix}$$

$R_2 + R_3 \rightarrow R_2$
$R_1 - R_2 \rightarrow R_1$

$$\begin{bmatrix} 1 & 0 & 3 & | & -1 \\ 0 & 1 & 0 & | & 1 \\ 0 & 0 & 1 & | & -1 \end{bmatrix}$$

$-3R_3 + R_1 \rightarrow R_1$

$$\begin{bmatrix} 1 & 0 & 0 & | & 2 \\ 0 & 1 & 0 & | & 1 \\ 0 & 0 & 1 & | & -1 \end{bmatrix}$$

31. $\begin{bmatrix} 1 & -3 & | & 5 \\ 3 & 5 & | & 1 \end{bmatrix}$

$-3R_1 + R_2 \rightarrow R_2$

$$\begin{bmatrix} 1 & -3 & | & 5 \\ 0 & 14 & | & -14 \end{bmatrix}$$

$(-1/14)R_2 \rightarrow R_2$

$$\begin{bmatrix} 1 & -3 & | & 5 \\ 0 & 1 & | & -1 \end{bmatrix}$$

$3R_2 + R_1 \rightarrow R_1$

$$\begin{bmatrix} 1 & 0 & | & 2 \\ 0 & 1 & | & -1 \end{bmatrix}$$

$(2, -1)$

33. $\begin{bmatrix} 1 & 2 & 1 & | & 8 \\ 1 & 4 & -1 & | & 12 \\ 1 & -2 & 1 & | & -4 \end{bmatrix}$

$R_1 - R_2 \rightarrow R_2$
$R_2 - R_3 \rightarrow R_3$

$$\begin{bmatrix} 1 & 2 & 1 & | & 8 \\ 0 & -2 & 2 & | & -4 \\ 0 & 6 & -2 & | & 16 \end{bmatrix}$$

$-(1/2)R_2 \rightarrow R_2$
$3R_2 + R_3 \rightarrow R_3$

$$\begin{bmatrix} 1 & 2 & 1 & | & 8 \\ 0 & 1 & -1 & | & 2 \\ 0 & 0 & 4 & | & 4 \end{bmatrix}$$

$-(1/4)R_3 \rightarrow R_3$

$$\begin{bmatrix} 1 & 2 & 1 & | & 8 \\ 0 & 1 & -1 & | & 2 \\ 0 & 0 & 1 & | & 1 \end{bmatrix}$$

$R_2 + R_3 \rightarrow R_2$
$-2R_2 + R_1 \rightarrow R_1$

$$\begin{bmatrix} 1 & 0 & 3 & | & 4 \\ 0 & 1 & 0 & | & 3 \\ 0 & 0 & 1 & | & 1 \end{bmatrix}$$

$$-3R_3 + R_1 \to R_1$$

$$\begin{bmatrix} 1 & 0 & 0 & | & 1 \\ 0 & 1 & 0 & | & 3 \\ 0 & 0 & 1 & | & 1 \end{bmatrix}$$

$$(1, 3, 1)$$

35. $\begin{bmatrix} 1 & -2 & 0 & | & -4 \\ 0 & 2 & 2 & | & 4 \\ 1 & 0 & 1 & | & 1 \end{bmatrix}$

$$R_1 + R_2 \to R_1$$
$$R_1 - R_3 \to R_3$$
$$(1/2)R_2 \to R_2$$

$$\begin{bmatrix} 1 & 0 & 2 & | & 0 \\ 0 & 1 & 1 & | & 2 \\ 0 & -2 & -1 & | & -5 \end{bmatrix}$$

$$2R_2 + R_3 \to R_3$$

$$\begin{bmatrix} 1 & 0 & 2 & | & 0 \\ 0 & 1 & 1 & | & 2 \\ 0 & 0 & 1 & | & -1 \end{bmatrix}$$

$$R_2 - R_3 \to R_2$$
$$-2R_3 + R_1 \to R_1$$

$$\begin{bmatrix} 1 & 0 & 0 & | & 2 \\ 0 & 1 & 0 & | & 3 \\ 0 & 0 & 1 & | & -1 \end{bmatrix}$$

$$(2, 3, -1)$$

37. $\begin{bmatrix} 1 & 1 & 1 & 1 & | & 7 \\ 2 & 1 & -1 & 2 & | & 6 \\ 3 & 2 & 1 & -1 & | & 3 \\ 1 & -1 & -1 & -2 & | & -10 \end{bmatrix}$

$$-2R_1 + R_2 \to R_2$$
$$-3R_1 + R_3 \to R_3$$
$$R_1 - R_4 \to R_4$$

$$\begin{bmatrix} 1 & 1 & 1 & 1 & | & 7 \\ 0 & -1 & -3 & 0 & | & -8 \\ 0 & -1 & -2 & -4 & | & -18 \\ 0 & 2 & 2 & 3 & | & 17 \end{bmatrix}$$

$$R_1 + R_2 \to R_1$$
$$-R_2 \to R_2$$
$$R_3 - R_2 \to R_3$$
$$2R_3 + R_4 \to R_4$$

$$\begin{bmatrix} 1 & 0 & -2 & 1 & | & -1 \\ 0 & 1 & 3 & 0 & | & 8 \\ 0 & 0 & 1 & -4 & | & -10 \\ 0 & 0 & -2 & -5 & | & -19 \end{bmatrix}$$

$$2R_3 + R_1 \to R_1$$
$$-3R_3 + R_2 \to R_2$$
$$2R_3 + R_4 \to R_4$$

$$\begin{bmatrix} 1 & 0 & 0 & -7 & | & -21 \\ 0 & 1 & 0 & 12 & | & 38 \\ 0 & 0 & 1 & -4 & | & -10 \\ 0 & 0 & 0 & -13 & | & -39 \end{bmatrix}$$

$$-(1/13)R_4 \to R_4$$

$$\begin{bmatrix} 1 & 0 & 0 & -7 & | & -21 \\ 0 & 1 & 0 & 12 & | & 38 \\ 0 & 0 & 1 & -4 & | & -10 \\ 0 & 0 & 0 & 1 & | & 3 \end{bmatrix}$$

$$7R_4 + R_1 \to R_1$$
$$-12R_4 + R_2 \to R_2$$
$$4R_4 + R_3 \to R_3$$

$$\begin{bmatrix} 1 & 0 & 0 & 0 & | & 0 \\ 0 & 1 & 0 & 0 & | & 2 \\ 0 & 0 & 1 & 0 & | & 2 \\ 0 & 0 & 0 & 1 & | & 3 \end{bmatrix}$$
$$(0, 2, 2, 3)$$

39. $f[g(x)] = \dfrac{3}{g(x) + 1}$

$= \dfrac{3}{\dfrac{x-1}{x} + 1}$

$= \dfrac{3 \cdot x}{\left(\dfrac{x-1}{x} + 1\right) \cdot x}$

$= \dfrac{3x}{x - 1 + x}$

$= \dfrac{3x}{2x - 1}$

$g[f(x)] = \dfrac{f(x) - 1}{f(x)}$

$= \dfrac{\dfrac{3}{x+1} - 1}{\dfrac{3}{x+1}}$

$= \dfrac{\left(\dfrac{3}{x+1} - 1\right)(x+1)}{\dfrac{3}{x+1} \cdot (x+1)}$

$= \dfrac{3 - (x+1)}{3}$

$= \dfrac{3 - x - 1}{3}$

$= \dfrac{2 - x}{3}$

41. $\dfrac{(x^{-2}y^{-2/3})^{1/2}}{9(x^{-1/4}y^{-1/6})^2}$

$= \dfrac{x^{-1}y^{-1/3}}{9x^{-1/2}y^{-1/3}}$

$= \dfrac{1}{9}x^{-1 - \left(-\frac{1}{2}\right)}y^{-1/3 - (-1/3)}$

$= \dfrac{1}{9}x^{-1/2}y^0$

$= \dfrac{1}{9x^{1/2}}$

1. $\begin{bmatrix} 1 & -3 \\ 5 & 2 \end{bmatrix} + \begin{bmatrix} 2 & 5 \\ 3 & 0 \end{bmatrix}$

$= \begin{bmatrix} 1 + 2 & -3 + 5 \\ 5 + 3 & 2 + 0 \end{bmatrix}$

$= \begin{bmatrix} 3 & 2 \\ 8 & 2 \end{bmatrix}$

3. $\begin{bmatrix} 1 & 3 \\ -5 & 0 \\ 2 & -7 \end{bmatrix} + \begin{bmatrix} -1 & -3 \\ 5 & 0 \\ -2 & 7 \end{bmatrix}$

$= \begin{bmatrix} 1 + (-1) & 3 + (-3) \\ -5 + 5 & 0 + 0 \\ 2 + (-2) & -7 + 7 \end{bmatrix}$

$= \begin{bmatrix} 0 & 0 \\ 0 & 0 \\ 0 & 0 \end{bmatrix}$

5. Not defined

7. $3\begin{bmatrix} 0 & 3 \\ 17 & 0 \end{bmatrix}$

$= \begin{bmatrix} 3(0) & 3(3) \\ 3(17) & 3(0) \end{bmatrix}$

$= \begin{bmatrix} 0 & 9 \\ 51 & 0 \end{bmatrix}$

9. $-A = -\begin{bmatrix} 2 & -3 \\ -1 & 0 \end{bmatrix}$

$= \begin{bmatrix} -2 & -(-3) \\ -(-1) & -0 \end{bmatrix}$

$= \begin{bmatrix} -2 & 3 \\ 1 & 0 \end{bmatrix}$

11. $\dfrac{2}{3}B = \dfrac{2}{3}\begin{bmatrix} 5 & 1 \\ -2 & 3 \end{bmatrix}$

$= \begin{bmatrix} \dfrac{10}{3} & \dfrac{2}{3} \\ -\dfrac{4}{3} & 2 \end{bmatrix}$

13. $A - B$

$= \begin{bmatrix} 2 & -3 \\ -1 & 0 \end{bmatrix} - \begin{bmatrix} 5 & 1 \\ -2 & 3 \end{bmatrix}$

$= \begin{bmatrix} 2 - 5 & -3 - 1 \\ -1 - (-2) & 0 - 3 \end{bmatrix}$

$= \begin{bmatrix} -3 & -4 \\ 1 & -3 \end{bmatrix}$

15. $2A - B$

$= 2\begin{bmatrix} 2 & -3 \\ -1 & 0 \end{bmatrix} - \begin{bmatrix} 5 & 1 \\ -2 & 3 \end{bmatrix}$

$= \begin{bmatrix} 4 & -6 \\ -2 & 0 \end{bmatrix} - \begin{bmatrix} 5 & 1 \\ -2 & 3 \end{bmatrix}$

$= \begin{bmatrix} -1 & -7 \\ 0 & -3 \end{bmatrix}$

17. $\begin{bmatrix} 3 & -3 & 1 \end{bmatrix}\begin{bmatrix} 5 \\ 1 \\ -2 \end{bmatrix}$

$= \begin{bmatrix} 3(5) - 3(1) + 1(-2) \end{bmatrix}$

$= \begin{bmatrix} 10 \end{bmatrix}$

19. $\begin{bmatrix} 0 & -1 & 1 & 2 \end{bmatrix}\begin{bmatrix} 1 \\ -1 \\ -2 \\ 2 \end{bmatrix}$

$= \begin{bmatrix} 0(1) - 1(-1) + 1(-2) + 2(2) \end{bmatrix}$

$= \begin{bmatrix} 3 \end{bmatrix}$

21. $\begin{bmatrix} 0 & -1 \\ 3 & 2 \end{bmatrix}\begin{bmatrix} -1 & 3 \\ 0 & -2 \end{bmatrix}$

$= \begin{bmatrix} 0(-1) + (-1)(0) & 0(3) + (-1)(-2) \\ 3(-1) + 2(0) & 3(3) + 2(-2) \end{bmatrix}$

$= \begin{bmatrix} 0 & 2 \\ -3 & 5 \end{bmatrix}$

23. Not defined

25. $\begin{bmatrix} 0 & 1 & -5 \\ -3 & 1 & 6 \end{bmatrix}\begin{bmatrix} -1 & 3 \\ 0 & -2 \\ 6 & 0 \end{bmatrix}$

$= \begin{bmatrix} 0(-1) + 1(0) + (-5)(6) & 0(3) + 1(-2) + (-5)(0) \\ -3(-1) + 1(0) + 6(6) & -3(3) + 1(-2) + 6(0) \end{bmatrix}$

$= \begin{bmatrix} -30 & -2 \\ 39 & -11 \end{bmatrix}$

27. $\begin{bmatrix} 1 & 0 & -2 \\ 3 & 1 & -1 \\ 2 & 0 & 1 \end{bmatrix} \begin{bmatrix} 1 & -3 & 0 \\ 0 & 2 & -2 \\ -6 & 0 & 5 \end{bmatrix}$

$= \begin{bmatrix} 1(1) + 0(0) + (-2)(-6) & 1(-3) + 0(2) + (-2)(0) & 1(0) + 0(-2) + (-2)(5) \\ 3(1) + 1(0) + (-1)(-6) & 3(-3) + 1(2) + (-1)(0) & 3(0) + 1(-2) + (-1)(5) \\ 2(1) + 0(0) + (1)(-6) & 2(-3) + 0(2) + 1(0) & 2(0) + 0(-2) + 1(5) \end{bmatrix}$

$= \begin{bmatrix} 13 & -3 & -10 \\ 9 & -7 & -7 \\ -4 & -6 & 5 \end{bmatrix}$

29. $\begin{bmatrix} 2 & 0 & -2 & 1 \\ -2 & 3 & -1 & 1 \end{bmatrix} \begin{bmatrix} 1 & 3 \\ 1 & 0 \\ -1 & 5 \\ 1 & -3 \end{bmatrix}$

$= \begin{bmatrix} 2(1) + 0(1) + (-2)(-1) + 1(1) & 2(3) + 0(0) + (-2)(5) + 1(-3) \\ -2(1) + 3(1) + (-1) + 1(1) & -2(3) + 3(0) + (-1)(5) + 1(-3) \end{bmatrix}$

$= \begin{bmatrix} 5 & -7 \\ 3 & -14 \end{bmatrix}$

31. $\begin{bmatrix} 1 & 0 & 0 \\ 0 & 1 & 0 \\ 0 & 0 & 1 \end{bmatrix} \begin{bmatrix} 3 & 1 & 0 \\ -2 & 4 & 6 \\ 3 & 8 & -10 \end{bmatrix}$

$= \begin{bmatrix} 1(3) + 0(-2) + 0(3) & 1(1) + 0(4) + 0(8) & 1(0) + 0(6) + 0(-10) \\ 0(3) + 1(-2) + 0(3) & 0(1) + 1(4) + 0(8) & 0(0) + 1(6) + 0(-10) \\ 0(3) + 0(-2) + 1(3) & 0(1) + 0(4) + 1(8) & 0(0) + 0(6) + 1(-10) \end{bmatrix}$

$= \begin{bmatrix} 3 & 1 & 0 \\ -2 & 4 & 6 \\ 3 & 8 & -10 \end{bmatrix}$

33. $CB = \begin{bmatrix} 2 & -2 \\ 4 & 1 \end{bmatrix} \begin{bmatrix} 3 & 2 \\ -5 & 0 \end{bmatrix}$

$= \begin{bmatrix} 2(3) + (-2)(-5) & 2(2) + (-2)(0) \\ 4(3) + 1(-5) & 4(2) + 1(0) \end{bmatrix}$

$= \begin{bmatrix} 16 & 4 \\ 7 & 8 \end{bmatrix}$

$$BC = \begin{bmatrix} 3 & 2 \\ -5 & 0 \end{bmatrix} \begin{bmatrix} 2 & -2 \\ 4 & 1 \end{bmatrix}$$

$$= \begin{bmatrix} 3(2) + 2(4) & 3(-2) + 2(1) \\ -5(2) + 0(4) & -5(2) + 0(1) \end{bmatrix}$$

$$= \begin{bmatrix} 14 & -4 \\ -10 & 10 \end{bmatrix}$$

$CB \neq BC$
False

35.　$B + C = \begin{bmatrix} 3 & 2 \\ -5 & 0 \end{bmatrix} + \begin{bmatrix} 2 & -2 \\ 4 & 1 \end{bmatrix}$

$$= \begin{bmatrix} 5 & 0 \\ -1 & 1 \end{bmatrix}$$

$$(B + C)A = \begin{bmatrix} 5 & 0 \\ -1 & 1 \end{bmatrix} \begin{bmatrix} 2 & 1 & 0 \\ 3 & 1 & -2 \end{bmatrix}$$

$$= \begin{bmatrix} 10 & 5 & 0 \\ 1 & 0 & -2 \end{bmatrix}$$

$$BA = \begin{bmatrix} 3 & 2 \\ -5 & 0 \end{bmatrix} \begin{bmatrix} 2 & 1 & 0 \\ 3 & 1 & -2 \end{bmatrix}$$

$$= \begin{bmatrix} 12 & 5 & -4 \\ -10 & -5 & 0 \end{bmatrix}$$

$$CA = \begin{bmatrix} 2 & -2 \\ 4 & 1 \end{bmatrix} \begin{bmatrix} 2 & 1 & 0 \\ 3 & 1 & -2 \end{bmatrix}$$

$$= \begin{bmatrix} -2 & 0 & 4 \\ 11 & 5 & -2 \end{bmatrix}$$

$BA + CA$

$$= \begin{bmatrix} 12 & 5 & -4 \\ -10 & -5 & 0 \end{bmatrix} + \begin{bmatrix} -2 & 0 & 4 \\ 11 & 5 & -2 \end{bmatrix}$$

$$= \begin{bmatrix} 10 & 5 & 0 \\ 1 & 0 & -2 \end{bmatrix}$$

$(B + C)A -= BA + CA$
True

37.　$$\left(\frac{1}{2} \begin{bmatrix} 3 & 5 \\ 2 & 4 \end{bmatrix} \right) \begin{bmatrix} 0 & 2 \\ 1 & -1 \end{bmatrix}$$

$$= \begin{bmatrix} \frac{3}{2} & \frac{5}{2} \\ 1 & 2 \end{bmatrix} \begin{bmatrix} 0 & 2 \\ 1 & -1 \end{bmatrix}$$

$$= \begin{bmatrix} \frac{5}{2} & \frac{1}{2} \\ 2 & 0 \end{bmatrix}$$

$$= \frac{1}{2} \left(\begin{bmatrix} 3 & 5 \\ 2 & 4 \end{bmatrix} \begin{bmatrix} 0 & 2 \\ 1 & -1 \end{bmatrix} \right)$$

$$= \frac{1}{2} \begin{bmatrix} 5 & 1 \\ 4 & 0 \end{bmatrix}$$

$$= \begin{bmatrix} \frac{5}{2} & \frac{1}{2} \\ 2 & 0 \end{bmatrix}$$

$$\begin{bmatrix} \frac{5}{2} & \frac{1}{2} \\ 2 & 0 \end{bmatrix} = \begin{bmatrix} \frac{5}{2} & \frac{1}{2} \\ 2 & 0 \end{bmatrix}$$

39.　AB
$\quad 5 \times 4 \quad 3 \times 5$

$\neq$
Not defined

BA

$3 \times 5 \quad 5 \times 4$

3×4

41.　Tom's Office Supply:

$$\begin{bmatrix} 7 & 6 & 2 \\ 4 & 10 & 9 \end{bmatrix} \begin{bmatrix} 230 \\ 65 \\ 18 \end{bmatrix}$$

$$= \begin{bmatrix} 2036 \\ 1732 \end{bmatrix}$$

Kuma's Office Supply:

$$\begin{bmatrix} 7 & 6 & 2 \\ 4 & 10 & 9 \end{bmatrix} \begin{bmatrix} 250 \\ 56 \\ 16 \end{bmatrix}$$

$$= \begin{bmatrix} 2118 \\ 1704 \end{bmatrix}$$

Tom's Office Supply is less expensive for the Sociology Department and Kuma's Office Supply is less expensive for the Political Science Department.

45. $(2x - 1)(x + 3) = (3x - 2)(x + 4)$

$2x^2 + 5x - 3 = 3x^2 + 10x - 8$

$0 = x^2 + 5x - 5$

$x = \dfrac{-5 \pm \sqrt{5^2 - 4(1)(-5)}}{2(1)}$

$= \dfrac{-5 \pm \sqrt{45}}{2}$

$= \dfrac{-5 \pm 3\sqrt{5}}{2}$

47. Center = midpoint

$= \left(\dfrac{-3 + 5}{2}, \dfrac{4 - 6}{2} \right)$

$= (1, -1)$

$r = \sqrt{(-3 - 1)^2 + [4 - (-1)^2]}$

$= \sqrt{16 + 25}$

$= \sqrt{41}$

$(x - h)^2 + (y - k)^2 = r^2$

$(x - 1)^2 + [y - (-1)]^2 = (\sqrt{41})^2$

$(x - 1)^2 + (y + 1)^2 = 41$

11.4 Exercises

1. $\begin{bmatrix} 3 & 2 \\ 2 & -1 \end{bmatrix} \begin{bmatrix} x \\ y \end{bmatrix} = \begin{bmatrix} 14 \\ 9 \end{bmatrix}$

$AX = B$

$X = A^{-1}B$

$= \begin{bmatrix} 4.57 \\ 0.14 \end{bmatrix}$

$x = 4.57, \ y = 0.14$

3. $\begin{bmatrix} 3 & -2 \\ 2 & 1 \end{bmatrix} \begin{bmatrix} x \\ y \end{bmatrix} = \begin{bmatrix} 15 \\ 10 \end{bmatrix}$

$AX = B$

$X = A^{-1}B$

$= \begin{bmatrix} 5 \\ 0 \end{bmatrix}$

$x = 5, \ y = 0$

5. $\begin{bmatrix} 8 & -6 \\ 2 & -3 \end{bmatrix} \begin{bmatrix} x \\ y \end{bmatrix} = \begin{bmatrix} 2 \\ 0 \end{bmatrix}$

$AX = B$

$X = A^{-1}B$

$= \begin{bmatrix} \dfrac{1}{2} \\ \dfrac{1}{3} \end{bmatrix}$

7. $\begin{bmatrix} 0.3 & -0.2 \\ 0.5 & 0.1 \end{bmatrix} \begin{bmatrix} x \\ y \end{bmatrix} = \begin{bmatrix} -1 \\ 7 \end{bmatrix}$

$AX = B$

$X = A^{-1}B$

$= \begin{bmatrix} 10 \\ 20 \end{bmatrix}$

$x = 10, \ y = 20$

9. $\begin{bmatrix} 2 & 1 & -1 \\ 1 & 1 & 2 \\ 1 & -1 & 1 \end{bmatrix} \begin{bmatrix} x \\ y \end{bmatrix} = \begin{bmatrix} 4 \\ 6 \\ 5 \end{bmatrix}$

$AX = B$

$X = A^{-1}B$

$= \begin{bmatrix} 3 \\ -\dfrac{1}{3} \\ \dfrac{5}{3} \end{bmatrix}$

$x = 3, \ y = -\dfrac{1}{3}, \ z = \dfrac{5}{3}$

11. $\begin{bmatrix} 3 & -2 & -4 \\ 4 & 3 & -5 \\ 6 & -5 & 2 \end{bmatrix} \begin{bmatrix} x \\ y \\ z \end{bmatrix} = \begin{bmatrix} -8 \\ -5 \\ -17 \end{bmatrix}$

$AX = B$
$X = A^{-1}B$

$= \begin{bmatrix} -2 \\ 1 \\ 0 \end{bmatrix}$

$x = -2, \quad y = 1, \quad z = 0$

13. $\begin{bmatrix} 1 & 0.2 & 1 \\ 0.3 & -1 & 0.2 \\ 1 & 1 & 2 \end{bmatrix} \begin{bmatrix} x \\ y \\ z \end{bmatrix} = \begin{bmatrix} 0.02 \\ 0 \\ 0.1 \end{bmatrix}$

$AX = B$
$X = A^{-1}B$

$= \begin{bmatrix} 1 \\ 0.1 \\ -1 \end{bmatrix}$

$x = 1, \quad y = 0.1, \quad z = -1$

15. x = number of shirts
y = number of blouses
z = number of skirts

$\begin{bmatrix} 0.1 & 0.2 & 0.2 \\ 0.3 & 0.4 & 0.5 \\ 0.08 & 0.06 & 0.05 \end{bmatrix} \begin{bmatrix} x \\ y \\ z \end{bmatrix} = \begin{bmatrix} 400 \\ 980 \\ 160 \end{bmatrix}$

$AX = B$
$X = A^{-1}B$

$= \begin{bmatrix} 966.7 \\ 683.3 \\ 833.3 \end{bmatrix}$

They should manufacture 966 shirts, 683 blouses and 833 skirts.

19. $\sqrt{3x + 1} + 3 = x$
$\quad\quad \sqrt{3x + 1} = x - 3$
$\quad\quad\quad 3x + 1 = (x - 3)^2$

$3x + 1 = x^2 - 6x + 9$
$\quad\quad 0 = x^2 - 9x + 8$
$\quad\quad 0 = (x - 8)(x - 1)$
$x - 8 = 0 \quad \text{or} \quad x - 1 = 0$
$\quad\quad x = 8 \quad \text{or} \quad\quad x = 1$

When you check $x = 1$, you find it to be extraneous. The only solution is $x = 8$.

21. $\quad\quad \dfrac{x}{x + 1} + 1 = \dfrac{x}{x + 2}$

$(x + 1)(x + 2)\left(\dfrac{x}{x + 1} + 1\right) = (x + 1)(x + 2)\left(\dfrac{x}{x + 2}\right)$

$x(x + 2) + (x + 1)(x + 2) = x(x + 1)$
$x^2 + 2x + x^2 + 3x + 2 = x^2 + x$
$\quad\quad\quad\quad\quad x^2 + 4x + 2 = 0$

$x = \dfrac{-4 \pm \sqrt{4^2 - 4(1)(2)}}{2(1)}$

$= \dfrac{-4 \pm \sqrt{8}}{2}$

$= \dfrac{-4 \pm 2\sqrt{2}}{2}$

$= -2 \pm \sqrt{2}$

11.5 Exercises

1. $\begin{vmatrix} 1 & 2 \\ 3 & 4 \end{vmatrix} = (1)(4) - (3)(2) = -2$

3. $\begin{vmatrix} 5 & 1 \\ -2 & 4 \end{vmatrix} = (5)(4) - (-2)(1) = 22$

5. $\begin{vmatrix} -3 & -1 \\ 2 & -2 \end{vmatrix} = (-3)(-2) - (2)(-1) = 8$

7. $\begin{vmatrix} 1 & 0 \\ 5 & 4 \end{vmatrix} = (1)(4) - (5)(0) = 4$

9. $\begin{vmatrix} 4 & 6 \\ 6 & 9 \end{vmatrix} = (4)(9) - (6)(6) = 0$

11.
$$\begin{vmatrix} 1 & 2 & -2 \\ 3 & -3 & 1 \\ -4 & 2 & -1 \end{vmatrix}$$

$$= 1\begin{vmatrix} -3 & 1 \\ 2 & -1 \end{vmatrix} - 2\begin{vmatrix} 3 & 1 \\ -4 & -1 \end{vmatrix} + (-2)\begin{vmatrix} 3 & -3 \\ -4 & 2 \end{vmatrix}$$

$$= (-3)(-1) - (2)(1) - 2[(3)(-1) - (-4)(1)] - 2[(3)(2) - (-4)(-3)]$$
$$= 1 - 2(1) - 2(-6)$$
$$= 11$$

13.
$$\begin{vmatrix} 1 & 2 & 3 \\ 2 & 4 & 6 \\ 1 & 1 & 1 \end{vmatrix}$$

$$= 1\begin{vmatrix} 2 & 3 \\ 4 & 6 \end{vmatrix} - 1\begin{vmatrix} 1 & 3 \\ 2 & 6 \end{vmatrix} + 1\begin{vmatrix} 1 & 2 \\ 2 & 4 \end{vmatrix}$$

$$= [(2)(6) - (4)(3)] - [(1)(6) - (2)(3)] + [(1)(4) - (2)(2)]$$
$$= 0 - 0 + 0$$
$$= 0$$

15.
$$\begin{vmatrix} -3 & 0 & 4 \\ 5 & 2 & -3 \\ 7 & 0 & 6 \end{vmatrix}$$

$$= 0\begin{vmatrix} 5 & -3 \\ 7 & 6 \end{vmatrix} + 2\begin{vmatrix} -3 & 4 \\ 7 & 6 \end{vmatrix} - 0\begin{vmatrix} -3 & 4 \\ 5 & -3 \end{vmatrix}$$

$$= 2[(-3)(6) - (7)(4)]$$
$$= -92$$

17.
$$\begin{vmatrix} 1 & 0 & 0 \\ 0 & 1 & 0 \\ 0 & 0 & 1 \end{vmatrix}$$

$$= 1\begin{vmatrix} 1 & 0 \\ 0 & 1 \end{vmatrix} - 0\begin{vmatrix} 0 & 0 \\ 0 & 1 \end{vmatrix} + 0\begin{vmatrix} 0 & 1 \\ 0 & 0 \end{vmatrix}$$

$$= 1[(1)(1) - (0)(0)]$$
$$= 1$$

19.
$$\begin{vmatrix} 2x & 4 \\ 3 & 5 \end{vmatrix} = 18$$

$$(2x)(5) - (3)(4) = 18$$
$$10x - 12 = 18$$
$$10x = 30$$
$$x = 3$$

21.
$$\begin{vmatrix} x^2 & 5 \\ x & 1 \end{vmatrix} = 14$$

$$(x^2)(1) - 5(x) = 14$$
$$x^2 - 5x - 14 = 0$$
$$(x - 7)(x + 2) = 0$$
$$x - 7 = 0 \quad \text{or} \quad x + 2 = 0$$
$$x = 7 \quad \text{or} \quad x = -2$$

23.
$$\begin{vmatrix} x & 3 \\ 2 & x + 2 \end{vmatrix} = 9$$

$$x(x + 2) - 2(3) = 9$$
$$x^2 + 2x - 6 = 9$$
$$x^2 + 2x - 15 = 0$$
$$(x + 5)(x - 3) = 0$$
$$x + 5 = 0 \quad \text{or} \quad x - 3 = 0$$
$$x = -5 \quad \text{or} \quad x = 3$$

25. $D = \begin{vmatrix} 9 & 2 \\ 5 & 1 \end{vmatrix} = (9)(1) - (5)(2) = -1$

$D_x = \begin{vmatrix} 1 & 2 \\ 0 & 1 \end{vmatrix} = (1)(1) - (0)(2) = 1$

$D_y = \begin{vmatrix} 9 & 1 \\ 5 & 0 \end{vmatrix} = (9)(0) - (5)(1) = -5$

$$x = \frac{D_x}{D} = \frac{1}{-1} = -1$$

$$y = \frac{D_y}{D} = \frac{-5}{-1} = 5$$

27. $D = \begin{vmatrix} 2 & -4 \\ -3 & 6 \end{vmatrix} = (2)(6) - (-3)(-4) = 0$

No unique solution

29. $\begin{cases} 3x - 5y = 2 \\ 4x - 3y = 10 \end{cases}$

$$D = \begin{vmatrix} 3 & -5 \\ 4 & -3 \end{vmatrix} = (3)(-3) - (4)(-5) = 11$$

$$D_x = \begin{vmatrix} 2 & -5 \\ 10 & -3 \end{vmatrix} = (2)(-3) - (10)(-5) = 44$$

$$D_y = \begin{vmatrix} 3 & 2 \\ 4 & 10 \end{vmatrix} = (3)(10) - (4)(2) = 22$$

$$x = \frac{D_x}{D} = \frac{44}{11} = 4$$

$$y = \frac{D_y}{D} = \frac{22}{11} = 2$$

31. $\begin{cases} 2x - 7y = -4 \\ -4x + 3y = 8 \end{cases}$

$$D = \begin{vmatrix} 2 & -7 \\ -4 & 3 \end{vmatrix} = (2)(3) - (-4)(-7) = -22$$

$$D_x = \begin{vmatrix} -4 & -7 \\ 8 & 3 \end{vmatrix} = (-4)(3) - (8)(-7) = 44$$

$$D_y = \begin{vmatrix} 2 & -4 \\ -4 & 8 \end{vmatrix} = (2)(8) - (-4)(-4) = 0$$

$$x = \frac{D_x}{D} = \frac{44}{-22} = -2$$

$$y = \frac{D_y}{D} = \frac{0}{-22} = 0$$

33. $\begin{cases} 6x - 5y = -7 \\ -3x + 4y = -5 \end{cases}$

$$D = \begin{vmatrix} 6 & -5 \\ -3 & 4 \end{vmatrix} = (6)(4) - (-3)(-5) = 9$$

$$D_x = \begin{vmatrix} -7 & -5 \\ -5 & 4 \end{vmatrix} = (-7)(4) - (-5)(-5) = -53$$

$$D_y = \begin{vmatrix} 6 & -7 \\ -3 & -5 \end{vmatrix} = (6)(-5) - (-3)(-7) = -51$$

$$x = \frac{D_x}{D} = \frac{-53}{9}$$

$$y = \frac{D_y}{D} = \frac{-51}{9} = \frac{-17}{3}$$

35. $D = \begin{vmatrix} 2 & -9 \\ 3 & 5 \end{vmatrix} = (2)(5) - (3)(-9) = 37$

$$D_x = \begin{vmatrix} 4 & -9 \\ 6 & 5 \end{vmatrix} = (4)(5) - (6)(-9) = 74$$

$$D_y = \begin{vmatrix} 2 & 4 \\ 3 & 6 \end{vmatrix} = (2)(6) - (3)(4) = 0$$

$$x = \frac{D_x}{D} = \frac{74}{37} = 2$$

$$y = \frac{D_y}{D} = \frac{0}{37} = 0$$

37. $D = \begin{vmatrix} 6 & 7 \\ 5 & 8 \end{vmatrix} = (6)(8) - (5)(7) = 13$

$$D_x = \begin{vmatrix} 3 & 7 \\ 9 & 8 \end{vmatrix} = (3)(8) - (9)(7) = -39$$

$$D_y = \begin{vmatrix} 6 & 3 \\ 5 & 9 \end{vmatrix} = (6)(9) - (5)(3) = 39$$

$$x = \frac{D_x}{D} = \frac{-39}{13} = -3$$

$$y = \frac{D_y}{D} = \frac{39}{13} = 3$$

39. $x - 24y = 12$
 $x - 24y = 18$

$D = \begin{vmatrix} 1 & -24 \\ 1 & -24 \end{vmatrix} = (1)(-24) - (1)(-24) = 0$

No unique solution

41. $D = \begin{vmatrix} 1 & 1 & 1 \\ 1 & -1 & 1 \\ -1 & 1 & 1 \end{vmatrix} = -4$

$D_x = \begin{vmatrix} 3 & 1 & 1 \\ 2 & -1 & 1 \\ 4 & 1 & 1 \end{vmatrix} = 2$

$D_y = \begin{vmatrix} 1 & 3 & 1 \\ 1 & 2 & 1 \\ -1 & 4 & 1 \end{vmatrix} = -2$

$D_z = \begin{vmatrix} 1 & 1 & 3 \\ 1 & -1 & 2 \\ -1 & 1 & 4 \end{vmatrix} = -12$

$x = \dfrac{D_x}{D} = \dfrac{2}{-4} = -\dfrac{1}{2}$

$y = \dfrac{D_y}{D} = \dfrac{-2}{-4} = \dfrac{1}{2}$

$z = \dfrac{D_z}{D} = \dfrac{-12}{-4} = 3$

43. $D = \begin{vmatrix} 3 & 4 & 2 \\ 2 & 3 & 1 \\ 6 & 1 & 5 \end{vmatrix} = -6$

$D_x = \begin{vmatrix} 1 & 4 & 2 \\ 1 & 3 & 1 \\ 1 & 1 & 5 \end{vmatrix} = -6$

$D_y = \begin{vmatrix} 3 & 1 & 2 \\ 2 & 1 & 1 \\ 6 & 1 & 5 \end{vmatrix} = 0$

$D_z = \begin{vmatrix} 3 & 4 & 1 \\ 2 & 3 & 1 \\ 6 & 1 & 1 \end{vmatrix} = 6$

$x = \dfrac{D_x}{D} = \dfrac{-6}{-6} = 1$

$y = \dfrac{D_y}{D} = \dfrac{0}{-6} = 0$

$z = \dfrac{D_z}{D} = \dfrac{6}{-6} = -1$

45. $D = \begin{vmatrix} 1 & 2 & 3 \\ 6 & 5 & 4 \\ 7 & 8 & 9 \end{vmatrix} = 0$

No unique solution

47. $D = \begin{vmatrix} 3 & -1 & 0 \\ 0 & 2 & 1 \\ 3 & 0 & 4 \end{vmatrix} = 21$

$D_x = \begin{vmatrix} 4 & -1 & 0 \\ 6 & 2 & 1 \\ 14 & 0 & 4 \end{vmatrix} = 42$

$D_y = \begin{vmatrix} 3 & 4 & 0 \\ 0 & 6 & 1 \\ 3 & 14 & 4 \end{vmatrix} = 42$

$D_z = \begin{vmatrix} 3 & -1 & 4 \\ 0 & 2 & 6 \\ 3 & 0 & 14 \end{vmatrix} = 42$

$x = \dfrac{D_x}{D} = \dfrac{42}{21} = 2$

$y = \dfrac{D_y}{D} = \dfrac{42}{21} = 2$

$z = \dfrac{D_z}{D} = \dfrac{42}{21} = 2$

49. $D = \begin{vmatrix} 6 & -2 & 4 \\ 12 & -4 & 8 \\ -3 & 1 & -2 \end{vmatrix} = 0$

No unique solution

51. $\begin{cases} x - y - z = 2 \\ -x + y + z = 3 \\ -x - y + z = 4 \end{cases}$

$D = \begin{vmatrix} 1 & -1 & -1 \\ -1 & 1 & 1 \\ -1 & -1 & 1 \end{vmatrix} = 0$

No unique solution

53. $D = \begin{vmatrix} 2 & 1 & 0 \\ -3 & 0 & 1 \\ 0 & 2 & 5 \end{vmatrix} = 11$

$D_a = \begin{vmatrix} 8 & 1 & 0 \\ -13 & 0 & 1 \\ 0 & 2 & 5 \end{vmatrix} = 49$

$D_b = \begin{vmatrix} 2 & 8 & 0 \\ -3 & -13 & 1 \\ 0 & 0 & 5 \end{vmatrix} = -10$

$D_c = \begin{vmatrix} 2 & 1 & 8 \\ -3 & 0 & -13 \\ 0 & 2 & 0 \end{vmatrix} = 4$

$a = \dfrac{D_a}{D} = \dfrac{49}{11}$

$b = \dfrac{D_b}{D} = \dfrac{-10}{11}$

$c = \dfrac{D_c}{D} = \dfrac{4}{11}$

55. $D = \begin{vmatrix} 68 & 85 \\ 920 & 743 \end{vmatrix} = (68)(743) - (920)(85) = -27676$

57. $D = \begin{vmatrix} 5 & -3 \\ 0.2 & 4 \end{vmatrix} = 20.6$

$D_x = \begin{vmatrix} -0.181 & -3 \\ 0.2482 & 4 \end{vmatrix} = 0.0206$

$D_y = \begin{vmatrix} 5 & -0.181 \\ 0.2 & 0.2482 \end{vmatrix} = 1.2772$

$x = \dfrac{D_x}{D} = \dfrac{0.0206}{20.6} = 0.001$

$y = \dfrac{D_y}{D} = \dfrac{1.2772}{20.6} = 0.062$

59. $D = \begin{vmatrix} 0.02 & 0.04 & -0.06 \\ 0.5 & 0.6 & -0.3 \\ 0.1 & -0.02 & 1 \end{vmatrix} = -0.00512$

$D_x = \begin{vmatrix} 12 & 0.04 & -0.06 \\ 16 & 0.6 & -0.3 \\ 15 & -0.02 & 1 \end{vmatrix} = 6.8672$

$D_y = \begin{vmatrix} 0.02 & 12 & -0.06 \\ 0.5 & 16 & -0.3 \\ 0.1 & 15 & 1 \end{vmatrix} = -6.304$

$D_z = \begin{vmatrix} 0.02 & 0.04 & 12 \\ 0.5 & 0.6 & 16 \\ 0.1 & -0.02 & 15 \end{vmatrix} = -0.8896$

$x = \dfrac{D_x}{D} = \dfrac{6.8672}{-0.00512} = 1341.25$

$y = \dfrac{D_y}{D} = \dfrac{-6.304}{-0.00512} = 1231.25$

$z = \dfrac{D_z}{D} = \dfrac{-0.8896}{-0.00512} = 173.75$

63. $\log_3 \dfrac{1}{81} = -4$

65. $\log_4 60 \approx 2.9534$
$\log_3 40 \approx 3.3578$
$\log_3 40$ is greater.

11.6 Exercises

1. $\begin{cases} x + y \le 6 \quad \text{(solid line)} \\ x + 2y \ge 3 \quad \text{(solid line)} \end{cases}$

Test point: $(3, 1)$

$3 + 1 \le 6 \qquad 3 + 2(1) \ge 3$
$\qquad 4 \le 6 \qquad\qquad\quad 5 \ge 3$
$\qquad\;\; \text{True} \qquad\qquad\quad\;\; \text{True}$

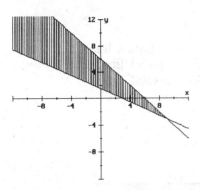

Test point: (3, 1)

$$3 + 1 \le 6 \qquad 3 + 2(1) \ge 3$$
$$4 \le 6 \qquad\qquad 5 \ge 3$$
$$\text{True} \qquad\qquad \text{True}$$

3. $\begin{cases} 2y + x \ge 6 & \text{(solid line)} \\ 3y + x \ge 9 & \text{(solid line)} \end{cases}$

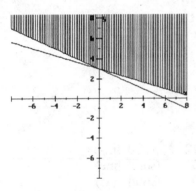

Test point: (1, 3)

$$2(3) + 1 \ge 6 \qquad 3(3) + 1 \ge 9$$
$$7 \ge 6 \qquad\qquad 10 \ge 9$$
$$\text{True} \qquad\qquad \text{True}$$

5. $\begin{cases} x - y \ge 2 & \text{(solid line)} \\ y - x > -1 & \text{(dashed line)} \end{cases}$

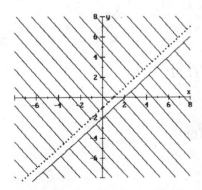

No solution, since the shadings do not overlap.

7. $\begin{cases} x + y \le 5 & \text{(solid line)} \\ 2x + y \le 8 & \text{(solid line)} \end{cases}$

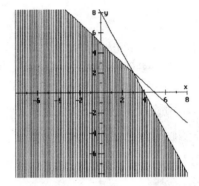

Test point: (0, 0)

$$0 + 0 \le 5 \qquad 2(0) + 0 \le 8$$
$$0 \le 5 \qquad\qquad 0 \le 8$$
$$\text{True} \qquad\qquad \text{True}$$

9. $\begin{cases} 3x + 2y \le 12 & \text{(solid line)} \\ y \le x & \text{(solid line)} \end{cases}$

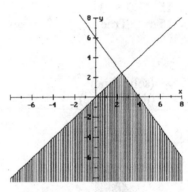

Test point: (0, -1)

$$3(0) + 2(-1) \le 12 \qquad -1 \le 0$$
$$-2 < 12 \qquad\qquad \text{True}$$
$$\text{True}$$

11. $\begin{cases} 3x + 2y \ge 12 & \text{(solid line)} \\ y \ge x & \text{(solid line)} \end{cases}$

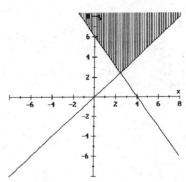

Test point: (0, 7)

$$3(0) + 2(7) \geq 12 \qquad 7 \geq 0$$
$$14 \geq 12 \qquad \text{True}$$
$$\text{True}$$

13. $\begin{cases} 5x - 3y \leq 15 & \text{(solid line)} \\ x < 3 & \text{(dashed line)} \end{cases}$

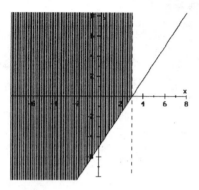

Test point: (0, 0)

$$5(0) - 3(0) \leq 15 \qquad 0 < 3$$
$$0 \leq 15 \qquad \text{True}$$
$$\text{True}$$

15. $\begin{cases} 5x - 3y \geq 15 & \text{(solid line)} \\ x > 3 & \text{(dashed line)} \end{cases}$

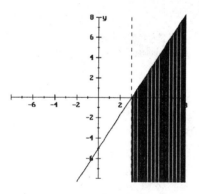

Test point: (4, 0)

$$5(4) - 3(0) \geq 15 \qquad 4 > 3$$
$$20 \geq 15 \qquad \text{True}$$
$$\text{True}$$

17. $\begin{cases} x - 2y < 10 & \text{(dashed line)} \\ x \geq 2 & \text{(solid line)} \\ y \leq 2 & \text{(solid line)} \end{cases}$

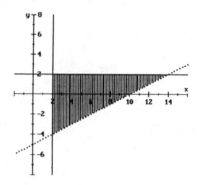

Test point: (3, 0)

$$3 - 2(0) < 10$$
$$3 < 10$$
$$\text{True}$$
$$3 \geq 2$$
$$\text{True}$$
$$0 \leq 2$$
$$\text{True}$$

19. $\begin{cases} 4x + 3y \leq 12 & \text{(solid line)} \\ x \geq 0 & \text{(solid line)} \\ y \geq 0 & \text{(solid line)} \end{cases}$

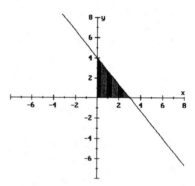

Test point: (1, 1)

$$4(1) + 3(1) \leq 12$$
$$7 \leq 12$$
$$\text{True}$$
$$1 \geq 0$$
$$\text{True}$$
$$1 \geq 0$$
$$\text{True}$$

21. $\begin{cases} x + 3y \geq 6 & \text{(solid line)} \\ 3x + 2y \leq 18 & \text{(solid line)} \\ y \leq 3 & \text{(solid line)} \\ x \geq 0 & \text{(solid line)} \end{cases}$

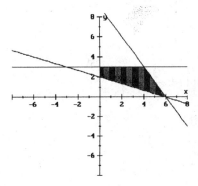

Test point: (1, 2)

$$1 + 3(2) \geq 6 \qquad 3(1) + 2(2) \leq 18$$
$$7 \geq 6 \qquad\qquad 7 \leq 18$$
$$\text{True} \qquad\qquad \text{True}$$

$$2 \leq 3 \qquad 1 \geq 0$$
$$\text{True} \qquad \text{True}$$

23. $\begin{cases} x + 2y \geq 4 & \text{(solid line)} \\ 2x - 4y \leq -8 & \text{(solid line)} \\ x \geq 0 & \text{(solid line)} \\ y \geq 0 & \text{(solid line)} \end{cases}$

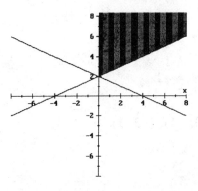

Test point: (1, 3)

$$1 + 2(3) \geq 4 \qquad 2(1) - 4(3) \leq -8$$
$$7 \geq 4 \qquad\qquad -10 \leq -8$$
$$\text{True} \qquad\qquad \text{True}$$

$$1 \geq 0 \qquad 3 \geq 0$$
$$\text{True} \qquad \text{True}$$

25. x = number of single mattresses
    y = number of double mattresses

$$\begin{cases} 80x + 120y \leq 8000 \\ 16x + 36y \leq 2000 \\ x \geq 0 \\ y \geq 0 \end{cases}$$

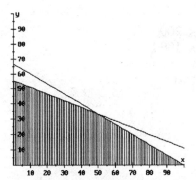

27. X = number of ounces of brand X
    Y = number of ounces of brand Y

$$\begin{cases} 10X + 13Y \geq 21 \\ 0.33X + 0.37Y \leq 1 \\ X \geq 0 \\ Y \geq 0 \end{cases}$$

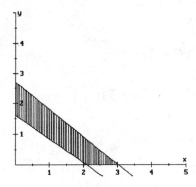

29. x = number of men's shoes
y = number of women's shoes

$$\begin{cases} x + \dfrac{3}{4}y \le 500 \\[2mm] \dfrac{1}{3}x + \dfrac{1}{4}y \le 150 \\[2mm] \qquad x \ge 0 \\ \qquad y \ge 0 \end{cases}$$

$$\begin{cases} 4x + 3y \le 2000 \\ 4x + 3y \le 1800 \\ \quad\; x \ge 0 \\ \quad\; y \ge 0 \end{cases}$$

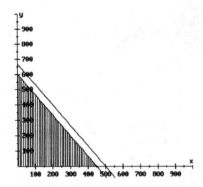

31. $\log_b \dfrac{\sqrt{x^3 y^4}}{5b}$

$= \log_b \dfrac{(x^3 y^4)^{1/2}}{5b}$

$= \log_b (x^3 y^4)^{1/2} - \log_b (5b)$

$= \dfrac{1}{2} \log_b (x^3 y^4) - \log_b (5b)$

$= \dfrac{1}{2} (\log_b x^3 y^4) - [\log_b 5 + \log_b b]$

$= \dfrac{1}{2} (3 \log_b x + 4 \log_b y) - (\log_b 5 + 1)$

$= \dfrac{3}{2} \log_b x + 2 \log_b y - \log_b 5 - 1$

33. $100 \text{ wk} = \dfrac{100}{52} \text{ yr}$

$= \dfrac{25}{13} \text{ yr}$

$A = P\left(1 + \dfrac{r}{n}\right)^{nt}$

$= 6000\left(1 + \dfrac{0.072}{52}\right)^{(52)\left(\frac{25}{13}\right)}$

$= \$6890.37$

11.7 Exercises

1. $\begin{cases} x^2 + y^2 = 10 \\ x^2 - y^2 = 8 \end{cases}$

$\begin{array}{r} x^2 + y^2 = 10 \\ \underline{x^2 - y^2 = 8} \\ 2x^2 \quad\;\; = 18 \\ x^2 = 9 \\ x = \pm 3 \end{array}$

$x = 3$:
$(3)^2 + y^2 = 10$
$y^2 = 1$
$y = \pm 1$
$(3, 1), \quad (3, -1)$

$x = -3$:
$(-3)^2 + y^2 = 10$
$y^2 = 1$
$y = \pm 1$
$(-3, 1), \quad (-3, -1)$

$(3, 1), \quad (3, -1), (-3, 1), \quad (-3, -1)$

3. $\begin{cases} x^2 + y^2 = 25 \\ x \; - y \; = 5 \end{cases}$

$x = 5 + y$
$(5 + y)^2 + y^2 = 25$
$25 + 10y + y^2 + y^2 = 25$
$2y^2 + 10y = 0$
$2y(y + 5) = 0$

$2y = 0 \qquad\qquad$ or $\; y + 5 = 0$
$y = 0 \qquad\qquad$ or $\qquad y = -5$
$x = 5 + 0 = 5 \qquad\qquad x = 5 + (-5) = 0$

$(5, 0), \quad (0, -5)$

5. $\begin{cases} 16x^2 - 4y^2 = 64 \\ x^2 + y^2 = 9 \end{cases}$

$4x^2 - y^2 = 16$
$\underline{x^2 + y^2 = 9}$
$5x^2 \qquad = 25$
$\qquad x^2 = 5$
$\qquad x = \pm\sqrt{5}$

$x = \sqrt{5}:$
$(\sqrt{5})^2 + y^2 = 9$
$\qquad y^2 = 4$
$\qquad y = \pm 2$
$(\sqrt{5}, 2), \ (\sqrt{5}, -2)$

$x = -\sqrt{5}:$
$(-\sqrt{5})^2 + y^2 = 9$
$\qquad y^2 = 4$
$\qquad y = \pm 2$
$(-\sqrt{5}, 2), \ (-\sqrt{5}, -2)$

$(\sqrt{5}, 2), \ (\sqrt{5}, -2), \ (-\sqrt{5}, 2), \ (-\sqrt{5}, -2)$

7. $\begin{cases} x^2 + y^2 = 9 \\ 2x - y = 3 \end{cases}$

$2x - y = 3$
$\qquad y = 2x - 3$
$x^2 + (2x - 3)^2 = 9$
$x^2 + 4x^2 - 12x + 9 = 9$
$5x^2 - 12x = 0$
$x(5x - 12) = 0$
$x = 0 \qquad\qquad \text{or} \ \ 5x - 12 = 0$
$y = 2(0) - 3 = -3 \qquad\qquad x = \dfrac{12}{5}$

$y = 2\left(\dfrac{12}{5}\right) - 3 = \dfrac{9}{5}$

$(0, -3, \ \left(\dfrac{12}{5}, \dfrac{9}{5}\right)$

9. $\begin{cases} x^2 + y^2 = 13 \\ 3x^2 - y^2 = 3 \end{cases}$

$x^2 + y^2 = 13$
$\underline{3x^2 - y^2 = 3}$
$4x^2 \qquad = 16$
$\qquad x^2 = 4$
$\qquad x = \pm 2$

$x = 2:$
$2^2 + y^2 = 13$
$\qquad y^2 = 9$
$\qquad y = \pm 3$
$(2, 3), \ (2, -3)$

$x = -2:$
$(-2)^2 + y^2 = 13$
$\qquad y^2 = 9$
$\qquad y = \pm 3$
$(-2, 3), \ (-2, -3)$

$(2, 3), \ (2, -3), \ (-2, 3), \ (-2, -3)$

11. $\begin{cases} x^2 - y = 0 \\ x^2 - 2x + y = 6 \end{cases}$

$x^2 \qquad - y = 0$
$\underline{x^2 - 2x + y = 6}$
$2x^2 - 2x \qquad = 6$
$2x^2 - 2x - 6 = 0$
$x^2 - x - 3 = 0$

$x = \dfrac{-(-1) \pm \sqrt{(-1)^2 - 4(1)(-3)}}{2(1)}$

$= \dfrac{1 \pm \sqrt{13}}{2}$

$x^2 - y = 0$
$\quad x^2 = y$

$x = \dfrac{1 + \sqrt{13}}{2}:$

$y = \left(\dfrac{2 + \sqrt{13}}{2}\right)^2$

$= \dfrac{1 + 2\sqrt{13} + 13}{4}$

$= \dfrac{14 + 2\sqrt{13}}{4}$

$= \dfrac{7 + \sqrt{13}}{2}$

$x = \dfrac{1 - \sqrt{13}}{2}:$

$y = \left(\dfrac{1 - \sqrt{13}}{2}\right)^2$

$\quad = \dfrac{1 - 2\sqrt{13} + 13}{4}$

$\quad = \dfrac{14 - 2\sqrt{13}}{4}$

$\quad = \dfrac{7 - \sqrt{13}}{2}$

$\left(\dfrac{1 + \sqrt{13}}{2}, \dfrac{7 + \sqrt{13}}{2}\right), \left(\dfrac{1 - \sqrt{13}}{2}, \dfrac{7 - \sqrt{13}}{2}\right)$

13. $\begin{cases} y = 1 - x^2 \\ x + y = 2 \end{cases}$

$x + (1 - x^2) = 2$

$0 = x^2 - x + 1$

$x = \dfrac{-(-1) \pm \sqrt{(-1)^2 - 4(1)(1)}}{2(1)}$

$\quad = \dfrac{1 \pm \sqrt{-3}}{2}$

No real solution

15. $\begin{cases} y = x^2 - 6x \\ y = x - 12 \end{cases}$

$\quad\quad x^2 - 6x = x - 12$

$x^2 - 7x + 12 = 0$

$(x - 4)(x - 3) = 0$

$x - 4 = 0 \quad\quad\quad \text{or} \quad x - 3 = 0$

$\quad\quad x = 4 \quad\quad\quad\quad\quad\quad x = 3$

$\quad\quad y = 4 - 12 = -8 \quad\quad\quad y = 3 - 12 =$

$(4, -8), \quad (3, -9)$

17. $\begin{cases} x^2 + y^2 = 16 \\ \quad x = y^2 - 16 \end{cases}$

$\quad\quad\quad x = y^2 - 16$

$\quad\quad x + 16 = y^2$

$x^2 + (x + 16) = 16$

$\quad\quad x^2 + x = 0$

$\quad\quad x(x + 1) = 0$

$x = 0 \quad \text{or} \quad x + 1 = 0$

$\quad\quad\quad\quad\quad\quad x = -1$

$x = 0:$

$0 + 16 = y^2$

$\quad 16 = y^2$

$\quad \pm 4 = y$

$x = -1:$

$-1 + 16 = y^2$

$\quad\quad 15 = y^2$

$\quad \pm\sqrt{15} = y$

$(0, 4), \quad (0, -4), \quad (-1, \sqrt{15}), \quad (-1, -\sqrt{15})$

19. $\begin{cases} x^2 + y^2 - 25 = 0 \\ x + y - 7 = 0 \end{cases}$

$x + y - 7 = 0$

$\quad\quad x = 7 - y$

$\quad\quad (7 - y)^2 + y^2 - 25 = 0$

$49 - 14y + y^2 + y^2 - 25 = 0$

$\quad\quad 2y^2 - 14y + 24 = 0$

$\quad\quad\quad y^2 - 7y + 12 = 0$

$\quad\quad\quad (y - 4)(y - 3) = 0$

$y - 4 = 0 \quad\quad\quad \text{or} \quad y - 3 = 0$

$\quad y = 4 \quad\quad\quad\quad\quad\quad y = 3$

$\quad x = 7 - 4 = 3 \quad\quad\quad x = 7 - 3 = 4$

$(3, 4), \quad (4, 3)$

21. $\begin{cases} x^2 - y^2 = 4 \\ 2x^2 + y^2 = 16 \end{cases}$

$\quad\quad\quad\quad\quad\quad y^2 = 16 - 2x^2$

$x^2 - (16 - 2x^2) = 4$

$\quad x^2 - 16 + 2x^2 = 4$

$\quad\quad\quad\quad 3x^2 = 20$

$\quad\quad\quad\quad x^2 = \dfrac{20}{3}$

$\quad\quad x = \pm\sqrt{\dfrac{20}{3}} = \pm\dfrac{2\sqrt{15}}{3}$

$x = \dfrac{2\sqrt{15}}{3}$

$y^2 = 16 - 2\left(\dfrac{2\sqrt{15}}{3}\right)^2$

$y^2 = \dfrac{8}{3}$

$$y = \pm\sqrt{\frac{8}{3}} = \pm\frac{2\sqrt{6}}{3}$$

$$x = \frac{-2\sqrt{15}}{3}$$

$$y^2 = 16 - 2\left(-\frac{2\sqrt{15}}{3}\right)^2$$

$$y^2 = \frac{8}{3}$$

$$y = \pm\sqrt{\frac{8}{3}} = \pm\frac{2\sqrt{6}}{3}$$

$$\left(\frac{2\sqrt{15}}{3}, \frac{2\sqrt{6}}{3}\right), \left(\frac{2\sqrt{15}}{3}, -\frac{2\sqrt{6}}{3}\right),$$

$$\left(-\frac{2\sqrt{15}}{3}, \frac{2\sqrt{6}}{3}\right), \left(-\frac{2\sqrt{15}}{3}, -\frac{2\sqrt{6}}{3}\right)$$

23. $\begin{cases} 9x^2 + y^2 = 9 \\ 3x + y = 3 \end{cases}$

$$y = 3 - 3x$$
$$9x^2 + (3 - 3x)^2 = 9$$
$$9x^2 + (3 - 3x)^2 = 9$$
$$9x^2 + 9 - 18x + 9x^2 = 9$$
$$18x^2 - 18x = 0$$
$$18x(x - 1) = 0$$
$18x = 0 \qquad\qquad$ or $\quad x - 1 = 0$
$x = 0 \qquad\qquad$ or $\quad x = 1$
$y = 3 - 3(0) = 3 \qquad y = 3 - 3(1) = 0$

$(0, 3), \quad (1, 0)$

25. $\begin{cases} x^2 + y = 9 \\ 3x + 2y = 16 \end{cases}$

$$y = 9 - x^2$$
$$3x + 2(9 - x^2) = 16$$
$$3x + 18 - 2x^2 = 16$$
$$0 = 2x^2 - 3x - 2$$
$$0 = (2x + 1)(x - 2)$$
$2x + 1 = 0 \qquad\qquad$ or $\quad x - 2 = 0$
$$x = -\frac{1}{2} \qquad\qquad\qquad x = 2$$
$$y = 9 - \left(-\frac{1}{2}\right)^2 = \frac{35}{4} \qquad y = 9 - 2^2 = 5$$

$\left(-\frac{1}{2}, \frac{35}{4}\right), \quad (2, 5)$

27. $\begin{cases} x = y^2 - 3 \\ x + y = 9 \end{cases}$

$$(y^2 - 3) + y = 9$$
$$y^2 + y - 12 = 0$$
$$(y + 4)(y - 3) = 0$$
$y + 4 = 0 \qquad$ or $\quad y - 3 = 0$
$y = -4 \qquad\qquad\qquad y = 3$
$x = (-4)^2 - 3 = 13 \qquad x = 3^2 - 3 = 6$

$(13, -4), \quad (6, 3)$

29. $\begin{cases} x^2 + 4x + y^2 - 6y = 7 \\ 2x + y = 5 \end{cases}$

$$y = 5 - 2x$$
$$x^2 + 4x + (5 - 2x)^2 - 6(5 - 2x) = 7$$
$$x^2 + 4x + 25 - 20x + 4x^2 - 30 + 12x = 7$$
$$5x^2 - 4x - 12 = 0$$
$$(5x + 6)(x - 2) = 0$$
$5x + 6 = 0 \qquad$ or $x - 2 = 0$
$$x = -\frac{6}{5} \qquad\qquad\qquad x = 2$$
$$y = 5 - 2\left(-\frac{6}{5}\right) = \frac{37}{5} \qquad y = 5 - 2(2) = 1$$

$\left(-\frac{6}{5}, \frac{37}{5}\right), \quad (2, 1)$

31. $\begin{cases} x^2 + y^2 = 10 \\ xy = 4 \end{cases}$

$$x = \frac{4}{y}$$

$$\left(\frac{4}{y}\right)^2 + y^2 = 10$$

$$\frac{16}{y^2} + y^2 = 10$$

$$16 + y^4 = 10y^2$$
$$y^4 - 10y^2 + 16 = 0$$
$$(y^2 - 8)(y^2 - 2) = 0$$
$y^2 - 8 = 0 \qquad$ or $\quad y^2 - 2 = 0$
$y^2 = 8 \qquad\qquad\qquad y^2 = 2$
$y = \pm\sqrt{8} = \pm2\sqrt{2} \qquad y = \pm\sqrt{2}$
$y = 2\sqrt{2} \qquad\qquad\qquad y = \sqrt{2}$

$x = \dfrac{4}{2\sqrt{2}} = \sqrt{2}$ $x = \dfrac{4}{\sqrt{2}} = 2\sqrt{2}$

$y = -2\sqrt{2}$ $y = -\sqrt{2}$

$x = \dfrac{4}{-2\sqrt{2}} = -\sqrt{2}$ $x = \dfrac{4}{-\sqrt{2}} = -2\sqrt{2}$

$(\sqrt{2}, 2\sqrt{2}),\ (-\sqrt{2}, -2\sqrt{2}),\ (2\sqrt{2}, \sqrt{2}),\ (-2\sqrt{2}, -\sqrt{2})$

33. $\begin{cases} x^2 - y^2 = 4 \\ xy = 2\sqrt{3} \end{cases}$

$$x = \dfrac{2\sqrt{3}}{y}$$

$$\left(\dfrac{2\sqrt{3}}{y}\right)^2 - y^2 = 4$$

$$\dfrac{12}{y^2} = y^2 = 4$$

$$12 - y^4 = 4y^2$$
$$0 = y^4 + 4y^2 - 12$$
$$0 = (y^2 + 6)(y^2 - 2)$$

$y^2 + 6 = 0$ $y^2 - 2 = 0$
$y^2 = -6$ $y^2 = 2$
No Solution $y = \pm\sqrt{2}$

$y = \sqrt{2}$ $y = -\sqrt{2}$

$x = \dfrac{2\sqrt{3}}{\sqrt{2}} = \sqrt{6}$ $x = \dfrac{2\sqrt{3}}{-\sqrt{2}} = -\sqrt{6}$

$(\sqrt{6}, \sqrt{2}),\ (-\sqrt{6}, -\sqrt{2})$

35. $\begin{cases} x^2 + y^2 = 25 \\ x^2 - xy + y^2 = 13 \end{cases}$

$\quad x^2 + y^2 = 25$
$\underline{-x^2 + xy - y^2 = -13}$
$\qquad\ xy = 12$

$$x = \dfrac{12}{y}$$

$$\left(\dfrac{12}{y}\right)^2 + y^2 = 25$$

$$\dfrac{144}{y^2} + y^2 = 25$$

$$144 + y^4 = 25y^2$$

$y^4 - 25y^2 + 144 = 0$
$(y^2 - 16)(y^2 - 9) = 0$

$y^2 - 16 = 0$ $y^2 - 9 = 0$
$\quad y^2 = 16$ $\quad y^2 = 9$
$\quad y = \pm 4$ $\quad y = \pm 3$

$y = 4$ $y = 3$
$x = \dfrac{12}{4}$ $x = \dfrac{12}{3} = 4$

$y = -4$ $y = -3$
$x = \dfrac{12}{-4} = -3$ $x = \dfrac{12}{-3} = -4$

$(3, 4),\ (-3, -4),\ (4, 3),\ (-4, -3)$

37. $\begin{cases} x^2 + y^2 = 29 \\ x - y = 3 \end{cases}$

$$x = 3 + y$$
$$(3 + y)^2 + y^2 = 29$$
$$9 + 6y + y^2 + y^2 = 29$$
$$2y^2 + 6y - 20 = 0$$
$$y^2 + 3y - 10 = 0$$
$$(y + 5)(y - 2) = 0$$

$y + 5 = 0$ $\qquad$ or $\ y - 2 = 0$
$\ y = -5$ $\qquad\qquad$ $y = 2$
$\ x = 3 + (-5) = -2$ $\qquad$ $x = 3 + 2 = 5$

$(-2, -5),\ (5, 2)$

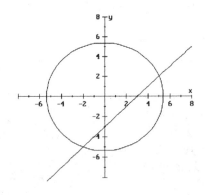

39. $\begin{cases} x^2 + y^2 = 29 \\ \quad y = x^2 - 9 \end{cases}$

$$y = x^2 - 9$$
$$y + 9 = x^2$$

$$y + 9 + y^2 = 29$$
$$y^2 + y - 20 = 0$$
$$(y + 5)(y - 4) = 0$$

$y + 5 = 0 \qquad$ or $\quad y - 4 = 0$
$\quad\quad y = -5 \qquad\qquad\quad y = 4$
$\quad\quad x^2 = -5 + 9 \qquad\quad x^2 = 4 + 9$
$\quad\quad x^2 = 4 \qquad\qquad\quad x^2 = 13$
$\quad\quad x = \pm 2 \qquad\qquad\quad x = \pm\sqrt{13}$

$(2, -5), \ (-2, -5), \ (\sqrt{13}, \ 4), \ (-\sqrt{13}, \ 4)$

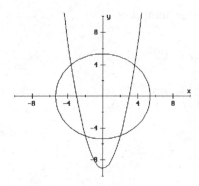

41. $\begin{cases} \quad x - y = 0 \\ 3y + x^2 = 4 \end{cases}$

$$x - y = 0$$
$$x = y$$

$$3(x) + x^2 = 4$$
$$x^2 + 3x - 4 = 0$$
$$(x + 4)(x - 1) = 0$$

$x + 4 = 0 \quad$ or $\quad x - 1 = 0$
$\quad x = -4 \qquad\qquad x = 1$
$\quad y = -4 \qquad\qquad y = 1$

$(-4, -4), \ (1, 1)$

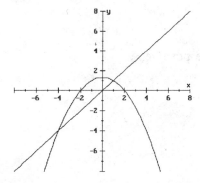

43. $\begin{cases} x^2 + y^2 = 5 \\ \quad y = 8x^2 - 6 \end{cases}$

$$y = 8x^2 - 6$$
$$4 + 6 = 8x^2$$
$$\frac{y + 6}{8} = x^2$$

$$\frac{y + 6}{8} + y^2 = 5$$

$$y + 6 + 8y^2 = 40$$
$$8y^2 + y - 34 = 0$$
$$(8y + 17)(y - 2) = 0$$

$8y + 17 = 0 \qquad$ or $\quad y - 2 = 0$
$\qquad y = -\dfrac{17}{8} \qquad\qquad\quad y = 2$

$-\dfrac{17}{8} + 6 = 8x^2 \qquad\quad 2 + 6 = 8x^2$

$\qquad \dfrac{31}{8} = 8x^2 \qquad\qquad\quad 8 = 8x^2$

$\qquad \dfrac{31}{64} = x^2 \qquad\qquad\quad \pm 1 = x$

$\qquad \pm\dfrac{\sqrt{31}}{8} = x$

$\left(\dfrac{\sqrt{31}}{8}, \ -\dfrac{17}{8}\right), \ \left(-\dfrac{\sqrt{31}}{8}, \ -\dfrac{17}{8}\right),$

$(1, 2), \ (-1, 2)$

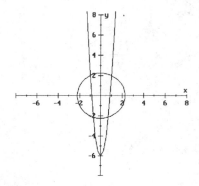

45. $\begin{cases} x^2 \quad\quad + y^2 = 1 \\ x^2 + 2x + y^2 = 0 \end{cases}$

$$-x^2 \quad\quad - y^2 = -1$$
$$\underline{x^2 + 2x + y^2 = 0}$$
$$2x \quad\quad = -1$$
$$x = -\frac{1}{2}$$

$$\left(-\frac{1}{2}\right)^2 + y^2 = 1$$

$$y^2 = \frac{3}{4}$$

$$y = \pm\frac{\sqrt{3}}{2}$$

$$\left(-\frac{1}{2}, \frac{\sqrt{3}}{2}\right), \left(-\frac{1}{2}, -\frac{\sqrt{3}}{2}\right)$$

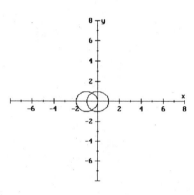

47. Let x = width
$\quad\quad y$ = length

$$\begin{cases} xy = 210 \\ 2x + 2y = 59 \end{cases}$$

$$x = \frac{210}{y}$$

$$2\left(\frac{210}{y}\right) + 2y = 59$$

$$\frac{420}{y} + 2y = 59$$

$$420 + 2y^2 = 59y$$
$$2y^2 - 59y + 420 = 0$$
$$(2y - 35)(y - 12) = 0$$

$2y - 35 = 0 \quad$ or $\quad y - 12 = 0$

$$y = \frac{35}{2} \quad\quad\quad y = 12$$

$$x = \frac{210}{\frac{35}{2}} \quad\quad\quad x = \frac{210}{12}$$

$$x = 12 \quad\quad\quad x = 17.5$$

The dimensions are 12″ by 17.5″.

49. Let x = rate going,
then $x + 2$ = rate returning

Let y = time going,
then $y - \frac{1}{4}$ = time returning

$$\begin{cases} xy = 60 \\ (x + 12)\left(y - \frac{1}{4}\right) = 60 \end{cases}$$

$$x = \frac{60}{y}$$

$$\left(\frac{60}{y} + 12\right)\left(y - \frac{1}{4}\right) = 60$$

$$60 - \frac{15}{y} + 12y - 3 = 60$$

$$-\frac{15}{y} + 12y - 3 = 0$$

$$-15 + 12y^2 - 3y = 0$$
$$12y^2 - 3y - 15 = 0$$
$$4y^2 - y - 5 = 0$$
$$(y + 1)(4y - 5) = 0$$
$$y + 1 = 0 \quad\quad 4y - 5 = 0$$
$$y = -1 \quad\quad\quad y = \frac{5}{4}$$

Since y must be positive, $y = \frac{5}{4}$.

$$y = \frac{5}{4}$$

$$y - \frac{1}{4} = \frac{5}{4} - \frac{1}{4} = 1$$

$$x = \frac{60}{\frac{5}{4}} = 48$$

$$x + 12 = 48 + 12 = 60$$

Going: rate = 48 mph; time $= \frac{5}{4}$ hr

Returning: rate = 60 mph; time = 1 hr

53. (a) $A(3) = 2000e^{0.0316(3)}$
        $= 2198.9$ bacteria

   (b) $5000 = 2000e^{0.0316t}$
           $2.5 = e^{0.0316t}$
        $\ln 2.5 = \ln e^{0.0316t}$
        $\ln 2.5 = 0.0316t$

$$\frac{\ln 2.5}{0.0316} = t$$

$$t \approx 29 \text{ hr}$$

55. $y = \left(\frac{2}{3}\right)^x$

x	y
-2	9/4
-1	3/2
0	1
1	2/3
2	4/9

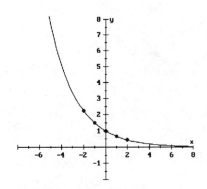

1. $\begin{cases} x + 2y - z = 2 \\ 2x + 3y + 4z = 9 \\ 3x + y - 2z = 2 \end{cases}$

$$\begin{array}{r} 4x + 8y - 4z = 8 \\ 2x + 3y + 4z = 9 \\ \hline 6x + 11y = 17 \end{array}$$

$$\begin{array}{r} 2x + 3y + 4z = 9 \\ 6x + 2y - 4z = 4 \\ \hline 8x + 5y = 13 \end{array}$$

$$6x + 11y = 17$$
$$8x + 5y = 13$$

$$\begin{array}{r} 24x + 44y = 68 \\ -24x - 15y = -39 \\ \hline 29y = 29 \\ y = 1 \end{array}$$

$$6x + 11(1) = 17$$
$$6x = 6$$
$$x = 1$$

$$1 + 2(1) - z = 2$$
$$-z = -1$$
$$z = 1$$

$$x = 1, \ y = 1, \ z = 1$$

3. $\begin{cases} 2x - 3z = 7 \\ 3x + 4y = -11 \\ 3y - 2z = 0 \end{cases}$

$$\begin{array}{r} 4x - 6z = 14 \\ -9y + 6z = 0 \\ \hline 4x - 9y = 14 \end{array}$$

$$4x - 9y = 14$$
$$3x + 4y = -11$$

$$\begin{array}{r} 16x - 36y = 56 \\ 27x + 36y = -99 \\ \hline 43x = -43 \\ x = -1 \end{array}$$

$$2(-1) - 3z = 7$$
$$-3z = 9$$
$$z = -3$$

$$3y - 2(-3) = 0$$
$$3y = -6$$
$$y = -2$$

$$x = -1, \ y = -2, \ z = -3$$

5. $\begin{cases} 3x + y - z = 2 \\ 3x + y - z = 5 \\ x - y + z = 1 \end{cases}$

$\begin{array}{r} -3x - y + z = -2 \\ \underline{3x + y - z = 5} \\ 0 = 3 \end{array}$

No solution

7. $\begin{bmatrix} 1 & 2 & | & 6 \\ 2 & 6 & | & 8 \end{bmatrix}$

$-2R_1 + R_2 \rightarrow R_2$

$\begin{bmatrix} 1 & 2 & | & 6 \\ 0 & 2 & | & -4 \end{bmatrix}$

$\tfrac{1}{2}R_2 \rightarrow R_2$
$R_1 - R_2 \rightarrow R_1$

$\begin{bmatrix} 1 & 0 & | & 10 \\ 0 & 1 & | & -2 \end{bmatrix}$

9. $\begin{bmatrix} 2 & 2 & 6 & | & 4 \\ 2 & 5 & 9 & | & -2 \\ 1 & 2 & 3 & | & -1 \end{bmatrix}$

$\tfrac{1}{2}R_1 \rightarrow R_1$
$R_1 - R_2 \rightarrow R_2$
$-2R_3 + R_1 \rightarrow R_3$

$\begin{bmatrix} 1 & 1 & 3 & | & 2 \\ 0 & -3 & -3 & | & 6 \\ 0 & -2 & 0 & | & 6 \end{bmatrix}$

$(-1/3)R_2 \rightarrow R_2$

$\begin{bmatrix} 1 & 1 & 3 & | & 2 \\ 0 & 1 & 1 & | & -2 \\ 0 & -2 & 0 & | & 6 \end{bmatrix}$

$2R_2 + R_3 \rightarrow R_3$
$R_1 - R_2 \rightarrow R_1$

$\begin{bmatrix} 1 & 0 & 2 & | & 4 \\ 0 & 1 & 1 & | & -2 \\ 0 & 0 & 2 & | & 2 \end{bmatrix}$

$\tfrac{1}{2}R_3 \rightarrow R_3$

$\begin{bmatrix} 1 & 0 & 2 & | & 4 \\ 0 & 1 & 1 & | & -2 \\ 0 & 0 & 1 & | & 1 \end{bmatrix}$

$-2R_3 + R_1 \rightarrow R_1$
$R_2 - R_3 \rightarrow R_2$

$\begin{bmatrix} 1 & 0 & 0 & | & 2 \\ 0 & 1 & 0 & | & -3 \\ 0 & 0 & 1 & | & 1 \end{bmatrix}$

11. $\begin{bmatrix} 1 & -2 & | & 1 \\ 2 & 3 & | & 9 \end{bmatrix}$

$-2R_1 + R_2 \rightarrow R_2$

$\begin{bmatrix} 1 & -2 & | & 1 \\ 0 & 7 & | & 7 \end{bmatrix}$

$\begin{cases} x - 2y = 1 \\ 7y = 7 \end{cases}$

$7y = 7$
$y = 1$

$x - 2(1) = 1$
$x = 3$

$(3, 1)$

13. $\begin{bmatrix} 1 & 2 & -1 & | & 5 \\ 1 & -1 & 1 & | & 0 \\ 1 & 1 & 2 & | & 1 \end{bmatrix}$

$R_1 - R_2 \rightarrow R_2$
$R_1 - R_3 \rightarrow R_3$

$\begin{bmatrix} 1 & 2 & -1 & | & 5 \\ 0 & 3 & -2 & | & 5 \\ 0 & 1 & -3 & | & 4 \end{bmatrix}$

$-3R_3 + R_2 \rightarrow R_3$

$$\begin{bmatrix} 1 & 2 & -1 & | & 5 \\ 0 & 3 & -2 & | & 5 \\ 0 & 0 & 7 & | & -7 \end{bmatrix}$$

$$\begin{cases} x + 2y - z = 5 \\ 3y - 2x = 5 \\ 7z = -7 \end{cases}$$

$7z = -7$
$z = -1$

$3y - 2(-1) = 5$
$ 3y = 3$
$ y = 1$

$x + 2(1) - (-1) = 5$
$ x = 2$

$(2, 1, -1)$

15. $\begin{bmatrix} 1 & -3 & | & -1 \\ 2 & 1 & | & 5 \end{bmatrix}$

$-2R_1 + R_2 \rightarrow R_2$

$\begin{bmatrix} 1 & -3 & | & -1 \\ 0 & 7 & | & 7 \end{bmatrix}$

$(1/7)R_2 \rightarrow R_2$

$\begin{bmatrix} 1 & -3 & | & -1 \\ 0 & 1 & | & 1 \end{bmatrix}$

$3R_2 + R_1 \rightarrow R_1$

$\begin{bmatrix} 1 & 0 & | & 2 \\ 0 & 1 & | & 1 \end{bmatrix}$

$x = 2, \quad y = 1$
$(2, 1)$

17. $5C = 5\begin{bmatrix} 2 & -1 \\ 0 & 5 \end{bmatrix}$

$ = \begin{bmatrix} 10 & -5 \\ 0 & 25 \end{bmatrix}$

19. $C - D$

$= \begin{bmatrix} 2 & -1 \\ 0 & 5 \end{bmatrix} - \begin{bmatrix} -3 & 0 \\ 2 & 1 \end{bmatrix}$

$= \begin{bmatrix} 2 - (-3) & -1 - 0 \\ 0 - 2 & 5 - 1 \end{bmatrix}$

$= \begin{bmatrix} 5 & -1 \\ -2 & 4 \end{bmatrix}$

21. AB

$= \begin{bmatrix} 1 \\ 0 \\ -2 \end{bmatrix} [2 \ -3 \ 1]$

$= \begin{bmatrix} 2 & -3 & 1 \\ 0 & 0 & 0 \\ -4 & 6 & -2 \end{bmatrix}$

23. CE

$= \begin{bmatrix} 2 & -1 \\ 0 & 5 \end{bmatrix}\begin{bmatrix} 1 & 5 & -2 \\ -1 & 0 & 6 \end{bmatrix}$

$= \begin{bmatrix} (2)(1) + (-1)(-1) & (2)(5) + (-1)(0) & (2)(-2) + (-1)(6) \\ (0)(1) + (5)(-1) & (0)(5) + (5)(0) & (0)(-2) + (5)(6) \end{bmatrix}$

$= \begin{bmatrix} 3 & 10 & -10 \\ -5 & 0 & 30 \end{bmatrix}$

25. $C + D$

$= \begin{bmatrix} 2 & -1 \\ 0 & 5 \end{bmatrix} + \begin{bmatrix} -3 & 0 \\ 2 & 1 \end{bmatrix}$

$= \begin{bmatrix} -1 & -1 \\ 2 & 6 \end{bmatrix}$

$(C + D)E$

$= \begin{bmatrix} -1 & -1 \\ 2 & 6 \end{bmatrix}\begin{bmatrix} 1 & 5 & -2 \\ -1 & 0 & 6 \end{bmatrix}$

$= \begin{bmatrix} 0 & -5 & -4 \\ -4 & 10 & 32 \end{bmatrix}$

27. $\begin{bmatrix} 2 & 3 \\ 3 & \frac{1}{2} \end{bmatrix} \begin{bmatrix} x \\ y \end{bmatrix} = \begin{bmatrix} 14 \\ 1 \end{bmatrix}$

$AX = B$

$X = A^{-1}B$

$= \begin{bmatrix} -\frac{1}{2} \\ 5 \end{bmatrix}$

$x = -\frac{1}{2}, \quad y = 5$

29. $\begin{bmatrix} 2 & -3 & 4 \\ 5 & 4 & -3 \\ 3 & -2 & 1 \end{bmatrix} \begin{bmatrix} x \\ y \\ z \end{bmatrix} = \begin{bmatrix} 18 \\ -4 \\ 12 \end{bmatrix}$

$AX = B$

$X = A^{-1}B$

$= \begin{bmatrix} 2 \\ -2 \\ 2 \end{bmatrix}$

$x = 2, \quad y = -2, \quad z = 2$

31. $\begin{vmatrix} 2 & 3 \\ 4 & -1 \end{vmatrix} = (2)(-1) - (4)(3) = -14$

33. $\begin{vmatrix} 1 & 3 & -2 \\ 2 & 4 & 3 \\ 5 & -1 & -3 \end{vmatrix}$

$= 1 \begin{vmatrix} 4 & 3 \\ -1 & -3 \end{vmatrix} - 3 \begin{vmatrix} 2 & 3 \\ 5 & -3 \end{vmatrix} + (-2) \begin{vmatrix} 2 & 4 \\ 5 & -1 \end{vmatrix}$

$= [-12 - (-3)] - 3[-6 - 15] - 2[-2 - 20]$

$= -9 + 63 + 44$

$= 98$

35. $D = \begin{vmatrix} 3 & 7 \\ 5 & 4 \end{vmatrix} = -23$

$D_x = \begin{vmatrix} 4 & 7 \\ 2 & 4 \end{vmatrix} = 2$

$D_y = \begin{vmatrix} 3 & 4 \\ 5 & 2 \end{vmatrix} = -14$

$x = \dfrac{D_x}{D} = \dfrac{2}{-23} = -\dfrac{2}{23}$

$y = \dfrac{D_y}{D} = \dfrac{-14}{-23} = \dfrac{14}{23}$

37. $D = \begin{vmatrix} 4 & 2 \\ 6 & 3 \end{vmatrix} = 0$

No unique solution

39. $D = \begin{vmatrix} 3 & 4 & 1 \\ 5 & -3 & 6 \\ 4 & -5 & -5 \end{vmatrix} = 318$

$D_x = \begin{vmatrix} 3 & 4 & 1 \\ 2 & -3 & 6 \\ 1 & -5 & -5 \end{vmatrix} = 192$

$D_y = \begin{vmatrix} 3 & 3 & 1 \\ 5 & 2 & 6 \\ 4 & 1 & -5 \end{vmatrix} = 96$

$D_z = \begin{vmatrix} 3 & 4 & 3 \\ 5 & -3 & 2 \\ 3 & -5 & 1 \end{vmatrix} = -6$

$x = \dfrac{D_x}{D} = \dfrac{192}{318} = \dfrac{32}{53}$

$y = \dfrac{D_y}{D} = \dfrac{96}{318} = \dfrac{16}{53}$

$z = \dfrac{D_z}{D} = \dfrac{-6}{318} = -\dfrac{1}{53}$

41. $\begin{cases} x^2 + y^2 = 8 \\ x + y = 4 \end{cases}$

$x = 4 - y$

$(4 - y)^2 + y^2 = 8$

$16 - 8y + y^2 + y^2 = 8$

$2y^2 - 8y + 8 = 0$

$y^2 - 4y + 4 = 0$

$(y - 2)^2 = 0$

$y - 2 = 0$

$y = 2$

$x = 4 - 2 = 2$

$(2, 2)$

43. $\begin{cases} 2x^2 - y^2 = 14 \\ 2x - 3y = -8 \end{cases}$

$$2x - 3y = -8$$
$$2x = 3y - 8$$
$$x = \frac{3}{2}y - 4$$

$$2\left(\frac{3}{2}y - 4\right)^2 - y^2 = 14$$

$$2\left(\frac{9}{4}y^2 - 12y + 16\right) - y^2 = 14$$

$$\frac{9}{2}y^2 - 24y + 32 - y^2 = 14$$

$$9y^2 - 48y + 64 - 2y^2 = 28$$
$$7y^2 - 48y + 36 = 0$$
$$(y - 6)(7y - 6) = 0$$

$y - 6 = 0$ or $7y - 6 = 0$

$y = 6$ $y = \dfrac{6}{7}$

$x = \dfrac{3}{2}(6) - 4 = 5$ $x = \dfrac{3}{2}\left(\dfrac{6}{7}\right) - 4 = -\dfrac{19}{7}$

$(5, 6), \quad \left(-\dfrac{19}{7}, \dfrac{6}{7}\right)$

45. $x^2 + y^2 = 4$
      $2x^2 + 3y^2 = 18$

$$\begin{aligned} -2x^2 - 2y^2 &= -8 \\ \underline{2x^2 + 3y^2} &= \underline{18} \\ y^2 &= 10 \\ y &= \pm\sqrt{10} \end{aligned}$$

$y = 10$:
$$x^2 + \left(\sqrt{10}\right)^2 = 4$$
$$x^2 + 10 = 4$$
$$x^2 = -6$$

No real solution

47. $\begin{cases} x + y \le 2 & \text{(solid line)} \\ 2x - 3y \le 6 & \text{(solid line)} \end{cases}$

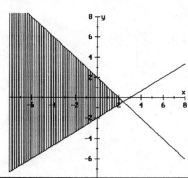

Test point: $(0, 0)$

$\begin{array}{ll} 0 + 0 \le 2 & 2(0) - 3(0) \le 6 \\ \quad 0 \le 2 & \quad\quad\quad 0 \le 6 \\ \quad \text{True} & \quad\quad\quad \text{True} \end{array}$

49. $\begin{cases} 2x + 3y \le 12 & \text{(solid line)} \\ \quad\quad y < x & \text{(dashed line)} \\ \quad\quad x \ge 0 & \text{(solid line)} \\ \quad\quad y \ge 0 & \text{(solid line)} \end{cases}$

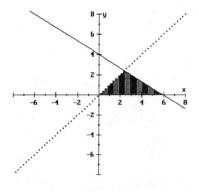

Test point: $(2, 1)$

$\begin{array}{ll} 2(2) + 3(1) \le 12 & 1 < 2 \\ \quad\quad 7 \le 12 & \text{True} \\ \quad\quad \text{True} & \end{array}$

$\begin{array}{ll} 2 \ge 0 & 1 \ge 0 \\ \text{True} & \text{True} \end{array}$

51. x = amount at 7.7%
      y = amount at 8.6%
      z = amount at 9.8%

$\begin{cases} \quad\quad\quad\quad\quad x + y + z = 20000 \\ \quad\quad\quad\quad\quad\quad\quad y + z = 3000 + x \\ 0.077x + 0.086y + 0.098z = 1719.10 \end{cases}$

$\begin{cases} \quad\quad x + y + z = 20000 \\ \quad -x + y + z = 3000 \\ 77x + 86y + 98z = 1719100 \end{cases}$

$$\begin{bmatrix} 1 & 1 & 1 \\ -1 & 1 & 1 \\ 77 & 86 & 98 \end{bmatrix} \begin{bmatrix} x \\ y \\ z \end{bmatrix} = \begin{bmatrix} 20000 \\ 3000 \\ 1719100 \end{bmatrix}$$

$$AX = B$$
$$X = A^{-1}B$$

$$= \begin{bmatrix} 8500 \\ 5200 \\ 6300 \end{bmatrix}$$

$8500 is invested at 7.7%, $5200 at 8.6% and $6300 at 9.8%.

53. x = width
y = length

$$\begin{cases} 2x + 2y = 41 \\ xy = 100 \end{cases}$$

$$x = \frac{100}{y}$$

$$2\left(\frac{100}{y}\right) + 2y = 41$$

$$\frac{200}{y} + 2y = 41$$

$$200 + 2y^2 = 41y$$
$$2y^2 - 41y + 200 = 0$$
$$(2y - 25)(y - 8) = 0$$
$$2y - 25 = 0 \qquad \text{or} \quad y - 8 = 0$$

$$y = \frac{25}{2} = 12.5 \qquad\qquad y = 8$$

$$x = \frac{100}{12.5} = 8 \qquad\qquad x = \frac{100}{8} = 12.5$$

The dimensions are 8 ft by 12.5 ft.

CHAPTER 11 PRACTICE TEST

1. $\begin{cases} 2x - 3y + 4z = 2 \\ 3x + 2y - z = 10 \\ 2x - 4y + 3z = 3 \end{cases}$

$$\begin{array}{r} 4x - 6y + 8z = 4 \\ \underline{9x + 6y - 3z = 30} \\ 13x \qquad + 5z = 34 \end{array}$$

$$\begin{array}{r} 6x + 4y - 2z = 20 \\ \underline{2x - 4y + 3z = 3} \\ 8x \qquad + z = 23 \end{array}$$

$$\begin{array}{r} 13x + 5z = 34 \\ 8x + z = 23 \end{array}$$

$$\begin{array}{r} 13x + 5z = 34 \\ \underline{-40x - 5z = -115} \\ -27x \qquad = -81 \\ x = 3 \end{array}$$

$$8(3) + z = 23$$
$$z = -1$$

$$\begin{array}{r} 2(3) - 3y + 4(-1) = 2 \\ -3y + 2 = 2 \\ -3y = 0 \\ y = 0 \end{array}$$

$$x = 3, \quad y = 0, \quad z = -1$$

2. (a) $\begin{bmatrix} 5 & 2 & | & 1 \\ 1 & 3 & | & 8 \end{bmatrix}$

$$-5R_2 + R_1 \rightarrow R_2$$

$$\begin{bmatrix} 5 & 2 & | & 1 \\ 0 & -13 & | & -39 \end{bmatrix}$$

$$\begin{array}{r} 5x + 2y = 1 \\ -13y = -39 \\ y = 3 \end{array}$$

$$\begin{array}{r} 5x + 2(3) = 1 \\ 5x = -6 \\ x = -1 \end{array}$$

$$(-1, 3)$$

(b) $\begin{bmatrix} 1 & -1 & 0 & | & 2 \\ 1 & 0 & 2 & | & 7 \\ -2 & 3 & 4 & | & 5 \end{bmatrix}$

$$R_2 - R_1 \rightarrow R_2$$
$$2R_2 + R_3 \rightarrow R_3$$

$$\begin{bmatrix} 1 & -1 & 0 & | & 2 \\ 0 & 1 & 2 & | & 5 \\ 0 & 3 & 8 & | & 19 \end{bmatrix}$$

$$-3R_2 + R_3 \rightarrow R_3$$

$$\begin{bmatrix} 1 & -1 & 0 & | & 2 \\ 0 & 1 & 2 & | & 5 \\ 0 & 0 & 2 & | & 4 \end{bmatrix}$$

$$\begin{cases} x - y & = 2 \\ y + 2z = 5 \\ 2z = 4 \end{cases}$$

$2z = 4$
$z = 2$

$y + 2(2) = 5$
$ y = 1$

$x - 1 = 2$
$ x = 3$

$(3, 1, 2)$

3. (a) $3A = \begin{bmatrix} 3 & 15 & -6 \\ -3 & 0 & 18 \\ 6 & 6 & 0 \end{bmatrix}$

$4B = \begin{bmatrix} 8 & -12 & 4 \\ 0 & 4 & -4 \\ 16 & 0 & 4 \end{bmatrix}$

$3A - 4B$

$= \begin{bmatrix} 3 & 15 & -6 \\ -3 & 0 & 18 \\ 6 & 6 & 0 \end{bmatrix} - \begin{bmatrix} 8 & -12 & 4 \\ 0 & 4 & -4 \\ 16 & 0 & 4 \end{bmatrix}$

$= \begin{bmatrix} -5 & 27 & -10 \\ -3 & -4 & 22 \\ -10 & 6 & -4 \end{bmatrix}$

(b) AB

$= \begin{bmatrix} 1 & 5 & -2 \\ -1 & 0 & 6 \\ 2 & 2 & 0 \end{bmatrix} \begin{bmatrix} 2 & -3 & 1 \\ 0 & 1 & -1 \\ 4 & 0 & 1 \end{bmatrix}$

$= \begin{bmatrix} (1)(2) + (5)(0) + (-2)(4) & (1)(-3) + (5)(1) + (-2)(0) & (1)(1) + (5)(-1) + (-2)(1) \\ (-1)(2) + (0)(0) + (6)(4) & (-1)(-3) + (0)(1) + (6)(0) & (-1)(1) + (0)(-1 + (6)(1) \\ (2)(2) + (2)(0) + (0)(4) & (2)(-3) + (2)(1) + (0)(0) & (2)(1) + (2)(-1) + (0)(1) \end{bmatrix}$

$= \begin{bmatrix} -6 & 2 & -6 \\ 22 & 3 & 5 \\ 4 & -4 & 0 \end{bmatrix}$

4. $\begin{bmatrix} 2 & 1 & -3 \\ 3 & 1 & -2 \\ 1 & 2 & 5 \end{bmatrix} \begin{bmatrix} x \\ y \\ z \end{bmatrix} = \begin{bmatrix} 5 \\ 1 \\ 6 \end{bmatrix}$

$AX = B$

$X = A^{-1}B$

$= \begin{bmatrix} -2.86 \\ 7.29 \\ -1.14 \end{bmatrix}$

$x = -2.86, \quad y = 7.29, \quad z = -1.14$

5. (a) $\begin{vmatrix} 2 & 3 \\ -1 & 4 \end{vmatrix} = (2)(4) - (-1)(3) = 11$

(b) $\begin{vmatrix} 5 & 0 & 2 \\ 2 & 3 & 1 \\ 1 & 1 & 2 \end{vmatrix}$

$= 0 \begin{vmatrix} 2 & 1 \\ 1 & 2 \end{vmatrix} + 3 \begin{vmatrix} 5 & 2 \\ 1 & 2 \end{vmatrix} - 1 \begin{vmatrix} 5 & 2 \\ 2 & 1 \end{vmatrix}$

$= 0 + 3(10 - 2) - 1(5 - 4)$

$= 23$

6. (a) $\begin{cases} 2x - 6y = -1 \\ 4x - 8y = 5 \end{cases}$

$D = \begin{vmatrix} 2 & -6 \\ 4 & -8 \end{vmatrix} = 8$

$D_x = \begin{vmatrix} -1 & -6 \\ 5 & -8 \end{vmatrix} = 38$

$D_y = \begin{vmatrix} 2 & -1 \\ 4 & 5 \end{vmatrix} = 14$

$x = \dfrac{D_x}{D} = \dfrac{38}{8} = \dfrac{19}{4}$

$y = \dfrac{D_y}{D} = \dfrac{14}{8} = \dfrac{7}{4}$

(b) $D = \begin{vmatrix} 5 & 1 & 0 \\ 3 & 0 & -1 \\ 0 & 1 & 3 \end{vmatrix} = -4$

$D_x = \begin{vmatrix} 9 & 1 & 0 \\ 3 & 0 & -1 \\ 8 & 1 & 3 \end{vmatrix} = -8$

$D_y = \begin{vmatrix} 5 & 9 & 0 \\ 3 & 3 & -1 \\ 0 & 8 & 3 \end{vmatrix} = 4$

$D_z = \begin{vmatrix} 5 & 1 & 9 \\ 3 & 0 & 3 \\ 0 & 1 & 8 \end{vmatrix} = -12$

$x = \dfrac{D_x}{D} = \dfrac{-8}{-4} = 2$

$y = \dfrac{D_y}{D} = \dfrac{4}{-4} = -1$

$z = \dfrac{-12}{-4} = 3$

7. x = number of $5-bills
y = number of $10-bills
z = number of $20-bills

$\begin{cases} 5x + 10y + 20z = 500 \\ \qquad\quad y \qquad\quad = 2z \\ \quad x + y + \quad z = 40 \end{cases}$

$5x + 10(2z) + 20z = 500$
$\quad x + \quad 2z + \quad z = 40$

$\quad 5x + 40z = 500$
$\underline{-5x - 15z = -200}$
$\qquad\quad 25z = 300$
$\qquad\qquad z = 12$

$x + 3(12) = 40$
$\qquad\quad x = 4$

$y = 2(12) = 24$

There are 4 $5-bills, 24 $10-bills and 12 $20-bills.

8. $\begin{cases} x^2 + 2y^2 = 6 \\ x - y = 3 \end{cases}$

$$x = y + 3$$
$$(y + 3)^2 + 2y^2 = 6$$
$$y^2 + 6y + 9 + 2y^2 = 6$$
$$3y^2 + 6y + 3 = 0$$
$$y^2 + 2y + 1 = 0$$
$$(y + 1)^2 = 0$$
$$y + 1 = 0$$
$$y = -1$$
$$x = -1 + 3 = 2$$

$(2, -1)$

9. $\begin{cases} y < 4 \quad \text{(dashed line)} \\ x \leq 2 \quad \text{(solid line)} \\ x + y > 3 \quad \text{(dashed line)} \end{cases}$

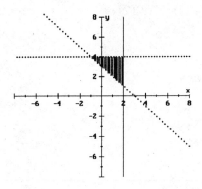

Test point: $(1, 3)$

$3 < 4$
True

$1 \leq 2$
True

$1 + 3 > 3$
$\quad 4 > 3$
$\quad\quad$ True

1. $\begin{cases} x + y + z = 5 \\ 2x + y - 2z = 6 \\ 3x - y + 3z = 10 \end{cases}$

$\begin{array}{r} x + y + z = 6 \\ \underline{3x - y + 3z = 10} \\ 4x \quad + 4z = 16 \\ x \quad + z = 4 \end{array}$

$\begin{array}{r} 2x + y - 2z = 6 \\ \underline{3x - y + 3z = 10} \\ 5x \quad + z = 16 \end{array}$

$\begin{array}{r} x + z = 4 \\ 5x + z = 16 \end{array}$

$\begin{array}{r} -x - z = -4 \\ \underline{5x + z = 16} \\ 4x \quad = 12 \\ x = 3 \end{array}$

$\begin{array}{r} 3 + z = 4 \\ z = 1 \end{array}$

$\begin{array}{r} 3 + y + 1 = 6 \\ y = 2 \end{array}$

$x = 3, \quad y = 2, \quad z = 1$

3. $\begin{cases} 3x - 4y + 5z = 1 \\ 2x - y + 3z = 2 \\ x - 2y + z = 3 \end{cases}$

$\begin{array}{r} 3x - 4y + 5z = 1 \\ \underline{-8x + 4y - 12z = -8} \\ -5x \quad - 7z = -7 \end{array}$

$\begin{array}{r} -4x + 2y - 6z = -4 \\ \underline{x - 2y + z = 3} \\ -3x \quad - 5z = -1 \end{array}$

$\begin{array}{l} -5x - 7z = -7 \\ -3x - 5z = -1 \end{array}$

$\begin{array}{r} 15x + 21z = 21 \\ \underline{-15x - 25z = -5} \\ -4z = 16 \\ z = -4 \end{array}$

$\begin{array}{r} -3x - 5(-4) = -1 \\ -3x = -21 \\ x = 7 \end{array}$

$\begin{array}{r} 7 - 2y + (-4) = 3 \\ -2y = 0 \\ y = 0 \end{array}$

$x = 7, \quad y = 0, \quad z = -4$

5. $\begin{cases} x - 2y + 3z = 4 \\ \dfrac{3}{2}x - 3y + \dfrac{9}{2}z = 6 \\ -3x + 6y - 9z = -12 \end{cases}$

$\begin{cases} x - 2y + 3z = 4 \\ 3x - 6y + 9z = 12 \\ -3x + 6y - 9z = -12 \end{cases}$

$\begin{array}{r} 3x - 6y + 9z = 12 \\ \underline{-3x + 6y - 9z = -12} \\ 0 = 0 \end{array}$

$\{(x, y, z) \,|\, x - 2y + 3z = 4\}$

7. $\begin{vmatrix} 3 & 2 \\ 4 & 5 \end{vmatrix} = (3)(5) - (4)(2) = 7$

9. $\begin{vmatrix} 1 & 2 \\ 2 & 4 \end{vmatrix} = (1)(4) - (2)(2) = 0$

11. $\begin{vmatrix} 4 & -1 & 2 \\ 2 & 1 & 0 \\ -1 & 2 & -3 \end{vmatrix}$

$= -2\begin{vmatrix} -1 & 2 \\ 2 & -3 \end{vmatrix} + 1\begin{vmatrix} 4 & 2 \\ -1 & -3 \end{vmatrix} - 0\begin{vmatrix} 4 & -1 \\ -1 & 2 \end{vmatrix}$

$= -2(3 - 4) + 1(-12 + 2) - 0$
$= -8$

13. $\begin{vmatrix} 5 & 4 & 3 \\ 2 & 0 & 0 \\ 3 & 1 & 1 \end{vmatrix}$

$= -2\begin{vmatrix} 4 & 3 \\ 1 & 1 \end{vmatrix} + 0\begin{vmatrix} 5 & 3 \\ 3 & 1 \end{vmatrix} - 0\begin{vmatrix} 5 & 4 \\ 3 & 1 \end{vmatrix}$

$= -2(4 - 3)$
$= -2$

15.
$$\begin{bmatrix} 1 & 2 \\ 2 & -3 \end{bmatrix} \begin{bmatrix} x \\ y \end{bmatrix} = \begin{bmatrix} 0 \\ 7 \end{bmatrix}$$

$$AX = B$$
$$X = A^{-1}B$$

$$= \begin{bmatrix} 2 \\ -1 \end{bmatrix}$$

$(2, -1)$

17. $D = \begin{vmatrix} 3 & 7 \\ 10 & 5 \end{vmatrix} = -55$

$$D_x = \begin{vmatrix} 2 & 7 \\ 11 & 5 \end{vmatrix} = -67$$

$$D_y = \begin{vmatrix} 3 & 2 \\ 10 & 11 \end{vmatrix} = 13$$

$$x = \frac{D_x}{D} = \frac{-67}{-55} = \frac{67}{55}$$

$$y = \frac{D_y}{D} = \frac{13}{-55} = -\frac{13}{55}$$

19. $D = \begin{vmatrix} 2 & 3 & 4 \\ 5 & 2 & -3 \\ 3 & -7 & 5 \end{vmatrix} = -288$

$$D_x = \begin{vmatrix} 3 & 3 & 4 \\ 2 & 2 & -3 \\ -7 & -7 & 5 \end{vmatrix} = 0$$

$$D_y = \begin{vmatrix} 2 & 3 & 4 \\ 5 & 2 & -3 \\ 3 & -7 & 5 \end{vmatrix} = -288$$

$$D_z = \begin{vmatrix} 2 & 3 & 3 \\ 5 & 2 & 2 \\ 3 & -7 & -7 \end{vmatrix} = 0$$

$$x = \frac{D_x}{D} = \frac{0}{-288} = 0$$

$$y = \frac{D_y}{D} = \frac{-288}{-288} = 1$$

$$z = \frac{D_z}{D} = \frac{0}{-288} = 0$$

21. $\begin{cases} x + y = 1 \\ x^2 + y^2 = 5 \end{cases}$

$$x = 1 - y$$
$$(1 - y)^2 + y^2 = 5$$
$$1 - 2y + y^2 + y^2 = 5$$
$$2y^2 - 2y - 4 = 0$$
$$y^2 - y - 2 = 0$$
$$(y - 2)(y + 1) = 0$$

$y - 2 = 0 \qquad\qquad$ or $\quad y + 1 = 0$
$\quad y = 2 \qquad\qquad\qquad\qquad y = -1$
$\quad x = 1 - 2 = -1 \qquad\quad x = 1 - (-1) = 2$

$(-1, 2), \quad (2, -1)$

23. $\begin{cases} y - x^2 = 4 \\ x^2 + y = 1 \end{cases}$

$$-x^2 + y = 4$$
$$\underline{x^2 + y = 1}$$
$$2y = 5$$
$$y = \frac{5}{2}$$

$$x^2 + \frac{5}{2} = 1$$

$$x^2 = -\frac{3}{2}$$

No real solution

25. $\begin{cases} x^2 + y^2 = 10 \\ 3x^2 - 4y^2 = 23 \end{cases}$

$$4x^2 + 4y^2 = 40$$
$$\underline{3x^2 - 4y^2 = 23}$$
$$7x^2 = 63$$
$$x^2 = 9$$
$$x = \pm 3$$

$x = 3$:
$$3^2 + y^2 = 10$$
$$y^2 = 1$$
$$y = \pm 1$$

$x = -3$:
$$(-3)^2 + y^2 = 10$$
$$y^2 = 1$$
$$y = \pm 1$$

$(3, 1), \quad (3, -1), \quad (-3, 1), \quad (-3, -1)$

27. $\begin{cases} x - y^2 = 3 \\ 3x - 2y = 9 \end{cases}$

$$x = 3 + y^2$$
$$3(3 + y^2) - 2y = 9$$
$$9 + 3y^2 - 2y = 9$$
$$3y^2 - 2y = 0$$
$$y(3y - 2) = 0$$

$y = 0$ or $3y - 2 = 0$

$x = 3 + 0^2 = 3$ $y = \dfrac{2}{3}$

$$x = 3 + \left(\dfrac{2}{3}\right)^2 = \dfrac{31}{9}$$

$(3, 0)$, $\left(\dfrac{31}{9}, \dfrac{2}{3}\right)$

29. $\begin{cases} x + y \le 6 & \text{(solid line)} \\ 2x + y \ge 4 & \text{(solid line)} \end{cases}$

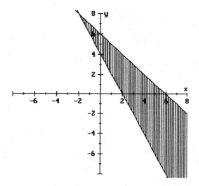

Test point: $(3, 1)$
$3 + 1 \le 6$ $2(3) + 1 \ge 4$
$4 \le 6$ $7 \ge 4$
True   True

31. $\begin{cases} 2x + 3y < 12 & \text{(dashed line)} \\ x < y & \text{(dashed line)} \end{cases}$

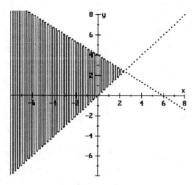

Test point: $(0, 1)$
$2(0) + 3(1) < 12$ $0 < 1$
$3 < 12$ True
True

33. $\begin{cases} x + y \le 4 & \text{(solid line)} \\ x - y \le 4 & \text{(solid line)} \\ \phantom{x - {}} x \ge 0 & \text{(solid line)} \end{cases}$

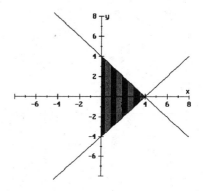

Test point: $(1, 0)$
$1 + 0 \le 4$
$1 \le 4$
True

$1 - 0 \le 4$
$1 \le 4$
True

$1 \ge 0$
True

35. $-3C = -3\begin{bmatrix} -1 & 2 \\ 5 & 0 \end{bmatrix}$

$ = \begin{bmatrix} 3 & -6 \\ -15 & 0 \end{bmatrix}$

37. $3C + 5D$

$= 3\begin{bmatrix} -1 & 2 \\ 5 & 0 \end{bmatrix} + 5\begin{bmatrix} -4 & 1 \\ 0 & 3 \end{bmatrix}$

$= \begin{bmatrix} -3 & 6 \\ 15 & 0 \end{bmatrix} + \begin{bmatrix} -20 & 5 \\ 0 & 15 \end{bmatrix}$

$= \begin{bmatrix} -23 & 11 \\ 15 & 15 \end{bmatrix}$

39. $CD = \begin{bmatrix} -1 & 2 \\ 5 & 0 \end{bmatrix} \begin{bmatrix} -4 & 1 \\ 0 & 3 \end{bmatrix}$

$= \begin{bmatrix} (-1)(-4) + (2)(0) & (-1)(1) + (2)(3) \\ (5)(-4) + (0)(0) & (5)(1) + (0)(3) \end{bmatrix}$

$= \begin{bmatrix} 4 & 5 \\ -20 & 5 \end{bmatrix}$

x	y
1/25	-2
1/5	-1
1	0
5	1
25	2

41. $\begin{bmatrix} 3 & 2 \\ ½ & 3 \end{bmatrix} \begin{bmatrix} x \\ y \end{bmatrix} = \begin{bmatrix} 10 \\ 1 \end{bmatrix}$

$AX = B$

$X = A^{-1}B$

$= \begin{bmatrix} 3.5 \\ -0.25 \end{bmatrix}$

$x = 3.5, \quad y = -0.25$

43. $f(x) = 3^{x+1}$

x	y
-3	1/9
-2	1/3
-1	1
0	3
1	9

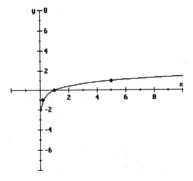

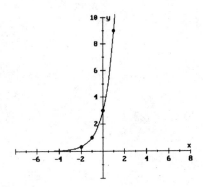

45. $f(x) = \log_5 x$

$y = \log_5 x$

$5^y = x$

47. $\log_2 64 = 6$

$2^6 = 64$

49. $\sqrt[3]{125} = 5$

$125^{1/3} = 5$

$\log_{125} 5 = \dfrac{1}{3}$

51. $\log_{27} 81 = \dfrac{4}{3}$

$27^{4/3} = 81$

53. $\log_3 81 = 4$

55. $\log_4 \dfrac{1}{16} = x$

$4^x = \dfrac{1}{16}$

$4^x = 4^{-2}$

$x = -2$

57. $\log_9 \dfrac{1}{3} = x$

$9^x = \dfrac{1}{3}$

$(3^2)^x = 3^{-1}$

$3^{2x} = 3^{-1}$

$2x = -1$

$x = -\dfrac{1}{2}$

59. $\log_b 1 = x$

$\quad\quad b^x = 1$

$\quad\quad b^x = b^0$

$\quad\quad x = 0$

61. $\log_b \sqrt[3]{5xy}$

$= \log_b (5xy)^{1/3}$

$= \dfrac{1}{3} \log_b (5xy)$

$= \dfrac{1}{3} \log_b 5 + \log_b x + \log_b y)$

$= \dfrac{1}{3} \log_b 5 + \dfrac{1}{3} \log_b x + \dfrac{1}{3} \log_b y$

63. $\log_3 \dfrac{x^2\sqrt{y}}{9wz}$

$= \log_3 \dfrac{x^2 y^{1/2}}{9wz}$

$= \log_3 (x^2 y^{1/2}) - \log_3 (9wz)$

$= \log_3 x^2 + \log_3 y^{1/2} - (\log_3 9 + \log_3 w + \log_3 z)$

$= 2 \log_3 x + \dfrac{1}{2} \log_3 y - (2 + \log_3 w + \log_3 z)$

$= 2 \log_3 x + \dfrac{1}{2} \log_3 y - 2 - \log_3 w - \log_3 z$

65. $\log 73{,}600 \approx 4.8669$

67. antilog $0.6085 \approx 4.0598$

69. $\log_9 384 = \dfrac{\log 384}{\log 9}$

$\quad\quad\quad\quad \approx 2.7083$

71. $\dfrac{1}{2} \log_8 x = \log_8 5$

$\quad \log_8 x^{1/2} = \log_8 5$

$\quad\quad x^{1/2} = 5$

$\quad\quad\quad x = 5^2$

$\quad\quad\quad x = 25$

73. $5^x = \dfrac{1}{25}$

$\quad 5^x = 5^{-2}$

$\quad\quad x = -2$

75. $\log_6 x + \log_6 4 = 3$

$\quad\quad \log_6 4x = 3$

$\quad\quad\quad 6^3 = 4x$

$\quad\quad\quad 216 = 4x$

$\quad\quad\quad 54 = x$

77. $\dfrac{4^{x^2}}{2^x} = 64$

$\dfrac{(2^2)^{x^2}}{2^x} = 2^6$

$\dfrac{2^{2x^2}}{2^x} = 2^6$

$2^{2x^2 - x} = 2^6$

$2x^2 - x = 6$

$2x^2 - x - 6 = 0$

$(2x + 3)(x - 2) = 0$

$2x + 3 = 0 \quad$ or $x - 2 = 0$

$x = -\dfrac{3}{2} \quad$ or $\quad\quad x = 2$

79. $\quad\quad 9^x = 7^{x+3}$

$\quad\quad \log 9^x = \log 7^{x+3}$

$\quad\quad x \log 9 = (x + 3) \log 7$

$\quad\quad x \log 9 = x \log 7 + 3 \log 7$

$x \log 9 - x \log 7 = 3 \log 7$

$x(\log 9 - \log 7) = 3 \log 7$

$\quad\quad x = \dfrac{3 \log 7}{\log 9 - \log 7}$

$\quad\quad x \approx 23.2288$

81. $A = P\left(1 + \dfrac{r}{n}\right)^{nt}$

$5000 = 3000\left(1 + \dfrac{0.08}{4}\right)^{4t}$

$\dfrac{5}{3} = 1.02^{4t}$

$$\ln \frac{5}{3} = \ln 1.02^{4t}$$

$$\ln \frac{5}{3} = 4t \ln 1.02$$

$$\frac{\ln \dfrac{5}{3}}{4 \ln 1.02} = t$$

$$t \approx 6.449$$

It will take 6.449 years.

83.
$$A = A_0 e^{rt}$$
$$2 = 20 e^{-0.002t}$$
$$0.1 = e^{-0.002t}$$
$$\ln 0.1 = \ln e^{-0.002t}$$
$$\ln 0.1 = -0.002t$$

$$\frac{\ln 0.1}{-0.002} = t$$

$$t \approx 1151$$

85.
$$pH = -\log \left[H_3O^+ \right]$$
$$8.2 = -\log \left[H_3O^+ \right]$$
$$-8.2 = \log \left[H_3O^+ \right]$$
$$10^{-8.2} = \left[H_3O^+ \right]$$
$$6.31 \times 10^{-9} = \left[H_3O^+ \right]$$

CHAPTERS 10 - 11
CUMULATIVE PRACTICE TEST

1. $\begin{cases} x + y + z = 6 \\ 3x + 2y - z = 11 \\ 2x - 4y - z = 12 \end{cases}$

$$\begin{array}{l} x + y + z = 6 \\ \underline{3x + 2y - z = 11} \\ 4x + 3y = 17 \end{array}$$

$$\begin{array}{l} x + y + z = 6 \\ \underline{2x - 4y - z = 12} \\ 3x - 3y = 18 \end{array}$$

$$\begin{array}{l} 4x + 3y = 17 \\ \underline{3x - 3y = 18} \\ 7x = 35 \\ x = 5 \end{array}$$

$$\begin{array}{l} 4(5) + 3y = 17 \\ 3y = -3 \\ y = -1 \end{array}$$

$$\begin{array}{l} 5 - 1 + z = 6 \\ z = 2 \end{array}$$

$$x = 5, \quad y = -1, \quad z = 2$$

3. (a) $D = \begin{vmatrix} 6 & 5 \\ 7 & 8 \end{vmatrix} = 13$

$$D_x = \begin{vmatrix} 13 & 5 \\ 26 & 8 \end{vmatrix} = -26$$

$$D_y = \begin{vmatrix} 6 & 13 \\ 7 & 26 \end{vmatrix} = 65$$

$$x = \frac{D_x}{D} = \frac{-26}{13} = -2$$

$$y = \frac{D_y}{D} = \frac{65}{13} = 5$$

(b) $D = \begin{vmatrix} 4 & -5 & 2 \\ 3 & 7 & -5 \\ 5 & -6 & 3 \end{vmatrix} = 28$

$$D_x = \begin{vmatrix} 17 & -5 & 2 \\ 2 & 7 & -5 \\ 21 & -6 & 3 \end{vmatrix} = 84$$

$$D_y = \begin{vmatrix} 4 & 17 & 2 \\ 3 & 2 & -5 \\ 5 & 21 & 3 \end{vmatrix} = -28$$

$$D_z = \begin{vmatrix} 4 & -5 & 17 \\ 3 & 7 & 2 \\ 5 & -6 & 21 \end{vmatrix} = 0$$

$$x = \frac{D_x}{D} = \frac{84}{28} = 3$$

$$y = \frac{D_y}{D} = \frac{-28}{28} = -1$$

$$z = \frac{D_z}{D} = \frac{0}{28} = 0$$

5. $\begin{cases} x - y \leq 5 & \text{(solid line)} \\ x - 2y \leq 0 & \text{(solid line)} \\ y \geq 0 & \text{(solid line)} \end{cases}$

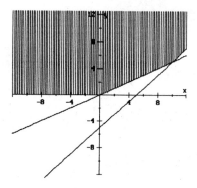

Test point: (0, 1)
$$0 - 1 \leq 5$$
$$-1 \leq 5$$
True

$$0 - 2(1) \leq 0$$
$$-2 \leq 0$$
True

$$1 \geq 0$$
True

7. $\begin{bmatrix} 9 & 4 & -2 \\ 5 & 2 & -1 \\ 2 & -7 & 3 \end{bmatrix} \begin{bmatrix} x \\ y \\ z \end{bmatrix} = \begin{bmatrix} 8 \\ 6 \\ 5 \end{bmatrix}$

$$AX = B$$
$$X = A^{-1}B$$

$$= \begin{bmatrix} 4 \\ 45 \\ 104 \end{bmatrix}$$

$$x = 4, \quad y = 45, \quad z = 104$$

9. $\log_8 \frac{1}{4} = -\frac{2}{3}$

$$8^{-2/3} = \frac{1}{4}$$

11. (a) $\log_b x \sqrt[3]{y}$
$$= \log_b xy^{1/3}$$
$$= \log_b x + \log_b y^{1/3}$$
$$= \log_b x + \frac{1}{3} \log_b y$$

(b) $\log_b \dfrac{x^3}{\sqrt{xy}}$

$$= \log_b \frac{x^3}{(xy)^{1/2}}$$

$$= \log_b x^3 - \log_b (xy)^{1/2}$$

$$= 3 \log_b x - \frac{1}{2} \log_b (xy)$$

$$= 3 \log_b x - \frac{1}{2}(\log_b x + \log_b y)$$

$$= 3 \log_b x - \frac{1}{2} \log_b x - \frac{1}{2} \log_b y$$

$$= \frac{5}{2} \log_b x - \frac{1}{2} \log_b y$$

13. $\log_8 985 = \dfrac{\log 985}{\log 8}$

$$\approx 3.3147$$

15. $\qquad A = P\left(1 + \dfrac{r}{n}\right)^{nt}$

$$10000 = 5000\left(1 + \frac{0.10}{4}\right)^{4t}$$

$$2 = 1.025^{4t}$$
$$\log 2 = \log 1.025^{4t}$$
$$\log 2 = 4t \log 1.025$$

$$\frac{\log 2}{4 \log 1.025} = t$$

$$t \approx 7.018 \text{ yr}$$